CAMBRIDGE LIBRARY COLLECTION

*Books of enduring scholarly value*

## Botany and Horticulture

Until the nineteenth century, the investigation of natural phenomena, plants and animals was considered either the preserve of elite scholars or a pastime for the leisured upper classes. As increasing academic rigour and systematisation was brought to the study of 'natural history', its subdisciplines were adopted into university curricula, and learned societies (such as the Royal Horticultural Society, founded in 1804) were established to support research in these areas. A related development was strong enthusiasm for exotic garden plants, which resulted in plant collecting expeditions to every corner of the globe, sometimes with tragic consequences. This series includes accounts of some of those expeditions, detailed reference works on the flora of different regions, and practical advice for amateur and professional gardeners.

## The Trees of Great Britain and Ireland

Although without formal scientific training, Henry John Elwes (1846–1922) devoted his life to natural history. He had studied birds, butterflies and moths, but later turned his attention to collecting and growing plants. Embarking on his most ambitious project in 1903, he recruited the Irish dendrologist Augustine Henry (1857–1930) to collaborate with him on this well-illustrated work. Privately printed in seven volumes between 1906 and 1913, it covers the varieties, distribution, history and cultivation of tree species in the British Isles. The strictly botanical parts were written by Henry, while Elwes drew on his extensive knowledge of native and non-native species to give details of where remarkable examples could be found. Each volume contains photographic plates as well as drawings of leaves and buds to aid identification. The species covered in Volume 7 (1913) include lime, box, willow, poplar and elm. The work's index appeared separately in 1913 but is now incorporated in this volume.

Cambridge University Press has long been a pioneer in the reissuing of out-of-print titles from its own backlist, producing digital reprints of books that are still sought after by scholars and students but could not be reprinted economically using traditional technology. The Cambridge Library Collection extends this activity to a wider range of books which are still of importance to researchers and professionals, either for the source material they contain, or as landmarks in the history of their academic discipline.

Drawing from the world-renowned collections in the Cambridge University Library and other partner libraries, and guided by the advice of experts in each subject area, Cambridge University Press is using state-of-the-art scanning machines in its own Printing House to capture the content of each book selected for inclusion. The files are processed to give a consistently clear, crisp image, and the books finished to the high quality standard for which the Press is recognised around the world. The latest print-on-demand technology ensures that the books will remain available indefinitely, and that orders for single or multiple copies can quickly be supplied.

The Cambridge Library Collection brings back to life books of enduring scholarly value (including out-of-copyright works originally issued by other publishers) across a wide range of disciplines in the humanities and social sciences and in science and technology.

# The Trees of Great Britain and Ireland

VOLUME 7

HENRY JOHN ELWES
AUGUSTINE HENRY

**CAMBRIDGE**
**UNIVERSITY PRESS**

University Printing House, Cambridge, CB2 8BS, United Kingdom

Published in the United States of America by Cambridge University Press, New York

Cambridge University Press is part of the University of Cambridge.
It furthers the University's mission by disseminating knowledge in the pursuit of
education, learning and research at the highest international levels of excellence.

www.cambridge.org
Information on this title: www.cambridge.org/9781108069380



This edition first published 1913
This digitally printed version 2014

ISBN 978-1-108-06938-0 Paperback

# THE TREES OF GREAT BRITAIN AND IRELAND

Photograph by Elliott & Fry Ltd. 55 Baker St. London W.

A. HENRY H. J. ELWES

# *The Trees of Great Britain & Ireland*

BY

Henry John Elwes, F.R.S.

AND

Augustine Henry, M.A.

VOLUME VII

Edinburgh: Privately Printed

MCMXIII

# CONTENTS

# Contents

# ILLUSTRATIONS

# TILIA

*Tilia*, Linnæus, *Hort. Cliff.* 204 (1735), *Sp. Pl.* i. 514 (1753), and *Gen. Pl.* 267 (1767); Bentham et Hooker, *Gen. Pl.* i. 236, 986 (1862-7); V. Engler, *Monog. Gatt. Tilia*, pp. 1-159 (1909).

Deciduous trees, belonging to the order Tiliaceæ, with tough fibrous inner bark. Leaves simple, long-stalked, alternate, arranged on the branchlets in two rows; unequal[1] and cordate or truncate at the base; acute or acuminate at the apex; serrate or toothed; venation pseudo-palmate, the midrib giving off secondary nerves pinnately on both sides, the lower two pairs of which arise together at the base, and give off tertiary nerves on the outer side only. Stipules ligulate, membranous, caducous.

Flowers white, fragrant, regular, perfect, in cymes; peduncle connate with the axis of a membranous elongated persistent bract, from the middle of which it apparently arises; inflorescence and bract springing from the axil of a leaf, alongside a bud, which develops into a branch in the following year.[2] Sepals five, distinct; petals five, imbricate in bud. Staminodes either absent or present as petaloid scales, one opposite each petal, and united with the base of the stamens. Stamens numerous, free, or in five clusters united together at the base. Filaments unbranched, or forked at the apex, with each branch bearing an extrorse half-anther. Ovary sessile, five-celled; style erect, dilated at the apex into five spreading stigmatic lobes; ovules two in each cell. Fruit nut-like, dry, indehiscent, one-celled, and one- to two-seeded by abortion. Seeds obovate, with fleshy albumen. Cotyledons reniform or cordate, palmately five-lobed, raised above ground on germination.

In winter the twigs are zig-zag and bear lateral buds, disposed alternately in two ranks; each bud with two to three scales visible externally, ovoid, obliquely displaced to one side of the semicircular leaf-scar, which is set on a prominent pulvinus. Stipule-scars small, linear or oblong, one on each side of the leaf-scar. Terminal bud absent; a circular scar at the apex of the twig, opposite the uppermost leaf-scar, indicating where the tip of the branchlet fell off in early summer. Base of the branchlet girt with a ring of scars, due to the fall of the bud-scales in the previous spring.

About twenty species of Tilia can be distinguished. These are widely distributed in the temperate regions of the northern hemisphere, extending southward in North America as far as the highlands of Mexico; but in the old world, while common in Europe and in northern and eastern Asia, no species is known in

[1] Cf. Van Tieghem, in *Ann. Sci. Nat.* (*Bot.*) iii. 378 (1906), on the peculiar asymmetry in the leaves and stipules of the lime.

[2] The occurrence of an inflorescence and a normal bud in one and the same axil is unusual; and is explained by the fact that the former is the result of the very early development of a flower-bud under the first scale of the normal bud, the other scales of the latter remaining dormant until the following season.

northern Africa or in the Himalayas. A large number of hybrid forms have arisen, some of which are common in cultivation. The following key, based on the characters of the branchlets and leaves, will serve to distinguish the species and hybrids which are cultivated in this country.

I. LEAVES GREEN BENEATH, WITH AXIL-TUFTS OF PUBESCENCE, BUT WITHOUT ANY STELLATE TOMENTUM.

(a) *Axil-tufts*[1] *present at the base of the leaf and elsewhere.*

* *Branchlets glabrous, or nearly so. Leaves glabrous, except for axil-tufts beneath.*

1. *Tilia mongolica*, Maximowicz. North China and Mongolia. See p. 1679.

Branchlets quite glabrous, reddish. Leaves 2½ in. wide, often trilobed, glaucous beneath with non-prominent tertiary veins, coarsely toothed.

2. *Tilia cordata*, Miller. Europe, Caucasus. See p. 1656.

Branchlets slightly pubescent at first, speedily becoming glabrous. Leaves 2 to 2½ in. wide, bluish green beneath with non-prominent tertiary veins, finely serrate.

3. *Tilia vulgaris*, Hayne. A hybrid, occasionally wild in Europe. See p. 1664.

Branchlets quite glabrous. Leaves 3 in. wide, dull green above, pale green beneath and with prominent tertiary veins, finely serrate with short points to the teeth.

4. *Tilia euchlora*, Koch. A hybrid, occasionally wild in the Caucasus. See p. 1674.

Branchlets usually quite glabrous. Leaves 2½ to 3 in. wide, dark shining green above, pale green beneath and with prominent tertiary veins, finely serrate with long points to the teeth.

** *Branchlets densely pubescent with long hairs.*

5. *Tilia platyphyllos*, Scopoli. Europe. See p. 1661.

Leaves 3 to 4 in. wide, upper surface with short scattered pubescence, lower surface covered with long hairs.

(b) *Axil-tufts absent at the base of the leaf, present elsewhere.*

6. *Tilia americana*, Linnæus. North America. See p. 1685.

Branchlets glabrous. Leaves 5 to 6 in. long and 3½ to 4½ in. wide, broadly ovate, cordate at the base, glabrous beneath and with numerous prominent parallel tertiary veins; margin with long-pointed coarse serrations.

7. *Tilia paucicostata*, Maximowicz. Western China. See p. 1680.

Branchlets glabrous. Leaves 2½ in. long and 2 in. wide; ovate, usually truncate at the base, glabrous beneath with few prominent irregular tertiary veins; margin with long-pointed fine serrations.

II. LEAVES GREEN OR GREYISH BENEATH, WITH SCATTERED STELLATE TOMENTUM.

(a) *Under surface of the leaves without axil-tufts, but with long hairs on the midrib.*

8. *Tilia Moltkei*, Schneider. See p. 1686.

A hybrid, with large leaves similar to those of *T. americana* in shape and serrations. Buds and branchlets glabrous.

[1] These are tufts of hairs at the junctions of the midrib and lateral nerves on the under surface of the leaf, which are now often termed *domatia*; they are the abodes of mites, and serve a useful purpose in the economy of the tree. They were fully studied and described by Lundström, in *Nov. Act. Reg. Soc. Sci. Upsala*, xiii. pt. 2, pp. 3-10 (1887). Cf. Lord Avebury, *Brit. Flowering Plants*, 123 (1905).

9. *Tilia spectabilis*, Dippel. See p. 1686.

A hybrid similar to *T. Moltkei*, but with smaller leaves, which have long hairs on the principal nerves, as well as on the midrib. Buds pubescent in their upper half. Branchlets with traces of stellate pubescence.

(b) *Under surface of the leaves with axil-tufts.*

10. *Tilia Michauxii*, Nuttall. North America. See p. 1689.

Leaves usually large,[1] 5 to 7 in. in length and 4 to 6 in. wide; ovate-cordate, very oblique at the base, with long-pointed large triangular serrations. Buds and branchlets glabrous.

III. LEAVES WHITE OR GREY BENEATH, AND COVERED WITH A DENSE STELLATE TOMENTUM.

(a) *Branchlets glabrous.*

* *Axil-tufts present.*

10A. *Tilia Michauxii*, Nuttall. In some forms of this species the leaves are densely greyish tomentose beneath. See above, No. 10.

* *Axil-tufts absent.*

11. *Tilia heterophylla*, Ventenat. North America. See p. 1688.

Leaves ovate-cordate, very oblique at the base, 4 to 5 in. long, 3 to 4 in. wide, covered beneath with a silvery white tomentum; serrations coarse and short-pointed.

12. *Tilia Oliveri*, Szyszylowicz. China. See p. 1681.

Leaves orbicular-ovate, 3 to 4 in. long, 2½ to 3 in. wide, silvery white beneath; serrations minute, crenate.

(b) *Branchlets pubescent.*

**Axil-tufts present.*

13. *Tilia Maximowicziana*, Shirasawa. Japan. See p. 1683.

Leaves orbicular-ovate, averaging 5 in. long and broad, covered beneath with a greyish tomentum; axil-tufts and tomentum on midrib and nerves brownish.

** *Axil-tufts absent.*

14. *Tilia tomentosa*, Moench. South-eastern Europe, Asia Minor. See p. 1675.

Leaves orbicular-ovate, 3 to 5 in. across, greyish or snowy white beneath, with stout or slender short petioles; serrations fine, regular, ending in short points. Buds, branchlets, and petioles grey tomentose.

15. *Tilia petiolaris*, Hooker. See p. 1677.

Possibly a sport of *T. tomentosa*, from which it differs in the pendulous habit of the tree, the long slender petioles, and the peculiar fruit.

16. *Tilia mandshurica*, Ruprecht and Maximowicz. Manchuria, North China, Korea. See p. 1682.

Leaves orbicular-ovate, 4 to 5 in. across, white beneath; margin often one- to two-lobed, with coarse serrations, ending in long awn-like points. Branchlets, buds, and stout petioles brown tomentose.

[1] In native specimens the leaves are smaller, averaging 4 to 5 in. long and 3 to 4 in. wide.

17. *Tilia Miqueliana*, Maximowicz. Cultivated in Japan. See p. 1684.

Leaves remarkably variable in shape, deltoid or ovate, usually much longer than broad, 3 to 4 in. long and 2 to 2½ in. wide, grey beneath; serrations irregular, ending in short points. (A. H.)

## TILIA CORDATA, Small-leaved Lime

*Tilia cordata*, Miller,[1] *Gard. Dict.* No. 1 (1768); Moench, *Verz. Ausl. Weissenst.* 135 (1785); Schneider, *Laubholzkunde*, ii. 372 (1909); V. Engler, *Monog. Gatt. Tilia*, 74 (1909).
*Tilia europæa*, Linnæus, *Sp. Pl.* 514 (1753) (in part); Loudon, *Arb. et Frut. Brit.* i. 364 (1838).
*Tilia ulmifolia*, Scopoli, *Fl. Carn.* i. 374 (1772); Sargent, in *Garden and Forest*, ii. 256, f. 111 (1889).
*Tilia parvifolia*, Ehrhart, *Beitr. Naturk.* v. 159 (1790); Willkomm, *Forstliche Flora*, 729 (1887); Mathieu, *Flore Forestière*, 29 (1897).
*Tilia microphylla*, Ventenat, in *Mem. Acad. Sc. Paris*, iv. 5 (1803).
*Tilia sylvestris*, Desfontaines, *Table Éc. Bot. Mus. Paris*, 152 (1804).

A tree, attaining 100 ft. in height and 20 ft. in girth. Bark smooth and grey on young trees; ultimately on old trunks divided by narrow longitudinal fissures into scaly ridges. Young branchlets green, slightly pubescent at first, speedily becoming glabrous, the pubescence, however, being often retained on short shoots; older branchlets dark brown. Leaves[2] (Plate 407, Fig. 8), membraneous, 2 to 2½ in. wide, usually broader than long, smooth and not wrinkled, cuspidate at the apex, cordate at the base; margin non-ciliate, regularly serrate, the teeth ending in short cartilaginous points; upper surface dark green, shining, glabrous except for occasional long hairs on the nerves; lower surface bluish or glaucous green, glabrous except for the conspicuous dense orange-brown axil-tufts at the base, and at the junctions of the midrib, primary, and secondary nerves; tertiary veins scarcely prominent on the under surface, and more irregular, and less straight and parallel than in *T. vulgaris* and *T. platyphyllos*; petiole about half as long as the blade, slender, glabrous, or with a few scattered hairs.

Cymes directed upwards, five- to seven-flowered; bract long-stalked, glabrous; pedicels glabrous or with a few scattered hairs; sepals pubescent; petals glabrous; stamens about thirty, longer than the petals; staminodes absent; ovary tomentose, style glabrous. Fruit globose, faintly ridged, apiculate at the apex, covered with long scattered tomentum; shell thin and fragile.

In winter the buds are more globose than those of *T. vulgaris* or *T. platyphyllos*, and appear to be composed of two external scales, though the

[1] E. G. Baker, in *Journ. Bot.* xxxvi. 318 (1898), states that Miller's specimen in the British Museum is *T. platyphyllos*; but there is no evidence that this is a type specimen. It is plain from Miller's statement that *T. cordata* "grows naturally in the woods in many parts of England," and from his identification of it with *Tilia foemina, folio minore*, C. Bauhin, *Pinax*, 426 (1671), that he meant the small-leaved lime.

[2] The leaves on coppice shoots in the first year are remarkably large. Mr. Riddelsdell sent us specimens from Glamorganshire, with leaves 5 to 7 in. long and nearly as broad, coarsely toothed, deeply and narrowly cordate at the base, ending at the apex in a long acuminate point, and on short petioles scarcely an inch in length. As the coppice shoots lengthen year by year, the leaves gradually assume their normal form, small in size, broader than long, and with long petioles. Lees, in *Bot. Worcestershire*, 16 (1867), argues from the variable appearance of the leaves of coppice shoots of *T. parvifolia*, that the common lime is only a variety of the latter. The coppice shoots of most broad-leaved trees have peculiar leaves, different from those in the adult state, and more alike in allied species, so that their discrimination is difficult.

pubescent tip of a third scale may be discerned at the apex of the bud; the first and second scales are shining, glabrous, ciliate.

This species is readily distinguished by the bluish tint of the under surface of the leaves, which are very different in their tertiary venation from the other common limes. The erect and not pendulous cymes of flowers are also a peculiar feature.

### VARIETIES

This species, as limited here to the European and Caucasian small-leaved lime, displays little variation in the wild state, the varieties[1] established by Schneider on the shape and size of the leaf and the amount of the pubescence on the fruit, being probably due to soil conditions, and not worth enumerating. A few peculiar sports have been noticed, none of which appear to be known in England:—

1. Var. *vitifolia*, Schneider, *op. cit.* Leaves three-lobed. Wild in Hungary.

2. Var. *aureo-variegata*, Schneider,[2] *op. cit.* Leaves variegated with yellow.

3. Var. *cucullata*, Henry (*var. nova*). Similar to the variety so named of *T. platyphyllos.* De Vries, *Species and Varieties*, 355, 669 (1906) and *Mutation Theory*, 470, fig. 106 (1910), draws attention to a tree with peltate and pitcher-like leaves, which is growing at Lage Vuursche, near Amsterdam.

### DISTRIBUTION

The small-leaved lime is a native of the greater part of Europe and of the Caucasus, the closely allied forms[3] in Siberia, Manchuria, and Japan being now regarded as distinct species. In Europe, it extends from northern Spain to the Ural range, attaining its maximum development in Russia, where it occasionally forms pure woods, but more usually, as is always the case elsewhere, growing as isolated trees or in small groups with other deciduous trees. It occurs as far north as the province of Volgoda, where it disappears after becoming a small shrub at lat. 62°. In the Ural, it reaches as far north as lat. 58° 50′. The finest lime woods

[1] Var. *Blockiana*, Schneider (*T. Blockiana*, Borbas), and var. *ovalifolia*, Spach, with leaves larger and less cordate than usual, are possibly of hybrid origin.

[2] *Tilia ulmifolia*, Scopoli, var. *foliis variegatis*, Petzold and Kirchner, *Arb. Musc.* 156 (1864), is another name for this variety.

[3] The Asiatic forms are distinguished as follows from the European *T. cordata*:—

A. *Tilia sibirica*, Bayer, in *Verh. Z. Bot. Ges. Wien.* xii. 23 (1862).

*Tilia cordata*, var. *sibirica*, Maximowicz, in *Bull. Ac. St. Petersb.* xxvi. 432 (1860). Indigenous in western Siberia. Not yet introduced. Differs mainly in the leaves, truncate or cuneate at the base, with sharper serrations, and long hairs on the nerves.

B. *Tilia amurensis*, Ruprecht, *Fl. Cauc.* 253 (1869).

*Tilia Maximowiczii*, Baker, in *Journ. Bot.* xxxvi. 319 (1898).

*Tilia cordata*, var. *mandshurica*, Maximowicz, in *Mél. Biol.* x. 584 (1880). Indigenous in Manchuria and Korea. Not yet introduced. Differs in the larger leaves, with fewer coarser serrations, which are tipped with long points.

C. *Tilia japonica*, Simonkai, in *Math. Term. Koezl.* xxii. 326 (1888).

*Tilia cordata*, var. *japonica*, Miquel, in *Ann. Mus. Lugd. Bat.* iii. 18 (1867); Sargent, in *Garden and Forest*, vi. 111 (1893), and *Forest Flora Japan*, 20 (1894); Shirasawa, *Icon. Ess. Forest. Japon*, i. text 115, t. 72, figs. 1-10 (1900). Indigenous in Japan, where it is a small tree, rarely higher than 60 ft. It was introduced into the Arnold Arboretum, U.S.A., in 1886, where it is hardy, producing flowers and fruit every year; but is said to be scarcely distinct from the European species. It appears to differ mainly in the flowers, which are 20 to 40 in each cyme, and possess staminodes. Specimens collected by Elwes at Asahigawa in Yezo show no difference in leaves and branchlets.

in Russia are in the region extending southwards from Kostroma to the edge of the steppe; and here both this species and *T. platyphyllos* grow together. In Norway it is found as a wild tree as far north as lat. 62° 9′ on the west coast, and in Sweden up to 63° 10′ in Angermanland; but according to Schübeler, it thrives when planted as far north as 67° 56′ in Norway, 65° 50′ in Sweden, and 63° in Finland.

It appears to be not a native tree in Belgium, Holland, Denmark, and north-western Germany; and is nowhere very common in central Europe at present, though it is supposed to have been more widely spread in ancient times, as the word linden is very prevalent in German and Slavonic names of places. It is rather a tree of the plains than of the mountains; but it ascends in Bohemia and Bavaria to 2000 ft., and in Switzerland and the Tyrol to 4000 ft. Bolle informed Sargent that very old and enormous trees[1] of this species, one being nearly 23 ft. in girth, exist at Paelitzaerder on the Paarestein lake near Eberswalde.

In France, it is met with in most of the forests of the plains and low hills, except in the departments bordering on the Mediterranean. It is occasionally treated as coppice, being used for firewood and making charcoal. Bast, which was formerly a product of some importance, is now only produced in the forest of Chantilly, nearly all the bast used being now imported from Russia. Mathieu mentions a tree, planted at Gerardmer in the Vosges, which measured 95 ft. in height and 19 ft. in girth, and was supposed to be at least 250 years old.

The small-leaved lime extends southwards to about lat. 41°, occurring in northern Spain, Italy, and the Balkan States; but is unknown in Greece and Sicily. Huffel[2] says that both it and *T. platyphyllos* are common in the forests of the hills of Dobrudja, Roumania, where they are the dominant trees.

The small-leaved lime is a native of England, ranging from Cumberland southward. It occurs in woods in rather inaccessible positions, where it is a rare tree, and more commonly in coppice, situations in which the indigenous vegetation has often been preserved. Ray[3] considered this species to be a true native; and in his time it was frequent and wild in woods and coppices in Essex, Sussex, Lincolnshire, and especially in Bedfordshire, "where there were thousands of lime trees." He adds that it was less common in the Forest of Dean, and rare in Cranborne Chase in Dorset. Many of the local floras give instances of its occurrence, as J. G. Baker[4] for Yorkshire, who states that it occurs "at Slip Gill near Rievaulx, where aboriginal woods composed principally of oak and hazel cover the steeply-sloping rocky banks of one of the loneliest and pleasantest glens in the eastern calcareous range." Ley[5] records it for different parts of Herefordshire. Murray[6] says it is abundant in the Leigh woods near Bristol. It is said[7] to be wild in several localities in Glamorganshire. Bromfield[8] mentions wild trees in one locality in the Isle of Wight, and in aboriginal woods on the chalk at Bordean Hill, near Petersfield, Hants.

Bromfield supposes that Lyndhurst, in the New Forest, owes its name to the

[1] Bean, in *Kew Bull.* 1908, p. 397, mentions a tree in the Grosse Garten, Dresden, branching close to the ground, where the trunk was about 8 ft. through.

[2] *Les Forêts de la Roumanie* (1890).

[3] *Syn. Meth.* 316 (1696) and *Philos. Letters*, 250 (1718).

[4] *Flora N. Yorks.* 274 (1906).

[5] *Flora Herefordshire*, 54 (1889).

[6] *Flora Somerset*, 64 (1896).

[7] Riddelsdell, in *Journ. Bot. Suppl.* 18 (1907).

[8] *Fl. Vect.* 83 (1856).

prevalence of this species there in ancient times. Limehouse, in London, according to Stowe, was originally called Limehurst, meaning a grove of linden trees in Saxon times. (A. H.)

Some doubt exists among botanists as to whether the small-leaved lime is truly native of England or not. It is not mentioned by Clement Reid as having been found in the fossil state in Britain; and by some it is supposed to have been introduced at an early epoch, perhaps by the Romans. But in some parts of the West Midlands it is found in woods remote from buildings, where one can hardly believe it was planted, so that it might fairly be considered a native but for one important fact. Notwithstanding many inquiries even in the districts where it now seems most at home, I have found no one who has seen a self-sown small-leaved lime. It seems hardly possible that a native tree should have lost its power of reproduction by seed, in a climate where it succeeds so well even as far north as Ross-shire; and in the north of France self-sown seedling limes are not uncommon, as I have myself observed in the Forêt de Retz. The tree has a remarkable power of persistence after repeated cutting, and of extending from stools to a considerable distance; so that in two old coppiced woods on my own property, it is now impossible to say where the stools originated. I have seen limes in remote rocky woods on the Wye valley near Moccas Court, whose stools had the appearance of very great age; and in the deep rocky gorge of Castle Eden Dene, on the coast of Durham, there are limes growing on such steep rocks that they could scarcely have been planted. But though rabbits will eat almost anything before they touch lime, I have searched in vain for seedlings in all these places. On the Carboniferous limestone rocks at Pen Moel near Chepstow, the residence of W. R. Price, Esq., I saw the tree growing in situations where it must have grown naturally from seed; and though Mr. Price has never found ripe fruit he has not the least doubt that it is indigenous here and elsewhere on the cliffs of the lower Wye valley.

E. Lees, in *Botany of Worcestershire*, 16 (1867) gives an excellent account of the occurrence of the lime in that county, where it is, in his opinion, "an undoubted native." He states that Shrawley wood, west of the Severn, which is about 500 acres in extent, is remarkable for a great part of it consisting of an undergrowth of lime, which is regularly cut as coppice-wood, and, therefore, is never in a flowering state.[1] On visiting this place, I agreed with Sir H. Vernon, of Hanbury Hall, near Droitwich, the owner of the wood, that the stools are in rows as though they had been planted; moreover there is not, so far as he knows, any lime in the adjoining woods. He says that this underwood used to be cut every seventeen years, and sent to the Potteries for making crates, but that this demand having ceased, it is now difficult to get rid of. It is now allowed to grow into poles, which are sometimes sold for copper-smelting in the Black Country, at about six or seven pounds per acre for twenty to twenty-five years' growth. In Sir H. Vernon's opinion, it would now pay better to grub the lime and plant larch in its place.

Lees[2] goes on to state that "Ockeridge wood, near Holt, though in a lesser

[1] Cf. p. 1656, note 2.

[2] Cf. also Lees' remarks in *Forest and Chace of Malvern*, abstracted in *Gard. Chron.* 1870, p. 1536.

degree, nourishes the same tree, as well as various coppices on the banks of the Severn between Ombersley and Hawford, where *T. grandifolia* exists in a naturalised state." He mentions "a very old and remarkable pollard tree of *T. parvifolia* at Hawford, on the ridge not far removed from the Severn. The base is more than 40 ft. round and six large boles arise from this in a semicircular manner. In fact, commencing with the border of Wyre forest and proceeding southward, the lime appears in numerous woods, coppices, and old hedgerows, to the very end of the Malvern range near Bromsberrow. The base of the round hills near Alberley, Ockeridge wood, the western base of the Berrow Hill near Martley, the banks of Leigh brook, Rosebury Rock on the Teme, the Old Storridge Hill, the country about Great Malvern, and ancient woods in the parishes of Castle Morton and the Berrow, may be particularly mentioned. Many of the old lime trees get pollarded, and then, in the course of years, put on a very grotesque appearance."

### Remarkable Trees

The small-leaved lime apparently never attains so great a height in England as the common lime, but is occasionally of great girth and is certainly long-lived. A tree (Plate 372), remarkable for its spreading habit, at Sprowston Hall, Norwich, was figured by Grigor in *Eastern Arboretum*, 200 (1841), where it is stated that it measured 24 ft. 7 in. "near the ground" and was believed to have been planted on 30th January 1649. It still survives in a shattered condition.

There is a remarkably fine tree of this species at The Hall, Thirsk, the seat of Reginald Bell, Esq., who has kindly sent us photographs. In 1904 the trunk in its narrowest part was 20 ft. in girth, and the spread of the branches was about 250 ft. in circumference.

One of the finest small-leaved limes is growing on a flat by the River Teme, at Oakly Park, Ludlow, which, in 1908, as nearly as I could measure it, was about 110 ft. by 14½ ft.

A fine tree, of weeping habit, at Hursley Park, Hants, the seat of Sir G. A. Cooper, Bart., measures about 80 ft. by 15½ ft. Close to it stands the hollow trunk of a much larger tree of the same species, which was blown down some years ago, and measures 19½ ft. in girth. The spread of its branches is said to have exceeded 100 ft.

At Arley Castle, Bewdley, a good specimen measured, in 1903, 85 ft. by 9 ft. 9 in. At Woburn Abbey, the largest tree of this species measures 76 ft. by 7 ft. 4 in., but appears to be still young, as the bark is comparatively smooth.

In Lincolnshire, the tree is not uncommon in parks and hedgerows. In Burghley Park there are several old trees, one of which measured, in 1908, 80 ft. by 11 ft. 4 in. At Casewick House, another was 82 ft. by 9 ft. 6 in. in the same year. At Syston Park there is a fine specimen, which measured 97 ft. by 11 ft. in 1906.

(H. J. E.)

## TILIA PLATYPHYLLOS, Large-leaved Lime

*Tilia platyphyllos*, Scopoli, *Fl. Carn.* i. 373 (1772); Sargent, in *Garden and Forest*, ii. 256, f. 109 (1889); Schneider, *Laubholzkunde*, ii. 376 (1909).
*Tilia grandifolia*, Ehrhart, *Beit.* v. 158 (1790); Willkomm, *Forstliche Flora*, 733 (1887); Mathieu, *Flore Forestière*, 33 (1897).
*Tilia pauciflora*, Hayne, *Arzn.* iii. 48 (1813).
*Tilia corallina*, Smith, in Rees, *Cycl.* xxxv. No. 2 (1819).
*Tilia mollis*, Spach, in *Ann. Sc. Nat.* ii. 336 (1834).
*Tilia europæa*, Linnæus, *Sp. Pl.* 514 (1753) (in part); Loudon, *Arb. et Frut. Brit.* i. 364 (1838).

A tree, attaining 130 ft. in height and upwards of 20 ft. in girth. Bark at first smooth and grey, ultimately on old stems with narrow shallow longitudinal fissures and ridges separating on the surface into small quadrangular scales. Young branchlets moderately covered with long white hairs; older branchlets glabrescent. Leaves (Plate 407, fig. 6) 3 to 4 in. in width and length, slightly uneven or wrinkled, ciliate in margin, regularly serrate, the serrations ending in short cartilaginous points; upper surface dull green, covered with short pubescence; lower surface lighter green, covered with long whitish pubescence, densest on the midrib, nerves, and veinlets, and forming dense axil-tufts at the base of the blade and at the junctions of the primary nerves with the midrib and with the secondary nerves; tertiary veinlets parallel and prominent on the under surface; petiole stout, shorter than the blade, whitish pubescent.

Flowers in pendulous, usually three-flowered cymes; about ½ in. in diameter, yellowish-white; sepals slightly pubescent externally, downy within; petals oblanceolate, longer than the sepals; stamens about thirty, longer than the petals; staminodes absent; ovary globose, tomentose; style glabrous. Fruit globose, pyriform, or ovoid, usually[1] with three to five prominent ribs, tomentose, apiculate at the apex; shell woody and hard.

In winter this species may be recognised by the twigs being slightly pubescent near the buds, which are minutely pubescent at the tip and show externally three glabrous ciliated scales.

### Varieties

This species in the wild state varies considerably in the amount of pubescence on the leaves, branchlets, and petioles; and has been subdivided into five subspecies by Schneider, who acknowledges, however, the great difficulty of limiting them clearly. The most pubescent forms occur in northern Germany, northern France, and Scandinavia; while nearly glabrous forms are found in southern France, Austria, and the Balkan States. V. Engler disagrees with Schneider's classification; and considers that the limes occurring in southern France, the Pyrenees, Italy, etc., should be united with *T. caucasica*; but this view is hardly tenable. The bract is stalked in most cases, but is occasionally sessile; and abnormal forms occur

[1] A tree at Kew of undoubted *T. platyphyllos*, bore fruit in 1907, on which no trace of ribs was perceptible.

(var. *multibracteata*) in which two bracts are borne on one peduncle. The fruit is remarkably variable, both in shape and in the prominence of the ribs.

The southern more glabrous forms are rarely cultivated in England, the only specimen which I have seen being a tree at Kew about 25 ft. high, which is named var. *obliqua*.[1] The branchlets are nearly glabrous; leaves very oblique and truncate at the base, glabrous above, with scattered pubescence below. It bears flowers similar to those of the type.

A large number of sports have arisen both under cultivation and in the wild state, the most noteworthy of which are:—

1. Var. *pyramidalis*, Simonkai, in *Math. Term. Koezl.* 334 (1888). Pyramidal in habit; leaves usually more or less cordate at the base. According to Schneider this is occasionally wild in south-eastern Europe.

2. Var. *tortuosa*, Bean, in *Kew Hand-list Trees*, 71 (1902). A peculiar sport, with all the twigs and branches twisted and curved. This[2] originated in the Royal Horticultural Society's garden at Chiswick in 1888 as a single specimen out of a bed of 500 large red limes. Grafts were sent to Kew from Chiswick in 1890, and three trees about 18 ft. high survive, in the Lime collection.

3. Var. *aurea*, Loudon (var. *aurantia*). Twigs golden yellow.

4. Var. *corallina*, Solander, in Aiton, *Hort. Kew.* ii. 229 (1789). Twigs bright red. Both these varieties are conspicuous in winter, and have been known for more than a century. According to Koch they were probably introduced from England to the Continent. The latter is the red-twigged lime of some English nurseries.

5. Var. *laciniata*, Loudon (var. *aspleniifolia*, var. *filicifolia*). Leaves smaller than in the type, deeply and irregularly cut and twisted. This never attains a large size,[3] and is only suitable for planting as a curiosity in gardens. It commonly throws out branches on which the foliage is normal.

6. Var. *vitifolia*, Simonkai, *op. cit.* Leaves lobulate or weakly three-lobed.

7. Forms with variegated leaves are known, as var. *albo-marginata*, Van Houtte.

8. Var. *cucullata*, Schneider (*T. cucullata*, Jacquin,[4] *Frag. Bot.* 19, t. 11, f. 3 (1800)). A form with small leaves, of which the edges of the two sides are joined together at the base, making the leaf pitcher-shaped. It is said to occur wild in southern Bohemia, where, according to Willkomm, there are some old trees, with all the leaves showing this peculiarity, at the monastery of Goldenkron, near Krumau.

## Distribution

This species is widely distributed throughout central and southern Europe, extending as far eastward as the Ural Mountains. Its northern limit as a wild tree is not known with certainty, and Willkomm considers it not to be indigenous in

[1] Probably identical with *T. obliqua*, Host, in Schmidt, *Oestr. Baumz.* iv. t. 224 (1822), and Host, *Fl. Austr.* ii. 62 (1831). The Kew tree agrees with a dried specimen collected in Host's garden in 1832.

[2] Cf. *Gard. Chron.* iv. 708 (1888).

[3] A. B. Jackson saw a tree at Blenheim, 40 ft. by $3\frac{1}{2}$ ft., in 1908.

[4] Jacquin figures leaves from trees in a cemetery at Sedlitz, near Kuttenberg in Bohemia. Leneck, in *Mitt. Nat. Ver. Univ. Wien*, 1893, pp. 19-29, figs. 1-11, gives an account of these abnormal leaves; and records a large-leaved lime growing at Leitmeritz in northern Bohemia, of which 20 to 30 per cent of the leaves were pitcher- or cowl-shaped. Cf. Just, *Bot. Jahresb.* xxii. pt. 2, p. 219 (1894).

northern Germany, Denmark, and the Baltic provinces of Russia. Bolle, however, states that it grows sparingly in these countries, and mentions small groups of wild trees growing in the islands on the west coast of Sweden, near Strömstad. It is most common in southern Russia, where in the provinces of Ukraine and Volhynia it often forms pure woods, though it is also seen in mixture with the small-leaved lime and *Quercus pedunculata*. It is also frequent in the southern states and Rhenish provinces of Germany, and ascends in the Bavarian Alps to 3300 ft. It is also widely spread in Austria, Hungary, and the Balkan States; and occurs in Italy and Spain, reaching its most westerly point in Asturias and New Castile. In France it is found scattered in the forests of the plain, except in the Mediterranean region, where, however, it has been observed as a rare tree in the Ravin des Arcs, 15 miles north of Montpellier.[1] More common in the hills and mountains, it attains its highest elevation, 4600 ft., in the Pyrénées-Orientales. It is replaced in Greece by the closely allied species *T. corinthiaca*, Bosc, and in the Caucasus, north Persia, and Armenia by *T. caucasica*, Ruprecht. Neither of these is in cultivation.

This species is a doubtful native of England, and was considered by Watson to be only a denizen. Bromfield[2] says that the broad-leaved lime, though partly naturalised in hedgerows, is nowhere indigenous in this country. Ley,[3] in 1889, however, considered it to be truly wild in rocky woods in the lower valley of the Wye, where, on the Great Doward and at Symonds Yat, it grows on bare limestone rock in company with the small-leaved lime. The occurrence here of *Pyrus latifolia* as an indigenous tree supports Ley's opinion. Baker[4] also considers it to be a native of Yorkshire, where it grows on the limestone scars of the lower part of Swaledale in a rocky wood, where no trees have ever been planted. Linton[5] also records it as growing wild on limestone cliffs in Derbyshire. (A. H.)

### Remarkable Trees

One of the oldest large-leaved limes in England is the famous tree[6] planted by Queen Elizabeth during her visit to Burghley Park, Stamford. This is now only about 60 ft. high, having lost many limbs in recent years, but it is 20 ft. in girth, and still bears foliage freely.

The tallest trees of this species which I have seen are those on the hill in the park at Longleat, where there are many from 120 to 130 ft. high, and some probably

[1] Elwes found a lime wild in the Forêt de Sainte Baume, near Aubagne (Var), which was identified by M. Mader of Nice with *T. platyphyllos*. It is recorded for this station by Albert and Jahandiez, *Plant. Vasc. du Var*, 84 (1908). Enormous trees of this species are said to have existed in France, one at Château Chaillé near Melle (Poitou) having measured 50 ft. in girth in 1804, when it was 538 years old. T. Hartig alludes to another at Saint Bonnet which was 55 ft. in girth. Cf. Kanngieser, in *Flora*, xcix. 428 (1909). Willkomm, *Forstliche Flora*, 736, note (1887), mentions also large limes in Germany, one at Staffelstein in Bavaria being 57 ft. in girth at three feet from the ground. We have not been able to confirm these records.

[2] *Flora Vect.* 83 (1856).

[3] *Flora Herefordshire*, 54 (1889). The late Rev. Augustin Ley, who kindly sent me specimens for examination, informed me in a letter, that "it occurs sparingly in aboriginal woodland, through Herefordshire, where there are nine stations in which the tree is native. It extends northwards into Shropshire, westwards into Radnor and Brecon, and southwards along the Wye valley into west Gloucester and Monmouth. In many of its stations it occupies crannies of limestone cliff, where it is physically impossible that it should be planted."

[4] *Flora North Yorkshire*, 274 (1906).

[5] *Flora Derbyshire*, 91 (1903).

[6] Figured in *Gard. Chron.* xvi. 400, fig. 78 (1881).

taller, all running up to a great height with clean straight stems. These are reputed to have been planted in 1690. At Revesby Abbey there is a fine old tree, about 100 ft. by 13 ft. 3 in., with the branches descending to the ground. The avenue at Poltimore, which is very fine, is composed of limes[1] of the species.

Mr. Renwick reports a large-leaved lime[2] at Ancrum, near Roxburgh, which in 1909 measured 26 ft. in girth at 6 ft. from the ground. (H. J. E.)

## TILIA VULGARIS, Common Lime

*Tilia vulgaris*, Hayne, *Arzn.* iii. 47 (1813); Sargent, in *Garden and Forest*, ii. 256, fig. 110 (1889).
*Tilia intermedia*, De Candolle, *Prod.* i. 513 (1824); Mathieu, *Flore Forestière*, 32 (1897).
*Tilia europæa*, Linnæus, *Sp. Pl.* 514 (1753) (in part); Loudon, *Arb. et Frut. Brit.* i. 364 (1838).
*Tilia cordata* × *platyphyllos*, Schneider, *Laubholzkunde*, ii. 374 (1909); V. Engler, *Monog. Gatt. Tilia*, 144 (1909).

A tree, attaining 130 ft. in height and 15 ft. in girth. Bark similar to that of *T. platyphyllos*. Young branchlets green, glabrous, becoming dark-brown with age. Leaves (Plate 407, fig. 4) larger than those of *T. cordata*, averaging 4 in. in length and 3 in. in width, slightly wrinkled or uneven, cuspidate or acuminate at the apex, truncate or cordate at the base; margin slightly ciliate and regularly serrate, the teeth ending in short points; upper surface dark green, glabrous; lower surface pale green, with brown axil-tufts at the base and the junctions of the midrib, primary and secondary nerves, and a few scattered long hairs on the nerves, elsewhere usually glabrous; tertiary nerves on the lower surface prominent, mostly straight and parallel; petiole green, glabrous, about half the length of the blade.

Cymes pendulous, five- to ten-flowered, glabrous; bract slightly pubescent, sessile or stalked; sepals, petals, stamens, ovary, and style as in *T. platyphyllos*. Fruit ovoid or globose, apiculate at the apex, not ribbed when mature, covered with a dense tomentum; shell thick and tough.

The buds are similar to those of *T. platyphyllos*, showing three external scales, which are glabrous, shining, and ciliate; but the glabrous branchlets will readily distinguish the common lime in winter, those of *T. platyphyllos* always being more or less pubescent.

This species, though the most common lime in cultivation, both in Britain and on the Continent, is extremely rare in the wild state. Mathieu says that it is occasionally seen in woods[3] in France; and Simonkai records reputed wild specimens from Upsala in Sweden and from Finland. Bolle informed Sargent that he had only once seen an indigenous specimen, a tree growing in the Tyrol.

[1] Cf. p. 1667, note 1, for other avenues of this species, and p. 1669.

[2] Cf. Christison, in *Trans. Bot. Soc. Edin.* xix. 494 (1893), who states that it was 20 ft. in girth at five feet up, the narrowest point, in 1877.

[3] In the Cambridge herbarium there is a specimen gathered by Vincent in 1847 in the wood of Champigneuille, near Nancy; and another gathered in Switzerland by J. Stuart Mill, labelled "Mountain side, near Altdorf, Canton Uri, completely wild and native."

It is now universally admitted to be a hybrid between *T. cordata* and *T. platyphyllos*. The distinctive marks of these species and the hybrid are :—

*T. platyphyllos.* Branchlets and leaves very pubescent with long hairs. Buds with three external scales. Cymes *pendulous*, usually three-flowered. Fruit with prominent ribs ; shell woody and hard.

*T. cordata.* Branchlets glabrous or nearly so. Leaves small, glabrous except for axil-tufts, bluish beneath with irregular and not prominent tertiary venation. Buds with two external scales. Cymes *erect*, five- to seven-flowered. Fruit faintly ridged ; shell thin and fragile.

*T. vulgaris.* Branchlets quite glabrous. Leaves larger than those of *T. cordata* ; under surface pale green, glabrous except for axil-tufts and a few hairs on the nerves, with parallel straight and prominent tertiary venation (as in *T. platyphyllos*). Buds with three external scales. Cymes *pendulous*, five- to ten-flowered. Fruit faintly ribbed ; shell woody and hard.

There are at least two distinct forms of the common lime in cultivation in England and elsewhere which require further study, one with leaves light green beneath, longer than broad, the form described above and considered by botanists to be typical *T. vulgaris*, Hayne ; and the following :—

1. *T. pallida*,[1] Wierzbicki, in Reichenbach, *Icon. Fl. Germ.* vi. 58, t. 315 (1844).

Leaves smaller, often not much larger than those of *T. cordata*, as broad as or broader than long, yellowish or bluish green beneath. It is readily distinguishable from *T. cordata* by its prominent tertiary venation, and has flowers and fruits like those of typical *T. vulgaris*. According to V. Engler, it is occasionally found in the wild state in Hungary.

Rarer hybrids also occur :—

2. *T. flavescens* and *T. floribunda*, A. Braun, in Doell, *Rhein Fl.* 672 (1843).

These peculiar trees, possibly hybrids of the same parentage as *T. vulgaris*, were noticed growing in an avenue at Carlsruhe in 1836. The leaves closely resemble those of *T. cordata*, but are larger and with paler axil-tufts. The cymes, with numerous flowers, in which staminodes[2] are developed, resemble in these respects those of *T. japonica*, the small-leaved lime of Japan. According to Koch,[3] the seed was sown, and produced pure *T. cordata* seedlings ; but two trees at Kew, labelled *T. flavescens*, presumably seedlings, have larger leaves than those of the common small-leaved lime, and are peculiar in their yellow branchlets and petioles. These, though young trees,[4] bear flowers, few in the cyme, without staminodes, in partly erect and partly pendulous cymes. One obtained from Späth in 1900 is about 20 ft. high ; the other, from Simon-Louis in 1902, is about 15 ft.

[1] Identified by Schneider, with *T. subparvifolia*, Borbas, in *Oest. Bot. Zeit.* xxxvii. 297 (1887). *T. vulgaris*, var. *pallida*, Sargent, *Bull. Pop. Inform.* No. 30 (1912), and in *Gard. Chron.* lii. 88 (1912), is the typical form of *T. vulgaris*, and not the tree described by Wierzbicki.

[2] On account of the staminodes, these trees are often supposed to be hybrids of *T. cordata* with *T. americana*, but they show no resemblance to the latter species in the shape, size, or serrations of the leaves.

[3] *Dendrologie*, i. 481 (1869).

[4] Sargent, *Bull. Pop. Inform.* No. 30, 1912, and in *Gard. Chron.* lii. 87 (1912), says that plants only a few feet high flower profusely.

3. *T. Beaumontia*,[1] which is sold in Späth's and Simon-Louis's nurseries, appears to be a form of the common lime, with pendulous branches.

The common lime is not indigenous in Britain, where it is never found except in plantations, avenues, and hedgerows, and rarely[2] produces natural seedlings. It comes into flower[3] about ten days later than *T. platyphyllos*, and a fortnight earlier than *T. cordata.*

In the common lime and allied species, the upper surface of the leaves is frequently found in summer to be sprinkled over with a viscid saccharine fluid, which is popularly known as honey dew. There has been great diversity of opinion as to whether this honey dew is always an exudation from the leaves, or is in some cases voided by aphides on the leaves. Sorauer,[4] the latest investigator of this subject, believes that the saccharine excretion originates without the assistance of aphides, and is the result of excessive transpiration, brought about usually by intense sunlight, a common occasion being when a cold night is followed by a hot morning sun. After the honey dew dries and thickens, it becomes the seat of growth of certain fungi, species of *Fumago*, which give the leaves a blackened appearance. Paths and garden seats situated under lime trees frequently show a disagreeable coating of this viscid exudation, which has fallen from the leaves.

The date of its introduction into England is uncertain, but this tree appears[5] to have been first planted on a large scale by Le Notre,[6] in the reign of Charles II., who used it for avenues, as was then the custom in France. The lime trees mentioned by Turner in 1562 as attaining a large size, and the old trees reported by Barrington[7] to be growing in 1769 in Moor Park in Hertfordshire and on the river Neath in Glamorganshire, were probably *T. cordata*, and of indigenous origin.

(A. H.)

CULTIVATION

The lime seems to ripen its seed more often than is generally supposed in warm summers in the south of England, and I have raised seedlings[8] from seeds gathered as far north as near Newark in 1904. In the same year Mr. A. C. Forbes sent me some of the common lime from Longleat, saying that very little, if any, of the seed of

[1] Cf. Schneider, *Laubholzkunde*, ii. 374 (1909), who considers it to be a hybrid between *T. euchlora* and *T. platyphyllos*.

[2] Mr. Anderson has found a few seedlings from trees planted on the edge of Lord Bathurst's deer park, just opposite the Royal Agricultural College at Cirencester.—H. J. E.

[3] Cf. Dr. Moss, in *Bot. Exchg. Club Rep.* 1910, p. 550.

[4] *Pflanzenkrank.* i. 412-414 (1909). The literature about honey dew on the lime is extensive. Boussingault's article in *Comptes Rendus*, lxxiv. 87 (1872), and Rivière's and Roze's articles in *Bull. Soc. Bot. France*, xiv. 12, 15 (1867), are abstracted in *Gard. Chron.* 1872, pp. 509, 609. See also various letters in *Gard. Chron.* 1873, pp. 920, 952, 1308, 1340, 1372, 1404, 1501, 1602. Buckton, *British Aphides*, i. 39-47 (1876), may also be consulted.

[5] Cf. Loudon, *Arb. et Frut. Brit.* i. 23, 24 (1838), who states, quoting Hasted, *Kent*, 562 (1769), that Sir John Speilman, in the reign of Elizabeth, brought over two lime trees from Germany, which were planted at Portbridge, near Dartford. These trees were cut down some time previous to 1769, and there is no means of determining what species of lime they belonged to.

[6] According to Chalmers, *Biog. Dict.* xxiii. 251 (1815), Andrew Le Notre, who was born in 1613 and died in 1700, laid out St. James's and Greenwich Parks in the reign of Charles II.

[7] In *Phil. Trans.* lix. 35 (1769).

[8] In the west court of the University Library, Cambridge, which is laid out in grass, secluded, and surrounded by high buildings, there were in 1912 several natural seedlings, arising from seed brought by winds or birds. These included *Betula pubescens* and *B. verrucosa*, elder, *Salix Caprea*, *Cratægus monogyna*, sycamore, and a solitary seedling, three or four years old, which was apparently *T. vulgaris*. It differed slightly in having slight pubescence on the branchlets and under surface of the leaves, thus showing a reversion to *T. platyphyllos*.—A. H.

the small-leaved lime was fertile, possibly because it flowers later than the other. A small proportion of this seed germinated in the first spring, but most of it lay dormant till the following year, and this has been my experience with many sowings of seed from abroad of the various European forms, as well as of the American lime. After germination they grow slowly for the first two years, and the young wood is liable to be killed back in winter. For this reason layering is the method adopted by nurserymen, though the varieties are usually grafted. I cannot say that my experience is as yet long enough to justify me in preferring seedlings.[1]

With regard to soil, the lime is not particular, but requires a good deep loam to bring it to perfection. It transplants very well, and may, if properly prepared by cutting round the roots two years previously, be safely moved when 20 to 30 ft. high.

Owing to the depreciated value of the timber, the lime cannot now be recommended except as an ornamental tree, the principal objection to it for this purpose being the early period at which its leaves wither and fall in autumn.

## Lime Avenues

The common lime is one of the most valuable avenue trees that we have; the fashion for planting them is, however, not very ancient, having apparently been introduced by Le Notre and other French landscape gardeners in the latter half of the seventeenth century, from which period most of our best avenues date.

Of these one of the finest is the avenue at Burghley Park, Stamford, the seat of the Marquess of Exeter. This is about 3000 yards long, with four rows of trees planted 6 yards apart in the row, 10 yards between the two outer rows and 20 yards between the inner ones. The trees are 100 ft. to 110 ft. in height on an average, and all appear to have been pollarded when young, though they have the upright habit which distinguishes most of the older lime avenues. I was informed by Mr. C. Richardson, of Stamford, that about fifty-five years ago—when lime wood was much more valuable than it now is, and made 5s. to 6s. per foot, single trees being sometimes sold at £40 to £50—an offer was made by a syndicate of London timber merchants to buy the whole of this avenue for £100,000. This story appears hardly credible, and I could obtain no verification; but, if made, the offer was refused, and there is no chance of such a price being paid for lime trees now.

Another beautiful avenue of fine tall limes is at Stratton Park, Hants, the seat of the Earl of Northbrook, whose late father informed me that it probably dates from about 1715. This avenue shows a common defect, which consists in the mass of spray that springs from some point usually near the root, though sometimes at 10 to 20 ft. up the trees, or even higher. I have searched in vain the works of Evelyn, Duhamel, Miller, Boutcher, and Loudon, for any reference to these abnormal growths,

[1] *T. platyphyllos* is usually imported from France as seedlings; and these appear to thrive in some cases better than plants of the common lime, which have been raised from layers. A young avenue of *T. platyphyllos* at Terling, Essex, with trees about 30 ft. high, is very thriving. The lime avenue at the back of Trinity College, Cambridge, which was celebrated by Tennyson, consists of two parts. That on the west side of the Cam now consists of 38 common limes, one half of which are very burry and much decayed, gaps showing where a few have died. These trees, which now average 6 ft. 9 in. in girth, were planted in 1671, at a cost of £10 : 6s., plus carriage from London amounting to £1 : 4s. On the east side of the Cam there are 20 trees, all but one of which are *T. platyphyllos*. These, which now average 6 ft. in girth, were planted in 1717, and look much healthier than the others. Cf. Willis and Clarke, *Archit. Hist. Univ. Camb.* ii. 641, 646 (1886).—A. H.

which, so far as I can learn, seldom appear on wild trees, or on any species but *T. vulgaris.* It is scarcely due to soil, since in some of the finest old lime avenues, as those at Cassiobury Park, Waldershare Park, and Newhouse Park, these growths appear in some trees only. They extend to a considerable height up some of the trees, which are much stunted, as it seems, from this cause. This may arise from the affected trees having been propagated by layers from inferior shoots, or having been planted later to fill gaps, and thus having to contend with trees already established.

It is, however, a point which deserves careful attention on the part of nurserymen; as, though these growths may be pruned off annually, they constantly reappear at the same spot, and not only take a great deal of trouble and time to remove, but eventually disfigure the trees; and limes which produce them rarely attain the same height or beauty as those which are free from them. The trees now sold by nurserymen, which are always propagated from layers, seem to be more subject to these growths than the limes planted two centuries ago; and I believe that limes raised from seed are rarely if ever affected.

The longest avenue of limes which I have seen is comparatively modern, and as I am informed by Mr. A. H. Elliott, agent for the Clumber estate, was planted by Henry, fourth Duke of Newcastle, about the year 1840. It is 1 mile and 1590 yards long, and consists of 1315 trees planted in a double row on each side of the drive at Clumber. The trees are 31 ft. apart each way, and the total width is 143 ft. The trees are fairly uniform in habit, but have spreading bushy tops, and when I saw them in 1906 did not exceed about 60 ft. in height by about 4½ ft. in girth. The soil is rather sandy, and the trees when planted were only 5 ft. high, and were not pruned sufficiently after planting to develop a good trunk, so that this avenue is never likely to rival those at Burghley, Ashridge, or Cassiobury. Mr. Elliott tells me that the trees have suffered considerably from the attacks of the following geometrid moths: *Cheimatobia brumata, C. boreata, Hybernia aurantiaria, H. defoliaria, H. progemmaria, Anisopteryx æscularia*; but this damage has been checked, if not entirely prevented, by putting grease bands on the trees, which arrest the female moths when they try to ascend the stems in the winter months, and by killing the pupæ in the soil in July with gas-lime. Besides this avenue there are at Clumber two much older ones, over 150 years old, running north and south on either side of the elm avenue leading to West Drayton. One of these is 385 yards long and 30 yards wide, the other 330 yards long and 55 ft. wide. The trees are planted 24 ft. apart. They were pollarded in 1888 in order to save their lives.

At Newhouse Park, near Mamhead, Devonshire, Sir Robert Newman showed me a fine avenue which seems to have been planted about 200 years ago in anticipation of a mansion which was never built. It is only 20 ft. wide and the trees 10 ft. apart; but favoured by a fine soil and climate, the trees, which seem to have been pollarded at 10 ft., have shot up to an immense height, averaging at least 120 ft., and several exceeding 130 ft. Two which I measured were 115 ft. by 5 ft. 9 in., and 135 ft. by 7 ft. 6 in. I have little doubt that they were seedlings. In a chestnut avenue at the same place the trees were much shorter and thicker, about 70 ft. to 80 ft. by 15 ft. to 18 ft. in girth.

At Cassiobury Park, Herts, there is a lime avenue supposed to have been planted by Le Notre, but some of the trees, which are very inferior to the rest in height and symmetry, are smothered in dense masses of small spray at 20 ft. to 30 ft. from the ground, and seem to have been planted later, possibly to replace dead trees in the original avenue. This avenue is 24 yards wide and the trees 8 yards apart. The best of the trees are 120 ft. to 130 ft. high, and one measures 13½ ft. in girth. There is a fine avenue, about half a mile long, at Denham Court, near Uxbridge.

At Braxted Park, Essex, the property of C. H. Du Cane, Esq., there is a lime avenue composed of three rows of trees on each side, which shows extraordinary variation in the growth of the trees, giving it a very irregular appearance. The tallest at the bottom of the hill are about 120 ft. high, and covered to an extent I have never seen elsewhere, with mistletoe growing in large bushes nearly up to their tops. At the top of the hill near the entrance gate, many of the trees are poor and stunted, with masses of spray at their root and higher up, and with gouty swellings on their branches which may be due to the mistletoe.

At Betchford Park, Surrey, there is a remarkable avenue of very old trees, some of which, when seen by Henry in 1906, were 130 ft. in height and 12 ft. to 13½ ft. in girth. This avenue was described as a very fine one by Dr. Aikin[1] in 1798.

At Doneraile Court, in Ireland, there is a fine avenue, one tree measuring 98 ft. by 10 ft. in 1907; most of the trees were covered with masses of spray.

An excellent article by Alfred Rehder of the Arnold Arboretum, on the lime as an avenue tree, in *Möller's Deutsche Gärtner Zeitung*, 1904, p. 188, should be consulted by those who think of planting. After giving the distinctive characters of the species, and describing their peculiarities of growth, he says that the choice must depend on the character of the soil and climate, and considers that the large-leaved lime, *T. platyphyllos*, is the best where the ground is deep and moist and where rapid growth and heavy shade are required. For drier soil he prefers *T. cordata*, which does not, however, make such a large or fine tree. *T. vulgaris* is in most of its characters intermediate, and this is the lime which is most generally used in England, though according to Rehder most, if not all, of the limes celebrated for their size and age in Germany are *T. platyphyllos*. He thinks that *T. euchlora*, Koch, is the best for town planting, because its smooth leaves do not hold the dust so much as those of other species, and because its leaves do not fall so early. He prefers *T. petiolaris*, the pendulous silver lime, for park avenues, and for single specimens where its branches can show their full beauty. Rehder does not give any observations as to the relative advantage of trees propagated from seed, from layers, or from grafts; but he rightly says that it is important that all the trees should be propagated from the same variety.

### Remarkable Trees

Among the most remarkable limes that I have seen is a walk at Ashridge Park. These trees are individually much larger than those at Burghley or Stratton, and

[1] In *Monthly Magazine* for 1798, quoted in *Gard. Chron.* 1841, p. 4.

though planted only 4 to 5 yards apart, average 120 ft. high and about 10 ft. in girth. They have the amount of variation in their leaves that one would expect to find in seedlings, and though 250 years old[1] only one out of forty is decayed. The bark of these trees is more like that of an elm than the usual bark of a lime, but this is perhaps owing to some peculiarity in the soil. A great lime at the end of the pinetum at Ashridge is of a totally different character, having very drooping branches and an immense spread. It has a trunk about 30 ft. high, and does not exceed 80 ft. to 90 ft. in total height, but the branches cover a circumference of 110 paces.

In Windsor Park, near Cranbourne Tower, there are some extremely tall and graceful limes growing with beech in a circle which were planted[2] in 1697. The best of these that I measured was 130 ft. by 14 ft. in girth, a beech close by it being 125 ft. by 10 ft.

A remarkable case of the tendency of the lime to layer which occurs at Rothamsted, is figured in the *Gardeners' Chronicle*, June 5, 1875. Here a row of fine old limes have dropped their branches on each side to the ground, and these have grown up in a thick mass, forming a shady corridor on each side of the trunks. There is another good example at Enville Hall, Stourbridge, where hundreds of young stems have arisen from layers, the whole mass measuring 140 paces round in 1904.

A curious instance of natural inarching of the lime is described and figured in *Gard. Chron.* xi. 277 (1879).

The branches of the lime sometimes spread laterally to a great distance. One of the best instances I have seen was shown me by Mr. Tudway at the Coombe, near Wells in Somersetshire. The tree is a large-leaved lime growing in a sheltered dell, about 100 ft. by 14 ft., and has three immense horizontal limbs 8 to 9 ft. in girth, one of which extends for 64 ft. from the trunk. In Stoneleigh Park, Warwickshire, there are some fine limes remarkable for their wide-spreading branches. In the Thames valley there are many fine lime trees, one at Osterley Park being about 120 ft. in height; while at Crowsley Park, near Henley, there are several in a clump, one of which was 118 ft. by 12½ ft. in 1908.

E. Lees, in his account[3] of the *Forest and Chace of Malvern*, speaking of the common lime, says:—"Some very fine trees now stand in a field about half a mile south of Bromsberrow Church, and by the side of the road leading from Ledbury towards Gloucester. Two of these, growing near each other, have become conjoined, both by the amalgamation of their arms, and by a lateral junction at the root.[4] The largest of these trees is 27 ft. in circumference at three feet from the ground, and is 36 ft. round the base; the other is 11 ft. 3 in. in girth at a yard from the ground, and 19 ft. in circumference at the base. The whole mass, if measured as one tree (and the interval between the boles where the connecting root joins them is only 19 in.) is full 48 ft. in circumference. In a field on the Priory Farm, Little Malvern, are several large

[1] In a book called *Chiltern and Vale Farming*, p. 153, published in 1745, it is stated by the anonymous author, who lived close to Ashridge, that they were planted in 1660, and in 1745 or thereabouts were near 3 ft. in diameter at the bottom.

[2] W. Menzies, *History of Windsor Great Park*, 44 (1864).

[3] Abstracted in *Gard. Chron.* 1870, p. 1536, figs. 264, 265, 266.

[4] I visited these trees in 1905, and found the largest now standing to be about 80 ft. high and 20 ft. 9 in. in girth. The fruit was fully formed on 18th July.

trees of *Tilia platyphyllos*, but these do not belong exactly to forest trees, having certainly been planted either by one of the Priors of Little Malvern or some of his lay successors to the Priory lands."

A very large lime formerly grew in Hagley Park, Worcestershire, the seat of Lord Cobham, which, according to Lees,[1] in 1874 measured 27 ft. in girth at 3 ft. from the ground, but this I am told was blown down about twenty-five years ago. At Arley Castle there is a fine tree, about 120 ft. high by 12 ft. 8 in. in girth.

The most curious instance of artificial layering I have seen anywhere is at Knole Park, where a lime described by Loudon as having covered nearly a quarter of an acre in 1820 still grows. The central stem no doubt originally dropped its branches on the ground in a circle of about 8 yards in diameter. These have grown up into trees 80 ft. to 90 ft. high, some of which are thicker than their parent. These again have layered themselves in a second concentric circle 20 yards in diameter, the trees in which are 20 ft. to 40 ft. high, and these are now rooting their outer branches in a third circle more irregular than the others, and 8 yards distant from it, so that the total diameter of the group is 36 yards. All the stems are more or less covered with spray, and the central one seems to have long ceased to grow.

Strutt paid less attention to the lime than it deserves. He figures only two, one of which, at Cobham, is in the same plate with a sycamore; the other at Moor Park, near Ricksmansworth, Herts. This stood at the end of a line of large limes, and was a very wide-spreading tree, with a trunk 17½ ft. at three ft. from the ground, and branches 120 yards in circumference. I am informed by Mr. Haynes, gardener at Moor Park, that this tree was blown down in 1860; but it is still alive, and some of the branches have taken root in the ground and have sent up stems about forty-five feet high. Within fifty yards of it another tree in the same row is now 21 ft. in girth.

A similar case of a lime having been blown down and the branches taking root occurs at Stratton Strawless. This tree, as Mr. Birkbeck tells me, was mentioned by Sir T. Browne in the reign of Charles I. It was blown down in 1895 and lay till 1900, when the roots were covered with a mound of sand. When I saw it in 1909, many branches were throwing up vigorous shoots, and the tree looked as if it might live for centuries. Its trunk was about 12 ft. in girth. A very fine tall red-twigged lime by the water at Gatton Park, in 1904, was 131 ft. by 12½ ft. A large spreading tree, at Osberton Grange, Notts, is about 80 ft. high by 19 ft. in girth.

At Dallam Tower, Westmoreland, there is an old lime in an exposed situation in the park which measures no less than 22 ft. 3 in. in girth, though not over 65 ft. in height; and, as showing the influence of situation on trees, I may say that, in a sheltered hollow close to the house at the same place, I measured a lime 128 ft. high, double the height of the first, but only 7 ft. 8 in. in girth.

There is a row of very large and apparently old limes at Hawsted, near Bury St. Edmunds, in the same field, and probably planted at the same time as the Oriental planes which we have described.[2] The largest is about 105 ft. by 20 ft. Another, whose trunk is covered with large burrs, is 16 ft. in girth. The leaves on the shoots from the base of these trees vary considerably in size and shape.

[1] *Gard. Chron.* i. 49 (1874).

[2] Cf. Vol. III. pp. 621, 622.

In the west garden at Hatfield House, Herts, there is a remarkable pergola of the common lime, about 280 yards long, and 10 ft. wide. The trees, which are pruned every year, stand 12 ft. apart in the row on each side, and are 7½ ft. high.

In some parts of England, especially in Essex and Herts, the lime is infested by mistletoe, which often kills the branches and causes irregular excrescences, which sometimes have an elongated gourd-like shape. A remarkable specimen of this, taken from a tree in front of Spains Hall, the seat of A. W. Ruggles-Brise, Esq., in Essex, is now in the Forestry Museum at Cambridge. It was stated in the *Gardeners' Chronicle*,[1] in answer to an inquiry by me, that such swellings are abundant on the limes at Hampton Court and at Dropmore, where Mr. Page states that a large tree was cut down on account of it being in a dying condition.

In Scotland[2] the lime is at many places almost as fine as in England. An immense tree growing at Kinloch, Meigle, is, as I am told by Sir John Kinloch, about 90 ft. by 21 ft., and spreads over an area of a hundred yards in circumference. A wide-spreading lime at Gordon Castle, which is known as the Duchess' tree, measures about 89 ft. high and 17 ft. 4 in. in girth; its layered branches form a dense mass of shoots and have not been trained into trees like the lime at Knole. Their total circumference is not less than 126 paces, so that it covers as large an area of ground as the Newbattle beech.[3] This tree is mentioned in *Old and Remarkable Trees* as having been, in 1867, 70 ft. by 16½ ft. at 3 ft., the circumference of the branches being 310 ft. There is a wide-spreading common lime at Pitfirrane, near Dunfermline, of which the gardener, Mr. Percy Brown, has sent us a photograph. It was, in 1912, 74 ft. high, and 13 ft. 3 in. in girth, the circumference around the branches being 298 ft. At Leny there are some fine limes, one of which I found to be 105 ft. by 12 ft. 3 in., which, at such a high elevation above the sea, is remarkable. At Ancrum, near Roxburgh, Mr. Renwick saw a common lime 17 ft. 3 in. in girth in 1909.

At Roseneath, near the great silver firs (see p. 729), is an old avenue of large-leaved limes, covered with such a mass of small spray that it was impossible to see the bark near the ground, and one of these, measured over the spray with the tape as tight as I could make it, was no less than 24 ft. There are some fine lime trees of great size in the park at Taymouth Castle.

In Ireland we have seen no limes of remarkable size, and the tree never seems to have been so generally planted as in England. There is an avenue of fair-sized trees at Muckross Abbey; and Loudon mentions a tree in the park at Charleville forest, Co. Meath, which was reported at that time to be 110 ft. high and 5½ ft. in diameter at 1 ft. from the ground. At Rossanagh a remarkable tree was growing in 1908 which, as I was informed by Mr. W. T. Tighe, was blown down about 1825. His grandfather had it pulled up into a leaning position, and placed a large boulder over the roots to keep it firm. It now leans at an angle of about 40°, and has grown into a flat trunk 6 ft. wide on the side but only 14 ft. 10 in. in girth. It is about 100 ft. in height and twice as large in girth as any other of the trees in the same line, which appear to have been planted at the same time.

[1] *Gard. Chron.* xli. 240, 257 (1907).

[2] Cf. Christison, in *Trans. Bot. Soc. Edin.* xix. 494 (1893).

[3] Vol. I. p. 23, Plates 8, 9.

TIMBER

The wood of the lime is pale yellow or white, light, soft, and close-grained, and is not liable to become worm-eaten. It was formerly valued by pianoforte-makers for sounding-boards; and cutting boards used by shoemakers, glovers, and harness-makers, were made of it. I am informed by Mr. Anderson that twenty-five years ago he was sometimes able to sell the best part of large trees at as much as 5s. or 6s. per foot; and at Longleat a large lime tree blown down in the park realised 4s. 6d. per foot; but its use has now been superseded by foreign imports from America and elsewhere, and from one to two shillings per foot is its usual value. Owing to its softness, consistency, and non-liability to split, it was preferred for wood-carving; and all the finest carvings by Grinling Gibbons are said to be done in lime wood.[1]

In northern Europe, especially in Russia, the inner bark of the small-leaved and large-leaved limes is largely used for making the bast mats which are used as dunnage in grain cargoes, and also imported for covering garden frames. The shoes worn by the Russian peasants are made from plaited lime bark, and Loudon says that ropes were made from it in his time in Devonshire and Cornwall; but this, like so many other rural industries, has now, I believe, quite died out.

I am indebted to Mr. J. Rose, of Messrs. Broadwood & Sons, for the following:—"Fine lime-tree was at one time very eagerly sought after in this country for the manufacture of pianoforte keys. When large and freely grown it is a beautifully straight and silky-grained wood, easily worked, not given to warping, very light in weight, and yet very tough. These qualities made it an admirable material for the purpose. But it became more and more difficult to obtain lime-tree of fine quality, and it was replaced by the importation of American basswood, a wood of similar character, easier to obtain in good sizes, free from knots, and straight in the grain, which is imported in the form of boards, or of glued-up and planed keyboards ready for the ivories. It has also been replaced, in part, for key-making by continental-grown pine, which has distinct advantages for the purpose."

A marked feature in the timber trade in recent years has been the importation of sawn timber, which has greatly affected the sale of home-grown timber. The manufacturer is now supplied with foreign timber ready sawn, seasoned to some extent, and often carefully graded; whereas home-grown timber has to be collected in comparatively small parcels, and its selection and handling require a great amount of knowledge and experience possessed by a very few persons. (H. J. E.)

[1] Evelyn (*Silva*, Hunter's ed. i. 205 (1801)), says: "Because of its colour and easy working, and that it is not subject to split, architects make with it models for their designed buildings; and the carvers in wood use it not only for small figures, but for large statues and entire histories in bass and high relieve; witness beside several more the festoons, fruitages, and other sculptures of admirable invention and performance, to be seen about the choir of St. Paul's and other churches, royal palaces, and noble houses in city and country; all of them the works and invention of our Lysippus, Mr. Gibbons, comparable, and for ought appears equal to anything of the antients. Having had the honour (for so I account it) to be the first who recommended this great artist to his Majesty Charles II., I mention it on this occasion with much satisfaction."

## TILIA EUCHLORA

*Tilia euchlora*, Koch, in *Wochenschr. Gärtn. u. Pflanzenk.* ix. 284 (1866), and *Dendrologie*, i. 473 (1869); Schneider, *Laubholzkunde*, ii. 374 (1909); V. Engler, *Monog. Gatt. Tilia*, 149 (1909).
*Tilia multiflora*, Simonkai, in *Math. Term. Közl.* xxii. 328 (1888) (not Ledebour).
*Tilia rubra*, var. *euchlora*, Dippel, *Laubholzkunde*, iii. 63 (1893).
*Tilia dasystyla*, Jack, in *Garden and Forest*, i. 332 (1888); Nicholson, in *Kew Hand-list Trees*, 45 (1894) (not Steven).[1]

A tree, attaining 50 ft. in height and 6 ft. in girth, but possibly larger in its native country. Bark grey and scaly. Young branchlets green, glabrous, the short shoots, however, being slightly pubescent. Leaves (Plate 407, Fig. 10) intermediate in size between those of *T. cordata* and *T. vulgaris*, averaging 2½ in. in width and length, orbicular-ovate, coriaceous, cuspidate at the apex, oblique and cordate at the base; upper surface dark shining green, glabrous; lower surface paler, almost glaucous, glabrous, except for brownish axil-tufts at the base and at the junctions of the midrib, primary and secondary nerves; margin regularly serrate, the teeth ending in long slender points; petiole glabrous, slender, more than half the length of the blade. Buds, with three external green glabrous ciliate scales.

Cymes glabrous, pendulous, exceeding the leaves in length, three- to seven-flowered; bract glabrous, tapering at both ends, shortly stalked; flowers similar to those of *T. platyphyllos*, but the ovary is tomentose with long hairs, which are continued on the base of the style, the upper ¾ of which is glabrous. Fruit ovoid, indistinctly five-ribbed, covered with dense short brownish grey tomentum, the base of the style persistent at the apex; shell thick and woody.

This species is remarkably distinct in appearance, owing to the dark green and remarkably glossy upper surface of the leaves. It comes into flower about the end of July at Kew, later in the season than most of the limes, and the flowers have a peculiar colour, owing to the distinctly yellow tinge of the petals and filaments.

It has been much confused with *T. dasystyla*,[2] Steven, a native of the Crimea, which has leaves quite different in colour and shape, and, as its name indicates, a densely tomentose long style.

*Tilia euchlora* is represented in the Kew herbarium by a wild specimen from Karabagh in Russian Armenia; but it is supposed by Schneider and V. Engler to be a hybrid; and, if this is the case, its parents are possibly *T. caucasica*,[3] Ruprecht, the common large-leaved lime in the Caucasus, and *T. cordata*. It is always propagated in nurseries by budding on the common lime; and seedlings of it appear to be unknown.

[1] Loudon, *Arb. et Frut. Brit.* i. 366 (1838), refers to Steven's species from the Crimea, and not to the tree called by us *T. euchlora*, Koch.

[2] *T. dasystyla* is represented in the Kew herbarium by Steven's type specimen and two other specimens, all from the Crimea.

[3] The Caucasian lime can scarcely be identified with *T. rubra*, De Candolle, in *Cat. Plant. Hort. Monsp.* 150 (1813), which is described as having leaves pubescent beneath as in *T. platyphyllos*, and evidently refers to the southern form of the latter species in Europe.

This species is perhaps the handsomest of all the limes, on account of its shining foliage, which is very late in falling in autumn, and seems to be free from insect and fungoid attacks and from honey dew. It is apparently quite as hardy as the common lime, and young trees at Kew are remarkably thriving and healthy. It was unknown in Loudon's time, and seems to have been introduced a short time before 1866, when it was first accurately distinguished by Koch. It is rather rare in cultivation in England, though it is planted in Berlin and other German cities,[1] and thrives[2] in the Arnold Arboretum in Massachusetts. (A. H.)

## TILIA TOMENTOSA, White Lime

*Tilia tomentosa*, Moench, *Verz. Ausl. Bäume Weissenst.* 136 (1785); V. Engler, *Monog. Gatt. Tilia*, 116 (1909); Schneider, *Laubholzkunde*, ii. 386 (1909).
*Tilia alba*, Aiton,[3] *Hort. Kew.* ii. 230 (1789), and iii. 300 (1811); Waldstein and Kitaibel, *Icon. Pl. Hung.* i. 2, t. 3 (1802); Loudon, *Arb. et Frut. Brit.* i. 372 (1838).
*Tilia pallida*, Salisbury, *Prod.* 367 (1796).
*Tilia rotundifolia*, Ventenat, *Mém. Inst. Paris*, iv. 12, t. 4 (1803).
*Tilia argentea*, Desfontaines, in De Candolle, *Cat. Pl. Hort. Monsp.* 150 (1813).
*Tilia petiolaris*, De Candolle, *Prod.* i. 514 (1824) (not J. D. Hooker).

A tree, attaining 100 ft. in height and 15 ft. in girth, usually with markedly ascending branches. Young branchlets covered with white stellate tomentum, more or less retained in the second year. Leaves (Plate 407, Fig. 3), about 3 to 5 in. across, nearly orbicular, cuspidate at the apex, cordate or truncate at the base; margin often lobulate, serrate or biserrate, the serrations ending in short blunt cartilaginous points; upper surface green, with scattered stellate pubescence; lower surface covered with a dense whitish tomentum, without axil-tufts; petiole stout or slender, less than half the length of the blade, stellate-pubescent. Buds, with three external grey tomentose scales.

Flowers, in seven- to ten-flowered pendulous tomentose cymes, which are shorter than the leaves; bract tomentose, sub-sessile; sepals tomentose, clothed with long hairs at the base within; petals glabrous, longer than the sepals; staminodes slender, spatulate, shorter than the petals; stamens, shorter than the staminodes, numerous, with the halves of each anther on a distinct short stalk; ovary ovoid, tomentose; style glabrous. Fruit ovoid, elongated, apiculate, slightly five-angled, grey tomentose, smooth or only indistinctly warty; shell woody.

This species shows in the wild state considerable variation in the shape of the

[1] Mr. Bean, in *Kew Bull.* 1908, p. 390, says that this species, which is so promising a tree for street planting, is abundant in the Boskoop nurseries, near Gouda, in Holland.

[2] *Garden and Forest*, i. 332 (1888).

[3] In Aiton, *Hort. Kew.* ii. 230 (1789), it is erroneously stated that the common white lime is a native of North America, an error which was rectified in the second edition of this work, iii. 300 (1811), where Hungary is correctly given. Some writers have supposed that *T. heterophylla* was the species referred to; but the type specimen in the British Museum, inscribed *Tilia alba* in Solander's handwriting, though bearing neither flowers nor fruit, is without doubt a branch of the common European lime, identical with var. *argentea*.

leaves—some specimens, distinguished by V. Engler, as var. *typica*, having suborbicular leaves, usually cordate at the base; whilst others, var. *petiolaris*,[1] V. Engler, have leaves usually broader than long, and more or less truncate or subcordate at the base. These are connected by numerous intermediate forms, and can scarcely be maintained as distinct varieties.

All the wild specimens which I have seen, are characterised by leaves, thin in texture, usually pale green above, greyish and not snowy-white beneath, and with slender petioles. Trees similar to the wild form are occasionally seen in cultivation; but the tree which is more commonly cultivated under the name *T. argentea*, and which possibly was the one described by Moench as *T. tomentosa*, and in that case technically the typical form of the species, differs considerably in having larger leaves, thick in texture, more or less orbicular, uneven on the surface, dark shining green above, snowy white beneath, margin often lobulate, petioles short and stout. It is convenient to distinguish this cultivated form, the origin of which is unknown to me, as var. *argentea*. It has always ascending branches, and is possibly a sport.

This species is a native of south-eastern Europe, and Asia Minor; but does not extend as far eastward as the Caucasus. Its northern limit is southern Hungary, where it is found in some parts of Banat, Slavonia, and Croatia. It is widely distributed throughout the Balkan peninsula, extending southwards as far as Laconia in Greece; and spreads eastward through Roumania and Moldavia to Bessarabia, Podolia, and the Crimea. In Asia Minor, it is limited to Bithynia and the island of Chios. It is usually a component of mixed deciduous woods, growing in valleys and mountain slopes at a low elevation, but occasionally forms pure woods of small extent. (A. H.)

## Cultivation

The white lime was introduced into England in 1767, and has been planted as an ornamental tree at many places; but I know no avenues of it in England.

This species should be multiplied either by seed, which here only ripens in hot seasons such as 1911, or by layering, as when grafted, as is often done, on the common lime, the scions grow thicker than the stock and produce an unsightly swelling at the point of junction. As an ornamental tree, and for use in towns, it is much superior to the common lime, on account of the freedom of the leaves from honey dew.

The only place where I have seen the silver lime growing wild in Europe is in the forests of Bosnia, near Maglai in the valley of the Bosna, at about 1000 ft. elevation. Here it was scattered in forests of oak, and other deciduous trees, and was so conspicuous when the silvery white undersides of the leaves were upturned by the wind, that at a distance of a mile or so I at first supposed it to be a tree covered with white flowers.[2] Probably some such experience must have induced a former owner of Highclere to plant it largely in that beautiful park, where, as Loudon

[1] To be carefully distinguished from *T. petiolaris*, Hooker (see p. 1677).

[2] In sunny weather, the leaves on the sunny side, especially at the ends of the branchlets, are reversed, turning their white sides to the light. This is a provision against excessive transpiration of water; and has been observed in the other silver limes and in *Quercus conferta*. See Kerner's remarks in his *Nat. Hist. Plants*, Eng. Trans. i. 338 (1898).—A. H.

remarks, its presence could be detected at some miles distance through the apparently dense forest by the white tops appearing at intervals among other trees.

The finest tree of this species that we have seen grows at Albury Park, Sussex, in front of the Duke of Northumberland's house, and measures 100 ft., or perhaps a little more, in height, by 13½ ft. in girth (Plate 373). The ascending branches seem characteristic of this species, in cultivation at least, but this tree has an unusually regular and perfect head. I was informed by the late Mr. Leach that at Albury the flowers of this tree are poisonous to bees,[1] whose dead or stupefied bodies are found lying on the grass below it in August, and this observation is confirmed by Mr. Comber at The Hendre and by other observers.

At Henham Hall, Suffolk, a tree measured 76 ft. by 8 ft. 10 in. in 1909. It is grafted on the common lime, and is of the typical upright habit. At Hewell Grange a large tree, with the bark decaying on one side, was 92 ft. by 9 ft. 8 in. in 1909. At Dropmore there are two trees in the avenue to the Taplow gate, the larger of which is about 70 ft. by 11 ft. 3 in. At Harpsden Rectory, Oxon., Henry saw two trees, the larger of which was 80 ft. by 6 ft. 8 in. in 1907. At Arley Castle, a round-headed tree of upright habit measured[2] 62 ft. by 6 ft. 9 in. in 1903. At Beauport, Sussex, there is a tall but slender tree in a rather crowded situation, which has been grafted at about 10 ft. from the ground. Dr. Masters reported[3] a tree of fine proportions and symmetry at Strathfieldsaye, 70 ft. by 6 ft. 2 in., in 1899. Mr. Bean[4] saw a tree 80 ft. high in the Royal Gardens at Sans Souci, Berlin. (H. J. E.)

## TILIA PETIOLARIS, Weeping White Lime

*Tilia petiolaris*, J. D. Hooker, *Bot. Mag.* t. 6737 (1884) (not De Candolle); Boissier, *Fl. Orientalis, Suppl.* 136 (1888).

*Tilia alba*, Koch, *Dendrologie*, i. 478 (excl. syn.) (1869) (not Aiton).

*Tilia tomentosa*, Moench, var. *petiolaris*, Kirchner, in Petzold and Kirchner, *Arb. Musc.* 162 (1864); V. Engler, in *Mitt. Deut. Dend. Ges.*, 1907, pp. 218-221.

*Tilia tomentosa*, Moench, var. *sphærobalana*, Borbas, in *Bot. Centralb.* xxxvii. 168 (1889); V. Engler, *Monog. Gatt. Tilia*, 121 (1909).

A tree, attaining 80 ft. in height, differing from the wild form of *T. tomentosa*, as follows: Branches and branchlets pendulous; leaves with long slender petioles, exceeding half the length of the blade; fruit globose, depressed at the summit, from which arises a short stout style, very warty on the greyish surface, and divided by five vertical furrows into as many lobes; seeds[5] often imperfect.

The leaves (Plate 407, Fig. 2) average about 3 in. across, and are obliquely orbicular, cordate or truncate at the base, cuspidate at the apex, flat on the surface and not wrinkled; margin finely and regularly serrate, the teeth ending in short points; under surface covered with a dense white tomentum; upper surface dark

[1] The flowers of the large tree of *T. petiolaris* at Kew are equally poisonous to bees. Cf. p. 1679, note 1.

[2] R. Woodward, *Hortus Arleyensis*, 25 (1907).

[3] In *Gard. Chron.* xxvi. 162 (1899).

[4] *Kew Bull.* 1908, p. 395.

[5] Engler, out of fifty fruits which he examined, found only three with good seed, 1 to 2 in each fruit.

green, with scattered stellate pubescence. The branchlets, buds, and flowers are identical with those of *T. tomentosa.*

This tree appears to be a sport of *T. tomentosa* of unknown origin. Schneider considers it to be a native of southern Hungary and the Balkan States; and no doubt, specimens of silver limes with longer petioles than usual occur in that region, but no one, so far as I am aware, has ever seen a tree in the wild state with the pendulous habit and the peculiar fruits of *T. petiolaris,* Hooker. V. Engler admits that it is only known in cultivation.

The history of this tree is obscure; but it seems to have been first accurately distinguished by Kirchner, who, in 1864, described it as *T. tomentosa,* var. *petiolaris,* and erroneously identified it with *T. petiolaris,* De Candolle.[1] Hooker, in *Bot. Mag.* t. 6737 (1884), following Kirchner, adopted De Candolle's name; but, though this was incorrect, it is convenient to retain the name as *T. petiolaris,* Hooker. This peculiar tree was not known in Loudon's time; but must have been introduced from the Continent soon afterwards, as the fine specimen in the Cambridge Botanic Garden was probably planted[2] in 1842. Before 1864 it was known in cultivation as *T. americana pendula*[3]; and Koch in 1869 considered it to be of American origin.

1. *Tilia orbicularis,* Carrière,[4] which originated in Simon-Louis' nursery at Plantières, is evidently a seedling of *T. petiolaris,* from which it differs mainly in being less pubescent. The leaves, which are dark glossy green above and dull grey beneath, also differ from those of *T. petiolaris* in the shorter petioles, which are, however, slender as in that species. The serrations are also slightly sharper and occasionally more irregular. Flowers as in *T. petiolaris,* but with the bract larger and nearly glabrescent, and the sepals and pedicels covered with a less dense and greyish tomentum. Fruit strictly globose, not depressed at the summit, and showing no furrows, but having the same warty surface as that of *T. petiolaris.*

This tree, which is not nearly so pendulous in habit as *T. petiolaris,* is reputed, on account of the dark glossy green of the upper surface of the leaves, to be a hybrid between *T. petiolaris* and *T. euchlora*; but, in the present state of our knowledge of hybrids, it is judicious to say nothing about its parentage until experimental sowings have been made on a large scale of the seed of *T. petiolaris.*[5]

Two small trees of *T. orbicularis,* obtained from Plantières in 1900, are thriving at Kew, one of which has already produced flowers and fruit. They retain their foliage late in the season. (A. H.)

[1] *T. petiolaris,* De Candolle, was founded in 1826 on a branch without flowers or fruit, now preserved in the Geneva Herbarium, which was taken from a tree cultivated in the Imperial Botanic Garden at Odessa; but Lange, in *Flora,* i. 233 (1827), who had seen this tree, states that it was identical in every respect with the ordinary form of *T. argentea* cultivated at Paris; and a drawing at Kew of De Candolle's specimen confirms Lange's opinion. The name *T. petiolaris,* De Candolle, thus disappears, being a mere synonym of *T. tomentosa.*

[2] The oldest herbarium specimen, which I have seen, is one in fruit from a tree growing in a street at Nancy, collected by Billot in 1861. According to Bunbury, *Arb. Notes,* 67 (1889), the fine tree at Barton was planted by his father; but no exact date can be now ascertained.

[3] Cf. Rehder, in *Mitt. Deut. Dend. Ges.* 1904, p. 209.

[4] *Ex* Beissner, in *Mitt. Deut. Dend. Ges.,* 1898, pp. 86 and 88. It appears, however, to have been first accurately described by Jouin in *La Semaine Horticole,* 1899, p. 335.

[5] A branch from a seedling of *T. petiolaris,* raised in the Arnold Arboretum and sent to me by Prof. Sargent in 1910, bears foliage identical with that of the parent. Sargent, *Bull. Pop. Inform.* No. 30 (1912), and in *Gard. Chron.* lii. 88 (1912), states, however, that plants raised in the Arnold Arboretum from the seeds of a tree of *T. petiolaris,* which was growing near *T. americana,* the two flowering at the same time, are identical with trees of *T. spectabilis.* See p. 1686.

*Tilia petiolaris* is a beautiful weeping tree, which has not been nearly so generally planted as it deserves to be. There are good examples in the Botanic Gardens at Kew, Cambridge, and Glasnevin. At Stowe, near Buckingham, there is a handsome tree, 74 ft. by 6 ft. 8 in. There are two fine trees, girthing 7 ft. 4 in. and 5 ft. 7 in., and about 80 ft. high, growing on the bank of the Thames near Cliveden on the Wharfe Estate, belonging to Lord Boston. A very similar one on the lawn at Barton, Suffolk, was measured by Henry as 83 ft. by 8 ft. 2 in. in 1908. Another at Chiswick House, which has the trunk decayed on one side, measured 77 ft. by 10½ ft. in 1903. At Bicton a handsome tree, near the house, grafted at seven feet from the ground, measured 80 ft. by 5½ ft. in 1906. At Hatherop Castle there is a beautiful specimen of moderate size on the lawn (Plate 374), which has layered naturally; and many plants have been propagated from it also by artificial layers. At Gunnersbury House there is a good tree, which in 1912 measured 56 feet high by 6 ft. in girth at 4 ft. from the ground. At Aldenham there is a tree on the lawn which ripened fruit in 1911 from which I raised seedlings. I noticed many dead bees[1] under it on August 20.

In Scotland it appears to be perfectly hardy at Durris; and Henry found, in 1905, a tree at Bargaly, 41 ft. by 4 ft. 8 in.

Mr. Bean[2] saw a fine specimen at Herrenhausen, in Hanover, which was 9 ft. 2 in. in girth in 1908. (H. J. E.)

## TILIA MONGOLICA

*Tilia mongolica*, Maximowicz, in *Mél. Biol.* x. 585 (1880), and *Enum. Pl. Mongol.* 118, t. 11 (1889); L. Henry, in *Rev. Hort.* 1902, p. 476, figs. 214, 215, 217; Rehder, in Sargent, *Trees and Shrubs*, i. 121, t. 61 (1903).

A small tree, scarcely exceeding 30 ft. in height, and flowering when only a few feet high. Young branchlets glabrous, reddish, becoming grey in the second year. Leaves (Plate 407, Fig. 7) about 2½ in. wide, acuminate at the apex, with one or two sharp-pointed lateral lobes; base truncate or cordate; coarsely serrate, with a few large triangular teeth, tipped with long callous points; upper surface dark green, shining, glabrous; lower surface glaucous, with pubescent tufts in the axils at the base and at the junctions of the primary and secondary nerves, elsewhere glabrous; petiole glabrous.

Flowers, six to twelve in a cyme; bract stalked; sepals erect, villous within, glabrous without; petals erect, longer than the sepals; staminodes five, obtuse; stamens, as long as the sepals, thirty-five to forty, in five bundles; style glabrous. Fruit ovoid, mucronulate, without ribs or only slightly ribbed, thick-walled, shortly tomentose.

This species is very distinct in appearance, the small coarsely serrate leaves resembling those of a birch, and opening with a reddish tint in spring.

[1] In 1908, the bodies of innumerable bees, poisoned by the flowers of a tree of *T. petiolaris* at Tortworth, had so much manured the ground under its outer branches, that a very green ring of turf was visible in the autumn following, and was noticed by the Earl of Ducie to be even more conspicuous in 1909.

[2] *Kew Bull.* 1908, p. 392.

This tree occurs in northern China, at Jehol, and on the Po-hua mountain, west of Peking; and has also been found in the Moni Ula range, north of Ordos, in Mongolia. It was introduced into cultivation by Dr. Bretschneider,[1] who sent seeds from Peking to the Museum at Paris in 1880, and to the Arnold Arboretum in 1882. The specimen in the Jardin des Plantes at Paris is about 20 ft. high. A small tree at Kew flowered, when only 5 ft. high, at the end of July 1907. This species has lately been introduced into the Coombe Wood Nursery by Mr. Purdom, who has been collecting for Messrs. Veitch in northern China. (A. H.)

## TILIA PAUCICOSTATA

*Tilia paucicostata*, Maximowicz, in *Act. Hort. Petrop.* xi. 82 (1890); Schneider, *Laubholzkunde*, ii. 371 (1909); V. Engler, *Monog. Gatt. Tilia*, 87 (1909).

*Tilia Miqueliana*, var. *chinensis*, Diels, in Engler, *Bot. Jahrb.* xxxvi., *Beibl.* No. 82, p. 75 (1905) (not Szyszylowicz).

A small tree. Young branchlets glabrous, green. Leaves about 2½ in. long and 2 in. wide, ovate, usually truncate and rarely cordate at the base, ending at the apex in a long non-serrate acuminate cusp; green and glabrous on both surfaces, except for minute axil-tufts of pubescence beneath, which are, however, absent at the base of the blade; tertiary veins on the lower surface few, irregular, not parallel, but anastomosing, more or less prominent; margin with regular fine serrations ending in long points; petiole about an inch long, green, glabrous.

Cymes erect, each with seven to fifteen flowers; bracts glabrous, stalked; staminodes present; style pilose at the base. Fruit globose, tomentose, faintly five-ribbed.

This species differs from *T. cordata* and its allies, in the prominent tertiary venation and the green and not glaucous under surface of the leaf, which is usually truncate at the base. It is a native of the provinces of Kansu, Shansi, and Szechwan, in western China, where it was collected by Potanin, Giraldi, and Wilson. The latter sent a living plant[2] in 1901 to Coombe Wood, which I consider to be probably of this species. From it many grafts have been taken, and it now produces coppice shoots with large leaves, which show at the base of the blade and on the adjoining end of the petiole a trace of scattered stellate pubescence. This is probably a juvenile character, disappearing on adult plants; and a young tree at Kew, one of the grafts, about 8 ft. high, bears leaves similar to those of the adult wild tree, though slightly larger, and in these the stellate pubescence has almost disappeared.

(A. H.)

[1] *Hist. Europ. Bot. Disc. China*, ii. 1050 (1898). The seedlings mentioned here as being alive at Kew in 1893 cannot be traced.

[2] This is probably the plant referred to *T. Miqueliana*, var. *chinensis*, in *Hortus Veitchii*, 381 (1906).

## TILIA OLIVERI

*Tilia Oliveri*, Szyszylowicz, in Hooker, *Icon. Plant.* ad t. 1927 (1890); Schneider, *Laubholzkunde*, ii. 387 (1909); V. Engler, *Monog. Gatt. Tilia*, 114 (1909).

*Tilia pendula*, V. Engler, *ex* Schneider, *Laubholzkunde*, ii. 387 (1909), and *Monog. Gatt. Tilia*, 113 (1909).

*Tilia mandshurica*, Szyszylowicz, in Hooker, *Icon. Plant.* ad t. 1927 (1890) (not Ruprecht and Maximowicz).

A tree, attaining in western China about 50 ft. in height. Young branchlets glabrous. Leaves variable in size, usually longer than broad, averaging 3 to 4 in. in length and 2½ to 3½ in. in breadth, orbicular-ovate, cordate at the base, cuspidate at the apex; margin regularly serrate, with shallow sinuses between the crenate teeth, which are very short and end in cartilaginous points; upper surface dark green, glabrous; lower surface covered with a dense white thin tomentum, without axil-tufts; petiole glabrous, one-half to three-fourths the length of the blade.

Cymes usually much longer than the leaves, each with about twenty flowers, which are similar to those of *T. tomentosa*, but are smaller in size and on short thickened pedicels. Fruit globose, grey tomentose and tuberculate on the surface, thick-shelled, apiculate, ⅓ in. in diameter.

This species, which promises to be a beautiful ornamental tree, is readily distinguished from the other limes with a pure white under surface to the leaves, by the glabrous branchlets and petioles, and the crenately serrate orbicular leaves.

This species is a native of central China, where it was discovered by me in 1888, in the mountains north of the Yangtze, in the Fang and Wushan districts of Hupeh. *T. Oliveri* was founded by Szyszylowicz on a branch (Henry, No. 7089) from a small shrubby tree, growing in a sunny exposure on high cliffs, and bearing in consequence small leaves, averaging 2 in. in length. Another specimen (Henry, No. 7452 B), gathered by me at no great distance, but in a shaded valley, bore leaves averaging 3½ in. in length, and was identified by Szyszylowicz with *T. mandshurica*, which is a native of northern China. This specimen has been made the type of a new species, *T. pendula*, V. Engler. A third specimen, collected since by Wilson (No. 2274), with leaves intermediate in size, is considered by Schneider to be *T. pendula*, but by Engler to be *T. Oliveri*. A careful examination of the whole material shows that all the specimens belong to one species.

*T. Oliveri* was introduced by Wilson, who sent seed from central China in 1900. It is now growing vigorously at Coombe Wood, and a small specimen is thriving at Kew. (A. H.)

## TILIA MANDSHURICA

*Tilia mandshurica*, Ruprecht et Maximowicz, in *Bull. Acad. St. Pétersb.* xv. 124 (1856); Maximowicz, *Prim. Fl. Amur.* 62 (1859), and in *Mél. Biol.* x. 586 (1880); Baker and Moore, in *Journ. Linn. Soc.* (*Bot.*) xvii. 380 (1879); Franchet, *Pl. David.* i. 60 (1884); Forbes and Hemsley, in *Journ. Linn. Soc.* (*Bot.*) xxiii. 94 (1886); Komarov, in *Act. Hort. Petrop.* xxv. 28 (1907),

*Tilia pekinensis*, Ruprecht, in *Bull. Acad. St. Pétersb.* xv. 125 (1856), and in Maximowicz, *Prim. Fl. Amur.* 469 (1859); Bayer, in *Verh. Zool. Bot. Ges. Wien*, xii. 49 (1862).

A tree, attaining in Manchuria about 60 ft. in height and 10 ft. in girth. Young branchlets and buds covered with brownish tomentum. Leaves, 4 to 5 in. in breadth and length, orbicular-ovate, usually cordate at the base, cuspidate at the apex; margin, often with one or two lobes, coarsely serrate, the teeth ending in long awn-like points; upper surface with a scattered stellate pubescence, which forms dense tufts at the base of the blade; lower surface densely covered with whitish stellate tomentum, but without axil-tufts; petiole half the length of the blade, stout, brown tomentose.

Flowers similar to those of *T. tomentosa*, but with bracts, pedicels, and sepals more densely covered with a brownish tomentum. Fruit globose, tomentose, and slightly warty, either without ribs or with five indistinct ribs towards the base.

This species is closely allied to *T. tomentosa*; but has larger leaves, with long-pointed serrations and different fruit. Young trees have usually lobed leaves, as is often the case in other species; and *T. pekinensis*,[1] founded on this character, cannot be retained even as a distinct variety.

*Tilia mandshurica* is widely spread throughout the whole of Manchuria, and also occurs in Korea, where it was found on the Diamond Mountains by Père Faurie. It is not uncommon in the mountains west and north of Peking. It occurs scattered or in groups throughout the broad-leaved forest of these regions. It is known to the Chinese, like all the other species of lime, as the *tuan* tree; and the bark is used for making ropes and sandals.

It was probably introduced by Maximowicz into the St. Petersburg botanic garden; but is extremely rare in cultivation, the only specimen which I have seen being a small tree at Kew, which was procured from Booth of Hamburg in 1871. As it comes into leaf very early in the spring, it is often cut by frost, and is not in a thriving condition. (A. H.)

[1] Var. *pekinensis*, Engler, *ex* Schneider, *Laubholzkunde*, ii. 384 (1909).

## TILIA MAXIMOWICZIANA

*Tilia Maximowicziana*, Shirasawa, in *Bull. Coll. Agric. Univ. Tokyo*, iv. 158, t. xviii (1900), and *Icon. Ess. Forest. Japon*, ii. t. 50 (1908); Schneider, *Laubholzkunde*, ii. 385 (1909).

*Tilia Miqueliana*, Sargent, in *Garden and Forest*, vi. 111, fig. 19 (1893), and *Forest Flora of Japan*, 19, t. 8 (1894) (not Maximowicz).

*Tilia Miyabei*, Jack,[1] in *Mitt. Deut. Dend. Ges.*, 1909, p. 285.

A tree, attaining in Japan 100 ft. in height and 10 ft. in girth. Young branchlets densely covered with a greyish brown tomentum. Leaves usually large,[2] about 5 in. in breadth and length, cordate at the base, cuspidate at the apex; margin ciliate, coarsely and regularly serrate, the serrations ending in blunt cartilaginous points; upper surface dark green, with scattered stellate tomentum on the surface between the nerves, and dense tomentum on the nerves, especially at the base of the blade; under surface greyish, densely covered with stellate tomentum, and with conspicuous brownish axil-tufts at the junctions of the nerves; petiole stout, less than half the length of the blade, covered with greyish brown tomentum. Buds covered with greyish or rusty brown tomentum.

Flowers, ten to eighteen in each pendulous tomentose cyme; bract shortly stalked, strongly veined and tomentose on both surfaces, the tomentum brown and very dense on the midrib; sepals lanceolate, acuminate, pubescent on both surfaces; petals keeled, glabrous; staminodes five, keeled, toothed at the apex, as long as but narrower than the petals; stamens sixty-five to seventy-five, united in five bundles, shorter than the petals, each half of the anther on a short stalk; ovary pubescent, ovoid; style glabrous. Fruit globose, about $\frac{1}{3}$ in. in diameter, with a thick woody grey pubescent five-ribbed shell.

This species is a native of Japan, ranging from central Hondo (lat. 36°) northwards to Hokkaido. In Hondo, according to Shirasawa,[3] who has observed it in the provinces of Kotsuke, Rikuchu, and Mutsu, it grows at altitudes of 800 to 1600 ft., in deep valleys, in mixed woods with *Alnus tinctoria*, *Populus Sieboldii*, and *Quercus grosseserrata*. It appears to be more common in Hokkaido, where it was seen by Elwes at Sapporo and Asahigawa, in virgin forest at 500 to 750 ft. above sea-level, where it attains about 100 ft. in height and forms a wide-spreading tree, with a stem 10 ft. in girth.[4] Shirasawa states that its wood is of little use except for firewood; but the bark, after steeping, is plaited by the Ainos into coarse cloth and mats.

There is a small tree of this species at Grayswood, Haslemere, which the late Mr. Chambers procured from the Yokohama Nursery Co. in 1894. It has not suffered from frost, but is rather slow in growth. Elwes has raised plants at Colesborne from

[1] Jack proposes this name on account of the earlier use of *T. Maximowiczii*, Baker, in *Journ. Bot.* xxxvi. 319 (1898), an untenable name for another species. Cf. p. 1657, note 3, B.

[2] Those on coppice shoots, gathered by Elwes in Hokkaido, are 8 to 10 in. long.

[3] Specimens collected by Shirasawa in different localities are now in the Kew Herbarium.

[4] Sargent says it attains 100 ft. in height in Hokkaido, where it frequently grows in company with *T. japonica*, which is much smaller in size. Cf. p. 1657, note 3, C.

seed sent by Shirasawa in the spring of 1905. These are 3 to 5 ft. in height, and differ from adult trees in having only traces of stellate pubescence on the leaves and branchlets. (A. H.)

## TILIA MIQUELIANA

*Tilia Miqueliana*,[1] Maximowicz, in *Bull. Acad. Imp. Sc. St. Petersb.* xxvi. 434 (1877), and *Mél. Biol.* x. 587 (1880); Shirasawa, *Icon. Ess. Forest. Japon*, i. text 116, t. 72, figs. 11-24 (1900), and in *Bull. Coll. Agric. Univ. Tokyo*, iv. 160 (1900); Schneider, *Laubholzkunde*, ii. 385 (1909); V. Engler, *Monog. Gatt. Tilia*, 111 (1909).

*Tilia mandshurica*, Miquel, *Prol.* 206 (1867) (not Ruprecht and Maximowicz); Franchet et Savatier, *Enum. Pl. Jap.* i. 67 (1875).

*Tilia Kinashii*, Léveillé and Vaniot, in *Bull. Soc. Bot. France*, li. 422 (1904).

*Tilia Franchetiana*, Schneider, in Fedde, *Repert.* vii. 201 (1909), and *Laubholzkunde*, ii. 386 (1909).

A tree, attaining in Japan about 40 ft. in height. Bark on young trees grey, smooth; on old trunks fissured longitudinally. Young branchlets with a minute grey stellate pubescence, very variable in amount. Leaves (Plate 407, Fig. 9) extremely variable in shape, but usually much longer than broad, averaging 3 to 4 in. in length and 2 to 2½ in. broad, ovate or deltoid; cordate at the base; acute, slightly acuminate or rounded, but rarely cuspidate at the apex; often lobulate and coarsely, irregularly, and sharply serrate, the teeth ending in incurved short callous points; upper surface shining green, with scattered slight pubescence, mostly on the nerves and at the base of the midrib; lower surface covered with a grey stellate tomentum, without axil-tufts; petiole short, less than half the length of the blade, with minute stellate pubescence. Buds grey tomentose.

Flowers, ten to twenty-two in a cyme; bract, peduncle, pedicels, and bracteoles tomentose; sepals ovate, acute, tomentose on both surfaces, shorter than the narrow obovate petals; stamens sixty to seventy-five, united in five clusters; staminodes more slender and shorter than the petals; ovary and base of the style covered with pale hairs. Fruit nearly globose, five-ribbed at the base, grey tomentose and warty.

This species is unknown in the wild state, and only occurs in Japan as a planted tree, most often seen in the courts of Buddhist temples, where it is regarded as sacred. It is reported by tradition to have been introduced from China by a Buddhist priest about the year 1190 A.D.; but it has not yet been found anywhere in that country.[2] Its extreme variability points to a hybrid origin. The species was founded by Maximowicz on two specimens, which he regarded as two forms: one with ovate-orbicular oblique leaves, long slender cymes, and broad bracts about 4 in. in length; the other with remarkably deltoid nearly symmetrical leaves, short cymes, and narrow bracts about 2½ in. long. It is impossible, however, following the opinions of Schneider and V. Engler, to retain those two forms as two distinct

[1] The tree described and figured under this name by Sargent, in *Garden and Forest*, vi. 111, fig. 19 (1893), and *Forest Flora Japan*, 19, t. 8 (1894), is *T. Maximowicziana*.

[2] *Tilia Miqueliana*, var. *chinensis*, Szyszylowicz, in Hooker, *Icon. Plant.* ad t. 1927 (1890), collected by me in Hupeh in central China, is referred by V. Engler, *Monog. Gatt. Tilia*, 130 (1909), to *T. chinensis*, Maximowicz, in *Act. Hort. Petrop.* xi. 83 (1890), a plant collected in Kansu. The Kansu and Hupeh plants do not seem to be quite identical; but neither can be considered the same as the Japanese *T. Miqueliana*. They have not been introduced into cultivation.

varieties or species; for the trees cultivated in Europe show great variation, no two specimens being alike, and both deltoid and ovate leaves are occasionally present on the same branch. Recent specimens from Japan, collected by Shirasawa, cannot be exactly matched with either of Maximowicz's specimens.

This species has lately been introduced into England, and the only specimens which we have seen are young trees at Kew, Aldenham, Casewick, and Glasnevin. The tree at Kew, obtained from Hesse's nursery in 1900, is now about 12 ft. high and appears to be thriving. It is also cultivated by Simon-Louis at Plantières, Metz.

(A. H.)

## TILIA AMERICANA, American Lime, Bass-Wood

*Tilia americana*, Linnæus, *Sp. Pl.* 514 (1753); Loudon, *Arb. et Frut. Brit.* i. 373 (1838); Sargent, *Silva N. Amer.* i. 52, tt. 24, 25 (1891), and *Trees N. Amer.* 671 (1905).
*Tilia caroliniana*, Miller, *Dict.* ed. 8, No. 4 (1768).
*Tilia latifolia*, Salisbury, *Prod.* 367 (1796).
*Tilia nigra*, Borkhausen, *Handb. Forstbot.* ii. 1219 (1800).
*Tilia glabra*, Ventenat, *Mém. Acad. Sc. Paris*, iv. 9 (1802).
*Tilia canadensis*, Michaux, *Fl. Bor. Am.* i. 306 (1803).
*Tilia stenopetala*, Rafinesque, *Fl. Ludovic.* 92 (1817).

A tree, occasionally attaining in America 130 ft. in height and 12 ft. in girth. Bark deeply fissured, with ridges broken on the surface into small thin scales. Young branchlets green, glabrous. Leaves (Plate 407, Fig. 1) broadly oval, averaging 5 in. to 6 in. long and 3½ in. to 4½ in. broad, turning yellow in autumn, cordate or truncate at the base, cuspidate at the apex; margin ciliate, with coarse triangular serrations, ending in long callous points; upper surface dull dark green, glabrous; lower surface[1] light green, glabrous, except for minute pubescent tufts in the axils at the junctions of the midrib, primary, and secondary nerves, but absent at the base of the blade; petiole glabrous, 1½ in. to 2 in. long.

Cymes pendulous, many-flowered; bract stalked, broad and rounded at the apex, glabrous except for slight stellate pubescence on the midrib beneath; peduncle glabrous; pedicels slightly pubescent; sepals ovate, acuminate, densely hairy within, slightly pubescent without, shorter than the lanceolate petals; staminodes present. Fruit globose or ovoid, without ribs, covered with thick rufous tomentum; shell thick.

The American lime is readily distinguished from the other species with glabrous branchlets and leaves, by the minuteness of the axil-tufts, which are, moreover, not present at the base of the leaf. In winter, the branchlets are shining, glabrous; buds with three external scales, glabrous except for tufted cilia at their tips.

The leaves are remarkable for their variation in size, an interesting account of which is given by Penhallow.[2] In Canada, some trees have tolerably uniform leaves, about 3½ in. in diameter. Other trees have in addition many leaves 4½ in. across. On vigorous shoots, especially on epicormic branches, the leaves are often 8 in.

[1] In *T. americana* the midrib and principal nerves are remarkably yellow, and this character is also seen in its reputed hybrids, *T. Moltkei*, *T. spectabilis*, and *T. Michauxii*.

[2] *Canadian Record of Science*, ix. 291 (1905).

wide. This lime coppices freely, and on shoots so produced the foliage is always enormous, reaching a maximum of 10 in. in diameter. This variation is due in part to the amount of nutrition available, which is greatest on coppice shoots, owing to their extensive root system; but the size of the leaves is also much influenced by shade. Trees with large leaves are often known in gardens as *var. mississippiensis*, but this name is erroneous, as there is no reason to suppose that such trees are limited to the Mississippi basin.

*Tilia americana*, since its introduction into Europe, has given rise to hybrids[1] with other species, especially with *T. tomentosa*, which is probably the other parent in the following:—

1. *Tilia Moltkei*, Schneider, *Laubholzkunde*, ii. 381 (1909).

A tree with leaves similar in shape and serrations to those of *T. americana*, but larger, 6 in. to 7 in. long and 5 in. to 6 in. wide, and differing in the lower surface, which is pale or greyish green, more or less covered with scattered stellate pubescence, without axil-tufts, but with occasional long hairs on the midrib. Buds, petioles, and branchlets glabrous, and identical with those of *T. americana*.

This tree, which is remarkable for its vigorous growth and handsome foliage, originated[2] in Späth's nursery at Berlin, and is named after the famous general, who planted a young tree in front of Herr Späth's house in 1888. Specimens are growing at Kew and Aldenham.

2. *Tilia spectabilis*, Dippel, *Laubholzkunde*, iii. 73 (1893).

A tree resembling *T. Moltkei* in the shape, colour, and serration of the leaves, but differing as follows:—Young branchlets apparently glabrous, but showing traces of stellate pubescence. Buds pubescent in their upper half. Leaves (Plate 407, Fig. 11), variable in size, 4 in. to 6 in. long and 3 in. to 5 in. wide; under surface with scattered stellate pubescence, very variable in amount,[3] and with long hairs on the principal nerves as well as on the midrib, but without axil-tufts; petiole glabrous or with a few scattered hairs. The flowers are intermediate in character between those of *T. americana* and *T. tomentosa*, and open at Kew three weeks earlier than those of the latter species. The stellate pubescence on the bract, peduncle and pedicels is like that of *T. tomentosa*, but less in amount. The shape of the bract and the size of the flowers are similar to *T. americana*.

This tree has probably been long in cultivation, but I have not been able to trace its origin.[4] It is often planted in botanic gardens, as at Kew, where there are several specimens, one nearly 40 ft. in height.[5] At Cambridge a tree, about 35 ft. by 3½ ft. in 1907, has long been labelled *T. heterophylla*; and reputed trees of the latter species usually turn out to be either *T. spectabilis* or *T. Michauxii*.

[1] In America it probably forms hybrids with *T. heterophylla*. See *T. Michauxii*, p. 1689.

[2] It is mentioned as a new plant, *Tilia americana Moltkei*, in Späth's *Catalogue*, No. 57, p. 71 (1883).

[3] In cases where the stellate pubescence on the under surface of the leaves becomes dense and greyish, trees of *T. spectabilis* are often confused with *T. tomentosa* and *T. Michauxii*. They are readily distinguishable from the latter by the absence of axil-tufts, and from the former by the glabrous branchlets. The long hairs on the midrib and nerves seem to be peculiar to *T. spectabilis* and *T. Moltkei*.

[4] Dippel found it in 1893 in Froebel's nursery under the name *T. alba spectabilis*, and in the Zoeschen arboretum as *T. Blechiana*; but it must have originated much earlier. Cf. *ante*, p. 1678, note 5.

[5] This tree has been labelled *T. argentea*, and resembles *T. tomentosa* in habit.

3. *Tilia viridis*, Simonkai, in *Math. Term. Koezl.* xxii. 320 (1888).

In addition to the usual form of *T. spectabilis* described above, there appears to be another hybrid of the same parentage, which is represented at Kew by a tree about 25 ft. high, obtained under the name *T. spectabilis* from Späth in 1899. It is closer to *T. tomentosa*, as the buds, branchlets, and upper surface of the leaves are covered with a scattered stellate pubescence, which is denser on the under surface of the leaves than in ordinary *T. spectabilis*. This tree closely resembles, if it is not identical with, a specimen[1] taken from a tree cultivated at Baden in 1835, which was identified by A. Braun with *T. argentea*, var. *virescens*, Spach, in *Ann. Sci. Nat.* ii. 344 (1834). Spach's description was based on a tree in the Trianon, which was said to have been raised from seeds of *T. tomentosa*; and if this account is correct, this tree would appear to have been the first cross observed between *T. tomentosa* and *T. americana*.

## Distribution

*Tilia americana* is widely spread, occurring in Canada from the valley of the Assiniboine River and the southern shores of Lake Winnipeg eastward to northern New Brunswick, and extending in the United States southwards to Virginia and along the Appalachian mountains to Georgia and Alabama, and ranging westward to Dakota, Nebraska, Kansas, Indian Territory, and eastern Texas. It grows in rich often moist soil, and formerly occurred as pure forest.[2] It attains its largest size in the alluvial lands of the lower Ohio river, where Ridgway records a tree 135 ft. in height and 17½ ft. in girth.[3]

According to Loudon, the American lime was cultivated[4] by Miller in 1752, but had not been extensively distributed. It is very rare at the present time in cultivation, and the tree at White Knights, mentioned by Loudon as 60 ft. in height, is no longer living. The only specimens which we have seen in England are small trees at Kew, Eastnor Castle, and Liphook. (A. H.)

## Timber

The wood of the American limes is very similar in character to that of the European species, and according to Sargent is largely used in the United States under the name of whitewood for the manufacture of cheap furniture, carriage panels, and woodenware. He states that, though one of the woods most largely used for making pulp, the quick decomposition of the sap makes it unfit for white paper. It is imported to some extent into Europe under the name of basswood, and has

[1] This specimen is considered by Bayer, in *Verhand. Zool. Bot. Ges. Wien*, xii. 50 (1862), to be a hybrid between *T. tomentosa* and *T. americana*, which he calls *T. argentea-nigra*. V. Engler (*op. cit.* 152) considers it to be *T. cordata* × *T. tomentosa*, but it shows no trace of *T. cordata* parentage.

[2] In *U.S. Forest Service, Circular* 63 (1907), which gives an account of this species, it is said to do well when planted in pure stands, and to be the most prolific of American trees in shoots from the stumps.

[3] Sargent, *Bull. Pop. Inform.* No. 30 (1912), and in *Gard. Chron.* lii. 87 (1912), says that it shows its greatest beauty in the forests of New Brunswick, northern New England, and the valley of the St. Lawrence. The leaves of planted trees in eastern Massachusetts are, especially in dry summers, made brown by the red spider, which, however, is easily controlled by spraying.

[4] In the *London Catalogue of Trees*, 1730, p. 81, the Carolina lime tree is mentioned as "*Tilia*, with leaf more longly mucronate. Seeds sent from Carolina by Catesby in 1726, hardy, and may be propagated as other limes."

partially displaced native limewood in the pianoforte trade on account of its cheapness. Owing to the facility with which large thin sheets can be turned off the log by rotating it against a knife-edge, it is coming into use for three-ply boards, a manufacture which, though only recently invented, is likely to grow rapidly for many purposes. (H. J. E.)

## TILIA HETEROPHYLLA

*Tilia heterophylla*, Ventenat, in *Anal. Hist. Nat. Madrid*, ii. 68 (1800), and *Mém. Acad. Sc. Paris*, iv. 16, pl. 5 (1803); Sargent, *Silva N. Amer.* i. 57, t. 27 (1891), and *Trees N. Amer.* 674 (1905).
*Tilia americana*, Linnæus, var. *heterophylla*, Loudon, *Arb. et Frut. Brit.* i. 375 (1838).

A tree, attaining in America 60 ft. in height and 12 ft. in girth. Young branchlets glabrous. Leaves similar to those of *T. Michauxii* in size and shape, but differing in being covered beneath with a silvery white tomentum, without axil-tufts, and having finer serrations with shorter straighter points. The flowers appear to differ in the bract[1] pubescent on both surfaces, and the stellate-pubescent peduncle and pedicels. There appears to be no constant difference in the fruit.

Both this species and *T. Michauxii* are readily distinguished from the other white limes by the glabrous branchlets, and the different shape of the leaves, which are usually very oblique at the base, and always longer than broad. *T. pubescens*, Aiton, the white lime of the Gulf States, is not in cultivation, and probably would not live in our climate. It has pubescent branchlets.

*T. heterophylla*, according to Sargent, is found on rich wooded slopes or near the banks of streams; and ranges from Ithaca, New York, southwards along the Alleghany mountains to northern Alabama, and westward to middle Tennessee, Kentucky, southern Indiana, and Illinois. It is most abundant and of its largest size on the high mountains of North Carolina and Tennessee.

Typical *T. heterophylla* is described by most authors as having the leaves covered beneath with a silvery white tomentum; but Sargent informs me in a letter that Rehder found trees on the mountains of West Virginia, with leaves nearly glabrous beneath,[2] which he considered to be undoubtedly *T. heterophylla*. Such trees are probably hybrids with *T. americana*; and a further study of the American limes is necessary, as the relationship of *T. Michauxii* and the glabrous forms of West Virginia with *T. heterophylla* in its typical form, is very obscure.

*T. heterophylla* was introduced, according to Loudon, in 1811; but he had seen no specimens except small trees in the Chiswick Garden; and these may have been *T. Michauxii*, which he did not distinguish as a separate species. Most of the trees which occur in cultivation under this name are either *T. Michauxii* or *T. spectabilis*; and we have seen no living specimens in England which can be identified with *T. heterophylla*. (A. H.)

[1] The bract is said by Schneider to be stalked in *T. heterophylla*, and sessile in *T. Michauxii*; but native specimens show it to be variable in both species.

[2] *T. eburnea*, Ashe, in *Bot. Gaz.* xxxiii. 231 (1902), usually considered to be a peculiar variety of *T. heterophylla* in the Carolinas and Georgia, is said to lose its pubescence in autumn occasionally.

## TILIA MICHAUXII

*Tilia Michauxii*, Nuttall, *N. Amer. Sylva*, i. 108 (1865); Sargent, *Trees N. Amer.* 673 (1905); Schneider, *Laubholzkunde*, ii. 387 (1909).
*Tilia alba*, Michaux, *Hist. Arb. Amer. Sept.* iii. 315, t. 2 (1813), and Nuttall, *N. Amer. Sylva*, iii. 84, t. 132 (1865). (Not Aiton, *Hort. Kew.*)

A tree, attaining in America 80 ft. in height and 9 ft. in girth. Young branchlets and buds glabrous. Leaves (Plate 407, Fig. 5) very variable in size, averaging[1] 4 to 5 in. in length, and 3 to 4 in. in breadth, always much longer than broad, ovate, very oblique and cordate or truncate at the base, cuspidate at the apex; margin with large triangular serrations ending in long incurved points; upper surface dark green, glabrescent; lower surface covered with a scattered or dense greyish tomentum, very variable in amount, and with conspicuous axil-tufts at the junctions of the midrib, primary and secondary nerves; petiole stout, short, glabrescent.

Cymes, ten- to twenty-flowered; bract tomentose above, glabrous below; peduncle glabrous; pedicels slightly pubescent; sepals lanceolate, pubescent on both surfaces; petals lanceolate, acute, twice as long as the sepals; staminodes present. Fruit globose, marked with five furrows and minutely tuberculate on the grey slightly tomentose surface.

This species was first described and figured by Michaux, who, however, considered it to be the same as the European white lime. Nuttall[2] based his name *T. Michauxii* on Michaux's description and figure. It appears to have much the same distribution as *T. heterophylla*, but is imperfectly known,[3] and in all probability is a hybrid of that species with *T. americana*, resembling the latter in the serrations, axil-tufts, and yellow midribs and nerves of the leaves. The characters of the flowers are intermediate between the two species mentioned. The extreme variability in the amount of the tomentum on the leaves points to a hybrid origin.

It is not very common in cultivation in England; but there are two fine trees at Beauport, which are both grafted; one of these measures 84 ft. by 9 ft. 1 in., and the other is nearly as large. There are also small trees at Kew and Tortworth. There is an ill-shaped tree at Castlemartyr in Ireland, which Elwes found in 1908 to be about 40 ft. by 3½ ft.; some of its leaves were no less than 12 in. long and 9 in. wide. It is cultivated under the name *T. americana rubra* in the Heatherside nursery; and appears to be known in continental nurseries under various names, as *T. americana pubescens*, *T. gigantea*, *T. macrophylla*, *T. hybrida superba*, etc.

(A. H.)

[1] In cultivated trees in Europe the leaves are usually larger, averaging 7 in. long and 5 to 6 in. broad; but this may be a juvenile character.

[2] Nuttall says: "The *Tilia alba* of Michaux, not being the *T. alba* of Kitaibel and Aiton, it is necessary to change the name, and we propose to call it *T. Michauxii*."

[3] I am unable to follow V. Engler's view, that all the trees in the wild state in America, supposed to be *T. Michauxii* by Sargent and others, are simply *T. heterophylla*. Engler restricts the hybrid *T. americana* × *T. heterophylla* to trees found in cultivation in Europe. Wild *T. Michauxii* in America, judging from a considerable number of specimens which I have seen, is readily distinguishable from the true *T. heterophylla*; and I see no reason for supposing that the trees referred by me to *T. Michauxii*, which are in cultivation in Europe, are not originally from America. A small tree at Kew, received from Biltmore, U.S.A., is undoubtedly *T. Michauxii*.

# TRACHYCARPUS

*Trachycarpus*, H. Wendland, *ex* Gay, in *Bull. Soc. Bot. France*, viii. 429 (1861); Bentham et Hooker, *Gen. Pl.* iii. 929 (1883); Beccari, in Martelli, *Webbia*, i. 41-73 (1905).
*Chamærops*, Martius, *Hist. Nat. Palm.* iii. 247 (1836-1850) (in part).

TALL unarmed trees, belonging to the order Palmaceæ, with flabellate leaves, without a rachis, deeply divided into numerous plicate segments, which are narrow, entire in margin, and bifid at the apex.

Flowers monœcious; spadices numerous, panicled, sessile between the leaves, and embraced by sheathing coriaceous spathes; individual flowers small; sepals three, ovate; petals three, broadly ovate, valvate; stamens six, with free filaments and short dorsifixed anthers, rudimentary in female flowers; ovary of three carpels, connate at the base, represented in the male flowers by three small subulate processes; stigmas three, recurved; ovules basilar, erect. Fruits, one to three, globose or oblong, with a terminal style; containing one erect seed, which is grooved on the ventral side, and has a small hilum at the base.

Three species[1] of Trachycarpus are known, natives of the Himalayas, Assam, Burma, and China.

## TRACHYCARPUS FORTUNEI, CHUSAN PALM

*Trachycarpus Fortunei*, H. Wendland, in *Bull. Soc. Bot. France*, viii. 429 (1861); Masters, in *Gard. Chron.* xxiv. 304, fig. 65 (1885).
*Trachycarpus excelsa*, H. Wendland, in *Bull. Soc. Bot. France*, viii. 429 (1861); Planchon, in *Flore des Serres*, xxii. 207, t. 2368 (1877); Beccari and J. D. Hooker, in J. D. Hooker, *Flora Brit. India*, vi. 436 (1892); Diels, in Engler, *Bot. Jahrb.* xxix. 233 (1900); Wright, in *Journ. Linn. Soc.* (*Bot.*) xxxvi. 168 (1903); Masters, in *Gard. Chron.* xxxv. 312, Supply. Illust. (1904); Beccari, in Martelli, *Webbia*, i. 41 (1905).
*Chamærops excelsa*, Martius, *Hist. Nat. Palm.* iii. 251, t. 125, figs. 2, 3 (1836-1850) (not Thunberg[2]); Miquel, *Anal. Bot. Ind.* ii. 21 (1851); Gay, in *Bull. Soc. Bot. France*, viii. 410 (1861); Houllet, in *Rev. Hort.* xl. 370, coloured plate (1868).
*Chamærops Fortunei*, W. J. Hooker, in *Bot. Mag.* t. 5221 (excl. figs. 6, 7) (1860).

A palm, attaining 30 to 50 ft. in height, with the trunk annulately scarred and clothed throughout with old leaf-sheaths. Leaves: lamina sub-orbicular or fan-

[1] The two species, not described by us, as they are scarcely hardy in cultivation, are: *T. Martiana*, Wendland, widely spread in the central and western Himalayas, Assam, and Burma; and *T. Takil*, Beccari, in *Webbia*, 52 (1905), which is a peculiar species occurring in Kumaon. The latter species is cultivated by Prof. Beccari in his garden at Florence. Bean, in *Kew Bull.* 1912, p. 291, quotes Beccari's description and gives a figure of this tree, which is readily distinguishable from *T. Fortunei* by its different habit and by the closer and neater fibre of the trunk.

[2] *Chamærops excelsa*, Thunberg, *Fl. Jap.* 130 (1784), is plainly from the description another palm, *Rhapis flabelliformis*, Aiton, *Hort. Kew.* iii. 473 (1789). Cf. Sims, *Bot. Mag.* t. 1371 (1811).

shaped, averaging 1½ ft. in length and breadth, deeply cut into numerous linear plicate segments, which are ½ to 1 in. broad, entire in margin, and bifid at the apex; petiole about 1½ ft. long, trigonous or nearly semicylindric in section, convex above, and flat beneath, with the two margins irregularly serrate or dentate towards the base; lower sheathing part furfuraceous, and separating into a network of brown fibres.

Staminate panicles[1] broad, thick, and obtuse, enveloped in a strong stout brownish spathe, at first erect, soon pendent, with arched ramifications; flowers small, numerous, close together, deep yellow-orange. Pistillate inflorescence smaller, less thick than the staminate one, clothed in an acuminate erect spathe, with distant spreading slender ramifications; flowers numerous, not close together, small, pale yellow or greenish. Drupe reniform, hollowed on one side.

This species is variable in habit, the leaves in some individuals being straight and stiff, whilst in others they droop at the tips; and the petioles show a varying amount of serration. There are no grounds, however, for supposing that there are two forms, corresponding to a supposed Chinese and Japanese origin. Plants which are sold as distinct, under the names *Chamærops excelsa* and *C. Fortunei*, are usually imported seedlings, which have been raised in the Riviera from the same batch of seed.

1. Var. *gracilis*, Carrière, in *Rev. Hort.* xlvii. 220 (1875). Petioles, longer and more slender than in the type, abruptly inflexed; laminæ more deeply divided with narrower segments. The variety was described from a plant in the Jardin des Plantes at Paris, which was obtained from Thibaut and Keteleer's nursery in 1862.

2. Var. *surculosa*, Henry (*var. nova*). Leaves smaller and stiffer than in the type. Dwarf in habit, and freely reproducing itself by suckers like *Chamærops humilis* of the Mediterranean region. This peculiar variety, which was obtained a few years ago from Japan by Sir E. G. Loder, is possibly a distinct species. It thrives at Leonardslee, where it has produced flowers; but no fruit has yet been noticed.

*T. Fortunei* is a native of the central provinces of China, where it is certainly wild on the hills and lower slopes of the mountains in Szechwan, Hupeh, Kiangsu, and Chekiang. It is everywhere much cultivated, so that it is impossible to define its original distribution in the wild state. It is commonly planted around Shanghai, where it lives unprotected through the severest winters.[2] It is also cultivated on the island of Chusan, whence the name given to it of Chusan palm by Fortune,[3] who saw it growing wild in great perfection near Yen Chow and on the Sung Lo mountain in Chekiang. It is known to the Chinese as *Tsung*, and is of great economic importance, owing to the large quantity of strong and useful fibre found on the stem at the bases of the leaf-stalks. This fibre, which is similar to the coir

[1] Inflorescences bearing both staminate and pistillate flowers have been noticed by Beccari. The germination of this species is described by Gatin, in *Ann. Sc. Nat.* (*Bot.*) iii. figs. 28, 29 (1906).

[2] It is unknown in North China, being killed when planted out at Chefoo and Tientsin. Cf. Gay, in *Bull. Soc. Bot. France*, viii. 412 (1861).

[3] Cf. Fortune, *Wanderings in China*, 53, 54 (1847); *Tea Countries*, 58, 88, 117, 318 (1852); *Residence among the Chinese*, 5, 145, 189 (1857); and in *Gard. Chron.* 1850, p. 757, and 1860, p. 170.

obtained from the cocoa-nut, is used for making ropes and cables, which are very durable under water; in the manufacture of hats and rain-cloaks[1] (called *so-i*), which are used by the peasants in wet weather; and in making brushes and mattresses.

This palm is largely cultivated in southern and central Japan, where it is known as *Shuro*, but it does not seem to be a true native,[2] though it is naturalised in some places.[3] It is recorded from Tamsui and Bankinsing in Formosa; but is probably only planted.[4]

This palm was first introduced by Siebold,[5] who sent seeds from Nagasaki in Japan to Leyden in 1830. Of the plants which were raised, one was sent to Kew in 1837, where it was long an inmate of the Palm House; but it died some years ago. The Chinese plant was introduced by Fortune,[6] who sent six living plants to Kew in 1849, two of which are still flourishing in the Temperate House, and are about 40 ft. and 50 ft. in height respectively; while two, which were planted out-of-doors, have remained stunted in growth and are only about 10 ft. high. Another of these plants, which was sent to Osborne in 1849, was planted out in front of the royal residence, and had attained[7] 14½ ft. high in 1881. It is now about 21 ft. high and 2 ft. 4 in. in girth at a foot from the ground. (A. H.)

The Chusan palm is perfectly hardy[8] in our climate, but grows slowly in the neighbourhood of London and to the northwards. It flourishes best in the south-west of England and in the south of Ireland, where it attains a height of 25 to 30 ft., and ripens seeds regularly. In the grounds of Tregothnan, I found in 1905 self-sown seedlings in the grass, and transferred some of them to my own garden at Colesborne, one of which has borne a temperature close to zero without injury. On the lawn at Williamstrip Park, Gloucestershire, an old specimen has flowered on several occasions. At Duncan House, Torquay, there is a group of about six trees, 14 to 25 ft. high in 1903, around which there are hundreds of young plants, which have come from their seed.[9]

Plate 375 shows a tree planted by the Rev. Hon. T. Boscawen at Lamorran in Cornwall, which was about 25 ft. high in 1905. At Riverhill, near Sevenoaks, the seat of J. T. Rogers, Esq., I saw one of the original seedlings introduced by Fortune, which was 27 ft. by 2 ft. 6 in. in 1911. At Powis Castle, I measured one in 1906, 25 ft. high, girthing 1 ft. 10 in. at the ground, but 3 ft. 4 in. at five feet up.

At Heckfield Place, near Winchfield, Hants, there are two good specimens[10] of different sexes, which were planted in 1869. They measured in January 1912, 17½ feet and 15½ feet respectively.

[1] A rain-cloak and hat, brushes, cordage, and other articles made from this fibre, are displayed in the museum at Kew.

[2] Makino, in *Tokyo Bot. Mag.* xviii. 20 (1904), divides the species into two forms, var. *typica* and var. *Fortunei*; but in this he simply follows Wendland, and gives no distinguishing characters; moreover, he states of both these supposed forms, that they are planted and not native in Japan.

[3] Cf. Franchet and Savatier, *Enum. Plant. Jap.* ii. 2 (1879).

[4] Cf. Matsumura, *Index Plant. Jap.* ii. 168 (1905). Elwes saw it planted in many places in Formosa up to about 4000 ft. elevation; and the fibre is used for the same purposes as in China.

[5] *Cat. Rais. Pl. Jap.* 1856, p. 7, note.

[6] Fortune also obtained seed near Tsee-Kee, north-west of Ningpo, in 1853, from which plants were raised in England.

[7] Cf. Smith, *Records of Kew Gardens*, 116 (1885).

[8] Cf. *Gard. Chron.* 1860, pp. 170, 362, where it is stated that this palm suffered much but did not die in Northamptonshire in the severe winter of 1859-1860; while at Swansea, where the temperature fell to 10° Fahr., they were quite uninjured.

[9] *Gard. Chron.* xxxiv. 329 (1903); one of these palms was figured in *Gard. Chron.* xxiv. 420, Supply Illust. (1898).

[10] Cf. *Gard. Chron.* xv. 12, fig. 4 (1881), and xxiv. 216 (1885).

In Surrey, this palm can be grown very successfully in light dry sandy soil, provided the plants are placed in a sheltered position. In these conditions, they grow slowly, attaining about 10 ft. high in thirty years; but they flower regularly, and are very ornamental.[1] There is a fine specimen, 25 ft. high and 4 ft. in girth, at Joldwynds, near Dorking, the seat of Sir P. Bowman, Bart.

At Wisbech, Lord Peckover has a very old specimen, probably one of the original seedlings, which was 25 ft. high in 1912. A younger tree, planted about thirty-five years ago, is nearly 20 ft. high, and flowers more profusely than the older tree.

In Scotland[2] there is a good specimen at Ardchapel, Shandon, on Gare Loch in Dumbartonshire, which was planted in 1866 by the late Prof. Swan of St. Andrews. In 1905, it was 11 ft. high, with a girth of 3 ft. 8 in. at five feet from the ground. It has produced flowers regularly since 1881. There are small but thriving specimens in Arran, at Craigard, Lamlash; Whitefarland; and Cromla, Corrie.

In Ireland, it is slow in growth and of no great height in the open at Glasnevin. Two trees at Narrow Water, Co. Down, were 12 ft. high in 1903. It grows well at Mount Usher in Wicklow, where it ripens fruit; and is represented at Fota by several fine specimens, one of which is 25 ft. high.

The Chusan palm was introduced[3] into France by seed sent from central China by M. de Montigny in 1851; and was first planted in Algeria in 1853. It exists in the climate of Paris, where it bore 16° Cent. below zero in 1870-1871; and it is hardier[4] than *Chamærops humilis* at Montpellier, where the temperature occasionally falls to the same point. There are fine specimens[5] in Brittany, notably one in the park of Porzantrez near Morlaix, which was planted in 1856, and had attained 27 ft. high in 1907. In Spain and Portugal there are fine specimens,[6] one at Lisbon being 40 ft. high. (H. J. E.)

[1] Cf. *The Garden*, 15th November 1902, p. 341.

[2] Cf. Rev. Dr. Landsborough, in *Trans. Bot. Soc. Edin.* xx. 510 (1896), and xxiii. 140, 141 (1908).

[3] Gay, in *Bull. Soc. Bot. France*, viii. 422 (1861).

[4] Planchon, in *Flore des Serres*, t. 2368 (1877). There is a small plant of *Chamærops humilis* in the bamboo garden at Kew; but it would not survive without protection in winter. Landsborough, in *Trans. Bot. Soc. Edin.* xxiii. 141 (1908), mentions a thriving plant of this species at Cromla in Arran.

[5] Pardé, in *Bull. Soc. Dend. France*, 1908, p. 20.

[6] *Ibid.* 1910, p. 106.

# ACANTHOPANAX

*Acanthopanax*, Miquel, in *Ann. Mus. Lugd. Bat.* i. 10 (1863); Bentham et Hooker, *Gen. Pl.* i. 938 (1863); Viguier, in *Ann. Sci. Nat.* (*Bot.*) iv. 33 (1906).

*Kalopanax*, Miquel, in *Ann. Mus. Lugd. Bat.* i. 16 (1863); Harms, in Engler and Prantl, *Pflanz. Fam.* iii. 8, p. 50 (1894).

*Brassaiopsis*, Seemann, in *Journ. Bot.* ii. 290 (1864) (not Decaisne and Planchon).

*Panax*, sub-genus *Acanthopanax*, Decaisne and Planchon, in *Rev. Hort.* 1854, p. 105.

Deciduous trees or shrubs, belonging to the order Araliaceæ. Branchlets with or without spines. Stipules inconspicuous or absent. Leaves either compound and three- to five-digitate or simple and palmately lobed.

Flowers small, in panicled or solitary umbels, polygamous or perfect, on pedicels that are not articulated; calyx-tube coherent with the ovary, limb dentate; petals five, rarely four, valvate in the bud; stamens as numerous as the petals and inserted with them on the margin of the disc, filaments filiform, anthers ovate or oblong; ovary usually two-celled, rarely five-celled; styles, two or five, more or less connate at the base, with the apex recurved and stigmatose internally. Fruit baccate, laterally compressed or globose, with a fleshy exocarp; nutlets two to five, crustaceous or cartilaginous, each containing one seed.

About eighteen species are known, natives of Eastern Asia, extending from Amurland and Manchuria, through Japan, Formosa, and China to the Himalayas. The following species is the only one attaining timber size.

## ACANTHOPANAX RICINIFOLIUM

*Acanthopanax ricinifolium*, Seemann, in *Journ. Bot.* vi. 140 (1868), and *Revis. Heder.* 86 (1868); Franchet and Savatier, *Enum. Pl. Jap.* i. 193 (1875); Hance, in *Journ. Bot.* 1885, p. 323; Hemsley, in *Journ. Linn. Soc.* (*Bot.*) xxiii. 340 (1888); Sargent, in *Garden and Forest*, vi. 234, fig. 36 (1893), and *Forest Flora Japan*, 45, fig. 16 (1894); Shirasawa, *Icon. Ess. Forest. Japon*, ii. t. 56, figs. 11-24 (1908); Beissner, in *Mitt. Deut. Dend. Ges.* 1909, p. 289; Schneider, *Laubholzkunde*, ii. 429 (1909).

*Acanthopanax acerifolium*, Schelle, in *Mitt. Deut. Dend. Ges.* 1908, p. 217, and 1909, p. 289.

*Panax ricinifolium*, Siebold and Zuccarini, *Fl. Jap. Fam. Nat.* i. 91 (1843), and in *Abh. Ac. Münch.* iv. 2, p. 199 (1845).

*Kalopanax ricinifolium*, Miquel, in *Ann. Mus. Lugd. Bat.* i. 16 (1863); Harms, in Engler, *Bot. Jahrb.* xxix. 488 (1900).

*Kalopanax* ? *sp.*, Zabel, in *Mitt. Deut. Dend. Ges.* 1904, p. 63.

*Brassaiopsis ricinifolia*, Seemann, in *Journ. Bot.* ii. 291 (1864).

A tree, attaining in Yezo 80 to 90 ft. in height, and 10 ft. or more in girth. Bark very thick,[1] dark in colour, and deeply fissured. Branchlets stout, reddish brown, glabrous, armed with short stout spines, which are enlarged at the base and are often glaucous. Leaves sub-orbicular, but broader than long, averaging 6 to 10 in. in diameter, sub-cordate at the base, palmately five- or seven-nerved; on adult trees, divided to one-third of the length of the blade by acute sinuses into five ovate-triangular acuminate lobes, serrate in margin, with callous tips to the teeth; upper surface dark green, shining, glabrous; lower surface with slight whitish tomentum on the nerves, forming axil-tufts at their base, where they unite; petioles long and slender, more or less covered with scattered tomentum, especially near the base.

Flowers on long slender glabrous pedicels in many-flowered umbels, arranged in terminal compound panicles, which are sometimes 2 ft. in diameter; small, white in colour, appearing in August and September. Fruit black, globose.

1. Var. *Maximowiczii*, Schneider, *Laubholzkunde*, ii. 429 (1909).

*Aralia Maximowiczii*, Van Houtte, in *Flore des Serres*, xx. 39, t. 2067-2068 (1874).

This varietal name is applied for the sake of convenience to young trees in cultivation, which only show juvenile foliage. The leaves differ from those characteristic of adult trees, in being deeply lobed, with the sinuses extending more than two-thirds the length of the blade. It is impossible to say whether, as a result of being perhaps propagated by grafts from young seedlings, these trees will preserve the seedling foliage indefinitely, or will, when older, develop normal leaves. Seedlings occasionally show both kinds of foliage; and Siebold's original specimens also bear both shallow- and deeply-lobed leaves, with all intermediate stages.

This species is a native of China, Korea, Japan, Saghalien, and the Liu Kiu Islands. In China, it is widely spread in the central provinces, from Szechwan and Hupeh to Chekiang and Fokien, usually forming a tree 40 to 50 ft. high in the mixed forests on the mountains at no great elevation. It is known to the Chinese as *tz'e ch'iu*;[2] but its timber is little valued. (A. H.)

In Japan, where the tree is called *hari-giri*, it attains its largest size in the forests of Hokkaido, where, Sargent states, it is often 80 feet in height and 4 or 5 ft. in diameter, with a tall straight stem, covered with furrowed bark, and giving off great limbs, which spread horizontally. In central Hondo, it is smaller in size. Mayr[3] who measured a tree in Hokkaido, which was 90 ft. high, states that this species is fast in growth and remarkable for its capacity of enduring shade in the forest. In Kiusiu I saw it in the forest, which covers the lower slopes of the volcano of Kireshima at 2000 to 3000 ft. altitude. Here it is not so fine a tree as in Hokkaido, the largest that I measured being about 60 ft. by 5 ft. 10 in.

*A. ricinifolium* was introduced into Europe by Van Houtte, who figured in 1874 a young plant, which had been raised from a single seed received from the

[1] The bark is well shown in a photograph of a tree growing in Japan, reproduced by Jack in *Mitt. Deut. Dend. Ges.* 1909, p. 285.

[2] Cf. Bretschneider, *Bot. Sinic.* ii. 344 (1892) and iii. 480 (1895).

[3] *Fremdländ. Wald- u. Parkbäume*, 437 (1906).

St. Petersburg botanic garden. A tree[1] in the forest garden at Münden, probably derived from Van Houtte's original specimen, was 21 ft. high in 1895.

This species is best raised from seed imported from Japan; but it can be propagated by cuttings of the half-ripened wood, which strike root if placed in gentle heat.[1] Mayr says that it bears transplanting badly, often dying to the ground, but afterwards producing an abundance of suckers from the collar of the root.

It ripens its young shoots well, and is perfectly hardy at Colesborne, where seedlings, that I raised in 1906, are now 4 to 6 feet high. The best specimen that I know in England is in the beautiful grounds of South Lodge, Horsham, the seat of F. D. Godman, Esq., which was, in 1905, 28 ft. high, fifteen years after planting. A solitary specimen at Kew has done badly, and appears to suffer from spring frost. At Westonbirt, there is a vigorous young tree, about 20 ft. high.[2]

At Grafrath, near Munich, this species has been planted in plots, and has thriven well for so far; but the trees are too young to give as yet any evidence as to their value in forestry.[3] M. Hickel states that there is a fine specimen in Messrs. Barbier's nursery at Beuvronne.[4]

## Timber

Mayr says that this species yields a light timber, which, though disagreeable in odour, is used in Japan for building generally, and for making shafts of spears. Sargent states that the wood is rather hard, straight-grained, and light brown, with a fine satiny surface. The timber has been largely imported into England during the last few years under the name *sen*, by which it is known in Hokkaido. It has been sold in some cases as Japanese ash, and mixed with the timber of *Fraxinus mandshurica*, which it somewhat resembles; but it is better fitted for making furniture than for the purposes for which ash is generally used. (H. J. E.)

[1] Rehder, in *Möller's Deut. Gart. Zeit.* 1st July 1897.

[2] In *Gard. Mag.* 1888, p. 526, mention is made of a plant then 12 ft. high in Messrs. Veitch's nursery at Coombe Wood. This was disposed of some years ago.

[3] Weiss, in *Mitt. Deut. Dend. Ges.* 1912, p. 3, states that it has been tried in groups in the forest at Augsburg, where the trees are about 20 ft. high, but show no sign of being useful.

[4] Cf. Hickel, in *Bull. Soc. Dend. France*, 1907, p. 153, and 1909, p. 230.

# ACACIA

*Acacia*, Adanson, *Fam. des Plantes*, ii. 319 (1763); Willdenow, *Sp. Pl.* iv. 1049 (1805); Bentham et Hooker, *Gen. Pl.* i. 594 (1865).

TREES, shrubs, climbers, and rarely undershrubs, belonging to the division Mimoseæ of the order Leguminosæ. Branchlets with or without stipular or infra-stipular spines. Leaves alternate, either compound and equally bipinnate with minute leaflets, or reduced to simple phyllodes,[1] which are equivalent to dilated and flattened petioles, with their surfaces placed vertically.

Flowers, yellow or white; in globose heads, cylindrical spikes, or panicles, which are solitary or fascicled in the axils of the leaves; or panicled and ending the branchlets; perfect or polygamous, small, regular: sepals five, four, or three, rarely absent, free or united; petals as many as the sepals, free or united; stamens numerous, free or slightly connate at the base; ovary sessile or stalked, with usually numerous (rarely only two) ovules; style long and slender, ending in a minute stigma. Pod linear or oblong; flat or nearly cylindrical; straight, falcate, or twisted; opening by two valves or indehiscent. Seeds, more or less flattened, usually marked on each face with an oval or horseshoe-shaped depression, oblique spot or ring; funicle filiform or thickened into a flat aril under or around the seed.

About 450 species of Acacia are known, natives of the warmer parts of the world, and occurring in Africa, Asia, America, Australia, and Polynesia.

A considerable number of species are cultivated under glass in this country; but in the south-west of England, and in the south of Ireland, four or five species, some of which are shrubs, have succeeded out-of-doors. Two of these species, representative of the two kinds of foliage which occur in the genus, will now be described.

## ACACIA DEALBATA, SILVER WATTLE

*Acacia dealbata*, Link, *Enum. Hort. Berol.* ii. 445 (1822); Loddiges, *Bot. Cab.* t. 1928 (1833); Loudon, *Arb. et Frut. Brit.* ii. 666 (1838); J. D. Hooker, *Fl. Tasm.* i. 111 (1860); Bentham, *Fl. Austral.* ii. 415 (1864); Gamble, *Indian Timbers*, 301 (1902); Rodway, *Tasm. Fl.* 43 (1903).
*Acacia irrorata*, Sieber, *ex* Sprengel, *Syst.* iii. 141 (1826).
*Acacia decurrens*, Willdenow, var. *dealbata*, Von Mueller *ex* Maiden, *Forest Flora N. S. Wales*, iii. 56 (1908).

An evergreen tree, attaining occasionally in Tasmania 100 ft. in height and

[1] These phyllodes are not leaves turned edgeways, as is shown by the fact that they are not twisted at the base. Moreover, phyllodes are occasionally produced, which bear bipinnate leaves at their ends. On seedlings of the species of Acacia with phyllode foliage, the first leaves are bipinnate; succeeding leaves have flattened stalks with fewer leaflets; ultimately only phyllodes are produced.

11 ft. in girth; but usually much smaller. Young branchlets angled, hoary, covered with a minute whitish pubescence. Stipules reduced to inconspicuous scales. Leaves equally bipinnate; rachis 3 to 4 in. long, hoary and pubescent like the branchlets, often glandular; pinnæ ten to twenty pairs, hoary and pubescent, each bearing thirty to forty pairs of crowded linear leaflets, which are about $\frac{1}{6}$ in. long, pubescent, glandular at the sessile base.

Flower heads, in axillary and terminal panicled racemes, globose, yellow, about $\frac{1}{6}$ in. in diameter; flowers twenty to thirty in a head, mostly pentamerous. Pods, straight or curved, flattened, 2 to 3 in. long, $\frac{1}{4}$ to $\frac{1}{2}$ in. wide, not or slightly constricted between the seeds, glaucous on both surfaces.

*A. dealbata*, which is known in Australia as the silver wattle, occurs in New South Wales, Victoria, and Tasmania. It is widely distributed in Tasmania, where it usually attains 50 ft. in height and 3 to 6 ft. in girth; and yields a timber of little value, which is used for making staves of cheap casks. The bark is not so rich in tannin as that of typical *A. decurrens*,[1] of which species *A. dealbata* is considered to be a variety by Maiden.

*A. dealbata* was introduced from Tasmania about 1820; and is now much cultivated on the Riviera for its flowers, which are sent in large quantities to Paris and London, under the popular name of "Mimosa." In France this species flourishes on the west coast as far north as Nantes, where, however, it is killed to the ground in severe winters, but nevertheless sends up shoots afterwards with increased vigour.[2] (A. H.)

In England,[3] it can be grown in the open air in the south-west; and has attained 50 ft. in height after seventeen years growth from seed at Abbotsbury, where it produces flowers annually in great abundance and good seed, from which plants have been raised. At Trebah in Cornwall there is a fine tree of about the same age.[4] In Ireland, the finest we know is at Derreen (Plate 376), and is believed by the Marquess of Lansdowne to have been planted about thirty-two years. When I measured it in 1910, it was 48 ft. high, with four stems measuring 2 ft. 1 in. to 2 ft. 7 in. in girth.

This species is now completely naturalized in the Nilgiris,[5] where it is useful for firewood. Gamble says that it is readily reproduced by suckers and coppice shoots. In France it does not thrive on soils containing lime.[6] (H. J. E.)

[1] Typical *A. decurrens* has branchlets and foliage, which are nearly glabrous and not hoary; and is known in Australia as Green or Black Wattle. A plant, imported from Johannesburg, has been growing since 1909 in the open air at Blackmoor, Liss, Hants, and is now about 4 ft. high. *A. decurrens*, var. *mollis*, Lindley, has tomentose foliage, but the pubescence assumes a golden yellow tinge on the branchlets.

[2] Cf. Maiden, *Forest Flora N.S. Wales*, iii. 60 (1908).

[3] A tree in the Temperate House at Kew is about 50 ft. high, but with a slender stem, about 6 in. in diameter; the bark is broken on the surface into small scales. In *Gard. Chron.* liii. 45 (1913) mention is made of a tree, 70 ft. high and 2 ft. 2 in. in girth, growing in the conservatory at Branksome Hall, Darlington, which was raised from a root-cutting twenty-five years ago.

[4] *Gard. Chron.* lii. 44 (1912).

[5] Brandis, in *Indian Forester*, viii. 26 (1882), quotes General Morgan's account of the remarkable change in the period of flowering of this species in the Nilgiris, where it was introduced by seed in 1845. The trees here flowered at first in October, which is the month in which the parents flower in Australia. In 1860, they began to flower in September; in 1870, they flowered in August; in 1878, in July; and in 1882, in June, which is the spring month in the Nilgiris corresponding with October in Australia.

[6] Cf. Mottet, in *Rev. Hort.* 1896, p. 503.

## ACACIA MELANOXYLON, Blackwood

*Acacia melanoxylon*, R. Brown, in Aiton, *Hort. Kew.* v. 462 (1813); Sims, *Bot. Mag.* t. 1659 (1814); Loudon, *Arb. et Frut. Brit.* ii. 663 (1838); J. D. Hooker, *Fl. Tasm.* i. 109 (1860); Bentham, *Fl. Austral.* ii. 388 (1864), and in *Trans. Linn. Soc.* xxx. 481 (1875); Gamble, *Indian Timbers*, 301 (1902); Rodway, *Tasm. Flora*, 42 (1903); Maiden, *Forest Flora N.S. Wales*, ii. 103, plate 57 (1907).

*Acacia latifolia*, Desfontaines, *Table École Bot. Paris*, 207 (1815).

*Acacia arcuata*, Sieber, *ex* Sprengel, *Syst.* iii. 135 (1826).

*Acacia brevipes*,[1] Cunningham, in *Bot. Mag.* t. 3358 (1834).

An evergreen tree, attaining occasionally, in Australia, 120 ft. in height, and 6 to 10 ft. in girth; but usually only about 80 ft. high. Young branchlets angled, minutely grey tomentose, or rarely glabrous. Stipules absent. Phyllodes coriaceous, glabrous, usually falcate, very variable in size, averaging 2½ to 4 in. long and ⅓ to ¾ in. wide; lanceolate or oblong, gradually tapering to an obtuse or acute apex, which is tipped with a cartilaginous point; much narrowed towards the base; with three or four conspicuous longitudinal veins, connected by anastomosing veinlets; margin entire, cartilaginous. True leaves often present erratically on young trees, bipinnate, with a tomentose rachis and secondary axes, the latter bearing oblong apiculate leaflets, about ⅕ to ¼ in. long.

Flower heads, three to four in an axillary raceme, globose, yellow, about ⅕ in. in diameter; peduncles ⅙ to ⅓ in. long, glabrous; flowers minute, thirty to fifty in a head, pentamerous, with coherent calyces and denticulate sepals; petals free. Pods linear, flat, often curved in a circle, 2 to 4 in. long, about ⅓ in. broad, with thickened cartilaginous margins, and glaucous coriaceous valves. Seed small, nearly orbicular, with a long dilated pale red funicle, encircling it in double folds.

*A. melanoxylon* is known in Australia as blackwood, and is widely distributed in Tasmania, Victoria, and New South Wales, extending into South Australia and Queensland. It is most common on rich soil in valleys or grassy ranges; and ascends in the mountains to considerable elevations. It yields a timber of the highest class, which Gamble compares to light-coloured walnut. It is hard and close-grained, taking a fine polish; and is very beautiful when figured. It is used in Sydney and Melbourne for making billiard tables, furniture, gunstocks, coaches, and railway carriages; and is imported by English pianoforte manufacturers.[2]

(A. H.)

*A. melanoxylon* was introduced into England about 1808, and is occasionally grown out-of-doors in the south-west. At Abbotsbury, raised ten years ago from seed, it is about 35 ft. high; and seedlings have been raised from it. At Tregothnan, I saw several trees flowering well in 1911, one of which was 35 ft. high by 1 ft. 8 in. in girth.

In the south of France,[3] it is the best of the acacias for avenues, as it forms a tall tree regular in habit; and at Hyères there are numerous natural seedlings. It was introduced in 1840 into the Nilgiris in India, where it is completely naturalised.

(H. J. E.)

[1] This is a variety, which appeared in cultivation at Kew, with longer and more falcate phyllodes, attaining 5 to 7 inches in length.

[2] Cf. Penny, *Tasm. Forestry*, 9 (1905).

[3] Pottier, in *Le Jardin*, xxii. 75 (1908).

# LAURELIA

*Laurelia*, Jussieu, in *Ann. Mus. Paris*, xiv. 134 (1809); Tulasne, in *Archiv. Mus. Hist. Nat. Paris*, viii. 414 (1855); A. de Candolle, *Prod.* xvi. pt. 2, p. 674 (1868); Bentham et Hooker, *Gen. Pl.* iii. 145 (1880); Perkins and Gilg, in Engler, *Pflanzenreich*, iv. 101, *Monimiaceæ*, 76 (1901); Perkins, in Engler, *op. cit.*, Suppl. 46 (1911).

*Pavonia*, Ruiz and Pavon, *Fl. Peruv. et Chil. Prod.* 127, t. 28 (1794) (not Cavanilles).

*Theyga*, Molina, *Sag. Chile*, 163 (1810).

*Thiga*, Molina, *Sag. Chile*, 297 (1810).

Evergreen trees, belonging to the order Monimiaceæ; with opposite coriaceous fragrant serrate penninerved leaves, which are without stipules.

Flowers polygamous or diœcious; in simple or panicled axillary cymes; perianth with six to twelve spreading lobes in two or three series. Male flowers, with a flat receptacle, and five to twenty stamens; filaments short, each with two lateral glands at the base; anthers with two oblong lateral cells, which dehisce by valves opening upwards. Female and perfect flowers, with a receptacle at first cup-shaped, bearing stamens, which are often reduced wholly or in part to staminodes, and numerous fusiform villous ovaries, which are tipped with long hirsute styles, and are one-celled, containing an erect anatropous ovule. Fruit, consisting of the enlarged and almost closed receptacle, which has become globose, ovoid, or tubular, and ultimately opens by splitting irregularly into three or four valves; on these are placed the ripened ovaries or achenes, which are ovoid, pilose, and end in a long plumose unbranched style. The valves of the receptacle remain firmly closed in moist air, but spread widely when dry.

Three species of Laurelia are known, two of which, not being in cultivation, need only be briefly mentioned:—

1. *Laurelia Novæ-Zelandiæ*, Cunningham, in *Ann. Nat. Hist.* i. 381 (1838). A tree attaining 150 ft. in height in New Zealand.

2. *Laurelia sempervirens*, Tulasne, in *Arch. Mus. Hist. Nat.* viii. 416 (1855).

*Laurelia aromatica*, Poiret, in Lamarck, *Encycl. Meth. Suppl.* iii. 313 (1813).

*Pavonia sempervirens*, Ruiz and Pavon, *Syst. Veg. Pl. Peruv. et Chil.* 253 (1798).

This species, with which *L. serrata* has been much confused, has a more northerly distribution, occurring in Peru, as well as throughout Chile, where it is called *laurel* by the inhabitants. It differs from *L. serrata*[1] in having the leaves undulate serrate, with rather distant appressed obtuse teeth; flowers in loose panicles with long pedicels; fruit receptacles fusiform or ovoid; seed plumose throughout, even to the tip of the style. It is a tall tree, very abundant in the evergreen forests, and yielding wood that is easily worked and much used.[2] (A. H.)

[1] *L. serrata* has sharply serrate leaves; flowers in short crowded panicles, with short pedicels; fruit receptacles globose; seed with the style not plumose at the tip.

[2] Cf. Castello and Dey, *Jeog. Vej. Rio Valdivia*, 51 (1908).

## LAURELIA SERRATA

*Laurelia serrata*,[1] Philippi, in *Bot. Zeit.* xv. 401 (1857); Castillo and Dey, *Jeog. Vej. Rio Valdivia*, 52, fig. 27 (1908); Stapf, in *Bot. Mag.* t. 8279 (1909); Perkins, in Engler, *Pflanzenreich*, iv. 101, *Monimiaceæ*, Suppl. 47 (1911).

*Laurelia aromatica*, Masters, in *Gard. Chron.* xxxvi. 401, fig. 172 (1904) (not Poiret).

A tree, attaining in Chile about 70 ft. in height. Bark greyish, smooth, with persistent brown lenticels, and resembling that of *Zelkova crenata.* Young branchlets, with two pubescent furrows, which have slightly projecting margins. Leaves coriaceous, in opposite pairs, narrowly elliptic or broadly lanceolate, averaging 4 to 5 in. long, and 1¼ to 1¾ in. wide; tapering to an acuminate apex, usually tipped with a cartilaginous point; cuneate at the base; lateral nerves, ten to twelve pairs, dividing and looping before reaching the margin, which is entire towards the base, but is elsewhere serrate, with sharp teeth directed towards the apex of the leaf or slightly spreading and tipped with a glandular thickening; both surfaces glabrous, except on the slightly pubescent prominent midrib above; petiole ⅛ to ¼ in. long, blackish, densely pubescent on the upper surface.

Flowers, three to nine, in simple or panicled silky cymes; pedicels about ⅛ in. long. Receptacle in the fruiting stage, globose but constricted at the apex, ultimately splitting irregularly, ashy-grey externally; achenes ovoid, nearly ½ in. long, densely hirsute with spreading hairs, except at the tip of the long and persistent style.

*L. serrata* is a native of the evergreen forests of southern Chile and northern Patagonia, occurring from Valdivia to the valley of the river Aysen. It is called *huan-huan* by the inhabitants, and seems to be a smaller and rarer tree than *L. sempervirens*; and is said by Castillo and Dey to have an inferior wood, which has an unsupportable odour when freshly cut, yet is much used.

*L. serrata* was probably introduced about 1860 by Pearce, who collected in Chile for Messrs. Veitch. It is one of the rarest trees in cultivation, and can only be propagated by seed or by layers, cuttings having always failed to take root at Kew. (A. H.)

The only specimens which we have seen are the following:—At Penjerrick, near Falmouth, the seat of R. Fox, Esq., there is a splendid tree in perfect health and looking as if it would become considerably larger, which was in flower in April, 1911, when it measured 47 ft. in height and 3 ft. 4 in. in girth. Another at Kilmacurragh, Co. Wicklow, is about 30 ft. by 3 ft.; but is not so handsome. This tree was procured by the late Mr. Acton from Messrs. Rollison of Tooting about 1868; and from it were drawn the figures cited above in the *Botanical Magazine* and in the *Gardeners' Chronicle.* (H. J. E.)

[1] *Laurelia serrata*, Bertero, in *Mercur. Chil.* 1829, manip. 15, p. 685, translated in *Amer. Journ. Sci.* xxiii. 89 (1833), is a name without any description.

# ILEX

*Ilex*, Linnæus, *Sp. Pl.* 125 (1753); Bentham et Hooker, *Gen. Pl.* i. 356 (1862); Loesener, in *Nova Acta Ac. Leop. Carol.* lxxviii. 8 (1901); Schneider, in *Gartenflora*, lii. 452 (1903), and *Laubholzkunde*, ii. 157 (1907).
*Prinos*, Linnæus, *Gen. Pl.* 153 (1754).
*Aquifolium*, Haller, *Enum. Stirp. Helv.* i. 296 (1758).

Trees or shrubs belonging to the order Aquifoliaceæ, mostly evergreen, rarely deciduous. Leaves alternate, simple, usually short-stalked; margin entire, crenate, serrulate, or with spiny teeth; stipules minute, deltoid, often deciduous.

Flowers axillary, solitary or cymose, normally diœcious, regular, usually tetramerous, rarely pentamerous or hexamerous; calyx gamosepalous, hypogynous, with a short tube and four to six lobes, imbricate in the bud; corolla rotate, with four to six petals, free or connate at the base; disc absent. Male flowers, with four to six stamens, alternate with the petals and adhering to them at the base; anthers introrse, with two cells opening longitudinally; ovary rudimentary, either without cells or with empty cells containing no ovules, stigma absent. Female flowers, with four to six staminodes, like the stamens, but sterile and small; ovary free, superior, with usually two to four (rarely nine to twenty-two) cells, each usually containing one pendulous ovule; stigma sessile, with as many lobes as there are cells in the ovary. Fruit drupaceous, with the calyx and stigma persistent; epicarp fleshy, containing two to four (or more) one-seeded nutlets. Seed pendulous, with a minute embryo at the apex of the copious albumen.

The genus comprises about 270 species, which are distributed throughout the greater part of the tropical and temperate regions of the world. About twenty-five exotic species are cultivated in Britain, either shrubs or small trees, not coming within the scope of our work; but of these the following four species will be briefly described on account of their relationship to the common holly.

I. *Ilex Perado*, Aiton, *Hort. Kew.* i. 169 (1789); Loudon, *Arb. et Frut. Brit.* ii. 519 (1838).

*Ilex maderiensis*, Lamarck, *Encyc.* iii. 146 (1789).

A small tree, attaining 30 ft. in height. Young branchlets glabrous. Leaves obovate-oblong, yellowish green, 2½ to 3 in. long, 1½ to 2 in. broad, flat on the surface, thick and coriaceous; apex rounded or minutely cuspidate, and tipped with a slender spine; base rounded or cuneate, decurrent on the glabrous petiole, which has two longitudinal depressions on the lower side; margin not undulate, entire or with a few minute serrations, mostly towards the apex, each tipped with a slender spine; under surface reticulate. Flowers reddish; fruit ellipsoid, reddish, with nutlets similar to those of *I. Aquifolium.*

*I. Perado,* in its typical form, is confined[1] to the Madeira Islands, whence it was introduced into England in 1760. It has usually been cultivated since in green-houses; but Loudon states[2] that in the Chiswick Garden and several other places near London it bore uninjured the winter of 1837-1838 without protection. It has, however, when tried out-of-doors at Coombe Wood, been always killed in ordinary winters. It is perfectly hardy in Wicklow, where at Kilmacurragh a healthy shrub was bearing fruit in July 1908; and at Powerscourt, where a fine specimen is about 20 ft. high, with a diameter of spreading branches of 24 ft.

It is doubtful if this species has been one of the parents of any of the hybrid hollies.[3]

II. *Ilex platyphylla,* Webb and Berthelot, *Phyt. Canar.* ii. 135, t. 68 (1836); W. J. Hooker, *Bot. Mag.* t. 4079 (1844).

A tree, attaining 40 ft. in height. Young branchlets minutely pubescent above the insertions of the leaves. Leaves thick and coriaceous, yellowish green, much larger as a rule than in *I. Perado,* 4 to 6 in. long, 2 to 4 in. broad, ovate-oblong, flat on the surface; base rounded or cuneate, decurrent on the glabrous petiole, which shows two longitudinal furrows on the lower side; apex shortly acuminate and tipped with a slender spine; margin not undulate, entire or with few or numerous minute serrations ending in slender spines; reticulate beneath. Flowers white; fruit sub-globose, reddish or blackish, with larger nutlets than in *I. Aquifolium.*

This is common in the Canary Islands, and also occurs in the Madeiras, in the latter case associated with *I. Perado.* From the latter species it differs mainly in the larger acuminate leaves. It does not seem ever to have been in cultivation[4] in England except at Kew[5] under glass; and there is no evidence that it was concerned in the origin of any of the hybrid hollies.

III. *Ilex balearica,* Desfontaines, *Hist. Arb.* ii. 362 (1809).

*Ilex maderensis,* Willdenow, *Enum. Pl. Hort. Berol. Suppl.* 8 (1813) (not Lamarck); Moore, in *Gard. Chron.* ii. 751 (1874).

*Ilex Aquifolium,* Linnæus, var. *balearica,* Lamarck, *Encycl.* iii. 145 (1789); Loudon, *Arb. et Frut. Brit.* ii. 516 (1838).

A small tree, attaining 30 or 40 ft. in height. Young branchlets stout, greenish, densely covered with a minute pubescence. Leaves thick and coriaceous, concave on the upper surface, 3 in. long, 2 in. broad, ovate, shortly acuminate at the apex, which ends in a slender spine; margin not undulate, either entire or with a few (three to ten) irregularly placed serrations, which end in slender spines; base of the blade rounded or cuneate, scarcely decurrent on the minutely pubescent petiole,

[1] *I. Perado,* var. *azorica,* Loesener, in *Nova Acta Acad. Leop. Carol.* lxxviii. 247 (1901), which is the holly of the Azores, differs from the typical form of the Madeiras in having smaller leaves.

[2] *Trees and Shrubs,* 161 (1842). Loudon, *Gard. Mag.* xiv. 226 (1838), records a shrub of *I. Perado* at Hendon Rectory which was 6½ ft. high.

[3] Cf., however, p. 1712, note 1.

[4] Plants commonly cultivated under this name are, in my opinion, either *I. balearica* or one of its hybrids with *I. Aquifolium.*

[5] A small specimen now growing in the Temperate House at Kew was sent from the Canaries by Dr. Perez.

which is convex on the lower side. Fruit and nutlets slightly larger than in *I. Aquifolium.*

This species occurs in southern Spain and in the Balearic Islands.[1] It is closely allied to *I. Perado*, but differs in having dense pubescence on the branchlets, and the base of the leaf is not decurrent on the petiole. It is very distinct from the common holly, which has leaves with very undulate margins and large sinuate teeth.

*I. balearica* is said to have been introduced into England in 1744, and was certainly cultivated in the Royal Garden at Versailles in 1789. It is usually propagated by budding or grafting upon the common holly, and is perfectly hardy at Kew, Cambridge, and Paris, but requires protection during winter in Germany.

This species regularly produces flowers[2] and fruit in England; and appears to have given rise, in conjunction with the common holly, to a series of hybrids, which began to be noticed about 1800, though they were considered at the time to be simply varieties of the latter species (see p. 1712).

IV. *Ilex opaca*, Aiton, *Hort. Kew.* i. 169 (1789).

A tree, attaining in America 50 ft. in height and 12 ft. in girth. Young branchlets minutely pubescent, becoming pale brown in their first year. Leaves elliptic, about 3 in. long and 1¼ in. broad, convex and dull green above, concave and conspicuously reticulate beneath; margin with a few irregular spreading spiny teeth. Fruit dull red, ¼ in. in diameter; nutlets four.

This species, which somewhat resembles the common holly in foliage, is readily distinguishable by the brown branchlets; and is a native of the United States from Massachusetts to Florida, and westward to Indiana and Texas. It was introduced in 1744 into England, and forms a small tree in our climate, which produces flowers and fruit regularly. Though often grown with the common holly in nurseries and botanic gardens, there is no evidence that this species has taken part in the origin of the hybrid hollies. (A. H.)

## ILEX AQUIFOLIUM, Common Holly

*Ilex Aquifolium*, Linnæus, *Sp. Pl.* 125 (1753); Loudon, *Arb. et Frut. Brit.* ii. 505 (1838); Willkomm, *Forstl. Flora*, 786 (1887); Mathieu, *Flore Forestière*, 58 (1897); Loesener, in *Nova Acta Ac. Leop. Carol.* lxxviii. 248 (1901); Schneider, *Laubholzkunde*, ii. 163 (1907).

A tree, occasionally attaining 50 to 70 ft. in height, and 9 to 12 feet in girth, often shrubby and then rarely over 30 ft. high. Bark smooth, greyish. Young branchlets green or purplish, minutely pubescent. Buds minute, with two acuminate outer scales. Stipules two, usually persistent as withered minute scales at the base of the petiole. Leaves persistent about fourteen months, coriaceous, thick, ovate,

[1] It is probably much more widely spread in the Mediterranean region, but I have been unable to study its distribution. *I. Aquifolium*, var. *platyphylloides*, Christ, in *Ber. Schweiz. Bot. Ges.* xiii. (1903), a tree 30 ft. high occurring on rocky cliffs on the west side of Lake Maggiore in Italy, is probably a form of *I. balearica.*

[2] In nurseries the male plant of *I. balearica* is sometimes called *I. maderensis*, following the erroneous view of Moore, in *Gard. Chron.* ii. 751 (1874), that it came from Madeira, while the name *I. balearica* is wrongly restricted to the female plant. *I. Aquifolium*, var. *platyphylla* of most nurseries, appears to differ in no respect from *I. balearica*; but see p. 1714.

oblong, or elliptic, 2½ to 4 in. long, 1 to 2 in. broad; acute or acuminate at the apex, usually broad and rounded at the base; dark green and shining above, paler and duller beneath; margin waved, encircled by a cartilaginous rim, usually with sinuate teeth, each ending in a sharp spine, which spread in different planes; but on the upper branches[1] of old trees entire, or with one or two spinescent teeth towards the apex; primary veins pinnate, six to eight pairs, dividing and looping before reaching the margin; secondary and tertiary veins indistinct; petiole glabrous, short, rounded and not sulcate on the lower side, the base of the blade not being decurrent upon it.

Flowers small, in short axillary cymose fascicles, normally diœcious,[2] rarely polygamous; sepals four, greenish; petals four, white, placed crossways, slightly connate at the base; stamens four; ovary four-celled, one ovule in each cell. Fruit globose, red, about ⅓ in. in diameter, crowned by the four-lobed stigma, usually containing four nutlets.

The holly is usually bisexual, all the flowers on a tree being either exclusively staminate or pistillate; but, in rare cases, a few perfect flowers, containing both good pollen and well-formed ovules, are produced in addition.[3] Dallimore states that female trees, which are isolated, often bear large quantities of berries; but in such cases most of the seed is infertile, and there appears to be no doubt that pollination is always effected by the pollen being brought from another tree by either insects or the wind.

The seeds when sown do not germinate for a long time, two or even three years elapsing before the appearance of the seedlings.[4] These have two ovate entire obtuse shortly stalked cotyledons, about ⅗ in. long, which are raised above ground on a glabrous caulicle, about an inch in length. The glabrous angled stem gives off in the first year three to five simple alternate leaves, similar to the adult leaves in form, but much smaller, and with small sinuate spiny teeth. These leaves have minute ovate acute black stipules, which are soon deciduous.

## Varieties

The common holly is variable in the wild state, there being two forms,[5] one with green and the other with purple branchlets; moreover, differences in habit, in the spinescence of the foliage, and in the colour of the fruit, are not uncommon. The peculiar geographical forms,[6] which occur wild in western Asia and in China, are not in cultivation. The horticultural varieties, that are usually given in nursery

[1] Kerner, *Nat. Hist. Plants*, Eng. Trans. 433 (1898), believes that the spiny leaves of holly are an adaptation against browsing by ruminants; and points to the fact that on adult trees leaves without spines are only produced on branches beyond the reach of these animals. Withering, *Arrange. Brit. Plants*, ii. 211, note (1796), long ago noticed the same fact, which was also referred to by the poet Southey. Mr. R. A. Phillips tells me that spineless leaves always commence about half-way up wild holly trees in Ireland; and Sir Herbert Maxwell mentions the same fact about wild holly in the wood between Murthly and Dunkeld.

[2] Darwin, *Forms of Flowers*, 297 (1892) states that the stamens in the female flowers, though quite destitute of pollen, are but slightly shorter than the perfect stamens in the male flowers. The male trees produce a greater number of flowers, and these have smaller corollas than occur in the other sex.

[3] Masters, in *Gard. Chron.* xxiii. 27, fig. 8 (1885) and iv. 358 (1888) describes polygamous flowers; and states that occasionally a tree which has hitherto only borne staminate flowers becomes covered with berries.

[4] Cf. Lubbock, *Seedlings*, i. 337, fig. 240 (1892).

[5] Cf. Bromfield, in *Phytologist*, iii. 536 (1849).

[6] Vars. *caspia* and *chinensis*, Loesener, in *Nova Acta Ac. Leop. Carol.* lxxviii. 262, 263 (1901).

catalogues, comprise both the hybrids, which we keep distinct as being of mixed origin; and the true varieties, due only to the common holly, which in many cases have arisen as branch sports, and been subsequently propagated by grafting. The latter usually revert, when old, in isolated branches on the tree, to the type of the common holly.

I. Differing from the type in habit.

1. Var. *pendula*, Loudon, *Trees and Shrubs*, 1113 (1842).

Branches pendulous; leaves as in the common holly. The original tree, from which this variety was propagated by Barron at Elvaston Castle, formerly existed in a private garden in Derby. Loudon also mentions [1] another weeping form, which was discovered about 1842 in Dalkeith Park. There are several good specimens of the weeping holly at Kew, which are clothed to the ground, and somewhat resemble the pendulous form of *Sophora japonica.* Dallimore states that as no leader is formed, it is necessary to keep a shoot tied up, in order to obtain a tall plant.

There are two variegated pendulous forms—var. *aurea pendula,* Waterer's weeping holly, and var. *argentea marginata pendula,* Perry's weeping holly.

2. Var. *fastigiata,* Loudon, *Gard. Mag.* xix. 442 (1843).

Branches erect. Loudon mentions two specimens, one in a garden in Edinburgh, and another in a garden in Derby. The fastigiate holly is rare, but is represented at Kew by a small shrub.

II. Differing from the type in the colour of the fruit.

3. Var. *fructu luteo,* Loudon, *Arb. et Frut. Brit.* ii. 509 (1838).

Vars. *chrysocarpa* and *xanthocarpa,* Koch, *Dendrologie,* ii. pt. i. 210, 216 (1872).

Berries yellow. This is one of the oldest known varieties.[2] Seedlings have been raised in Waterer's and Paul's nurseries, which have produced orange-coloured fruit, and are supposed to have originated from crossing between the yellow and red-fruited forms.[3] Loudon mentions a variety [4] with white berries, and another variety [5] with black berries, neither of which we have seen.

III. Differing from the type in foliage.

A. *Leaves, with both marginal and superficial spines.*

4. Var. *echinata,* De Candolle, *Prod.* ii. 14 (1825).

Var. *ferox,* Aiton, *Hort. Kew.* i. 169 (1789); Loudon, *Arb. et Frut. Brit.* ii. 507 (1838).
*Ilex echinata,* Miller, *Gard. Dict.* ed. 8, No. 2 (1768).

Leaves ovate, about 2 in. long, with their edges rolled backwards; upper surface covered more or less with sharp prickles; margin with irregular large spines. This, which is called the Hedgehog Holly, is the oldest known variety, as it was mentioned by Parkinson [6] in 1640, and was cultivated at Fulham about 1700. It occurs now only as a male plant; but Parkinson and Martyn refer to it as bearing berries, so that a female form was in all probability formerly in cultivation. Dalli-

[1] Loudon, *Gard. Mag.* xix. 442 (1843).

[2] In Cole's *History of Plants,* published in 1657, the yellow-berried holly is mentioned as having been found wild near Wardour Castle.

[3] Moore, in *Gard. Chron.* ii. 520 (1874).

[4] Koch, *Dendrologie,* ii. part i. 212 (1872) calls this var. *leucocarpa,* and speaks of it as common.

[5] Cf. *Arb. et Frut. Brit.* iv. 2545 (1838).

[6] *Theat. Bot.* 1486 (1640).

more[1] considers this variety to have originated from the typical form directly as a branch sport, which was subsequently propagated by grafting, and which by further sports has given rise to other varieties, like *crispa.*

Two variegated forms of this variety are known :—

Var. *ferox argentea,* Loudon, *op. cit.* 509. Spines and margin cream-coloured; known as the Silver-Striped Hedgehog Holly.

Var. *ferox aurea,* Loudon, *op. cit.* 509. Margin green; centre of the blade with a triangular yellow blotch; known as the Gold-blotched Hedgehog Holly.

B. *Leaves deformed; marginal spines abnormal.*

5. Var. *latispina,* Goeppert, in *Gartenflora,* iii. 318 (1854); Moore, in *Gard. Chron.* ii. 812, fig. 164 (1874).

Leaves thick in texture, broadly oval or quadrangular, with an acuminate apex ending in a long decurved spine; averaging 2 in. long and 1½ in. broad; margin with a few coarse spines, variable in number and direction. Dallimore considers that this variety originated probably as a branch sport from var. *crispa.*

6. Var. *trapeziformis,* Moore, in *Gard. Chron.* ii. 812, fig. 164 (1874).

This variety mainly differs from the last-mentioned in the slightly smaller leaves, which have the margin entire or with only one or two erratically placed spines.

7. Var. *monstrosa,* Goeppert, in *Gartenflora,* iii. 318 (1854); Moore, in *Gard. Chron.* ii. 750, fig. 147 (1874).

Leaves ovate-oblong, 3½ in. long, 1 to 1½ in. broad (exclusive of the spines), resembling var. *latispina* in the long acuminate apex with a deflexed spine, but with a longer blade and more numerous marginal spines, which project in all directions. Dallimore states that this originated in the Handsworth Nurseries.

8. Var. *albo-picta,* Loudon, *Arb. et Frut. Brit.* ii. 508 (1838).

Var. *argentea medio-picta,* Moore, in *Gard. Chron.* iv. 688 (1875).

Leaves similar to var. *monstrosa,* but slightly smaller and variegated; the centre of the blade having an irregular blotch of creamy white, the margin remaining green. This variegated holly is usually called "milkmaid" or "silver milkmaid," and has been known from an early period.

9. Var. *hastata,* Moore, in *Gard. Chron.* ii. 687, fig. 138 (1874).

Var. *kewensis,* Loesener, in *Nova Acta Ac. Leop. Carol.* lxxviii. 266 (1901).

Leaves dark green, coriaceous, ¾ to 1¼ in. long, with two or three pairs of long marginal spines towards the base; the upper half of the blade forming a large entire acuminate or emarginate triangular lobe. This is a dwarf shrub, with purple branches, which originated in the Handsworth Nurseries.

10. Var. *Beetii,* Moore, in *Gard. Chron.* ii. 520, fig. 107 (1874).

Leaves nearly orbicular, about 1½ in. in diameter, dark green, with a thickened margin, and a few (usually five or six) large spines, pointing in various directions. This originated in the Handsworth Nurseries.

[1] *Holly, Yew, and Box,* 68, 69 (1908).

11. Var. *crassifolia*, Aiton, *Hort. Kew.* i. 169 (1789); Loudon, *Arb. et Frut. Brit.* ii. 508 (1838); Moore, in *Gard. Chron.* ii. 752, fig. 152 (1874).

Leaves very thick in texture, lanceolate-oblong, 1¾ to 2 in. long, with the apex recurved; margin with triangular teeth, ending in coarse points. This, which is called the Leather-leaf or Saw-leaved Holly, is one of the oldest known varieties,[1] and is remarkably constant in character; no case of a branch reverting to the common holly having been noticed.[2] It is shrubby with purple branchlets; and always bears pistillate flowers, which ripen into peculiarly flattened fruit.

C. *Leaves small, less than 2 in. long, with regular marginal spines.*

12. Var. *lineata*, Moore, in *Gard. Chron.* ii. 752 (1874).

The form of the common holly with the smallest leaves, which are narrowly lanceolate, ¾ in. long, flat, with four or five minute spines on each side. This is a small compact bush with green branchlets. Var. *microphylla*, Moore, *loc. cit.*, appears to differ only in having purple branchlets.

13. Var. *recurva*, Aiton, *Hort. Kew.* i. 169 (1789): Loudon, *Arb. et Frut. Brit.* ii. 507 (1838): Moore in *Gard. Chron.* ii. 687, fig. 138 (1874).

Leaves ovate, acuminate, ending in a long spine, about 1½ in long, curved from base to apex, and usually twisted to one side above the middle; spines divaricate and resembling those of the typical form. This is a dense shrub, and bears staminate flowers.

14. Var. *serratifolia*, Loudon, *Arb. et Frut. Brit.* ii. 507 (1838); Moore, in *Gard. Chron.* ii. 687, fig. 138 (1874).

Var. *myrtifolia*, Goeppert, in *Gartenflora*, iii. 320 (1854); Moore, in *Gard. Chron.* ii. 687, fig. 138 (1874).

Leaves ovate-lanceolate, 1½ in. long, ½ to ¾ in. broad, with divaricate spines similar to those of the type. This forms a dense pyramidal shrub.

Several variegated small-leaved forms are known, which may be mentioned here:—

Var. *myrtifolia aureo-maculata*; centre of the leaf with deep yellow blotches.

Var. *myrtifolia aureo-marginata*; centre of the leaf mottled, margin pale yellow.

Var. *myrtifolia elegans*; leaf with a green centre and a narrow golden edge; and

Var. *Ingrami*; leaf mottled with irregular white streaks, margin pinkish.

15. Var. *angustifolia*,[3] Loudon, *Arb. et Frut. Brit.* ii. 507 (1838); Moore, in *Gard. Chron.* ii. 752, fig. 154 (1874).

Leaves lanceolate, about 1½ in. long and ½ in. broad, often ending in a long acuminate entire apex; spines numerous, regular, slender. This is a female tree, and is very distinct in its narrow erect almost fastigiate habit, and its elegant small leaves. At Kew, Tortworth, and elsewhere, there are specimens 30 to 40 ft. high,

[1] First described by Hanbury, *Complete Book of Gardening* (1770), as the saw-leaved holly.

[2] Cf. Dallimore, *Holly, Yew, and Box*, 67, 90 (1908).

[3] Var. *angustifolia*, Hohenacker, in *Bull. Soc. Nat. Mosc.* iii. 319 (1838), was found wild in woods near Lenkoran, on the south-west coast of the Caspian Sea; and is a form of var. *caspia*, Loesener, differing from the horticultural variety here described.

as straight as a lance, and always very pointed at the top. It bears small, but not numerous berries.

Variegated forms of this are known :—

Var. *angustifolia albo-marginata* ; a small shrub ; leaves as in the green variety, but creamy white on the margin ; and

Var. *angustifolia aureo-maculata*, centre of the leaf unevenly marked with yellow.

16. Var. *handsworthensis*, Moore, in *Gard. Chron.* ii. 520, fig. 108 (1874).

Leaves ovate, acuminate, 1½ in. long, ¾ in. broad, with numerous regular strong spines, pointing towards the apex, the margin only slightly undulate. This has neat and elegant foliage, and originated in the Handsworth Nurseries.

17. Var. *ciliata*, Loudon, *Arb. et Frut. Brit.* ii. 507 (1838); Moore, in *Gard. Chron.* ii. 752, fig. 153 (1874).

Leaves ovate-lanceolate or ovate, 1½ to 2 in. long, ½ to 1 in. wide; margin scarcely undulate, with regular slender spines. This is a male tree at Kew, with purple branchlets and dark foliage.

18. Var. *ovata*, Moore, in *Gard. Chron.* ii. 752, fig. 149 (1874).

*Ilex ovata*, Goeppert, in *Gartenflora*, iii. 321 (1854) ; Koch, *Dendrologia*, ii. pt. i. 216 (1872).

Leaves ovate, with a broad truncate or rounded base, 1½ to 2 in. long, 1¼ in. broad ; margin slightly undulate with small sinuate teeth, tipped with minute spines. This remarkably distinct-looking holly is only known in the staminate form, and has purple branchlets. It is slow in growth. It is undoubtedly a sport of the common holly, as branches with ordinary leaves of the latter are often present on old specimens.

D. *Leaves 2 to 4 in. long, with regular marginal spines.*

19. Var. *Foxii*, Moore, in *Gard. Chron.* ii. 752, fig. 150 (1874).

Leaves broadly ovate, 2 to 2½ in. long, 1¼ to 1½ in. broad, with distant sinuate spinous teeth of moderate size on the slightly undulate margin. This variety, which is only known in the staminate form, is very distinct in appearance.

20. Var. *whittingtonensis*, Moore, in *Gard. Chron.* ii. 687, fig. 138 (1874).

Leaves lanceolate or narrowly elliptic, 2½ to 3 in. long, 1 in. broad, dense shining green on the upper surface ; margin undulate with numerous stiff divaricate spines. This is a distinct variety, bearing staminate flowers.

21. Var. *Fisheri*, Moore, in *Gard. Chron.* ii. 520, fig. 105 (1874).

Leaves ovate, with a broad base and an entire triangular apex, 3 in. long, 1½ to 2 in. broad; dark shining green above ; margin slightly undulate, with large sinuate spiny teeth, variable in number, one to eight on each side. This is a vigorous handsome tree, which originated in the Handsworth Nurseries. It has green branchlets and is a male.

22. Var. *aurea regina*, Moore, in *Gard. Chron.* v. 44 (1876).

Leaves similar in shape to var. *Fisheri*, but variegated ; margin broad, well-defined, deep golden yellow ; centre of the blade mottled with different shades of green. This beautiful holly is usually called "Golden Queen," and bears staminate flowers.

23. Var. *Lichtenthalii*, Simon-Louis, *Cat.* 1880-1881, p. 50.

Leaves oblong or narrowly elliptic, 4 in. long, 2 in. broad; dark dull green above; midrib beneath yellow; margin slightly undulate, and with large regular sinuate spiny teeth. This is a distinct large-leaved form, with purplish branches, and is possibly of hybrid origin.

24. Var. *alcicornis*, Moore, in *Gard. Chron.* ii. 433, fig. 90 (1874).

Var. *Robinsoniana*, Dallimore, *Holly, Yew, and Box*, 76 (1908).

Leaves narrowly elliptic, mostly with an entire cuneate base, about 3 in. long and 1 in. broad (exclusive of the spines); margin undulate, with numerous large spiny teeth, ½ in. or more long, variously directed. This variety, which is distinct in appearance, has green branchlets, and is a free grower. It was originally sent out by Lawson.

E. *Leaves variable in margin, some quite entire, others with spiny teeth.*

25. Var. *donningtonensis*, Moore, in *Gard. Chron.* ii. 687, fig. 138 (1874).

Leaves lanceolate, 1½ to 2 in. long, ½ in. broad, very variable, either entire in margin or with a few irregular divaricate spines; usually recognisable by some of the non-spiny leaves bearing one or two peculiar curved lobes at the base. This variety, which has dark purple branchlets, always bears staminate flowers; and originated in the Handsworth Nurseries.

26. Var. *Smithiana*, Moore, in *Gard. Chron.* ii. 520, fig. 106 (1874).

Leaves lanceolate or elliptic, 2 to 2½ in. long, ¾ in. wide, similar to the last in having leaves both entire and with a few irregular spines, but considerably larger and not showing the peculiar lobes at the base. This always has staminate flowers.

27. Var. *heterophylla*, Aiton, *Hort. Kew.* i. 169 (1789); Loudon, *Arb. et Frut. Brit.* ii. 506 (1838).

This name was originally given to the wild form of the common holly, in which spiny leaves occur near the ground and entire leaves in the upper branches of the tree; but it is now often applied to the next variety, and to intermediate forms.

Two variegated varieties occur:—

Var. *heterophylla aureo-marginata*; leaves of two kinds, entire and spiny, with an irregular golden margin; and var. *heterophylla aureo-picta*, leaves mostly entire and blotched with yellow in the centre.

28. Var. *integrifolia*, Goeppert, in *Gartenflora*, iii. 320 (1854); Moore, in *Gard. Chron.* ii. 812, fig. 164 (1874).

Leaves mostly ovate, entire; flat, but slightly twisted at the apex; about 2 in. long and 1 in. broad; acute, acuminate, or rounded at the apex. Leaves bearing a few spines are usually present. This variety occurs in both sexes.

29. Var. *Watereriana*, Moore, in *Gard. Chron.* vi. 232 (1876).

Leaves oval, 1 to 1¾ in. long, either entire or with a few irregular small spines; edged with a broad irregular band of golden yellow. Waterer's Golden Holly is a staminate form with green branchlets, which are striped with yellow. It is a neat, dense, slow-growing variety.

F. *Leaves all (or nearly all) entire in margin; spines absent, except at the apex, which usually terminates in a spiny point.*

30. Var. *laurifolia*, Loudon, *Arb. et Frut. Brit.* ii. 507 (1838); Moore, in *Gard. Chron.* ii. 812 (1874).

Leaves narrowly elliptic or oblong-lanceolate, entire, flat, about 2 to 3 in. long, and 1 in. broad; rarely a few leaves are present bearing one or two spines. This is known only as a male plant.

Several variegated forms are known:—

Var. *laurifolia sulphurea*; mottled green in the centre, with a broad band of sulphur yellow along the margin.

Var. *laurifolia aurea*, with a narrow bright golden yellow margin; and

Var. *laurifolia aureo-picta*, blotched in the centre with deep golden yellow.

31. Var. *marginata*, Loudon, *Arb. et Frut. Brit.* ii. 507 (not Moore).[1]

Var. *scotica*, Moore, in *Gard. Chron.* ii. 812, fig. 164 (1874).

Leaves obovate or oblong, rounded and spineless, or with a slight cuspidate spine at the apex, 2 to 2½ in. long, 1 to 1½ in. broad; margin entire, undulate, strongly thickened; upper surface often marked near the apex with a cup-like depression.

This peculiar holly, which is occasionally cultivated under the erroneous name *I. Dahoon*,[2] is now generally named var. *scotica*, but seems to be without doubt the plant appropriately called var. *marginata* by Loudon. It is considered by Dallimore to have arisen as a branch sport from var. *crispa*, and often shows leaves approaching in character those of the latter variety; but in opposition to this view may be mentioned the fact that var. *marginata* is always female, while var. *crispa* is a staminate form.

Two variegated forms occur, which are known as:—

Var. *scotica aureo-picta*, a handsome variety which originated in the Cheshunt Nursery; it has leaves variegated with golden yellow in the centre; and

Var. *scotica aurea*, a dwarf form, having leaves with a broad golden edge.

32. Var. *crispa*, Loudon, *Arb. et Frut. Brit.* ii. 507 (1838).

Var. *calamistrata*, *revoluta*, and *contorta*, Goeppert, in *Gartenflora*, iii. 319 (1854).
Var. *tortuosa*, Waterer, *ex* Moore, in *Gard. Chron.* ii. 812, fig. 164 (1874).
Var. *marginata*, Moore, in *Gard. Chron.* ii. 812, fig. 164 (1874) (not Loudon).

Leaves spirally twisted and variously folded, about 2 in. long and 1 in. broad, shining green; margin much thickened, undulate, entire, or with one or two erratically placed spines; apex rounded or prolonged into a long stout decurved spine.

This variety, which is called the Screw-leaved Holly, is always a male tree, and is supposed by Dallimore to be a branch sport of var. *echinata*, as it occasionally bears leaves like those of the latter in having spines on the upper surface.

Var. *crispa aureo-picta* has leaves variegated with gold blotches. It often produces branches with green leaves, and occasionally has leaves with superficial spines.

[1] Var. *marginata*, Moore, is var. *crispa*, Loudon.

[2] *I. Dahoon*, Walter, is a native of the southern United States, and is apparently now not in cultivation. It was killed at Kew in the great frost of 1895. Cf. *Kew Bull.* 1896, p. 10.

G. *Variegated Hollies.*

Many of the variegated hollies have been mentioned above, where they are placed with the green varieties, which they resemble. Of these the most useful for ornament are var. *Watereriana* (see No. 29), var. *aurea regina* (see No. 22), and var. *argentea marginata pendula* (see No. 1). A few remain to be noticed.

33. Var. *aureo-marginata*, Loudon, *Arb. et Frut. Brit.* ii. 508 (1838). Leaves like the common holly, but yellow in margin, with the centre of the blade showing various shades of green. This includes a considerable number of sub-varieties, the most noteworthy of which are var. *aureo-marginata angustifolia* and var. *aureo-marginata bromeliæfolia.*

34. Var. *albo-marginata*, Loudon, *Arb. et Frut. Brit.* ii. 508 (1838). Leaves like the common holly, but with a silvery white or cream-coloured margin. Of the numerous sub-varieties, the best are var. *argentea regina*, "Silver Queen," and var. *handsworthensis argenteo-variegata*, "Handsworth New Silver Holly."

35. Var. *aureo-picta*, Loudon, *Arb. et Frut. Brit.* ii. 509 (1838).

This variety, which is usually known as var. *aurea medio-picta*, or "Gold Milkmaid," has leaves like the common holly in shape, but with their centre irregularly marked with a large golden yellow blotch.

36. Var. *lutescens*, Petzold and Kirchner, *Arb. Musc.* 350 (1864).

Var. *flavescens*, Moore, in *Gard. Chron.* vi. 616 (1876).

Leaves like the common holly in size and shape, but differing in having a soft yellow tinge when young, which usually lasts throughout the season, but is best marked on the side of the tree which is most exposed to the light. This beautiful variety is known as Moonlight Holly, the dark central mass of green foliage with light yellow terminal shoots giving the effect of a shrub seen by moonlight.

## Hybrids

The following, which are usually considered to be varieties of the common holly, are of hybrid origin, the parents being probably *I. balearica* and *I. Aquifolium.*[1] These hybrids are not known to produce branches which revert to the type of the common holly. They are vigorous trees, characterised by large leaves, with the margins flat or much less undulate than in the case of *I. Aquifolium*, the spinous teeth being also less sinuate, in these characters approaching *I. balearica.*

I. *Leaves conspicuously reticulate beneath.*

1. *Ilex Wilsoni*, Fisher, *ex Proc. Hort. Soc.* 1899, p. cxix.; Dallimore, *Holly, Yew, and Box*, 143 (1908).

*Ilex Aquifolium*, var. *latifolia*, Loudon, *Arb. et Frut. Brit.* ii. 507 (1838).
*Ilex Aquifolium*, var. *princeps*, Moore, in *Gard. Chron.* xiii. 45, fig. 8 (1880).

---

[1] Chambers, *Vestiges of Creation*, 310 (1851), quotes from the *Gardener's and Farmer's Journal*, 1848, p. 164:—"The following was related to us by Mr. M'Nab (of the Edinburgh Botanic Garden): he had sown the seeds of *I. balearica*, from which he had produced the common holly. He had also raised from the seeds of the tender Madeira holly (*I. Perado*) a variety identical with that known as Hodgins's holly; and although the offspring of a tender parent, yet like Hodgins's variety, it was also quite hardy." From this it would appear that *I. Perado* may have taken part in the origin of some of the hybrid hollies.

Leaves broadly ovate or oval, 3 to 5 in. long, 2½ in. wide, flat or slightly concave and shining above; conspicuously reticulate beneath; margin with numerous regular spiny teeth directed towards the apex, ¼ in. long, and lying in the plane of the blade.

This is one of the most ornamental hollies, and is a pistillate form, producing abundance of large berries. *I. Wilsoni* was exhibited in 1899 by the Handsworth Nurseries (where there is now a specimen 13 ft. high) as a new kind of seedling origin; but is apparently identical with var. *princeps*, from the same firm, described by Moore[1] a few years earlier. It seems also to be the same as var. *latifolia*, sold by Lawson in Loudon's time.

2. *Ilex Mundyi*, Dallimore, *Holly, Yew, and Box*, 142 (1908).

(?) *Ilex Aquifolium*, var. *nigra*, Moore, in *Gard. Chron.* ii. 433 (1874).

This differs mainly from *I. Wilsoni* in the dull and not shining upper surface of the leaf, the marginal spines of which are directed outwards rather than towards the apex. This is a male plant, which was sent out about twenty years ago by the Handsworth Nurseries, where the tallest specimen is now about 10 ft.

3. *Ilex Lawsoniana*, Dallimore, *Holly, Yew, and Box*, 141 (1908).

*Ilex Aquifolium*, var. *Lawsoniana*, Moore, in *Gard. Chron.* v. 624, fig. 110 (1875).

A variegated form, like *I. Mundyi* in the shape and texture of the leaf, which is, however, often sub-entire in margin; the centre of the blade is marked with broad irregular blotches of yellow. It is one of the handsomest of the golden hollies,[2] but often reverts to the green state.

II. *Leaves not conspicuously reticulate beneath.*

4. *Ilex altaclerensis*, Dallimore, *Holly, Yew, and Box*, 139 (1908).

*Ilex Aquifolium*, var. *altaclerensis*, Loudon, *Arb. et Frut. Brit.* ii. 507 (1838); Moore, in *Gard. Chron.* ii. 752 (1874).

Leaves broadly ovate, 3 to 3½ in. long, and 2 to 2½ in. wide, sub-entire, or with short spiny sinuate teeth variable in number and mostly near the apex of the blade; margin slightly undulate; petiole purplish, often ¾ in. long, with the base of the blade decurrent on it for a short distance.

This is a very fine male tree with purplish branchlets which originated at Highclere in the early part of the nineteenth century.

5. *Ilex Hodginsii*, Dallimore, *Holly, Yew, and Box*, 140 (1908).

*Ilex Aquifolium*, var. *Hodginsii*, Moore, in *Gard. Chron.* ii. 433 (1874).
(?) *Ilex Aquifolium*, var. *Shepherdii*, Goeppert, in *Gartenflora*, iii. 318 (1854) (not Moore[3]).

Leaves ovate or elliptic, 3 to 4 in. long, 2¼ to 3 in. broad, dark green and very shining above; margin undulate and with rather distant regular large triangular

[1] Moore states that var. *princeps* was raised from "the seed of *I. Aquifolium nigrescens* crossed with a male seedling from *I. balearica*."

[2] Gumbleton, in *Gard. Chron.* iii. 595 (1888), states that *I. Lawsoniana* originated in "Hodgen's nursery at Cloughjordan in Tipperary," whence grafts were sent to Messrs. Lawson at Edinburgh. This is an error, and should read "Hodgins's nursery at Dunganstown."

[3] *Ilex Aquifolium*, var. *Shepherdii*, Moore, in *Gard. Chron.* ii. 751 (1874), judging from the description and Moore's specimen in the Kew Herbarium, was identical with *I. Hendersoni*, or only differed from it in the yellowish green leaves being shining and not dull above.

divaricate spiny teeth. This has purplish branchlets, and is very vigorous in growth. It is a staminate form, and is said[1] to be the best holly for planting in towns, being less affected by smoke than the other kinds.

Vars. *nobilis* and *belgica*, Moore, in *Gard. Chron.* ii. 433 (1874), are scarcely distinct from *I. Hodginsii.* A form of the latter, called *I. Hodginsii*, "King Edward VII.," with leaves mottled in the centre, and having a broad yellow margin, was put in commerce in 1898 by Messrs. Little and Ballantyne. Var. *nobilis variegata* has smaller leaves than *I. Hodginsii*, with a yellow blotch in the centre and a broad green margin.

*I. Hodginsii* is sold under the name *I. Shepherdii*, by the Handsworth Nurseries, who inform us that Mr. Shepherd, past curator of the Liverpool Botanic Garden, received two varieties of seedling hollies from Hodgins, nurseryman in the early part of the nineteenth century at Dunganstown near Wicklow.[2] The late Mr. Holmes obtained for the Handsworth Nurseries a stock of these varieties, one of which he named *I. Shepherdii* and the other *I. Hendersoni*, the latter name after a friend of the curator. Fine specimens of the original grafted plants still exist at Handsworth. Mr. Holmes always asserted that it was to the variety *I. Shepherdii* that other nurserymen subsequently applied the name *I. Hodginsii.*[3]

6. *Ilex nigricans*, Henry.

*Ilex Aquifolium*, var. *nigricans*, Goeppert, in *Gartenflora*, iii. 319 (1854).
*Ilex Aquifolium*, vars. *atrovirens* and *nigrescens*, Moore, in *Gard. Chron.* ii. 751, 752 (1874).

Leaves ovate, 2½ to 3½ in. long, 1½ to 2 in. broad, dark shining green above; differing from *I. Hodginsii* in the margin being almost flat or only slightly undulate, with smaller sinuate teeth, ending in long slender spines, often very regular and numerous. Under this name may be included perhaps two or three distinct forms, sold as *maderensis*, *maderensis atrovirens*, *platyphylla*, etc., the nomenclature of which is at present confused in different nurseries.

7. *Ilex Hendersoni*, Dallimore, *Holly, Yew, and Box*, 140 (1908).

*Ilex Aquifolium*, var. *Hendersoni*, Moore,[4] in *Gard. Chron.* ii. 752, fig. 148 (1874).

Leaves elliptic or ovate, 2½ to 3 in. long, 1½ to 2 in. broad, dull yellowish green above; mostly entire or sub-entire, or with a few short sinuate spinous teeth, the margin being only slightly undulate. This is a female tree, which produces large red fruit, but in no great abundance.

8. *Ilex camelliæfolia*, Henry.

*Ilex Aquifolium*, var. *camelliæfolia*, Koch, *Dendrologie*, ii. pt. i. 210 (1872); Moore, in *Gard. Chron.* ii. 812, fig. 164 (1874).

Leaves ovate-oblong, about 4 in. long and 2 in. broad, entire or with a few small sinuate spiny teeth, the margin being slightly undulate; upper surface deep green and very shining; petioles and branchlets purplish. This is a very vigorous tree, with dense foliage, and is a female, bearing larger and darker coloured fruits than the common holly.

[1] Hibberd, in Robinson, *Eng. Flower Garden*, 468 (1893).
[2] Cf. Loudon, *Arb. et Frut. Brit.* i. 116 (1838).
[3] It is preferable to use *I. Hodginsii*, as it is now more commonly known, appearing as an alternative name in the Handsworth Catalogue. *I. Shepherdii* is ambiguous, having been applied to both varieties.
[4] Cf. p. 1713, note 3.

*I. Aquifolium*, var. *Marnocki*, Dallimore, *Holly, Yew, and Box*, 84 (1908), differs mainly from the preceding in the leaves being peculiarly twisted about the middle. Messrs. Fisher, Son, and Sibray tell us that it arose at Handsworth as a chance seedling about forty years ago, and state in their catalogue that it "bears immense berries on the two- and three-year-old wood, forming sprays of vermilion red colour from 1 to 2 ft. long." Their largest specimen is about 16 ft. high.

## Distribution

The common holly is a native of western and southern Europe, and extends in two or three geographical forms into Asia Minor, the Caucasus, northern Persia, and China. It cannot live in regions[1] where the winter temperature is low, its distribution in Europe being limited to the north and eastward by the January isothermal line of 32° F., so that it is unknown in Russia and Sweden and in the eastern parts of Germany and Austria. It occurs in Norway only on the islands[2] and fjords along the west coast from Christianssund (lat. 63° 10′) southwards, and is also met with in west Jutland and on the island of Rugen. It exists here and there in Germany, west of a line drawn from Mecklenburg to Bonn, and also in the Black Forest. It is more abundant in the Alps, where it ascends to 4000 feet elevation in Switzerland and in the Tyrol. In France[3] it is most common in the central and southern departments, growing, according to Fliche, on all soils that are not marshy, but preferring those deficient in lime; it is usually shrubby, but attains large dimensions in Vendée and Corsica.[4] Throughout its area it is most often seen as underwood or small trees in the shade of the broad-leaved forest; but it also grows in many places amidst scrub or in rocky situations in full sunlight.

(A. H.)

## Cultivation

Though the holly is usually looked upon and treated as a shrub, yet in many parts of England it attains the dimensions of a forest tree, which on account of its beautiful foliage and berries has always been one of the greatest ornaments of our natural woodlands and hedgerows. As it is a most useful nurse to oaks, beeches, and other valuable timber trees, and forms excellent shelter for game, it should be encouraged and planted in all places where the soil suits it.

Holly is hardy in all parts of Great Britain; and though it will grow on almost any soil, thrives especially on deep sandy loam and on soils with cool subsoil, grows well on chalky and limy soils, and in very moist climates on thin rocky hillsides. It sows itself freely where rabbits are kept down, but grows slowly at first.

The berries should be gathered in winter, and mixed with sand or light soil in which they decay slowly; but as the seeds rarely germinate until a year has elapsed,

[1] Cf. A. de Candolle, *Geog. Bot. Rais.* i. 147, 162, pl. 1 (1855), who gives much information concerning the distribution of the holly.

[2] Holly grows on some of the islands near Bergen, in Hardanger and Sogn fjords; and attains on Amuglen over 30 ft. in height and 5 ft. in girth.

[3] Cf. vol. iii. p. 560, note 4.

[4] I have two specimens living at Colesborne, which I brought home in 1903 from the Sila mountains in Calabria, where holly grows at 4000 to 5000 ft. altitude.—H. J. E.

it is better to keep them in sand until the following spring, as is done with haws. When sown the beds should be covered with leaves, fern, or branches, to keep out the frost and drought until they begin to germinate. A rich light soil is best to encourage rapid growth when young, and after two years in the seed-bed the seedlings should be transplanted in the month of May with as little damage as possible to the tender roots. I have found that the seedlings grow faster under a wall with a north aspect than in the full sun. At four years old the seedlings should again be transplanted, and the strongest of them, which may then be 2 to 3 ft. high, can be planted out, either in early autumn if the soil is cool and moist enough, or just before growth commences in May. In the latter case a mulch will be desirable, as they suffer from drought after transplanting, and in my experience it is never wise to transplant[1] hollies between November and the end of March.

In forming holly hedges care must be taken to have the ground thoroughly clean and deeply dug beforehand, and to keep the young plants free from grass and weeds, which often choke the young trees. For want of this precaution, and even in spite of it, deaths occur on dry or poor ground which may permanently ruin the regularity of the hedge; and a great deal of money and time are wasted by planting holly hedges and not attending to them afterwards. Holly bears pruning well, and requires attention when young to make a regular and even hedge, and as the growth of the different varieties is very variable, it is important that all the plants should be from the same source.

When variegated hollies are required, they are budded or grafted; and as this is an operation requiring time, experience, and suitable stocks, it is better to purchase the varieties from a nursery. Messrs. Fisher, Son, and Sibray, of the Handsworth Nurseries near Leeds, have long been celebrated for their hollies, which may be transplanted successfully, if due care is taken, up to 5 or 6 ft. high.

The holly, though an indigenous species, suffers from cold in very severe winters, as in 1838, when the thermometer fell at Chiswick on 20th January to $-4\frac{1}{2}°$ Fahr. Lindley[2] states that the holly in this year "was extensively affected in several places in the middle and north of England; this plant, however, offered very different powers of resisting cold, some of the varieties proving much hardier than others, and, according to Mr. M'Intosh, those which are variegated more so than the plain kinds." In 1837-1838 *I. balearica* was not in the least hurt about London.[2] Moreover, in 1905, when the temperature at Kew fell to 2° Fahr. on 8th February, none of the varieties of *I. Aquifolium* seems to have suffered.[3]

## REMARKABLE TREES

The holly occurs in woods and copses throughout the greater part of the British Isles, ascending[4] to 1000 feet in the Highlands, but seems to be commonest and of

[1] Cf. *Gard. Chron.* 1848, p. 99.
[2] In *Trans. Hort. Soc.* ii. 226, 275 (1842).
[3] Cf. *Kew Bull.* 1896, p. 9.
[4] Moss, in Tansley, *Brit. Veg.* 126 (1911), states that "the holly is found in almost every oak wood on the Pennines, but although it sometimes produces flowers it rarely fruits." This is due no doubt to the altitude and prevalent low temperature in these woods.

its largest size in the moister and milder parts of Ireland, Wales, and southern England. It is a very characteristic tree of the New Forest, and is not uncommon in parts of the Chiltern Hills, where large specimens may be seen on the roadside between Wyfold and Reading.

There are few places in Great Britain where the holly grows in greater size and abundance, and forms such an important feature in the scenery, than in the New Forest, where, since the red deer were killed off, it has increased very fast; and in some of the old woods of oak and beech forms almost impenetrable thickets, which not only add to the beauty of the scenery, but protect many young oaks and beeches from being eaten off by the ponies. I am inclined to think that but for these hollies the number of saplings which are coming up would be so small that the timber trees would in time disappear, and though deer, sheep, and rabbits all browse on and bark hollies in winter, they bear the shade of oak very well. In Mark Ash some of the hollies are over 50 ft. high, with straight trunks, and the annual growths here are over a foot long.

In all the beautiful country about Midhurst and Haslemere the holly grows very well, and is cut by gypsies for whip-sticks, which when straight and slender are the best in England, being light and elastic. I have not seen here, however, any trees so remarkable for size as those which grow in Hertfordshire and Bedfordshire on sandy and gravelly loam. At Russells, near Watford, in 1907, I saw a group of trees in a thick shrubbery from 70 to 75 ft. high, but crowded by beech trees, and not well shaped. In Rod's Wood, Teppingley, near Ampthill, Henry measured in 1909 a fine specimen, 60 ft. high and 11½ ft. in girth, at a foot from the ground, just below the point where it divided into about seven stems.

On the oolite of the Cotswold Hills it is common on old downs and in hedgerows, but though often forming trunks of 4 to 5 ft. in girth does not attain as great height as on better soil.

On the coast of Suffolk the holly grows remarkably well on sandy soil in Orwell Park, where in a natural wood of oak and holly I saw many 50 ft. high and more, with clean stems 15 to 20 ft. high, and over 5 ft. in girth. Mrs. Rivis tells us that part of Staverton Wood, near Butley, consists of numerous old holly trees crowded together, and with their stems clear of branches to a considerable height.

At Rougham Hall, Norfolk, in 1907, I measured a splendid weeping silver holly 50 ft. by 4 ft. 11 in., with a bole 15 ft. high, the finest of the sort I have seen.

On the Steiperstone hills, south of Shrewsbury, a natural forest of hollies was said to have existed in which trees of great size were found; one is mentioned 14 ft. in girth;[1] but when I visited this place in 1909 I found that part of the land was now planted, and on the open part the hollies were injured by cattle, and by being lopped for Christmas decorations, so that I could find no old ones more than 20 to 30 ft. high, and few seedlings coming up.

At Doddington Hall, near Lincoln, Lord Kesteven in 1907 measured a splendid holly about 50 ft. high, and 9½ ft. in girth at 4 feet from the ground.

[1] *Notes and Queries*, ser. v., xii. 508.

At Mount Edgecumbe, Henry saw in 1911 a fine holly about 60 ft. high, with a clean stem of 25 ft., measuring 6 ft. in girth.

The longest holly hedge in England is probably one bounding the park at Tyntesfield, near Bristol, the seat of G. A. Gibbs, Esq. Planted on a bank 3 feet high it extends by the side of the public road for about two miles. It is about 4 ft. high and 3 feet thick, and is very dense and even. At Kew there is a fine holly hedge surrounding the shrub nursery,[1] which is 315 yards long, the greater part of it 9 feet high and 4 feet wide, but one portion as much as 12 feet high and 7 feet in width. A remarkable holly hedge[2] of great length at Keele Hall, Staffordshire, is 25 feet high and 30 feet in thickness. Near Bagshot a fine holly hedge[3] around a private garden is 100 yards long and 40 feet high.

At Gorddinog, Llanfairfechan, North Wales, there is a remarkable avenue of yews, originally planted alternately with hollies, which the owner Colonel Henry Platt, C.B., considers to be very old, one yew tree bearing the date 1654 cut into the bark, with the figures stretched by age to a foot in length. Most of the hollies have died of old age, the largest one, which was 42 ft. high and 8 ft. 8 in. in girth, succumbing last year. A few remain, about the same height, all very decayed, and none exceeding 5½ ft. in girth.

The finest golden variegated holly that we know of is growing in the Isle of Man at Kirby Park, the seat of G. Drinkwater, Esq. In January 1913 it was 46 ft. high and 6 ft. 8 in. in girth at one foot from the ground, above which it divides into three stems, forming a beautiful narrow pyramid of foliage.

In Scotland the holly is quite as much at home as in England, and attains as great a size. Hutchison, who gives a complete list[4] of the remarkable holly trees in Scotland, states that they are most abundant in Morayshire and Aberdeenshire, in the basins of the Findhorn, Spey, and Dee, where the climate is mild, and there are numerous woods of ancient date. In Darnaway Forest, Morayshire, there are thousands of hollies, many of large dimensions, the finest, measured by the forester in 1891, being 42 ft. high with a clean bole of 16 ft., girthing 9 ft. 4 in. at five feet from the ground. These grow amidst oaks in a soil of reddish clayey loam, and are supposed to be 200 years old. The most remarkable collection of varieties of the holly is at Ochtertyre, Perthshire. Some of the oldest holly trees in Scotland appear to be at Glenkill, near Lamlash, in the Isle of Arran, the finest measuring in 1891 50 ft. in height and 8 ft. 3 in. in girth at three feet from the ground. An ash-tree, 30 ft. high and 2 ft. in girth, was growing, naturally grafted, on one of these hollies, the junction being about three feet from the ground, where the holly was 8 ft. 1 in. in girth.

Hunter records[5] a grand holly tree at Gourdiehill, Perthshire, which had a stem

[1] Dallimore, *Holly, Yew, and Box*, 32 (1908).

[2] Cf. *Gard. Chron.* xiii. 10, fig. 5 (1893).

[3] Described in *Gard. Chron.* xxvi. 424 (1899). A photograph of this is reproduced by Robinson, *English Flower Garden*, 467 (1893).

[4] *Trans. Highland and Agric. Soc. Scotland*, iv. 80-94 (1892).

[5] *Woods of Perthshire*, 495 (1883).

28 ft. long, as clean and smooth as a pillar, and holding a girth of 6½ ft. throughout its length.

Christison records[1] a holly at Fullarton House, near Troon, 30 to 40 ft. high, with a bole[2] 6 ft. long and 11 ft. 9 in. in girth, in 1891. Loudon[3] mentions a holly at Blair Drummond which grew in sandy loam, and measured 59 ft. by 8 ft., but it is no longer living.

Sir Joseph Sabine in *Trans. Hort. Soc.* vii. 194 (1830), gives a long account of some remarkable holly hedges in Scotland, of which those at Tyninghame, East Lothian, the seat of the Earl of Haddington, extended altogether to a length of 2952 yards. The most striking were those on both sides of a grass walk 36 feet wide, extending from the North Berwick road to the mansion. This walk was 743 yards long, and the hedges 15 feet high, and 11 feet broad at the base; another was 170 yards long, 25 feet high, and 13 feet wide. Most of these were planted by Thomas, the sixth Earl, in 1712; but when I visited the place in 1905 I found that they had become old and ragged, many of the bushes having died. The largest tree that Sabine mentions here in 1830 was 54 ft. by 5 ft. 3 in. at three feet from the ground, but I measured one no less than 71 ft. by 4 ft. 9 in. drawn up among beech trees, but not a handsome specimen.

The most remarkable groups of hollies I have seen anywhere are on the holly bank at Gordon Castle, growing on a moist gravelly old red sandstone soil facing west. These are in clumps, and many have evidently sprung from the same stool. Sabine counted seventy-three groups, containing 508 trees, of which eighty-seven had clean trunks from 8 to 14 ft. long. The largest he measured, which grew at the bottom of the bank, were 52 ft. by 5 ft. 7 in. with a bole of 10½ ft., and 43 ft. 9 in. by 4 ft. 9 in. with a bole of 8½ ft. He mentions one clump which then contained fifty-five trees growing on an area 134 ft. in circumference, and from 1½ to 3½ ft. in girth, which I believe to be the same that I saw in April 1904, not knowing that it had been previously described. I counted fifty-four trees 30 to 40 ft. high, averaging about 3 ft. in girth, and containing about 6 to 7 feet of timber, so that on an area of about a quarter of an acre there must have been over 300 feet of timber.[4] Plate 377 gives a very good idea of this wonderful group, which appeared to me, as it did to Sir J. Sabine, to have been the work of nature, but Mr. Webster, the gardener, could give me no record of their age and origin. The Duke of Richmond, however, tells me that they were flourishing in 1760, as they are alluded to in an account of Gordon Castle written at that time.

At Colinton House, Midlothian, the seat of J. Erskine Guild, Esq., there are some holly hedges, supposed to have been planted between 1670 and 1680, which are still in good health; and as I am informed by the gardener, Mr. Bruce, are

[1] *Trans. Bot. Soc. Edinburgh*, xx. 387 (1896). This tree was photographed by Paxton, who presented in 1894 to the library of the Edinburgh Botanic Garden a book of photographs and measurements of thirty remarkable trees in Ayrshire.

[2] Renwick, in *Trans. Nat. Hist. Soc. Glasgow*, vii. pt. iii. 263 (1904), gives the girth in 1903 as 11½ ft. at five feet from the ground.

[3] *Gard. Mag.* xvii. 507 (1841), where a list of measurements in 1836 of all the remarkable trees at Blair Drummond is given.

[4] In *Gard. Chron.* iii. 51 (1875), Mr. Webster says that one of the trees here measures 7 ft. 9 in. at three feet, and 7½ ft. at ten feet, and another 8½ ft. in the narrowest part of the trunk two feet from the ground.

now in better condition than they were twenty-five years ago. Some parts of them are solid green, and quite thick from base to top. The height of the tallest is 40 ft. after clipping, which is done once in two years. The broadest part at the base is 21 ft. through, and where protected from rabbits the leaves touch the ground. They grow on light dark loamy soil, with a sandy or gravel subsoil, but hollies here thrive equally well on clay with a rocky subsoil. These are supposed to be the oldest and the tallest holly hedges in Scotland, and perhaps in the whole of Great Britain.

The holly, according to Mr. R. A. Phillips, is distributed throughout Ireland, but is more abundant in the non-calcareous districts of the west, south, and south-east than elsewhere. It also occurs on the islands[1] off the west coast. In Ireland it formerly attained an enormous size, the most famous tree being one on Innisfallen Island, Killarney, which Hayes recorded[2] as 15 ft. in girth in 1794; but I could find no trace of it in 1909. There were also remarkable woods in which holly grew nearly pure, and produced valuable timber in quantity. The late Earl Annesley informed Henry that out of a wood of this kind by the lake at Castlewellan his brother sold in 1871 more than £500 worth of holly timber; but in 1906 the largest tree which remained was scarcely 6 ft. in girth. Near Mount Usher, in Wicklow, Henry measured in a wood in 1904 a tree 70 ft. in height and 6 ft. in girth.

## Timber, etc.

The wood of the holly is white, hard, and heavy; and has a fine close grain, being very homogeneous in texture. It takes a good polish, and is used for making mathematical instruments, for inlaying, and for turnery. Pulley-blocks for ships were formerly made of holly. When dyed black, it is a cheap substitute for ebony. Dallimore states[3] that snuff-boxes were, in the early part of the last century, made out of the knots and burrs that are sometimes found on the trunk. Young straight quickly grown shoots are used for making walking-sticks and whip handles.

Bird-lime is made out of the mucilaginous bark of the young shoots.

(H. J. E.)

[1] Praeger, in *Proc. R. I. Acad.* xxxi. pt. 10, pp. 19, 21, 25, records it for Clare Island.

[2] *Treatise on Planting*, 143 (1794).

[3] *Holly, Yew, and Box*, 20 (1908).

# BUXUS

*Buxus*, Linnæus, *Syst. Nat.* 9 (1735), *Gen. Pl.* 284 (1737), and *Sp. Pl.* 983 (1753); Baillon, *Monog. Buxac.* 58 (1859); Müller, in De Candolle, *Prod.* xvi. pt. i. p. 13 (1869); Bentham et Hooker, *Gen. Pl.* iii. 266 (1880); Pax, in Engler and Prantl, *Nat. Pflanzenfam.* iii. 5, p. 133 (1892); Van Tieghem, in *Ann. Sci. Nat.* v. 289 (1897); Hutchinson, in *Kew Bulletin*, 1912, p. 52.

EVERGREEN trees or shrubs belonging to the family Buxaceæ. Young branchlets quadrangular. Leaves opposite, coriaceous, simple, entire, penninerved, shortly stalked, without stipules. Flowers monœcious, without petals, in axillary clusters, which are composed of several staminate flowers, and usually a central single pistillate flower. Staminate flower: sepals four, in two opposite pairs; stamens four, each opposite a sepal, with a thick filament inserted beneath the rudimentary ovary, and with a two-celled anther, which dehisces longitudinally. Pistillate flower; sepals four to seven, occasionally ten, some of which may represent bracts, imbricated; ovary free, three-celled, crowned by three peripheral styles; ovules, two in each cell. Fruit,[1] a capsule, crowned by the three persistent styles, three-celled, each cell containing two seeds, or occasionally only one by abortion; capsule, when ripe, splitting longitudinally, through the three styles and the dorsal sutures, thus producing three two-horned valves; the endocarp afterwards opens down six lines, suddenly expelling the seeds, which are trigonous, shining, black, and smooth.

The above description is restricted to the section *Eu-buxus* of the genus,[2] which comprises about ten species, two, natives of tropical Africa and Socotra, and about eight species, natives of the Mediterranean region, Caucasus, Persia, Himalayas, China, and Japan. The following synopsis briefly deals with those which are in cultivation:—

I. *Branchlets glabrous, or occasionally with a few minute hairs near the nodes.*

1. *Buxus japonica*, Müller, in De Candolle, *Prod.* xvi. 1, p. 20 (1869); Shirasawa, *Icon. Ess. Forest. Japon*, ii. t. 38, figs. 16-32 (1908).

*Buxus sempervirens*, Linnæus, var. *japonica*, Makino, in *Tokyo Bot. Mag.* ix. 281 (1895), and xv. 169 (1901); Hayata, in *Journ. Coll. Sci. Tokyo*, xx. 3, p. 82, t. vi. c. (1904).

A shrub or small tree, with glabrous branchlets. Leaves coriaceous, very similar to those of *B. sempervirens* in size and appearance, but usually more oval and occasionally almost orbicular, ¾ to 1 in. long and ½ in. broad, emarginate and rounded

[1] Cf. Marshall Ward, *Trees*, iv. 154, fig. 147 (1908).

[2] The other sections of the genus, which are sometimes regarded as distinct genera, are *Buxella*, which includes four species, natives of Madagascar, tropical and south Africa; and *Tricera*, comprising about a dozen species, natives of the West Indies.

at the apex; upper side of the short broad petiole pubescent, the leaf elsewhere being glabrous. Flowers of both sexes sessile, similar to those of *B. sempervirens*, but with the rudimentary ovary of the staminate flower much enlarged and as long as the inner sepals.

This is wild in the mountains of Japan, where it has given rise in cultivation to several peculiar forms, one of which with almost orbicular leaves, puckered and uneven on the surface, has been introduced into Europe. The typical form is rare; but is represented at Kew by a shrub about 3 ft. high.[1]

2. *Buxus microphylla*, Siebold and Zuccarini, in *Abh. Ac. München*, iv. 2, p. 142 (1845).

> *Buxus sempervirens*, var. *microphylla*, Blume, *ex* Miquel, in *Ann. Mus. Lugd. Bat.* iii. 128 (1867); Hayata, in *Journ. Coll. Sci. Tokyo*, xx. 3, p. 83, t. vi. D. (1904).
> *Buxus japonica*, var. *microphylla*, Müller, in De Candolle, *Prod.* xvi. 1, p. 20 (1869).

A low shrub, with very slender branchlets, which are glabrous or with a few minute hairs above the nodes. Leaves thin and membranous, spatulate, lanceolate or elliptic, about ½ in. long and ⅙ to ¼ in. broad, emarginate at the apex, slightly pubescent on the upper side of the very short broad petiole. Flowers, as in *B. japonica*.

This species[2] occurs wild in Japan, in the provinces of Shimosa, Awa, and Tosa. It is a very distinct plant, with small thin leaves, and has lately been introduced into Kew.

3. *Buxus Harlandi*,[3] Hance, in *Journ. Linn. Soc. (Bot.)* xiii. 123 (1873).

A small shrub, attaining 1 to 2 ft. in height, with densely crowded foliage. Young branchlets very slender, glabrous or occasionally with a few minute hairs above the nodes. Leaves oblanceolate, very narrow in proportion to their length, ¾ to 1¼ in. long, ⅙ to ¼ in. wide, emarginate at the apex, very tapering at the base, quite glabrous throughout, or with slight pubescence on the upper side of the petiole. Flowers similar to those of *B. japonica*, sessile in both sexes, with the rudimentary ovary of the staminate flower as long as the inner sepals; but with the style of the pistillate flower as long as the ovary.

This shrub, which is very distinct in appearance, occurs in central and southern China, where it is found growing in the shingly or rocky beds of rivers and streams. It has been in cultivation about ten years at Kew, where it is a dwarf shrub.[4]

II. *Branchlets minutely puberulous.*

4. *Buxus balearica*, Lamarck, *Encyc.* i. 511 (1753); Willdenow, *Sp. Pl.* iv. 337 (1805).

> *Buxus sempervirens*, var. *gigantea*, Loiseleur, in Duhamel, *Traité des Arbres*, i. 82, t. 23 (1801).

---

[1] *B. Fortunei rotundifolia*, of some French nurseries, is probably a prostrate form of *B. japonica*.

[2] *B. stenophylla*, Hance, in *Journ. Bot.* vi. 331 (1868), wild in the province of Fukien, in China, is closely allied to, if not identical with *B. microphylla*. *B. sempervirens*, var. *riparia*, Makino, in *Tokyo Bot. Mag.* xxvi. 293 (1912), a shrub 3 ft. high, growing beside the river Yoshino, in the province of Tosa, is said to be intermediate between *B. japonica* and *B. microphylla*.

[3] This species was founded by Hance, on specimens gathered in Hong-Kong; but his sheet, No. 322, so-named, includes two different plants, one, the true *B. Harlandi*, and the other, resembling *B. japonica* in the shape of the leaves, but differing in having pubescent branchlets.

[4] *B. Fortunei*, Carrière, in *Rev. Hort.* xlii. 519 (1871), also sometimes known as *B. longifolia*, Hort., judging from specimens cultivated under this name in France, is identical with *B. Harlandi*.

A shrub or small tree. Young branchlets with a minute pubescence, only visible with a strong lens. Leaves light green, coriaceous, larger than those of the common box, but similar in shape, elliptic, 1 to 1½ in. long, ½ to ⅝ in. wide, emarginate at the apex, minutely pubescent on the upper side of the short petioles, but glabrous elsewhere. Staminate flowers shortly stalked; pistillate flowers with the styles as long as the ovary.

This species[1] occurs in the Balearic Isles, and in the province of Granada, in Spain, on the coast at Nerja, near Malaga, and on the Sierra de Almijara, at 2000 feet elevation. Loudon[2] states that it was introduced into England in 1780. It is perfectly hardy, the oldest specimen at Kew being about 23 ft. high.

III. *Branchlets densely pubescent.*

5. *Buxus sempervirens*, Linnæus. See p. 1724.

6. *Buxus Wallichiana*, Baillon, *Monog. Buxac.* 63 (1859); Hayata, in *Journ. Coll. Sci. Tokyo*, xx. 3, p. 84, t. vi. E. (1904).

*Buxus sempervirens*, J. D. Hooker. *Fl. Brit. Ind.* v. 267 (1887) (not Linnæus); Gamble, *Indian Timbers*, 592 (1902).

*Buxus sempervirens*, Linnæus, var. *liukiuensis*, Makino, in *Tokyo Bot. Mag.* ix. 279 (1895), and xv. 169 (1901).

*Buxus liukiuensis*, Makino, in *Tokyo Bot. Mag.* xvi. 179 (1902); Schneider, *Laubholzkunde*, ii. 140 (1907).

A shrub or small tree. Branchlets densely pubescent. Leaves lanceolate, 1 to 2¼ in. long, ¼ to ½ in. wide, rounded or emarginate, with a short point at the apex; petiole densely pubescent, the pubescence spreading along the midrib on the upper side. Flowers of both sexes sessile; rudiment of the ovary in the male flower very short, dilated and slightly four-lobed at the apex; styles in the pistillate flower as long as the ovary, persisting elongated and erect on the fruit.

*B. Wallichiana* is widely spread in Asia, occurring in India in the Suliman and Salt ranges, and in the Himalayas from Kumaon to Bhutan (but not in Sikkim), at 4000 to 8000 feet. It is common in the mountainous districts of central and southern China, and is the only species known at present in the Liu Kiu Islands and Formosa. Its distribution[3] in the Himalayas is local and peculiar; but it mainly grows along the banks of streams in moist and sheltered places, preferring a northerly aspect; and is met with on sandstone as well as on limestone. It often occurs in considerable quantity, attaining a large size, trees being recorded as much as 5 ft. in girth; but is slow in growth, averaging, according to Gamble, about twenty rings per inch of radius. In China,[4] the box is known as *huang-yang*, and is of consider-

[1] *B. longifolia*, Boissier, *Diag. Fl. Orient.* 107 (1853), and *Fl. Orient.* iv. 1144 (1879), of which I have seen no specimen, is said to occur in the mountains of Syria; and appears to be closely allied to *B. balearica*. It is entirely distinct from *B. longifolia*, Hort, which appears to be a synonym of *B. Harlandi*. Cf. p. 1722, note 4.

[2] *Arb. et Frut. Brit.* iii. 1341 (1838). His account of the size and distribution of this species, which has been followed by Dallimore, *Holly, Yew, and Box*, 229 (1908), is erroneous. So far as I can learn, *B. balearica* is a small tree or shrub, not exceeding thirty feet in height; and does not occur wild except in the localities mentioned above. Nyman, *Consp. Fl. Europ.* i. 647 (1878), records it for Sardinia; but this is doubtful.

[3] A dwarf box with very small leaves occurs in the Himalayas at high elevations; and has been referred to *B. iaponica*, var. *microphylla*, Müller, by Hooker, *Flora Brit. India*, v. 267 (1887); but is much more pubescent than the Japanese plant, and is probably a distinct species.

[4] The species of box in China require further study. *Buxus Henryi*, Mayr, *Fremdländ. Wald- u. Parkbäume*, 451 (1906), described from specimens (No. 3387) collected by me on cliffs near Ichang, in central China, has glabrous branchlets,

able economic importance in the mountains north of Ichang, in Hupeh, where it is a tree of considerable dimensions.

*B. Wallichiana* is very rare in cultivation; but thrives at Kew, where specimens, about 6 or 7 feet high, produce flowers and fruit regularly. (A. H.)

## BUXUS SEMPERVIRENS, Common Box

*Buxus sempervirens*, Linnæus, *Sp. Pl.* 983 (1753); Loudon, *Arb. et Frut. Brit.* iii. 1333 (1838); Baillon, *Monog. Buxac.* 59 (1859); Müller, in De Candolle, *Prod.* xvi. 1, p. 19 (1869); Willkomm, *Forstl. Flora*, 802 (1887); Mathieu, *Flore Forestière*, 306 (1897).
*Buxus arborescens*, Miller, *Dict.* ed. viii. No. 1 (1768).

A shrub or small tree, attaining about 30 ft. in height and 3 ft. in girth. Young branchlets densely pubescent with short white hairs, which are more or less retained in the second and third years. Leaves persistent five or six years, coriaceous, opposite, oval or elliptic, averaging 1 to 1¼ in. long and ½ in. broad, rounded and usually emarginate at the apex; shining and dark green above; duller and yellowish green below; secondary nerves pinnate, often forked, conspicuous on the upper surface; margin entire; tapering at the base to a very short petiole, which is pubescent like the branchlets, the pubescence extending along the midrib on the upper surface, and on the edges of the base of the blade.

Flowers small, white; both sexes sessile; rudimentary ovary of the staminate flower scarcely half as long as the inner sepals; styles of the pistillate flower short, about half as long as the ovary. Capsule ovoid, longer than broad, brown when ripe, crowned by short spreading styles; seeds trigonous, smooth, shining, black. The seedling[1] has two oblong obtuse glabrous cotyledons about ½ in. long, raised above ground on a glabrous caulicle about 1½ in. long; primary leaves opposite, decussate, elliptic, shortly stalked.

### Varieties

A considerable number of varieties of the common box are in cultivation, most of which have arisen in gardens and nurseries—the variation in the wild state[2] being slight.

1. Var. *angustifolia*, Loudon, *Arb. et Frut. Brit.* iii. 1333 (1838).

*Buxus angustifolia*, Miller, *Gard. Dict.* ed. viii. No. 2 (1768).

Leaves oblong-lanceolate, about 1 in. long and ⅓ in. wide. This is said to occur

lanceolate leaves as much as 3 in. long, and ¾ in. broad; staminate flowers on long pedicels, with a minute and linear rudimentary ovary. This species, which is allied to *B. balearica*, has remarkably fine foliage and conspicuous flowers; but has not yet been introduced. It has lately been fully described and figured by Dümmer, in *Gard. Chron.* lii. 423, fig. 182 (1912).

[1] Cf. Lubbock, *Seedlings*, ii. 481, fig. 639 (1892).

[2] The typical form of the species was distinguished as var. *arborescens*, by Linnæus, *Sp. Pl.* 983 (1753), a name kept up by Loudon, *Arb. et Frut. Brit.* iii. 1333 (1838). Var. *grandifolia*, Müller, in De Candolle, *Prod.* xvi. 1, p. 19 (1869), wild in Spain, Greece, and the Caucasus, is scarcely distinguishable, though occasionally the leaves are longer and more lanceolate than in the type.

wild in Algeria; and is commonly shrubby. An upright tall-growing form of this is known as var. *salicifolia elata.*

2. Var. *myrtifolia*, Loudon, *Arb. et Frut. Brit.* 1333 (1838).

*Buxus myrtifolia*, Lamarck, *Encyc. Meth.* i. 511 (1783).

Leaves dark green, oblong-lanceolate, smaller than in the last variety, about ¾ in. long and ¼ in. wide. This is a low shrub, which was described by Lamarck, from specimens obtained from M. Cels' nursery at Paris, where it probably originated.

3. Var. *myosotifolia*, Dallimore, *Holly, Yew, and Box*, 227 (1908).

Leaves resembling those of var. *myrtifolia*, but still smaller, about ½ in. long and ⅙ in. broad, lanceolate, dark green. This is a dwarf shrub.

4. Var. *rosmarinifolia*, Baillon, *Monog. Buxac.* 61 (1859).

Leaves lanceolate or spatulate, more slender and thinner in texture than the last variety, about ½ in. long and ⅛ in. wide, marked beneath with whitish dots and tubercles. This variety, which is also known as var. *thymifolia*, grows to be a bush 5 or 6 ft. high.

5. Var. *suffruticosa*, Linnæus, *Sp. Pl.* 983 (1753).

*Buxus suffruticosa*, Miller, *Gard. Dict.* ed. viii. No. 3 (1768).

Leaves oval or obovate, ½ in. long and ¼ in. wide. This is the well-known dwarf variety, which is used for edging beds in gardens. It has been in cultivation for several centuries at least, and is occasionally called var. *nana* or var. *humilis*.

6. Var. *latifolia*, Dallimore, *Holly, Yew, and Box*, 226 (1908).

Under this name are included several forms, in which the leaves are broader than usual, averaging 1 in. long and ¾ in. wide. In var. *latifolia bullata*, Späth, the leaves are uneven with peculiar swellings. Var. *handsworthensis*, Fisher, with broadly oval leaves, is vigorous in growth, and suitable for making hedges.

7. A considerable number of variegated forms are in cultivation. Those with leaves normal in size, or nearly so, are:—Var. *argentea* or var. *argenteo-marginata*, leaves white in margin; var. *aureo-marginata*, leaves yellow in margin; and var. *aureo-maculata*, leaves spotted with yellow. Var. *elegantissima*, Koch, *Dendrologie*, ii. pt. i. 477 (1872), is a distinct form with small oval leaves, variegated with white, and with many of the leaves deformed.

8. Var. *pendula*, Simon-Louis, *Cat.* 1869, p. 21. Tree-like in habit, and graceful in outline, the secondary branches and branchlets being pendulous.

9. Var. *pyramidalis*, Simon-Louis, *Cat.* 1869, p. 21. Pyramidal in habit, with upright branches.

### Distribution

The common box is a native of western Europe, the Mediterranean region, the Caucasus, and northern Persia. It is probably a true native of England; and in the south and east of France[1] is widely spread in the Jura, Dauphiné, Languedoc,

[1] Chatin, in *Bull. Soc. Bot. France*, viii. 364 (1861), considers the box to be naturalised in many localities in France, which are mainly in the neighbourhood of abbeys and castles that date from the middle ages. He mentions the forest of Marly, Vaux-de-Cernay, Neauphle-le-Château, Arthieul near Magny, Roche-Guyon, Chantilly, Nemours, Provins, and Jaux near Compiègne. He adds that it is abundant on the millstone grits and sandstones of Vaux-de-Cernay, near Paris, and on granite at Mauves-sur-Loire.

Provence,[1] and Pyrenees,[2] and occurs in a few scattered spots in the north of the departments of Meuse and Meurthe-et-Moselle, extending across the frontier into Belgium. It is recorded for Germany from one locality in Baden. Farther south it is common in the southern Tyrol, Istria, Dalmatia, the Balkan States, and Greece. It is also met with in Italy, Spain, and Portugal; and is a rare plant in the cedar forests of Algeria. In Europe, it is usually a gregarious shrub, often growing on arid hills or mountains on limestone[3] soil in the Mediterranean region. It is cultivated in Norway as far as lat. 67° 56′ on the coast, and in Sweden up to lat. 59° 7′.

The box attains its largest size in the Caucasus,[4] where it is very common in the coast region along the Black Sea from Sotschi to Batoum, at elevations between sea-level and 4000 ft. It also grows wild in Talysch, but in other parts of the Caucasus is a doubtful native, though it is often planted, being considered a holy tree by the natives. The finest trees grow inland from the Black Sea, at about 2500 ft. altitude, where they are commonly 30 to 40 ft. in height, with stems 8 to 12 inches in diameter. Formerly still larger trees were known, Köppen[4] mentioning 50 ft. in height and 2 ft. in diameter as the maximum size. Many large trees are of enormous age, 500 to 600 annual rings having been counted by Medwejew; but nearly all these are rotten at the heart. The box grows usually in the Caucasus in narrow bands along rivers and streams as undergrowth in the great forests of oak, ash, and beech; and is rarely found mixed with conifers. It thrives best in moist, rainy, sheltered, and shaded spots. The woods of box tree in Georgia are mentioned[5] by Marco Polo. Until about 1890 the export of boxwood from the Black Sea to England, France, and Turkey, was enormous, averaging 2340 tons annually for the years 1883-1887.

Consul Stevens, in his *Trade Report*[6] *for Batoum* for 1895, states: "Although all the private forests of boxwood have been exhausted, the Government up to the present still refuse to sell or allow boxwood to be cut in their extensive forests throughout Abkhasia; consequently the total exports from the Caucasus have not exceeded 1200 tons."

The box appears to be very common in the Elburz mountains in northern Persia, especially in the forests of Mazanderan, whence the export of boxwood in 1906 amounted to 125,864 pieces, weighing about 1560 tons.[7]

Whether the box is a native of England or not is doubtful; but it is certainly

[1] Tansley in *Gard. Chron.* lii. 113, fig. 51 (1912) describes the peculiar shrubby vegetation, which grows on the southern slopes of the hilly regions of Provence, at about 4500 ft. elevation. This consists of isolated bushes, with bare rock between them, of box, lavender, and *Genista cinerea.* The north slopes are covered with beech woods, in which there are holly and box, the latter being very abundant inland.

[2] On the south side of the Pyrenees, above Venasque at 5000 ft. altitude, I saw, in 1912, much box, forming dense scrub on sunny slopes; and it is the prevailing undergrowth in many of the valleys of the western Pyrenees on the French side. In the forest near Esterencuby, south of St. Jean-Pied-de-Port, box attained 30 ft. high on limestone, and was being felled for making prayer-beads.

[3] It is most commonly found on limestone; but occurs frequently on other soils, as in the cases mentioned (p. 1725, note 1) by Chatin; and according to De Candolle, *Geog. Bot. Rais.* i. 426 (1855), it grows on schist in the Pyrenees, on granite in Brittany, and on volcanic soil in Auvergne. Gèze, in *Bull. Soc. Bot. France*, lv. 464 (1908), states that at Villefranche it is not calcicole.

[4] Cf. Radde, *Pflanzen-verb. Kaukasusländ.* 145, 182, 201 (1899), and Köppen, *Holzgewächse Europ. Russlands*, ii. 1-9 (1889).

[5] Yule, *Marco Polo*, i. 50, 54 (1871).

[6] *Foreign Office Ann. Series*, No. 1717, p. 27.

[7] *Consular Report for Resht*, No. 3864, p. 25.

naturalised, if not truly indigenous, in a few localities. It was cultivated[1] in Britain by the Romans; and as it seeds itself freely in the south of England, it may have spread from abandoned villas. It was well known in Anglo-Saxon times, the earliest mention, I was informed by the late Dr. Skeat, being in the "Corpus Glossary" of Latin and English words, which was written about 750 A.D. In this work *box* is given as the Anglo-Saxon equivalent of the Latin word *buxus*. The word *box* begins to appear early in place-names,[2] the oldest example known to Dr. Skeat being *Box-ōra* (*i.e.* box-bank), the old name of Boxford, Berks; and there must have been box trees here at an early date. In the thirteenth century, numerous names of places occur with *box*, as La Boxe, Le Boxe, Hundred de Boxe, Boxen, Boxford, Boxhale, Boxhey, Boxore, Boxley, Boxland, Boxstead. These place-names show that box was well known in former times; but whether it was wild or cultivated, there is no means of determining.

Professor Babington[3] believed that the following extract from the beginning of Asser's *Life of King Alfred* showed that it was plentiful in Berkshire 1000 years ago: "Berrocscire; quæ paga taliter vocatur a Berroc sylva, ubi *buxus* abundantissime nascitur." Gough's *Camden*, 155 (1789), says: "The last remains of Boxgrove[3] in Sulham parish, near Reading, whence the county probably took its name, were grubbed up about forty years ago."

Gerard,[4] writing in 1597, says: "The box tree groweth upon sundry waste and barren hills in England." Ray[5] in 1696 records it growing at Boxhill near Dorking, at Boxley in Kent, and at Boxewell in the Cotswolds.

At the present day, box is apparently wild in several places in the south of England, the most famous being Box Hill in Surrey, where many acres on the western slopes are covered with a mixture of yew, box, and other trees. The occurrence of the box-tree here was first recorded by Merrett[6] in 1666. Count Solms-Laubach suggests[7] that the box and yew trees of Box Hill might probably be the remains of a native forest which originally clothed the North Downs. He urges the unlikelihood of such a soil as that of Box Hill being planted at all, and the improbability of any one hitting upon such a combination as yew and box for the purpose. Manning and Bray, *History of Surrey*, i. 560 (1804), give the following account: "The Downs, which rise from the opposite bank of the Mole, are finely chequered with Yew and Box trees of great antiquity, to a considerable height. Of the latter of these in particular there was formerly such abundance that that part of the Downs which is contiguous to the stream, and within

[1] Clement Reid, at a meeting of the Linnean Society, London, on 2nd December 1909, said that Box leaves have been found in three different rubbish heaps in the Roman remains at Silchester. The branches may have been used for wreaths.

[2] The names of places with box, given by Spelman, *Villare Anglicum* (1653) and by Adams, *Index Villaris* (1680) are Box (Wilts), Boxend (Beds), Boxford (Berks and Suffolk), Boxley (Kent), Boxted (Essex and Suffolk), Boxwell (Gloucester), and Boxworth (Berks and Cambridge). Some of these, as we know from the old spelling, do not indicate the box tree; thus Boxworth near Cambridge means the "farm of the buck." In some cases these places seem to be connected with the Roman occupation of Britain, as Boxmoor House (Herts), near which a Roman dwelling-house was discovered in 1851; but where the places occur on the chalk downs, the presumption is that the tree is indigenous.

[3] *Phytologist*, iv. 873 (1853). Cf. Stevenson, Asser's *Life of King Alfred*, pp. 1, 156, 157 (1904), who states that Berroc Wood was identified with Boxgrove by Francis Wise in 1738.

[4] *Herball*, 1225 (1597).

[5] *Syn. Meth.* ii. 310 (1696). Box does not appear to be growing wild at the present time near Boxley in Kent.

[6] *Pinax. Rerum Nat. Brit.* 18 (1666).

[7] Cf. article by G. R. M. Murray, in *Journ. Bot.* xxxix. 27 (1901).

the precinct of this Maner, hath always been known by the name of Box Hill. Here was formerly also a Warren with its Lodge; in a lease [1] of which from Sir Matthew Brown to Thomas Constable, dated 25th August 1602, the Tenent covenants to use his best endeavours for preserving the Yew, Box, and all other trees growing thereupon; and in an account of the rents and profits for one year to Michaelmas, 1608, the receipt for Box trees cut down upon the Sheep Walk on the hill is £50. I have seen also an account of this Maner, taken in 1712, in which it is supposed that as much had been cut down [2] within a few years before as amounted to £3000."

E. S. Marshall [3] says: "He must be very sceptical who doubts it being native on the steep slopes of Box Hill, above Burford Bridge. I have also seen it growing rather plentifully a mile or more away towards Betchworth." Bromfield [4] says: "Box is profusely abundant on most parts of Sidon Hill, in Highclere Park (Hants), scattered over its shelving sides as if quite spontaneous, and said to disperse itself freely by seed;" but he avers that it was certainly planted here. Thus the natural appearance now of the box on Box Hill, Surrey, is no sure proof of its being indigenous there.

Another locality where the box occurs apparently wild is the Chiltern Hills; as on the chalk downs [5] between Ashridge and Berkhampstead, where there are some very old-looking trees. Near here, on the top of Dunstable Downs, there is a place named Boxstead. On the Chequers Court estate, about half a mile from Ellesborough Church, near Wendover, box has all the appearance of being indigenous [6] over a considerable area. Mr. Raffety of High Wycombe tells me that box is thickest here in two valleys, known locally as Ellesborough Warren and Kimble Warren, the bushes being about 20 ft., with numerous seedlings of all ages. It extends up over the adjoining spur of the Chiltern Hills to an altitude of 500 feet, the chalk subsoil being near the surface, and the exposure almost due south-west. Messrs. Sprague and Hutchinson state [7] that some of the stems are 8 in. in diameter, and that the only tree which has obtained a place amongst the box thickets is the elder.

There is a place named Boxe in Domesday Book for Herts, section xxviii.; but the village [8] so-called no longer exists, being now part of Wymondley. Mr. H. Clinton-Baker tells me of a field at the Priory, Wymondley,[9] around which is a broad

[1] Barrington, in *Phil. Trans.* lix. 23 (1769) quotes from *A Journey through England*, printed in 1722: "Box Hill was first planted by that famous antiquary the Earl of Arundel, with box wood, designing to have a house there; but want of water made him alter his resolution and build one at Albury hard by." This is erroneous, as the Earl of Arundel was only sixteen years old in 1602, when Box Hill was already covered with box and yew, according to the lease cited above.

[2] Ellis, *Timber Tree Improved*, 103 (1745), says that "great quantities of box were felled off the Chalky Downs near Dorking in 1716, which paid its owner several hundred pounds."

[3] In *Journ. Bot.* xlv. 346 (1907).

[4] In *Phytologist*, iii. 817 (1850).

[5] Cf. W. G. Smith, in *Journ. Bot.* xxxix. 73 (1901).

[6] Loudon, *Derby Arboretum*, 50 (1840), says: "There are extensive native woods of the box tree on the estate of Sir Robert Russell at Chequers in Buckinghamshire."

[7] *Gard. Chron.* lii. 404 (1912). The southern end of Ellesborough Warren is Velvet Lawn, a favourite meet of the hounds.

[8] Chauncy, *Hist. Antiq. Herts.* ii. 126 (1826), says: "The Vill or Parish of Box was situated between the parishes of Stevenage, Chivesfield, and Walkerne; and this parish was called Box from a great wood which retains the name to this day." Boxbury Farm and Box Wood are in the parish of Walkern; but there are no box trees in these places at the present time.

[9] It is worth recording here that a chestnut tree, mentioned in Domesday book, still survives at Wymondley. Mr. H. Clinton-Baker tells me that it is a mere shell, no less, however, than 19 ft. in diameter, and still bearing fruit.

belt (about 10 yards in width) of box trees, which average 20 ft. in height, some being 2 feet in girth. (A. H.)

### Remarkable Trees

Though a native of countries much hotter than Britain, and known to most people only as a bush, box is capable of attaining large dimensions in England, and may under favourable circumstances become a small tree. The largest I have seen in this country are in the Hermitage Road in Hitchin, where there is a row of about forty trees, many of which attain 25 to 30 feet in height, and 2 to 2½ ft. in girth, the largest being 35 feet by 2 ft. 10 in. Mr. Seebohm informed me that this line of trees formerly grew on private property[1] which he bought, and opened a new road on which they now stand between an old wall and iron railings which protect them.

At Beckford Hall, Gloucestershire, there is a walk 3 yards wide on each side of which a line of box trees grow which are from 30 to 31 feet high and 20 to 30 inches in girth. A monastery once stood here, and the trees are supposed to be about 800 years old, which is probably twice or three times their real age.

At Boxwell Court,[2] Gloucestershire, the seat of the Rev. O. Huntley, a wood of box trees exists which must be of very great age. Mr. Huntley showed me in the will of his ancestor, Henry Huntley, dated 1556, the following passage: "I will that it shall be lawful for the said Anne my wife, to cut and fell all my boxe, reserving the young store, at any time or times at her pleasure within the space of the said five years." It is thus clear that 350 years ago this wood was looked on as a valuable possession; and that the trees were coppiced, as they are now. It is possible that they were planted by some monk or returned Crusader, and are not, as some have supposed, wild. This wood lies on a steep slope just below the level plateau of the Cotswolds at an elevation of 500 to 600 feet and extends for about 800 yards along the slope which faces south. They grow very thickly and form a dense shade under which nothing grows, and show every sign of having been regularly cut over and reproduced from the stool; in one place fresh plants have been planted in rows, 5 feet apart, which now form tall slender poles clean to 10 or 15 feet high, many of which have died from overcrowding. In other parts the shoots average 20 to 25 ft. high, and though the average girth is not above 6 to 8 in. yet there are a few stems of 2 ft. and over, the thickest that I measured being 2 ft. 10 in. and 3 ft. 4 in. in girth. Mr. Huntley tells me that, though an uncle of his is reputed to have sold a

[1] W. Wilshere, M.P., who owned this property, states in Loudon, *Gard. Mag.* xv. 236 (1839), that these box trees were, in 1839, sixty in number, forming a hedge about 180 ft. long. They then averaged 36 ft. in height, and 3 ft. 3½ in. in girth at 2 ft. from the ground; and were very old, thin, and ragged. They were supposed to have been planted as a border, and allowed to grow up through neglect; but their exact history was unknown. Cf. J. E. Little, in *Journ. Hitchin Nat. Hist. Club*, May 1891, who adds that in the park at Hexton, near Hitchin, box is very luxuriant, forming tall hedges along the drives.

[2] Rudge, *History of the County of Gloucester* (1803), states: "Boxwell, anciently Boxewelle. This name is derived from a box wood of about 16 acres, within a warren of 40 acres, from which arises a considerable spring. This is the most considerable wood of the kind in England, excepting Boxhill in Surrey; and from the name, which has now been on record for more than seven centuries, it must have been of long standing."

Ray, *Syn.* ii. 310 (1696), speaking of box trees, says: "At Boxwel in Coteswold in Gloucestershire, and at Boxley in Kent, there be woods of them.—*Mr. Aubrey's Notes.*" Cf. p. 1727, note 5.

quarter of an acre for £70, between 1858 and 1863, yet now he can only get £1 a ton for the wood, and at that low price no one seems to want it. In parts of the wood where rides have been cut, or openings made by the falling of trees, seedlings spring up abundantly; but the growth of the shoots from the stool seems very slow, owing perhaps to the rabbits, which are hard to keep out.

Mr. Cedric Bucknall describes[1] another wood of box trees, between Wotton-under-Edge and Alderley, clothing the hill-side for a considerable distance, and with abundance of natural seedlings.

In Ireland perhaps the best specimens of box are those growing in the grounds of the Earl of Rosse at Birr Castle.

## Timber

The wood of the box tree is dense and homogeneous, with a very fine grain; and is said to be the nearest approach to ivory of any wood known. Boxwood is unrivalled for wood-engraving, and is used for turnery and inlaying, and for making rules, scales, and other mathematical instruments. It is also employed in making shuttles and rollers that are used in textile factories. A good account of boxwood, with information about the best modes of felling, seasoning, and shipping, is given by Gamble,[2] who quotes largely from a letter written by Messrs. J. Gardner and Sons, Liverpool. Boxwood from the Caucasus, whence formerly the main supply was drawn, is now being replaced, except for the very best articles, on account of its increasing cost, by "West Indian boxwood,"[3] by *Buxus Macowani* from South Africa, and by other woods, belonging to different and often quite unallied genera.

(H. J. E.)

[1] In *Journal of Botany*, xxxix. 29 (1901). Cf. also J. H. White, *Flora of Bristol*, 523 (1912).

[2] *Manual of Indian Timbers*, 592-593 (1902).

[3] Sir David Prain, Director of Kew Gardens, informs me that "West Indian boxwood" is not really shipped from the West Indies, but from Venezuelan ports; and that its botanical origin is still unknown. It was erroneously stated in *Kew Bull.*, 1904, p. 11, to be *Tabebuia pentaphylla*, Bentham and Hooker, a Bignoniaceous tree, which is known in the West Indies as "white cedar." H. Stone, *Timbers of Commerce*, 169, plate xii. fig. 105, gives an account of the so-called "West Indian boxwood," which he confuses with "white cedar," although he rightly questions the accuracy of the determination of *Kew Bull.*, 1904, p. 11. "West Indian boxwood" is used for making parasol and umbrella handles, shuttles, rulers, thermometers, etc.—A. H.

# CRATÆGUS

*Cratægus*, Linnæus, *Sp. Pl.* 475 (1753); Bentham et Hooker, *Gen. Pl.* i. 626 (1865); Koehne, *Deut. Dend.* 227 (1893); Sargent, *Trees N. Amer.* 363 (1905); Schneider, *Laubholzkunde*, i. 766 (1906).

*Oxyacantha*, Medicus, *Phil. Bot.* i. 150 (1789).

*Azarolus*, Borkhausen, *Forst. Bot.* ii. 1224 (1803).

*Mespilus*, sub-genus *Cratægus*, Ascherson and Graebner, *Syn. Mitteleurop. Flora*, vi. pt. ii. 12 (1906).

TREES or shrubs, belonging to the order Rosaceæ, usually armed with simple or branched spines, which are either axillary accompanying a bud, or terminate a short shoot. Leaves usually deciduous, alternate, simple, stalked, usually lobed, serrate; stipules often foliaceous and persistent on the long shoots. Buds small, globose, with numerous imbricated scales.

Flowers, in corymbs, which are terminal on short lateral leafy branchlets; with quickly deciduous linear bracts and bractlets; pedicellate, regular, perfect; calyx superior, with an urceolate, campanulate, or obconic calyx-tube, and five lobes, which are reflexed after the flower opens and either fall off or persist enlarged on the fruit; petals five, inserted with the stamens on the edge of a disc lining the calyx-tube; stamens 5, 10, 15, 20, or 25, with filaments broad at the base and incurved; ovary composed of one to five carpels, concealed in the bottom of the calyx tube and adnate to it; styles, one to five, free, with dilated truncate stigmas; ovules two in each cell, erect. Fruit, a false berry or haw, usually umbilicate at the apex, and often crowned by the marcescent calyx-teeth, composed of the fleshy calyx-tube, which encloses one to five stones or nutlets, each containing one seed, the other ovule having aborted.

This genus is widely spread in the extratropical regions of the northern hemisphere, occurring in Europe, Asia Minor, Siberia, Himalayas, China, and Japan; and with numerous species in North America. Schneider admits about 150 species in all; but Sargent and other American botanists have already described over 500 species in America alone, most of which may be regarded as varieties or hybrid forms. At least sixty species are in cultivation, all of which are either shrubs or small trees, not coming within the scope of our work. The native Whitethorn, which is described below, is now usually considered to comprise two species.

Cratægus is closely allied to Mespilus, of which it has been made a section by some botanists. The following hybrids, one of which is doubtful, between the two genera are worthy of brief mention.

1. *Mespilus grandiflora*, Smith, *Exot. Bot.* i. 33, t. 18 (1814).

*Mespilus lobata*, Poiret, in Lamarck, *Encyc. Meth. Suppl.* iv. 71 (1816); W. J. Hooker, in *Bot. Mag.* t. 3442 (1835).
*Mespilus Smithii*, De Candolle, *Prod.* ii. 633 (1825); Loudon, *Arb. et Fruit. Brit.* ii. 878 (1838).
*Cratægus lobata*, Bosc, *Nouv. Cours. Agric.* ii. 223 (1821).
*Cratægus grandiflora*, Koch, in *Verh. Ver. Bef. Gartenb.* i. 227 (1853).
*Cratægus oxyacantho-germanica*, Gillot, in *Bull. Soc. Bot. France*, xxiii. p. xiv. (1876).
*Pyrus lobata*, Nicholson, *Kew Hand-list Trees*, 195 (1894) (not Koch).
*Cratæmespilus grandiflora*, Camus, in *Journ. de Bot.* xiii. 326 (1899).

A tree, attaining about 25 ft. in height. Branchlets pubescent, with short aborted spines. Leaves very variable in shape, entire or three- to five-toothed near the apex, or five- to seven-lobed and finely serrate, pubescent beneath. Flowers large, white, fragrant like a hawthorn, solitary or two to five in a corymb; calyx segments lanceolate, soon reflexed; petals five; stamens twenty; styles one to four; disc lobed. Fruit ovoid or globose, reddish brown, $\frac{1}{2}$ in. in diameter, crowned by the sepals, with usually two or three nutlets, which are sterile.

This tree, the origin of which is unknown, is considered by most botanists to be an accidental hybrid between *Cratægus oxyacantha* and *Mespilus germanica*; but Koehne[1] considers it to be an independent species, possibly native of the Caucasus. Five apparently wild shrubs were found in 1875 at Saint-Sernin-du-Bois, near Autun, in Seine-et-Loire, in a hedge around the ruins of an old priory, by Dr. Giliot,[2] whose interesting article should be studied. This remarkable tree, of which there is a good specimen[3] at Kew, near the Director's office, was in cultivation at Paris about 1800; and possibly earlier in England, as Loudon mentions old trees at Syon and other places near London.

2. Two very remarkable graft hybrids[4] originated about 1885 in the garden of M. Dardar, at Bronvaux, near Metz; and have been propagated by Simon-Louis. On a very old medlar tree, that had been grafted on a stock of hawthorn, two peculiar branches[5] were observed to arise just beneath the graft. One of these branches, from which has been propagated the form known as *Cratægo-Mespilus Dardari*,[5] differed from the medlar in the branches being spiny, and the flowers in corymbs; while the leaves and fruit were like those of the medlar but smaller. The other branch, which has been propagated as *Cratægo-Mespilus Asnieresi*,[6] was more like the hawthorn, the leaves being lobed and the flowers like *Cratægus monogyna* in form and arrangement; but the branchlets and leaves were pubescent as in the medlar. These two graft hybrids, which are now in cultivation at Kew, are said by Mr. Bean[7] to be very different in appearance. The *Asnieresi* form has remained true to type, and is a small tree of great elegance and beauty. The *Dardari* form, according

[1] *Deut. Dend.* 230 (1893).

[2] Cf. *Rev. Hort.* lxxi. 470 (1899), where Dr. Gillot states that it resembles *Mespilus* more than *Cratægus*, and is of undoubted hybrid origin.

[3] There are several trees in the Green Park, London, and a fine one at Tortworth.

[4] The history of these graft hybrids has been given by Simon-Louis and by Bellair in *Revue Horticole*, lxxi. 403, 482, 530 (1899); and by Koehne, in *Gartenflora*, l. 628 (1901). R. P. Gregory, in *Gard. Chron.* l. 185, fig. 86 (1911), gives Baur's explanation of their anatomical structure.

[5] A third branch was subsequently produced on the original tree at Bronvaux, also from the junction of the stock and scion; but on the opposite side to that occupied by the first two branches. It had at its base pure hawthorn; but was transformed towards the extremity into the *Asnieresi* form.

[6] Jouin, in *Le Jardin*, 1899, p. 22.

[7] Bean, in *Kew Bull.* 1911, p. 268, figs. 1 and 2.

to Mr. Bean, behaves to some extent like *Laburnum Adami*, and bore at Kew in 1911 three distinct kinds of foliage and flowers on the same specimen. One of its branches was like the *Asnieresi* form; another branch was a pure medlar; and all the other branches were the *Dardari* form. Neither of the two hybrids has as yet shown a branch of pure hawthorn. (A. H.)

## CRATÆGUS MONOGYNA, Hawthorn, Whitethorn

*Cratægus monogyna*, Jacquin, *Fl. Austr.* iii. 50, t. 292, fig. 1 (1775); Willkomm, *Forstl. Flora*, 835 (1887); Mathieu, *Flore Forestière*, 162 (1897); Schneider, *Laubholzkunde*, i. 781 (1906).

*Cratægus oxyacantha*,[1] Linnæus, *Sp. Pl.* 477 (1753) (in part); Druce, *List of British Plants*, 26 (1908).

*Cratægus oxyacantha*, Linnæus, var. *monogyna*, Loudon, *Arb. et Frut. Brit.* ii. 834 (1838).

*Cratægus oxyacantha*, Linnæus, sub-species *monogyna*, Hooker, *Student's Flora*, 127 (1870).

*Mespilus monogyna*, Allioni, *Fl. Pedem.* ii. 141 (1785); Willdenow, *Enum. Pl. Hort. Berol.* i. 524 (1809); Ascherson and Graebner, *Syn. Mitteleurop. Flora*, vi. 2, p. 27 (1906).

A shrub or small tree, attaining occasionally about 40 ft. in height. Bark greyish, broken on the surface into small scales. Young branchlets glabrous or with scattered pubescence. Leaves, variable in shape and size, usually broadly ovate, averaging $1\frac{3}{4}$ in. long, and $1\frac{1}{2}$ in. wide at the broad almost truncate or occasionally cuneate base; pinnatifid with five or seven deep lobes, separated by narrow sinuses; margin serrate; upper surface with scattered white hairs; lower surface pale green, with similar pubescence mainly on the midrib and nerves; petiole slender, $\frac{1}{2}$ to 1 in. long, slightly pubescent.

Flowers, variable in number in the corymb; calyx-tube and pedicel pubescent; sepals five, triangular, soon reflexed, persistent on the fruit; petals five, usually white, but occasionally pink, even in the wild state;[2] stamens fifteen or twenty; style one.[3] Fruit, ellipsoid or ovoid, reddish, with one stone, which is either smooth or marked with shallow furrows.

The seeds, when sown, do not germinate till the second year. The seedling[4] has two obovate-oblong cotyledons, which are about $\frac{2}{5}$ in. long, $\frac{1}{4}$ in. broad, shortly stalked, glabrous, obscurely three-nerved, and raised above ground by a glabrous caulicle about 1 to $1\frac{1}{2}$ in. long. The pubescent stem bears alternate serrated leaves, the first three of which are small, cuneate, and three-lobed; those succeeding becoming larger and deeply five-lobed.

[1] *C. oxyacantha*, Linnæus, included both species; but this name was early limited to the two-styled species by Jacquin, who separated the one-styled species as *C. monogyna*. All botanists until lately have followed Jacquin's nomenclature of the two species, which is adopted by us. Recently much confusion has been caused by one or two writers, who restrict the name *C. oxyacantha*, Linnæus, to the one-styled species. These authors are obliged to use another name, *C. oxyacanthoides*, Thuillier, for the two-styled species.

[2] Briggs, *Flora of Plymouth*, 143 (1880), records a bush with pink flowers, and another with deep red flowers. In hedges near Cambridge shrubs with pink flowers are not uncommon.

[3] *C. kyrtostyla*, Fingerhut, in *Linnæa*, iv. 372, t. iii. fig. 1 (1829), is a form of *C. monogyna*, in which the flowers have a peculiar curved or deflexed style. F. A. Lees, *Flora of W. Yorkshire*, 231 (1888), states that most old gnarled thorns in parks and pastures show this peculiarity.

[4] Cf. Lubbock, *Seedlings*, i. 500, fig. 324 (1900), where the seedling of this species is described under the name *C. oxyacantha*.

This species is more pubescent than *C. oxyacantha*, has a differently shaped leaf, and is readily distinguishable, apart from the single style, in the flower and solitary stone in the fruit. It is apparently much more variable in the wild state than the other species; but some of the supposed varieties may be due to hybridising. Certain forms, which are plainly intermediate in various ways between the two species, are recognised as hybrids by continental botanists[1] under the names: *C. media*, Bechstein, *Diana*, i. 88 (1797); and *C. intermixta*, Beck, *Fl. Niederr. Oesterr.* ii. 1, p. 706 (1892).

The commonest hybrid form in England has leaves like those of *C. monogyna*, but with flowers having a glabrous calyx-tube and pedicel. White states[2] that there are several trees of this kind near Bristol, one of which on Leigh Down was noticed to be the last to bloom in 1881, a year remarkable for the abundant blossom of the hawthorn.

In south-eastern and southern Europe there are peculiar races or allied species, which are not in cultivation in this country, and need not be more than alluded to.[3]

The following varieties have arisen in cultivation :—

A. *Differing from the type in habit.*[4]

1. Var. *flexuosa*, Dippel, *Laubholzkunde*, iii. 459 (1893).

*Cratægus oxyacantha*, var. *flexuosa*, Loudon, *Arb. et Frut. Brit.* ii. 835 (1838).

Branches spinescent, dense, flexuose, twisted or like a corkscrew. This peculiar variety originated in Smith's nursery at Ayr, and is represented at Kew by a shrub about 15 ft. high.

2. Var. *salisburifolia*, Nicholson, *Kew Hand-list Trees*, 205 (1894).

*Cratægus oxyacantha*, var. *salisburiæfolia*, Späth, *Cat.*, No. 59, p. 61 (1884).

Branches similar to those of var. *flexuosa*, but without spines. Leaves with few and obtuse lobes somewhat like those of a Ginkgo tree in shape. This is represented at Kew by a shrub about 5 ft. high, which was planted in 1885.

3. Var. *pendula*, Dippel, *Laubholzkunde*, iii. 459 (1893).

*Cratægus oxyacantha*, var. *pendula*, Loudon, *Arb. et Frut. Brit.* ii. 832 (1838).

Branchlets pendulous. Several forms are known. One is said by Loudon to have been picked out of a bed of seedlings at Somerford Hall. Anderson, curator of the Chelsea Botanic Garden in 1830, obtained several pendulous varieties by grafting shoots which were taken from the witches' brooms, that are occasionally

[1] Focke, who describes the hybrid in Koch, *Syn. Deutsch. Flora*, i. 859 (1892), says that it is common in northern and central Germany, where it is more generally distributed in hedges and plantations than the true species.

[2] *Flora of Bristol*, 300 (1912).

[3] *C. azarella*, Grisebach, *Spicil. Fl. Rum.*, i. 88 (1843) is a very pubescent form, which occurs in the Balkan States, Hungary, and Transylvania. *C. hirsuta*, Schur, *Enum. Pl. Trans.* 206 (1866), is very similar and widely distributed in the Mediterranean region. The following are peculiar local forms :—*C. Insegnæ*, Bertolini, *Fl. Ital.* vii. 629 (1847), a native of Sicily; *C. granatensis*, Boissier, *Elenchus*, 41 (1838), wild on the Sierra Nevada in Spain; and *C. brevispina*, Kunze, in *Flora*, 1846, p. 737, a native of southern Spain.

[4] In *Rev. Hort.* 1899, p. 489, it is stated that a form without spines was found in 1893 as a seedling in M. Hémeray-Proust's nursery at Orleans; but I have not seen it in commerce. A compact dwarf spineless variety (var. *inermis compacta*) is advertised by Simon-Louis.

found as conglomerations of slender branches on old trees.[1] A weeping tree[2] at Edinburgh, reputed to be over 300 years old and a favourite of Queen Mary, survived till 1836; and from it was propagated a form known as var. *reginæ*, or Queen Mary's Thorn.

Var. *pendula variegata*, a weeping form with variegated leaves; and var. *pendula rosea*, a weeping shrub with pink flowers, are also in cultivation.

4. Var. *ferox*, Schneider, *Laubholzkunde*, i. 781 (1906).

*Cratægus oxyacantha*, var. *ferox*, Carrière, in *Rev. Hort.* 1859, p. 348.
*Cratægus oxyacantha*, var. *horrida*, Carrière, in *Flore des Serres*, xiv. 201, t. 1468 (1861); Regel, in *Act. Hort. Petrop.* i. 119 (1870); Lynch, in *Gard. Chron.* xxiv. 13, fig. 5 (1898).

Branches pendulous, and armed with tufts of several spines. Carrière, who was unaware of the origin of this variety, states that seedlings raised from its seeds in the Jardin des Plantes at Paris reverted to the ordinary form.

5. Var. *stricta*, Nicholson, *Kew Hand-list Trees*, 205 (1894).

*Cratægus oxyacantha*, var. *stricta*, Loudon, *Arb. et Frut. Brit.* ii. 832 (1838).

Fastigiate in habit, with upright branches. This was discovered in 1826 in a bed of seedlings in Ronalds's nursery at Brentford, and was said by Loudon to resemble the Lombardy poplar in shape. A tree so named at Kew, about 20 ft. high, is pyramidal, with ascending branches, but is not fastigiate. The truly fastigiate form is sold by Smith of Newry and by Späth of Berlin.

6. Var. *ramulis aureis*, Nicholson, *Kew Hand-list Trees*, 205 (1894).

Var. *xanthoclada*, Zabel, *ex* Späth, *Cat.* No. 148, p. 91 (1911-1912).

Branchlets of a bronze colour, conspicuous in winter. A shrub of this at Kew, obtained from Simon-Louis in 1885, is about nine feet high.

B. *Differing from the type in foliage.*

7. Two variegated forms, mentioned by Loudon,—var. *foliis argenteis*, leaves mottled with white, and var. *foliis aureis*, leaves variegated with yellow,—are still in cultivation, but are of little ornamental value.

8. Var. *laciniata*, Dippel, *Laubholzkunde*, iii. 459 (1893).

*Cratægus oxyacantha*, var. *laciniata*, Loudon, *Arb. et Frut. Brit.* ii. 830 (1838); Regel, in *Act. Hort. Petrop.* i. 119 (1870).

Leaves deeply pinnatifid, with irregularly serrated lobes. This occurs in hedgerows, in company with the typical form. F. A. Lees states[3] that in Yorkshire, Worcestershire, Lincolnshire, and Berkshire, it flowers much less freely than the other forms.

9. Var. *filicifolia*, Koehne, *Deut. Dend.* 238 (1893).

*Cratægus oxyacantha*, var. *filicifolia*, Van Houtte, *Flore des Serres*, xx. 51, t. 2076 (1874).

Leaves broad, fan-shaped, deeply divided into numerous curled segments. This beautiful variety, which resembles *Adiantum farleyense* in foliage, does not

[1] Cf. Loudon, *Gard. Mag.* ix. 596 (1833).
[2] Figured by Loudon, *Arb. et Frut. Brit.* ii. 833, fig. 556 (1838).
[3] *Flora W. Yorkshire*, 231 (1888). It is recorded for Warwickshire by Bagnall, *Flora of Warwickshire*, 107 (1891).

seem to be in cultivation[1] in England. There are several other varieties[2] with peculiar cut leaves, none of which I have seen.

C. *Differing from the type in flowers.*

10. Var. *semperflorens*, Dippel, *Laubholzkunde*, iii. 460 (1893).

*Cratægus oxyacantha*, var. *semperflorens*, André, in *Rev. Hort.* liv. 354 (1882) and lv. 140, fig. 26 (1883).

A low bushy shrub, which numerous flowers, which appear more or less continuously throughout the season from May to autumn. This was found about 1879 at Poitiers by M. Bruant, in a bed of seedlings of the common hawthorn, and was subsequently propagated by grafting. A shrub at Kew is about 2 ft. high.

11. Var. *præcox*, Dippel, *Laubholzkunde*, iii. 459 (1893). Glastonbury Thorn.[3]

*Cratægus oxyacantha*, var. *præcox*, Loudon, *Arb. et Frut. Brit.* ii. 833 (1833).

Flowers, usually appearing at Christmas or early in January, and not ripening into fruit, the leaves being somewhat later; a second crop of flowers, which produce fruit, is borne in May and June. The original tree,[4] which grew at Glastonbury, was mentioned by Turner in 1562, and by Gerard in 1597, and had the appearance of a very old tree, when it was seen by Withering[5] in 1793. From this tree the variety was propagated. A specimen, growing near the Temperate House at Kew, is irregular in the time of flowering, which depends upon the nature of the season. Bean states[6] that, with a mild November and December, the tree at Kew will flower about Old Christmas Day (6th January); but if cold weather sets in before New Year, the flowers may not open till March or April. In 1908, owing to the unusual warmth of the autumn, it was in full blossom in the first week of November, before the leaves had fallen, so that it was carrying flowers, fruit (derived from the flowers of the preceding May), and foliage simultaneously.

D. *Differing from the type in fruit.*

12. Var. *eriocarpa*, Dippel, *Laubholzkunde*, iii. 460 (1893).

*Cratægus oxyacantha*, var. *eriocarpa*, Loudon, *Arb. et Frut. Brit.* ii. 831 (1838).

---

[1] Var. *filicifolia* is advertised by Späth, *Cat.* No. 148, p. 91 (1911-12).

[2] Var. *pinnatiloba*, Dippel, *Laubholzkunde*, iii. 458 (1893), is identified by Schneider, *Laubholzkunde*, i. 785 (1906), with *C. microphylla*, Koch, *Die Weissdorn*, 68 (1884), which is a Caucasian species, not apparently in cultivation, though it is given in the Kew Hand-list. Koehne, *Deut. Dendr.* 238 (1893), however, considers it to be a hybrid between *C. monogyna* and *C. oxyacantha.*

[3] Chevallier, in *Ann. Soc. Agric. Sci. Départ. Indre et Loire*, xxx. 70 (1850), describes a similar sport of *Prunus spinosa*, to which is attached a similar legend. At the Château du Chabrol, St. Patrice, on the Loire, midway between Saumur and Tours, there is a large blackthorn, called *l'épine miraculeuse*, which flowers every year in the last week in December, even in the severest seasons. The legend is that St. Patrick, while on his way to Tours in A.D. 395, reposed one night in winter under the shade of this tree, which burst forth into flowers and leaves to shield him from the cold. The tree did not appear to be very old in 1850, but is now of considerable size, judging from a photograph sent me by M. Hickel. This curious variety, which may be named *Prunus spinosa*, var. *præcox*, does not seem ever to have been propagated.

[4] Parkinson, *Theat.* 1025 (1640), mentions other trees of the same kind at Romney Marsh and at Nantwich. Plot, *Nat. Hist. Oxfordshire*, 159 (1705), mentions a very old tree, which flowered at Christmas, in Lord Norrey's park in Oxfordshire; but was uncertain whether it was a graft from the Glastonbury tree or an original specimen of the variety.

[5] *Arr. Brit. Pl.* 459 (1793). Cf. Loudon, *Gard. Mag.* ix. 122 (1833). However, James Howel, *Dodona's Grove*, 55 (1644), implies that the original thorn was destroyed by Puritan fanatics, one of whom "was wel served for his blind zeale, who, going to cut doune an ancient white Hawthorne tree, which, because she budded before others, might be an occasion of Superstition, had some of the prickles flew into his eye and made him Monocular."

[6] *Kew Bull.* 1908, p. 452. Cf. also J. W. White, *Flora of Bristol*, 302 (1912), who mentions a specimen at Ipswich, which flowered 14th November 1885, and another at Evesham, which was in flower on 26th November, 1899.

Fruit woolly pubescent. This variety is rare; but is occasionally found in the wild state, as in the Isle of Wight, where it is recorded by Bromfield.[1] It is said[2] to occur near Breslau, in Silesia.

13. Var. *maurianensis*, Didier, in *Bull. Soc. Dauph.* ix. 385 (1882).

Fruits very large, $\frac{1}{2}$ to 1 in. long, and $\frac{2}{5}$ in. or more wide. This variety was described from specimens found in Savoy, and occurs also in hedges near Toulon.[3]

Mr. J. W. White records[4] two trees, with branches bending down under the weight of numerous large haws, which were found in 1909 near Bristol, one on a low cliff near Walton-by-Clevedon, the other in Chelvey Batch wood. The fruits were very handsome, and four times larger than the typical form, averaging $\frac{3}{5}$ in. long and $\frac{1}{2}$ in. wide.

As both *C. monogyna* and *C. oxyacantha* have been much confused, it is impossible to give an accurate account of their separate distribution. The common hawthorn, comprising both species, is a native of Europe, and of the mountains of Algeria and Morocco; and extends from Asia Minor and the Caucasus, through Armenia, Persia, and Afghanistan to the western Himalayas, where it grows between 6000 and 9000 feet elevation.[5] It grows wild in Norway as far north as lat. 62° 55′, in Sweden as far as Upsala, lat. 59° 52′, and in Finland to lat. 61° 30′. In Russia, it is common in Livland, Kazan, and Orenburg, and throughout the southern provinces except in the Steppes. It occurs mainly in hedges, waste places, and on the margins of woods, ascending in the Alps to about 3000 feet altitude.

*C. monogyna* is by far the commoner of the two species in Britain, where it is found in hedges and woods from Moray and Islay southwards; and it is met with in all districts in Ireland. It is the most commonly planted hawthorn either for hedges or for ornament; and most of the large trees in parks are referable to this species, though some of them look as if of hybrid origin.

The hawthorn lives to a great age, probably to three or four hundred years.[6] Old trees often grow irregularly, so that ribs are formed upon their stem, which assume a vertical or a spiral direction. As years go on, these rib-like projections become larger, and the intervening channels deeper. Ultimately, when decay begins at the heart and spreads outwards, the projecting parts become separated and appear to be a number of subordinate stems, which are, however, united at the base, and bear on their inner surface, instead of bark, remains of the decayed heart-wood.[7]

(A. H.)

[1] In *Phytologist*, iii. 288 (1848).

[2] Ascherson and Graebner, *Syn. Mitteleurop. Flora*, vi. pt. ii. 30 (1906).

[3] Described as var. *macrocarpa*, Reynier, *ex* Albert and Jahandiez, *Plantes Vasc. du Var*, 185 (1908).

[4] *Flora of Bristol*, 300 (1912), where these two trees are assigned to var. *splendens*, Druce, a name which cannot be retained, as var. *splendens*, Koch, *Dendrologie*, i. 159 (1869), much earlier in date, was applied to forms with pink and scarlet flowers. Mr. White identifies this large-fruited variety with *Oxyacanthus folio et fructu majore* from Oxfordshire of Merrett's *Pinax* (1667). Ray, *Syn.* 454 (1724), states that it was also found by Sherard in Northamptonshire.

[5] The forms in southern Europe, northern Africa, and Asia, are apparently very distinct varieties or allied species; and require further study. Cf. Schneider, *Laubholzkunde*, i. 782 (1906).

[6] Lees, in *Gard. Chron.* iii. 688 (1875), states that he counted over 300 annual rings in the stem of a tree about 1 ft. in diameter, that grew on the Malvern Hills.

[7] Cf. Purchas, in *Journ. of Bot.* iii. 366 (1865), who points out that stems growing close together, which have commenced as independent trees in a hedge, are surrounded on all sides by bark, and are thus readily distinguishable from the peculiar stems described above. Lees, in *Gard. Chron.* iii. 688, figs. 142, 143 (1875), gives illustrations of trees with divided stems at Garnstone, Herefordshire, and at Upper Wyck, near Worcester.

## Cultivation

The hawthorn, though it grows on almost all soils, succeeds best in a rich loam, and does well on strong clay. It is propagated by seeds, which as they lie over for a year, should be mixed with ordinary soil or sand in pits or heaps, where they are kept until the second year following, when they may be sown in February or March. Young plants should be removed from the seed bed at the end of the first year, and after standing in nursery lines for two years, be planted out. The varieties are budded or grafted on seedlings of the common species.

The most important use of the hawthorn is as a hedge plant. Though hedges appear to have been in use in England from the time of the Romans, they were not generally planted to enclose ordinary fields and meadows till about the end of the seventeenth century. Dr. Walker states [1] that the first hedges in Scotland were planted by Cromwell's soldiers in East Lothian and Perthshire.

Fine old hawthorns, with trunks of great girth and wide-spreading branches, exist in many parks throughout the country; and tall slender specimens are occasionally seen, drawn up in woods. We may mention a few, remarkable for age, though doubtless there are many quite as large that we have not heard of.

An immense old thorn, at Hethel in Norfolk, was first mentioned by Marsham, who, in a letter to the Bath Society about 1740, made its girth 9 ft. 1 in. at four feet from the ground. Grigor [2] states on the authority of H. Gurney, Esq., that the first Sir Thomas Beevor, who owned the place towards the end of the eighteenth century, put a railing round it, which was subsequently repaired, and the spreading limbs propped up by Mr. Gurney. Grigor says that the trunk measured 12 ft. 1 in. at one foot, and 14 ft. 3 in. at five feet, whilst the branches, though several large ones had been lost, spread over an area 31 yards round. Both the trunk and the large branches were then hollow, but the wood sound and hard. Sir Hugh Beevor in 1895 found it to be 13½ ft. at eighteen inches from the ground where the girth was least. Miss Eaton sent me a photograph in 1903 which showed branches supported by numerous props. Mr. Edwards, who photographed the tree in September 1912, tells me that the tree now consists of several stems formed by the splitting of the original trunk. The branches, which now measure 37 yards round, are sound and covered with leaves and fruit, though bearing many tufts of mistletoe. It is protected from cattle by a rail; and the branches are supported by numerous props (Plate No. 378).

In the park at Holwood House, Kent, Mr. A. D. Webster records [3] a tree, which in 1888 was 14 ft. 6 in. in girth at three feet from the ground, above which it divided into six limbs, measuring at a yard from the fork, 4 ft. 2 in., 4 ft., 5 ft. 8 in., 2 ft. 8 in., 4 ft. 4 in., and 3 ft. 5 in. respectively. Its height was 42 ft., with a spread of branches 63 ft. in diameter. This tree was growing in strong clayey loam, and was in perfect health.

Mr. Edwin Lees described [4] a remarkable hawthorn at Lenchford, in Worcester-

[1] *Essays on Nat. Hist.* 54 (1812).

[2] *Eastern Arboretum*, 282 (1841).

[3] *Trans. Roy. Scot. Arbor. Soc.* xii. 311 (1889).

[4] *Gard. Chron.* iii. 688, figs. 141, 146 (1875).

shire, which was 60 ft. high and 9 ft. in girth in 1875; and another in Downton Park, Herefordshire, which was over 50 ft. in height. He also mentioned others of considerable size in the same counties.

At Chideock Manor, Dorset, there is a remarkable thorn, about 25 ft. high and 6 ft. in girth, with wide-spreading branches and very pendulous branchlets, the diameter of the spread being about 40 ft.

In Lilford Park, Northamptonshire, where there are many fine thorns, I measured, in 1906, a tree no less than 51 ft. in height, but only 5 ft. in girth.

In the pleasure grounds of Hatfield House, Herts, there is a tree, which Henry measured in 1911 as 48 ft. by 3 ft. 8 in. In Ware Park, in the same county, Mr. H. Clinton Baker found in 1910 a tree 30 ft. by 12 ft. 5 in.

Within the walls of Rothesay Castle, in the island of Bute, there was growing in 1878 a remarkable thorn which Mr. James Kay described[1] as follows: "Though the tree was blown down thirty-nine years ago, it is still vigorous and healthy. The extreme length of the tree as it now lies, measuring from the original surface of the root, is 47 ft.; present vertical height, 28 ft.; circumference, three feet up, 6 ft. 8½ in.; four and a half feet up, 6 ft. 6½ in.; six feet up, 6 ft. 10 in." This tree dates from some time after 1685, when the Castle was burnt; and its age in 1878 did not probably exceed 190 years.

### Timber

The wood of the hawthorn is white, often tinged with red; and is hard, heavy, and difficult to work, but with a fine grain and susceptible of a good polish. It is not much used, as it seldom can be obtained of sufficient size, and is usually spoiled by defects or knots. It is occasionally employed by turners, and was formerly found suitable for teeth of mill-wheels. Mr. W. G. Smith, who made many woodcuts for the *Gardeners' Chronicle*, states[2] that hawthorn wood is quite as good for engraving as ordinary boxwood, and possesses a far better colour. The best box, however, cuts a little smoother, as it has a somewhat closer grain. (H. J. E.)

## CRATÆGUS OXYACANTHA, Hawthorn, Whitethorn

*Cratægus oxyacantha*,[3] Linnæus, *Sp. Pl.* 477 (1753) (in part); Jacquin, *Fl. Austr.* iii. 50, t. 292, fig. 2 (1775); Loudon, *Arb. et Frut. Brit.* ii. 829 (1838); Willkomm, *Forstl. Flora*, 838 (1887); Mathieu, *Flore Forestière*, 163 (1897); Schneider, *Laubholzkunde*, i. 780 (1906).
*Cratægus oxyacantha*, var. *vulgaris*, De Candolle, *Prod.* ii. 628 (1825).
*Cratægus oxyacantha*, sub-species *oxyacanthoides*, Hooker, *Student's Flora*, 127 (1870).
*Cratægus oxyacanthoides*, Thuillier, *Fl. Envir. Paris*, 245 (1799).
*Mespilus oxyacantha*, Crantz, *Stirp. Austr.* i. 39 (1763); Allioni, *Fl. Pedem.* ii. 141 (1785); Willdenow, *Enum. Pl. Hort. Berol.* i. 524 (1809); Ascherson and Graebner, *Syn. Mitteleurop. Flora*, vi. 2, p. 25 (1906).

A shrub or small tree, similar to *C. monogyna* in bark and habit. Young branchlets glabrous. Leaves obovate or ovate, usually with three shallow lobes, the terminal

[1] *Trans. Roy. Scot. Arbor. Soc.* ix. 76 (1879).
[2] In *Gard. Chron.* iii. 689, note (1875), it is stated that figures 142 and 143 on p. 688 were engraved on hawthorn wood.
[3] Cf. p. 1733, note 1.

one being often lobulate; margin irregularly serrate; upper surface dark green, shining, glabrous; lower surface pale green, glabrescent or slightly pubescent on the midrib; petiole slender, usually glabrous.

Flowers, variable in number in the corymb; peduncle, pedicel, and calyx-tube glabrous; sepals five, triangular, spreading, persistent on the fruit; petals five, white; stamens fifteen or twenty; styles usually two, rarely in some flowers one or three. Fruit ovoid or globose; stones two or three, rarely one, flattened on the inner surface, convex with deep longitudinal furrows on the outer surface.

This species is very distinct in appearance, and in England as a rule comes into flower a fortnight earlier than *C. monogyna.*

## Varieties

This species is apparently much less variable in the wild state [1] than *C. monogyna*, but has given rise to some remarkable garden forms.

1. Var. *multiplex*, Loudon, *Arb. et Frut. Brit.* ii. 832 (1838).

Var. *flore pleno albo*, Rodigas, in *Flore des Serres*, xv. t. 1509, fig. 2 (1862).
*Cratægus monogyna*, var. *alba plena*, Rehder, in Bailey, *Cyc. Amer. Hort.* i. 396 (1900).

Flowers white, double, produced in great profusion and dying off a beautiful pink colour.[2] This variety, the origin of which is unknown, differs little from the type in other respects, having two-styled flowers with glabrous pedicels and calyx-tubes, and glabrescent three-lobed leaves.[3]

2. Var. *rosea*, Loudon, *Arb. et Frut. Brit.* ii. 832 (1838).

Petals pink, with white claws; in other respects similar to the type. This is a single pink variety, which is occasionally found wild.

3. Var. *punicea*, Loddiges, *Bot. Cat.* t. 1363 (1828); Loudon, *Arb. et Frut. Brit.* ii. 832 (1838).

Var. *flore puniceo*, Rodigas, in *Flore des Serres*, xv. t. 1509, fig. 2 (1862).

Petals larger than in var. *rosea*, dark red and without white on the claws. This is the handsome single crimson variety, which was first raised in Scotland, and afterwards propagated by Loddiges, who budded it upon the common whitethorn.

4. Var. *Gumpperii bicolor*, Van Houtte, in *Fl. des Serres*, xvi. t. 1651 (1866).

Flowers single; petals white, edged with a pink margin. This originated at Stuttgart about 1860, and is probably of hybrid origin, as the flowers, while having a glabrous calyx-tube, bear only one style.

[1] Var. *integrifolia*, Wallroth, *Sched. Crit.* 219 (1822), a name given to shrubs having obovate leaves with three very shallow lobes, can scarcely be retained as a variety, as such leaves are common in the typical form of the species. Similarly, var. *auriculata*, Dippel, *Laubholzkunde*, iii. 457 (1893), said to have persistent large stipules, is doubtfully distinct, as the retention of the stipules depends on the vigour of the branches, and is common enough in the ordinary form of the species.

[2] Späth, *Cat.* No. 148, p. 91 (1911-1912), advertises var. *candido-plena*, a new variety with double flowers that remain pure white.

[3] M'Nab, in *Trans. Bot. Soc. Edin.* vi. 284 (1860) gives an account of a tree with double white flowers in the Edinburgh Botanic Garden, which retained most of its leaves during the preceding winter, some remaining green till 12th May 1859. It then flowered and produced normal single flowers.

5. Var. *punicea flore pleno*, Loudon, *Arb. et Frut. Brit.* ii. 832 (1838), and *Trees and Shrubs*, 377 (1842).

Var. *flore rubro pleno*, Rodigas, in *Flore des Serres*, xv. t. 1509, fig. 3 (1862).

Flowers double, pink. This is said by Loudon to have been imported about 1832 by Masters of Canterbury, and to bear flowers not so brilliant in colour as the single crimson variety. It differs from typical *C. oxyacantha* in the pubescent leaves and calyx-tube, and is possibly a hybrid.

6. Var. *coccinea flore pleno*, Paul, in *Florist and Pomologist*, vi. 117 (1867).

Var. *floribus coccineis plenis*, Lemaire, in *Illust. Hort.* t. 536 (1867).
*Cratægus monogyna*, var. *Pauli*, Rehder, in Bailey, *Cyc. Amer. Hort.* i. 396 (1900).

Flowers double, deep scarlet. This originated about 1858 in Mr. Christopher Boyd's garden near Waltham Cross, as a single branch on a tree of the double pink variety, which was about 25 years old and nearly 30 feet high. This branch was observed to bear flowers of a deep scarlet colour, year after year, whilst all the other branches on the tree continued to produce flowers of the original pink colour. It was propagated by Messrs. Paul, who showed it at the International Horticultural Exhibition in 1866, under the name "Paul's New Double Scarlet Hawthorn," by which it is still known.

7. Var. *Gireoudi*, Späth, *Cat.* No. 104, p. 89 (1899-1900).

Young shoots pink, bearing new leaves, which are mottled with white and pink. This is represented at Kew by a spreading bush about 6 ft. high, received from Späth in 1899.

8. Var. *aurea*, Loudon, *Arb. et Frut. Brit.* ii. 831 (1838).

Var. *xanthocarpa*, Lange, *Rev. Spec. Cratægi*, 71 (1897).

Fruits yellow. This variety, which has been in cultivation over a hundred years, bears freely and is very showy.

9. Var. *leucocarpa*, Loudon, *Arb. et Frut. Brit.* ii. 831 (1838).

Fruit white. Plot, *Nat. Hist. Oxfordshire*, 158 (1705), mentions[1] a tree with white haws in a hedge near Bampton; but it does not appear to have been propagated, and this variety was unknown to Loudon.

The following is either a geographical form or a closely allied species:—

10. *Cratægus polyacantha*, Jan, *Elench. Hort. Parm.* 8 (1826).

*Cratægus oxyacantha*, sub-species *monogyna*, var. *polyacantha*, Nicholson, *Kew Hand-list Trees*, 343 (1902).

Leaves small, about ¾ in. long and broad, tri-lobed. Young branchlets, petioles, peduncles, pedicels, and calyx-tubes densely covered with white woolly pubescence. This is a small shrub, native of Sicily and Calabria. It is in cultivation at Kew.

[1] Cf. Ray, *Syn. Meth.* 453 (1724). A form with fruit larger than usual, occurring in the south of France and Switzerland, has been distinguished as var. *macrocarpa*, Le Grand, *Stat. Bot. Forez*, 119 (1873), *ex* Rouy and Camus, *Flore de France*, vii. 4 (1901), identical with *C. macrocarpa*, Hegetschweiler, *Fl. Schweiz*, 464 (1840).

### Distribution

*C. oxyacantha* is widely spread throughout Europe, occupying nearly the same territory as *C. monogyna*, but being much less common. Willkomm believes it to be more prevalent in the north than in the south of Europe. In France, according to Mathieu, it does not grow in the region of the olive; while elsewhere it is comparatively rare, and scarcely ever attains the dimensions of a tree. I saw it in 1912 in the interior of the Forest of Orleans, forming a large shrub, and growing in shade, as it does in the Gamlingay Wood, near Cambridge.

It is doubtfully wild in Scotland and Ireland; and is apparently indigenous only in the midland, eastern, and south-eastern counties of England. It is recorded for many stations in Kent, Surrey, Middlesex, Oxfordshire, and Cambridgeshire. It is not mentioned as a native plant for the northern or western counties in any of the published floras in which the two species are distinguished. In England, as far as we know, it is usually a shrub; and we have no records of any large trees of this species. (A. H.)

# SALIX

*Salix*, Linnæus, *Sp. Pl.* 1015 (1753); Forbes, *Salic. Woburn.*, 1-294 (1829); Andersson, *Monog. Salic.* 1-180 (1863), and in De Candolle, *Prod.* xvi. 2, p. 191 (1868); Bentham et Hooker, *Gen. Pl.* iii. 411 (1880); Buchanan White, in *Journ. Linn. Soc.* (*Bot.*) xxvii. 333-457 (1890); Camus, *Saules d'Europe*, 9-40 (1904); Schneider, *Laubholzkunde*, i. 23 (1904).

Trees or shrubs, with scaly bark, and slender terete branchlets, which are often easily separated at the joints. True terminal buds are not developed,[1] the top of the branchlet dying off in summer, and leaving a minute scar close to the uppermost axillary bud, which prolongs the branch in the following season; buds apparently covered by one scale, which is composed of two scale-like leaves fused together, as indicated by their keeled margins. Leaves deciduous,[2] alternate, rarely sub-opposite, simple, variable in shape, penni-nerved, stalked; stipules oblique, serrate, either small and early deciduous, or large, leafy, and persistent, often conspicuous on barren vigorous young branches. On branches from which the leaves have fallen the leaf-scars are crescentic and 3-dotted, and accompanied on each side by a minute stipular scar.

Flowers, appearing in some species before the leaves, in others after the leaves, diœcious, in catkins, each of which terminates a short shoot, and bears numerous flowers on a slender axis; each flower, with one or two honey-glands, placed front and back at its base, and subtended by a scale, which is usually entire in margin. Staminate flowers, with two or three to twelve stamens, inserted on the base of the scale, with slender filaments, free or more or less connate, and two-celled anthers opening longitudinally. Pistillate flowers; ovary free, stalked or sessile, one-celled, with four to eight ovules on each of the two placentæ; crowned by a style, which is often extremely short or obsolete, with two stigmas, which are either entire or bifid. Fruit, an acuminate capsule, separating when ripe into two recurved valves. Seeds minute, narrowed at the ends, dark brown or nearly black, furnished with long silky hairs.

About 160 species of Salix are known, distributed from the Arctic regions southwards to the Andes of Chile in the New World, and to South Africa, Madagascar, Himalayas, Burma, Malay Peninsula, Java, and Sumatra in the Old World. Most of the species are shrubs or small trees; and in the following account we have only dealt with the few species in cultivation which attain a large size, including *S. Caprea*, on account of its interest to foresters. These may be distinguished as follows:—

[1] However, in § *Chamætia*, Dumortier, in *Bijdr. Natuurk. Wetensch.* i. 56 (1826), which includes Arctic and Alpine under-shrubs, like *S. reticulata*, Linnæus, there are true terminal buds, giving rise to catkins in the following year. The species of this section in some respects are intermediate between *Salix* and *Populus*.

[2] *S. Bonplandiana*, H. B. K., a native of Mexico, is said to have persistent leaves. Cf. Dode, in *Bull. Soc. Dend. France*, 1909, p. 151.

I. *Leaves ovate or oval.*

1. *Salix Caprea*, Linnæus. See p. 1745.

Leaves oval, elliptic or ovate, 2 to 3 in. long, 1½ in. broad, irregularly crenate; lower surface bluish grey, tomentose.

2. *Salix pentandra*, Linnæus. See p. 1747.

Leaves fragrant when bruised, ovate to ovate-lanceolate, about 3 in. long and 1¼ in. broad, abruptly acuminate, minutely serrate, glabrous on both surfaces.

II. *Leaves lanceolate.*

A. *Mature leaves not ciliate in margin, green above, green or glaucous beneath.*

* *Leaves broadly lanceolate, with coarse serrations, which are not close together.*

3. *Salix speciosa*, Host. See p. 1756.

Leaves, broader than in *S. fragilis*, 4 to 6 in. long, 1¼ to 1½ in. broad, glabrous except for a few scattered hairs on the slightly glaucous under surface. Young branchlets pubescent or glabrescent.

4. *Salix fragilis*, Linnæus. See p. 1754.

Leaves about 4 in. long, ¾ to 1 in. broad, glabrous except for a few scattered hairs on the glaucous under surface. Young branchlets pubescent.

** *Leaves narrowly lanceolate, with fine close serrations.*

(a) *Young branchlets crimson, glabrous; clay-coloured in the second year.*

5. *Salix decipiens*, Hoffmann. See p. 1755.

Leaves, 2 to 3 in. long, ½ to ¾ in. broad, similar to those of *S. fragilis*, but smaller and green beneath.

(b) *Young branchlets green, becoming a brilliant yellow in winter and the following year.*

6. *Salix vitellina*, Linnæus. See p. 1768.

Leaves, 2 to 2½ in. long, ⅜ to ½ in. broad, green above, glaucous beneath; with scattered appressed hairs, sparse above, more abundant beneath. See No. 11A.

(c) *Young branchlets green, becoming olive green or brownish grey in the second year.*

7. *Salix babylonica*, Linnæus. See p. 1749.

Leaves, 2½ to 4 in. long, ½ to ¾ in. wide, tapering to a long slender filamentous apex; when mature, glabrous and glaucous beneath. Branchlets pendulous, always injured in England by spring frost.

8. *Salix elegantissima*, Koch. See p. 1751.

Leaves similar to those of *S. babylonica*, but broader, and more coriaceous, and less glaucous on the lower surface. Branchlets pendulous, uninjured in England by spring frost.

9. *Salix Salamonii*, Carrière. See p. 1750.

Leaves similar to those of *S. babylonica*, but with scattered appressed long hairs on both surfaces. Branches ascending, with pendulous branchlets, which are not injured by frost in England.

B. *Mature leaves ciliate in margin, covered more or less on both surfaces with appressed silky hairs.*

10. *Salix alba*, Linnæus. See p. 1759.

Leaves 2 to 2½ in. long, ⅜ to ½ in. wide, covered with silky pubescence, densest on the whitish under surface. A wide-spreading male or female tree, with pendulous or spreading branchlets; ovary and fruit sessile.

11. *Salix cœrulea*, Smith. See p. 1763.

Leaves similar to those of *S. alba*, but thinner in texture, more translucent, less pubescent, the lower surface being bluish grey and not white. A pyramidal female tree, with ascending branches and erect terminal branchlets; ovary and fruit shortly pedicellate.

11A. *Salix vitellina*, Linnæus. See No. 6.

Leaves occasionally ciliate till autumn, with both surfaces more or less appressed-pubescent. This is readily distinguishable by the bright yellow branchlets in winter. (A. H.)

## SALIX CAPREA, Sallow, Goat Willow

*Salix Caprea*, Linnæus, *Sp. Pl.* 1020 (1753); Loudon, *Arb. et Frut. Brit.* iii. 1561 (1838); Andersson, *Monog. Salic.* 75 (1863), and in De Candolle, *Prod.* xvi. 2, p. 222 (1868); Willkomm, *Forstliche Flora*, 487 (1887); Buchanan White, in *Journ. Linn. Soc.* (*Bot.*) xxvii. 385 (1890); Mathieu, *Flore Forestière*, 465 (1897); Camus, *Monog. des Saules*, 202 (1904).

A small tree, occasionally attaining 40 ft. in height. Bark smooth and greenish at first, ultimately ridged and fissured. Young branchlets, with a minute tomentum, becoming more or less glabrescent in the second year. Buds ovoid-conic, minutely tomentose or glabrous. Leaves oval or elliptic, about 2 to 3 in. long and 1¼ to 1¾ in. broad; apex acuminate, with the point usually directed to one side; base broadly cuneate, often unequal; upper surface light green, slightly shining, glabrous except for slight pubescence on the midrib and nerves; lower surface bluish grey, reticulate, covered with a whitish tomentum; margin irregularly crenulate, undulate, or rarely almost entire; petiole more or less tomentose; stipules oblique, reniform or half cordate, dentate.

Flowers appearing early, before the leaves; catkins sub-sessile, with scale-like leaves at the base, very silky, on account of the spatulate scales, brown or blackish towards the apex, and pubescent with long hairs. Staminate flowers; stamens two, free, glabrous. Pistillate flowers; ovary tomentose on a long pedicel, much surpassing the gland in length; style short, with two stigmas, which are usually emarginate or rarely bifid. Capsules on long pedicels, narrow, elongated, covered with greyish silky hairs.

### Varieties and Hybrids

I. *S. Caprea* is variable in the wild state. The catkins, which are usually sessile or sub-sessile, are occasionally provided with leafy peduncles. The branchlets and buds, which are normally glabrous, are occasionally pubescent. Varieties, differing in the leaves, have been distinguished:—

1. Var. *orbiculata*, Kerner, in *Verh. Zool. Bot. Ges. Wien*, x. 248 (1860). This is the typical form of the species, in which the leaves are broadly oval, slightly cordate at the base, and with the apex bent to one side.

2. Var. *elliptica*, Kerner, *loc. cit.* Leaves elliptic and tapering at both ends. This is said to be the more common form in northern and mountainous districts.

3. Var. *sphacelata*, Wahlenberg, *Fl. Carpat.* 319 (1814).

Var. *alpina*, Gaudin, *Fl. Helv.* vi. 240 (1830).

*Salix sphacelata*, Smith, *Brit. Bot.* iii. 1066 (1805), *Eng. Bot.* t. 2333 (1812), and *Eng. Flora*, iv. 224 (1828); Loudon, *Arb. et Frut. Brit.* iii. 1563 (1838).

Leaves small, tomentose on both surfaces. This is an alpine form, which occurs in the Highlands of Scotland.

4. Var. *pendula*, Petzold and Kirchner, *Arb. Musc.* 576 (1864).

*Salix Kilmarnocki*, Nicholson, *Kew Hand-list Trees*, ii. 213 (1896).

Pendulous in habit, usually grafted on a stock about 4 ft. high, and forming a weeping shrub, which is known as the Kilmarnock Willow. This was discovered[1] in 1840 on the banks of the river Ayr, and was propagated by Lang, nurseryman at Kilmarnock. The original tree died of old age about 1884.

II. *S. Caprea* is closely related to both *S. cinerea* and *S. aurita*, which are bushy species. Intermediate forms, which are often difficult to discriminate, have been referred to the following hybrids:—

5. *Salix Reichardtii*, Kerner, in *Verh. Zool. Bot. Ges. Wien*, x. 249 (1860).

This hybrid between *S. Caprea* and *S. cinerea* is rare, only a few examples from Perthshire, Fifeshire, Worcestershire, and Kent being recorded by Buchanan White. Both parents are common; but they do not flower at the same time, and are rarely met with together, as *S. Caprea* is most frequent in woods, whilst *S. cinerea* grows chiefly on river banks.

6. *Salix capreola*, Kerner, *ex* Andersson in De Candolle, *Prod.* xvi. 2, p. 223 (1868).

This hybrid between *S. Caprea* and *S. aurita* is also rare, as the periods of flowering of the two species are not identical. There are pistillate specimens at Kew from Derbyshire and Surrey; and Buchanan White mentions other localities in Perth, Worcestershire, and Somerset.

## Distribution

The common sallow or goat willow is widely distributed, occurring in Europe, the Caucasus, Siberia, Amurland, Manchuria, and Korea. In Europe it exists in every country, extending as far north as Iceland and Lapland, and as far south as southern Spain, Italy, and Greece. It is most commonly found in woods on the plains and lower hilly regions, but reaches 3300 feet elevation in Norway and 4500 feet in the Carpathians. It grows on almost all soils, on those which are moist, marshy, or even peaty, as well as in dry rocky or stony ground. It attains its largest

[1] Cf. Rev. Dr. Landsborough, in *Ann. Kilmarnock Glenfield Ramblers' Soc.*, 1893-1894, p. 20.

size in East Prussia, Lithuania, and the Russian Baltic provinces, where, according to Willkomm, it is often a fine tree, 30 to 50 ft. in height. It is naturally regenerated by seed, and when cut down produces vigorous coppice shoots. It can be propagated by cuttings and by sets. (A. H.)

The sallow is common in all parts of the British Isles, growing in woods and copses, and in waste places, but rarely attains a large size. It is often 20 to 30 ft. in height, and sometimes produces a trunk a foot in diameter, but seems to be a short-lived tree. The finest which I have seen is growing by the roadside two miles below the lodge at Guisachan, Inverness-shire, and measured in 1910 about 50 ft. by 6 feet. It is the only willow which commonly grows from seed, and in some of my plantations is so abundant that it may be called a forest weed.

It is usually looked upon by foresters as a useless tree; but it has proved valuable in fixing loose and shifting soil on river embankments and similar situations, as it is so readily propagated by cuttings. Mitchell[1] says: "It is the best underwood that we have. It makes good fences, and sheep hurdles made of it will always last a year or two longer than those made of hazel; and no soil or situation comes wrong to it, wet or dry." In Sussex it is used for making truck-baskets and handles of rakes, and also for fencing, as it is light and tough, and splits easily.[2]

Though the wood is of a nice pinkish colour, it is too small, as a rule, to have any recognised value. In northern Russia (as well as formerly in Scotland) the bark, which contains 7 per cent of tannin, is sometimes used for tanning leather. (H. J. E.)

## SALIX PENTANDRA, Bay Willow

*Salix pentandra*, Linnæus, *Sp. Pl.* 1016 (1753); Loudon, *Arb. et Frut. Brit.* iii. 1503 (1838); Andersson, *Monog. Salic.* 35 (1863), and in De Candolle, *Prod.* xvi. 2, p. 206 (1868); Willkomm, *Forstliche Flora*, 475 (1887); Buchanan White, in *Journ. Linn. Soc.* (*Bot.*) xxvii. 359 (1890); Mathieu, *Flora Forestière*, 449 (1897); Camus, *Monog. des Saules*, 84 (1904).

*Salix fragrans*, Salisbury, *Prod.* 393 (1796).

A tree, occasionally attaining 30 to 40 ft. in height, but often shrubby in the wild state. Young branchlets glabrous, dark brown, shining as if varnished. Buds ovoid, pointed, dark brown, shining, viscid. Leaves fragrant when bruised, ovate, ovate-oblong, or ovate-lanceolate, averaging 3 in. long and 1¼ in. broad, glutinous when young, coriaceous when fully grown; rounded at the base, abruptly acuminate at the apex; glabrous on both surfaces, dark green and very shining above, pale and dull beneath; margin closely and finely serrate, the serrations tipped with dark red glands; petiole, about ¼ inch long, glabrous, with two or three glands near its junction with the blade, and expanded at its origin from the branchlet, where there are one or two glands probably representing stipules.

Catkins, appearing with or after the leaves, terminating a branchlet which bears four or five leaves, spreading; axis pubescent. Staminate catkins 1½ in. long, densely

[1] *Dendrologia*, 56 (1827). Cf. Smith, *Eng. Flora*, iv. 227 (1828), who states that the wood and branches make the best hurdles, being tough, flexible, and durable. The wood was also used for the cutting-boards of shoemakers.

[2] *Gard. Chron.* xlvi. 19 (1909).

crowded with flowers; scales pubescent at the base; stamens usually five, rarely four to ten, unequal in length, with long hairs on the base of the filaments; and with two glands, the one beneath the scale often three-lobed, the other quadrate or broadly crescentic. Pistillate catkins, 1½ in. long; scales about half as long as the ovary, fringed with long hairs; ovary stalked, glabrous, about $\frac{1}{5}$ in. long, narrowly conic-subulate, with a short style, dividing into two arms, each of which is bifid; glands[1] two, one minute beneath the scale, the other quadrate and half the length of the pedicel. Capsules, ¼ inch long, on distinct pedicels, glabrous, narrow and elongated.

This species varies in the size and shape of the leaf, broad-leaved and narrow-leaved forms being distinguished by Andersson as vars. *latifolia* and *angustifolia;* but intermediate forms are common. Buchanan White states that in Britain, *S. pentandra* is a bushy shrub in the wild state, but that when cultivated, it becomes a tree with broader and larger leaves than those of the wild plant.

The bay willow is a very distinct-looking species, on account of its broad glabrous shining leaves, which resemble those of a Prunus; but is readily recognised to be a willow by its buds with a single scale. The flowers are fragrant, with an odour similar to that of the bay or true laurel (*Laurus nobilis*); and the leaves exhale the same fragrance, especially when bruised.

The following hybrids have *S. pentandra* as one of the parents:—

1. *Salix Meyeriana*, Rostkovius, *ex* Willkomm, *Berlin Baumz.* 427 (1811).

*Salix cuspidata*, Schultz, *Prod. Fl. Starg. Suppl.* 47 (1819); Borrer, in Smith and Sowerby, *Eng. Bot. Suppl.* v. tt. 2961-2962 (1863); Buchanan White, in *Journ. Linn. Soc.* (*Bot.*) xxvii. 360 (1890).

*Salix tinctoria*, Smith, in Rees, *Cycl.* xxxi. No. 13 (1815).

Leaves similar to those of *S. pentandra*, but narrower and more cuspidate at the apex. Pistillate catkins more slender and more tapering, and bearing narrower and more cylindrical capsules with longer pedicels. Stamens usually four.

This hybrid, of which the parents are supposed to be *S. pentandra* and *S. fragilis*, is said to be somewhat common on the Continent; but in England is rare in the wild state. It is recorded as a pistillate plant in Shropshire by Buchanan White. There are also specimens in the Kew Herbarium collected in 1895 on Wybunbury Bog in Cheshire by Linton. Dr. Moss adds to this distribution Herefordshire and Westmoreland in England, and Co. Mayo and Co. Kildare in Ireland; and tells me that it is occasionally planted in osier beds in Cambridgeshire and Suffolk. There is a fine tree in the Glasnevin Botanic Garden, about 40 feet high by 4 ft. in girth in 1912. Sir F. Moore states that it was obtained from Smith of Worcester, as the Purple King Willow, a name given to it on account of its purplish shoots. He adds that it is a quick grower, flowering early in spring, with beautiful large staminate catkins, the twigs being useful for house decoration.

2. *Salix hexandra*, Ehrhart, *Beit.* vii. 138 (1791); Buchanan White in *Journ. Linn. Soc.* (*Bot.*) xxvii. 361 (1890).

*Salix Ehrhartiana*, Smith, in Rees, *Cycl.* xxxi. No. 10 (1815).

Leaves lanceolate, long acuminate at the apex, silky pubescent at first, but

[1] Fraser, *Proc. Roy. Hort. Soc.* xxxv. p. cxv (1909), showed specimens of flowers of *S. pentandra*, with proliferation of the posterior gland, giving rise to two or three additional pistils.

soon becoming glabrous; dark green and very shining on the upper surface; minutely glandular-serrate; petiole slender, nearly ½ in. long. Stamens four to six.

This hybrid, the parents of which are supposed to be *S. pentandra* and *S. alba*, is known on the Continent in the staminate form. Buchanan White refers to it two specimens: one, a shrub with pistillate catkins, growing near Duddingston, Edinburgh, and the other, a barren specimen from a bush at Restennet, near Forfar. It has recently been found[1] in the Lake District in Cumberland and Westmoreland.

*S. pentandra* is a native of nearly the whole of Europe—except the extreme south—the Caucasus, and northern Asia, as far east as Kamtschatka and Amurland, and apparently extending to the province of Yunnan, in China. It grows mainly on river banks, and in marshy places, ascending in peat mosses in the Alps to 4000 ft. elevation. It grows in similar situations in Britain[2] from Argyle and Moray southwards to Derbyshire, ascending in Northumberland[3] to 1300 ft. In Ireland,[4] it is frequent and native in the north, becoming less common southwards, till in Kerry and Cork it appears only as an introduction. (A. H.)

The bay willow makes a handsome tree, with very distinct foliage, but is rather slow in growth. The finest specimens we have seen are:—one at Kew, on the lawn near the Palm House, which measured, in 1907, 58 ft. by 9 ft. 8 in. at eighteen inches above the ground, its trunk forking at three feet. Another at Woburn was 50 ft. by 5 ft. in 1908, with a short bole. A handsome tree at Beauport, Sussex, measured in 1911 about 35 ft. by 6 ft., with a bole 6 ft. long, another near the keeper's lodge not being quite so large. (H. J. E.)

## SALIX BABYLONICA, Weeping Willow

*Salix babylonica*, Linnæus, *Sp. Pl.* 1017 (1753); Loudon, *Arb. et Frut. Brit.* iii. 1507, and iv. 2588 (1838); Andersson, *Monog. Salic.* 50 (1863), and in De Candolle, *Prod.* xvi. 2, p. 212 (1868); Willkomm, *Forstliche Flora*, 471 (1887); Mathieu, *Flore Forestière*, 453 (1897); Burkill, in *Journ. Linn. Soc.* (*Bot.*) xxvi. 526 (1899); Camus, *Monog. des Saules*, 65 (1904).

*Salix pendula*, Moench, *Meth.* 336 (1794).

*Salix propendens*, Seringe, *Saules de la Suisse*, 73 (1815).

*Salix Napoleonis*, Schultz, *Arch. Fl.* 239 (1856).

A tree, attaining 30 to 40 ft. in height, with rough ridged bark, and usually with a short trunk, and wide-spreading branches, the ultimate branchlets being very long and pendulous. Branchlets slender, glabrous except near the nodes. Leaves linear-lanceolate, about 2½ to 3½ in. long, and ½ to ¾ in. broad, tapering at the apex into a long slender acuminate thread-like point, cuneate at the base, slightly pubescent when young, perfectly glabrous when fully grown, bright green and shining above, pale and covered with a glaucous bloom beneath; margin finely serrate, the serrations often ending in minute sharp incurved points; petiole ⅙ in. long, without glands, glabrous or slightly pubescent; stipules early deciduous.

[1] *Journ. Bot.* xxviii. 229 (1900).

[2] Hooker, *Student's Flora*, 355 (1878).

[3] Baker and Tate, *Fl. Northumberland and Durham*, 248 (1868), state: "Frequent in damp woods and by stream sides, ascending in Coquetdale to Harbottle, in Allendale to 450 yards, and in Teesdale to the junction of the Whey Sike with Harwood Beck."

[4] Praeger, in *Proc. Roy. Irish Acad.* vii. 282 (1901).

Only female trees are known in cultivation. Catkins terminating branchlets with one, two, or three leaves, very slender, green, curved, about 1 in. long; axis pubescent; scale about ⅔rds the length of the ovary, pilose at the base, ciliate, ovate-acuminate; ovary sub-sessile, about $\frac{1}{10}$ in. long, ovate, glabrous, ending in a very short style, which is divided into two stigmatic arms, each of which is bilobed; gland posterior, quadrate, emarginate or bilobed.

### Variety and Hybrids

I. The following variety is known in cultivation:—

1. Var. *annularis*, Ascherson, *Fl. Brandenburg*, 630 (1864).

Var. *crispa*, Loudon, *Arb. et Frut. Brit.* iii. 1514 (1838).
*Salix annularis*, Forbes, *Sal. Woburn.* 41, t. 21 (1829).

Leaves folded and rolled up, so as to form a ring or spiral, otherwise as in the type. This remarkable variety, which is called the ring-leaved willow, is of unknown origin;[1] but undoubtedly is a sport of *S. babylonica*, and like it is a female tree. W. Masters of Canterbury stated[2] that his father gave fifteen shillings for a plant 6 in. high, but had no clue as to where or how the variety had originated. He mentions an instance of a ring-leaved willow, which after being planted twenty years, produced a single branch with leaves of the ordinary form, which continued to be borne for years afterwards on the same branch and its ramifications.

There is a fine specimen on the lawn near the Palm House at Kew, which was, in 1912, 56 ft. high with a trunk 11 ft. in girth at three and a half feet from the ground, above which it divides into two stems. Lord Kesteven measured in 1906 a fine specimen, 57 ft. by 7 ft., on a farm near Caythorpe, Grantham.

II. The following trees, often considered to be varieties of *S. babylonica*, are probably hybrids:—

2. *Salix Salamonii*, Carrière, in *Rev. Hort.* xl. 463 (1869) and xlix. 444 (1877).

*Salix babylonica Salamonii*, Simon-Louis, *Cat.* 1869, p. 85; Carrière, in *Rev. Hort.* xliv. 115 (1872).

A tree with a tall straight stem, and ascending branches, forming when young a pyramidal crown; ultimate branchlets pendulous, but not so long as in *S. babylonica.* Young branchlets pubescent near the nodes, becoming glabrous. Leaves similar in shape, size, and colour to those of *S. babylonica*, but pubescent with scattered appressed long hairs on both surfaces. Only pistillate trees are known; catkins similar to those of *S. babylonica*, but with the axis more pubescent and the scales furnished with long cilia.

This remarkable tree, which is supposed to be a cross[3] between *S. babylonica* and *S. alba*, is very distinct in habit, forming when young a handsome pyramidal tree, which grows with astonishing vigour and is not injured in our climate by frost

[1] Dode, in *Bull. Soc. Dend. France*, 1909, p. 153, believes that this may have originated from a witches' broom; and states that he obtained a similar sport as a cutting, which was taken from an abnormal growth on a *Salix alba.*

[2] *Gard. Chron.* 1855, p. 726. Cf. Darwin, *Animals and Plants under Domestication*, i. 408 (1890).

[3] Schneider, *Laubholzkunde*, i. 36, note (1904), supposes it to be identical with *S. sepulcralis*, Simonkai, in *Termes. Füz.* xii. 157 (1890), said to have been found in Hungary. Camus, *Monog. Saules*, 235 (1904), identifies with the last-named, *S. alba*, var. *tristis*, Trautvetter, *Fl. Alt.* iv. 255 (1833), described from an Altai specimen.

in spring, as is almost invariably the case with the common weeping willow. Though not so pendulous in habit as the latter, it has graceful drooping branchlets.

This tree originated on the estate of Baron de Salamon, near Manosque (Basses Alpes) some time before 1869, when it was put on the market by Simon-Louis of Metz. Carrière recommended this tree, on account of its vigour,[1] for the production of timber; and stated that it grew on all soils, even on dry soils and on limestone, where the weeping willow refused to grow, or remained stunted and yellow. *S. Salamonii* comes nearly as early into leaf and retains its foliage almost as late in the season as *S. babylonica.*

There are several old trees of *S. Salamonii* on the borders of the lake at Kew. The exact age of these is unknown; but they much surpass in size the true weeping willows beside them.[2] At Casewick, where Lord Kesteven has planted *S. Salamonii* as a park tree, it thrives well, and has attained 35 ft. in height at eighteen years old.

3. *Salix pendulina*, Wender, in *Schrift. Natf. Ges. Marburg*, ii. 235 (1831).

*Salix blanda*, Andersson, *Monog. Salic.* 50 (1863), and in De Candolle, *Prod.* xvi. 2, p. 212 (1868); Camus, *Monog. Saules*, 232 (1904).

*Salix elegantissima*, Koch, in *Wochschrf. Ver. Bef. Gartb.* xiv. 380 (1871), and *Dendrologie*, ii. pt. i. 505 (1872).

Under these names are possibly included three slightly different hybrids of the same parentage, *S. babylonica* and *S. fragilis.* They are wide-spreading trees, with long and pendulous branchlets, differing only slightly in habit from *S. babylonica,* but much more hardy than this species, and on that account often cultivated in Germany where the true weeping willow is killed by severe frost. I have not been able to study the three forms; but a tree at Kew, labelled *S. elegantissima,*[3] obtained from Dieck in 1889, has leaves more coriaceous and slightly broader than those of *S. babylonica*; both surfaces glabrous, slightly glaucous beneath. Catkins occasionally androgynous,[4] usually only pistillate, shorter than in *S. babylonica*; axis pubescent; ovary shortly pedicellate, about $\frac{1}{10}$ in. long, slightly pubescent at the base, with two stylar arms, each arm bilobed; scale two-thirds the length of the ovary, very pilose; glands irregular, usually two,—one narrowly oblong between the scale and the ovary,—the other, posterior, nearly quadrate, and occasionally bilobed.

Koch states that this tree, which on the Continent is often sold in commerce as *S. Sieboldii,* is fast in growth, with very long branchlets, almost reaching to the ground. There is a thriving specimen at Glasnevin, about 25 ft. high, which is labelled *S. blanda.* In the Botanic Garden at Leyden there is a handsome tree, about 40 ft. by 3 ft. in 1912, which is labelled *S. Petzoldii pendula.*[5]

[1] In *Garden and Forest*, x. 497 (1897), it is said to be the fastest-growing willow in California, where, at the Chico Forestry Station, stems cut back in February 1896 to 2 ft. from the ground, were 31 to 32 ft. high in August 1897.

[2] A vigorous young tree on the bank of the Cam, Trinity College, Cambridge, was 35 ft. high by 4 feet in girth in 1912. Its pyramidal crown with ascending upper branches contrasts much with the older but lower adjacent weeping willows, which have broad flattened crowns and spreading branches.

[3] Both *S. blanda* and *S. elegantissima* were described as having perfectly glabrous ovaries; but notwithstanding this, the Kew tree, received from Dieck, is probably *S. elegantissima.* Ascherson and Graebner, *Syn. Mitteleurop. Flora*, iv. 73 (1908), correctly describe *S. elegantissima* as having a pubescent ovary; and on that account doubt its being a hybrid between *S. babylonica* and *S. fragilis.* Following Koch, they believe that it was introduced from Japan by Siebold; and if this is the case, *S. fragilis* could scarcely have been one of the parents.

[4] As Camus points out, androgynous catkins are common in the hybrids *S. blanda* and *S. sepulcralis* which he describes.

[5] Cf. Lauche, *Haupt-Katalog Muskauer Baumschulen*, 1905, p. 32.

### Distribution

*S. babylonica* was the name given by Linnæus [1] to the common weeping willow cultivated in Europe, which he erroneously supposed to have been identical with the trees growing by the rivers of Babylon, which are mentioned in Psalm cxxxvii. 2. The latter are, without doubt,[2] a species of poplar, *Populus euphratica*. The original home of the weeping willow appears to be central and southern China, where it is commonly found on the banks of rivers and canals, as well as in gardens. Fortune [3] observed this tree, identical in all respects with that cultivated in Europe, in the neighbourhood of Shanghai and Canton, and also near Ningpo, where it is sometimes planted around graves.

The typical form of the species, that long cultivated in Europe and prevalent in central and southern China, is always a female tree, with long pendulous branchlets. In the neighbourhood of Peking [4] a variety is more common, which may be distinguished as var. *pekinensis*, Henry. This is an upright tree, with ascending branches, which is known in both sexes. The foliage is practically identical [5] with that of the typical form; but there is a marked difference, not only in habit, but in the pistillate catkins, which are extremely short and compact in this variety, not exceeding ½ in. in length, often on leafless peduncles; ovaries wider, ending in two short undivided stigmas; scales nearly glabrous. Var. *pekinensis* is represented at Kew by a tree, about 15 ft. high, originally from Peking, which was obtained from the Arnold Arboretum in 1905. This tree, like the common weeping willow, is injured regularly by spring frosts.

From China the weeping willow was early introduced into Japan, where it is now cultivated and naturalised in many places.[6] During the Middle Ages it was probably carried westward to Persia, Asia Minor, and Turkey; and it is now also "cultivated in Baluchistan, Northern India, and the Punjab, and less commonly in the plains farther east, and also in Kurdistan." [7]

The first mention of the tree in European literature is by Petiver,[8] who refers to a specimen gathered in China by Cunningham in 1701, but which cannot now be found in the British Museum. The first mention of the tree in the Levant was by Tournefort [9] in 1719; and it is possible that either he or Wheler, who travelled in Asia Minor and Greece in 1675-1676, introduced it into western Europe.

[1] First described by Linnæus, in *Hort. Cliff.* 454 (1737).

[2] Cf. Koch, *Dendrologie*, ii. pt. i. 507 (1872). *S. babylonica* is not now found in Babylonia. See p. 1771, note 4.

[3] *Wanderings in China*, 118, 136 (1847), and *Residence Among the Chinese*, 52 (1857). The weeping willow is apparently depicted by Nieuhoff, *L'Embassade des Provinces Unies vers China*, i. 189 (1665), in a view of Tonnau, a village on the river Wei in the province of Shantung.

[4] Cf. Bretschneider, in *Journ. N. C. Branch, R. Asiat. Soc.* xv. 15, 30 (1880), and *Bot. Sinic.* ii. 359 (1892). The name of *S. babylonica* in China is *yang-liu*. The weeping form is rare at Peking, where it is distinguished as *ch'ui-yang-liu*.

[5] The serrations of the leaves in var. *pekinensis* are more distantly placed, and without the cilia, which are present in the typical form. These slight differences may be due to the different ages of the specimens.

[6] Franchet et Savatier, *Enum. Pl. Jap.* i. 459 (1875).

[7] Brandis, *Indian Trees*, 637 (1906).

[8] *Mus. Petiv. Cent.* 997 (1703), where it is referred to as follows: "*Yang-diu chinensibus. Arbor salicis folio ramulis pendulis.* Frequently painted on Japan work. Of the wood they make arrows."

[9] *Corollarium*, 41 (1719), where it is described as follows: "*Salix orientalis, flagellis deorsum pulchre pendentibus, hujus etiam meminit Wheler Itin.*" Wheler, *Journey to Greece and Asia Minor*, 217 (1682), saw near Brusa, and not far from Mt. Olympus in Asia Minor, a tree which appears from his description to have been a weeping willow.

The weeping willow was introduced into England some time before 1730, as it was on sale in London, according to a catalogue[1] published in that year. Collinson states[2] that it was introduced by Mr. Vernon, Turkey merchant at Aleppo, who planted it at his seat in Twickenham Park, where Collinson saw it in 1748. The latter says that this tree was the original of all the weeping willows in England; and adds that he measured one in 1765 at Mr. Snelling's at Godalming, which, though only fifteen years old, was 6 ft. in girth.

There was a famous weeping willow, planted by Pope in front of his villa at Twickenham, which was felled[3] in 1801, when the story was given in *St. James's Chronicle*, August 25-27, of that year,[4] that this tree was the first one planted in England, having been introduced as a withy round a package from Spain; but doubtless Pope's tree was a cutting from Mr. Vernon's willow.

Another celebrated tree was the weeping willow in St. Helena, which was planted in 1810, and was a favourite of Napoleon. After his death cuttings were brought to England, and planted in many places, where they were called Napoleon's willow, but differed in no respect from the ordinary form.[5] There is a weeping willow at the Fountain Pond, Cassiobury, which formerly bore a plate[6] stating that it had been a cutting from the tree in St. Helena. This tree fell and sustained considerable damage; but it has been replanted, and, according to Mr. Daniel Hill, who measured it in 1912, is 36 ft. high and 4 ft. 7 in. in girth. (A. H.)

The weeping willow strikes freely from cuttings, and grows rapidly in good soil beside water; but is very liable to have the young shoots killed by frost, and is not nearly so hardy as the hybrid *S. Salamonii*. It is one of the earliest trees to come into leaf, and the latest in retaining its foliage, being frequently green in December.

The finest trees known to Loudon were those at various places on the banks of the Thames, which were 50 to 60 ft. high in 1838; but it is doubtful if any of these now survive, as it is not a long-lived tree. The best that I have seen is perhaps a tree (Plate 379) on the Promenade, Cheltenham, which was planted about 1860, and is still thriving, although its limbs have been supported by iron rods for some years. It measured in 1911 about 75 ft. by 9 ft., and on 24th November 1911, after a severe frost, it still retained most of its leaves.

There are several picturesque trees,[7] but of no great height, growing on the banks of the Cam, behind the Colleges of Cambridge. These are exceeded in size by one

[1] Miller, *Catalogus Plantarum, a Catalogue of Trees, Shrubs, Plants, and Flowers, which are propagated for sale in Gardens near London*, p. 71 (1730), where it is mentioned as "*S. orientalis*, T. *Cor.* 41. The weeping willow *vulgo*."

[2] Dillwyn, *Hortus Collins.* 48 (1843), quoted by Loudon, *Gard. Mag.* xix. 64 (1843). Aiton, *Hort. Kew.* v. 356 (1813), states that the weeping willow was first cultivated in 1692 in the Royal Garden at Hampton Court, but gives no authority for this. If true, the introduction into Western Europe was probably made by Wheler. Aiton quotes Plukenet, *Phytographia*, t. 173, fig. 5, which is not the weeping willow.

[3] Corbett, in *Mem. Twickenham*, 285 (1872), states that this tree perished and fell to the ground in 1801. The wood was worked up by an eminent jeweller into all sorts of trinkets and ornaments, which had an extensive sale. The Empress of Russia took cuttings from Pope's willow in 1789 for the gardens at St. Petersburg.

[4] Phillips, *Sylva Florifera*, ii. 263 (1823).

[5] Loudon, *Arb. et Frut. Brit.* iv. 2588 (1838), and *Trees and Shrubs*, 758 (1842). Forbes, *Salic. Woburn.* 43 (1829) states that a plant raised from a cutting of the St. Helena tree was identical with the common weeping willow.

[6] D. Hill, in *Trans. Herts. Nat. Hist. Soc.* xiv. pt. ii. 132 (1911).

[7] These trees are comparatively young, and have replaced the original trees, which were planted in 1760. Cf. Willis and Clarke, *Archit. Hist. Univ. Camb.* ii. 646 (1886).

in the Fellows' Garden of King's College, which was about 45 ft. high and 10 ft. in girth in 1912, with the trunk decayed and mended with cement.

The weeping willow attains[1] a much greater size and beauty in warm countries than it does in England. I have seen none finer than in Chile, where it is often planted by the sides of the irrigation canals, and enjoys a long and warm summer.

(H. J. E.)

## SALIX FRAGILIS, Crack Willow

*Salix fragilis*, Linnæus, *Sp. Pl.* 1017 (1753); Smith, *Fl. Brit.* iii. 1051 (1804), *Eng. Bot.*[2] t. 1807 (1808), and *Eng. Flora*, iv. 184 (1828); Andersson, *Monog. Salic.* 41 (1863), and in De Candolle, *Prod.* xvi. 2, p. 209 (1868); Willkomm, *Forstliche Flora*, 472 (1887); Buchanan White, in *Journ. Linn. Soc.* (*Bot.*) xxvii. 368 (1890); Mathieu, *Flora Forestière*, 450 (1897); Camus, *Monog. des Saules*, 76 (1904).

A tree, attaining about 70 ft. in height. Bark rough, strongly ridged, and divided into broad deep fissures. Young branchlets slightly pubescent, glabrous in the second year. Buds appressed to the branchlet, compressed, shining, glabrous. Leaves lanceolate, about 4 in. long and ¾ in. broad, gradually tapering above into a long caudate oblique acuminate apex, cuneate at the base, silky when young; upper surface in summer glabrous, shining; lower surface glaucous or glaucescent, with scattered silky appressed hairs; margin coarsely serrate, each serration tipped with a conspicuous reddish brown gland; petiole ⅜ in. long or more, pubescent, usually with two glands at the insertion of the blade.

Catkins appearing with the leaves, terminating short branchlets, which bear three or four usually entire leaves; axis densely pubescent. Male catkins, about 2½ in. long; scales concave, oblong, truncate or cuspidate, glabrous within, pubescent without, margin fringed with long hairs; stamens two, filaments as long as the scale, slightly pubescent at the base, anthers yellow; glands two, the posterior transversely oblong, the anterior half its size and narrowly oblong. Female catkins about 2 in. long; scales concave, lanceolate, pilose at the base, ciliate, with long hairs; ovary distinctly stalked (pedicel $\frac{1}{25}$ in. long), one-third longer than the scale, fusiform, glabrous, ⅙ in. long, gradually narrowing to the apex, which ends in a short style, divided into two arms, each of which is bilobed; only one gland,[3] which is posterior, usually present, quadrate, entire in margin, much shorter than the pedicel. Fruiting catkins about 3 in. long, with a pubescent axis, and distinctly stalked capsules.

*S. fragilis* is called crack willow, on account of the ease with which the branchlets disarticulate, especially in spring. It is readily distinguishable by its

[1] *Salix Safsaf*, Dode, in *Bull. Soc. Dend. France*, 1906, p. 62, fig. (not Forskal), received from Palermo, where it was introduced by Schweinfurth from Egypt, is indistinguishable from *S. babylonica*; but is remarkably fast in growth and may be a hybrid. A specimen in the nursery of the Jardin des Plantes at Paris, which I saw in 1912, though only three years old from a small cutting, was 20 ft. high and 15 in. in girth. M. Dode states that this is perfectly hardy at Paris, none of the shoots being damaged by frost.

[2] One of the male flowers in *Eng. Bot.* t. 1807, has three stamens; and it is possible that the male plant figured is a hybrid like *S. speciosa*.

[3] The presence of two glands in the pistillate flower of *S. fragilis* is very rare, and abnormal.

nearly glabrous coarsely glandular-serrate large leaves, ending in a long point directed to one side. When adult the stem is covered with a much rougher bark than that of *S. alba*, the depressions between the ridges being broad and deep; on this account buyers of willow timber usually designate it as the "open-bark willow." It is of no value for making cricket bats, as its wood lacks the necessary strength, lightness, and elasticity.

### Varieties and Hybrids

True *S. fragilis* does not seem to be a very variable plant; but peculiar forms, more or less resembling it, are known, which are supposed to be of hybrid origin. These are described below. The twigs in some trees become dull grey in the second year, whilst in others they assume a brilliant orange colour. Varieties founded on the size of the leaf or on the colour of its lower surface are not sufficiently distinct to be worth naming; and the two forms, vars. *britannica* and *genuina*, Buchanan White, in *Journ. Linn. Soc.* (*Bot.*) xxvii. 368 (1890) based on the relative length of the bract and flower, are untenable, as this depends simply on the age at which individual flowers are observed.

The following are supposed to be hybrids:—

1. *Salix decipiens*, Hoffmann, *Hort. Sal.* ii. 9, t. 31 (1791); Smith, *Eng. Bot.* t. 1937 (1808); Forbes, *Sal. Woburn.* t. 29 (1829); Buchanan White, in *Journ. Linn. Soc.* (*Bot.*) xxvi. 348 (1890).

*Salix fragilis*, var. *decipiens*, Koch, *Syn. Fl. Germ.* 643 (1837).
*Salix cardinalis*, Veitch, *Cat. Trees and Shrubs*, 1910, p. 75.

A shrub or small tree. Young branchlets crimson or dark red, glabrous; becoming in the second year polished and yellowish white or of a clay colour. Leaves glabrous, smaller than those of *S. fragilis*, 2 to 3 in. long, $\frac{1}{2}$ to $\frac{3}{4}$ in. broad, more oblong, more parallel-sided, usually less narrowed and often rounded at the base; under surface pale dull green, scarcely glaucous; serrations finer, sharper, and closer together than in *S. fragilis*. This is supposed to be a hybrid between *S. fragilis* and *S. triandra*, and occurs in both sexes. The catkins of the female plant are intermediate between these two species; while the number of stamens in the male plant is usually two, rarely three.

*S. decipiens*, according to Buchanan White, is widely spread, but not abundant in Britain; and is usually found growing in company with *S. triandra* and *S. fragilis*. He mentions various localities from Perth to Somerset. There are specimens in the Kew Herbarium from Dorset, Sussex, and Essex. This species was said by Forbes[1] to be cultivated for basket-work, producing when cut crimson-coloured annual shoots, which are very remarkable in appearance. It appears[2] to be largely used at the

[1] *Sal. Woburn.* 57, t. 29 (1829). Cf. also Loudon, *Arb. et Frut. Brit.* iii. 1515 (1838). Smith, *Eng. Bot.* t. 1937 (1808), states that Crowe found it in several osier beds in Norfolk and Cambridgeshire, where it was known as white Welsh osier.

[2] Cf. Ellmore and Okey, in *Journ. Board Agric.* xviii. 915 (1912), who refer to it as *Salix alba*, var. *cardinalis*.

present time for the same purpose in Belgium; and when imported is called "Belgian red willow." It is occasionally planted for ornament, appearing in some nursery catalogues under the name *S. cardinalis.*

2. *Salix speciosa,* Host, *Sal.* 5, t. 17 (1828); Buchanan White, in *Journ. Linn. Soc.* (*Bot.*) xxvii. 353 (1890).

*Salix fragilis,* var. *latifolia,*[1] Andersson, in De Candolle, *Prod.* xvi. 2, p. 209 (1868).

A large tree, similar in bark and habit to *S. fragilis.* Young branchlets slightly pubescent, glabrous in the second year. Leaves longer and broader than those of *S. fragilis,* up to 6 in. long and 1¾ in. wide; broadly lanceolate, with a long acuminate often curved apex; pubescent when young; when mature, upper surface dark green, shining, glabrous; lower surface pale, glabrescent, or with scattered appressed hairs; margin coarsely glandular-serrate; petiole ¾ in. long, often with one to four glands at the junction with the blade.

Staminate catkins, 2½ in. long; flowers crowded on the densely pubescent axis; scale, shorter than the stamens, and covered with long white silky hairs; stamens usually two, rarely three, glabrous or pilose at the base; glands two, variable, entire or lobed. Pistillate flowers not seen.

The above description is drawn up from an old tree[1] on the side of the lake at Kew, which has been named *S. triandra* × *fragilis,*[2] by Linton. It is identical with some trees, which have been called "open-bark willow," by Carter, and is perhaps not very rare in Britain.

3. *Salix viridis,* Fries, *Nov. Pl. Suec.* 283 (1828); Andersson, in De Candolle, *Prod.* xvi. 2, p. 210 (1868); Buchanan White, in *Journ. Linn. Soc.* (*Bot.*) xxvii. 364, 371 (1890).

(?) *Salix rubens,* Schrank, *Baier Fl.* i. 226 (1789)
*Salix excelsior,* Host, *Sal.* 8, tt. 28, 29 (1828).
*Salix palustris,* Host, *Sal.* 7, tt. 24, 25 (1828).
*Salix montana,* Forbes, *Sal. Woburn.* 37, t. 19 (1829).
*Salix fragilis-alba,* Wimmer, *Denkschr. Schles. Ges.* 156 (1853).

A tree similar in size to the reported parents, *S. alba* and *S. fragilis,* with spreading branches, and usually more or less pendulous branchlets. Young branchlets variable, glabrous or pubescent. Leaves intermediate between these two species, variable in size, shape, colour, and pubescence; darker green, less shining, more finely serrated, and less oblique at the apex than those of *S. fragilis*; longer and broader than those of *S. alba,* and soon becoming glabrescent.

Staminate catkins narrower and larger than in *S. fragilis.* Pistillate catkins more slender than in that species; ovary shortly but distinctly pedicellate; capsules intermediate in size between *S. alba* and *S. fragilis.*

[1] Specimens which were sent from Kew to Buchanan White in 1887 and 1888, and are now preserved in his herbarium at Perth, show that this tree was then labelled *S. fragilis,* var. *latifolia,* a name accepted by White.

[2] *S. alopecuroides,* Tausch, *Ind. Hort. Canal.* (1821), *ex* Andersson, in De Candolle, *Prod.* xvi. 2, p. 203 (1868); A. Kerner, in *Verh. Z. B. Ges. Wien,* 69 (1860); Koch, *Dendrologie,* ii. pt. i. 516 (1872), is the name given to the hybrid *S. triandra* × *fragilis,* which occurs on the continent. This, judging from Tausch's specimen, preserved in the Cambridge Herbarium, is quite distinct from *S. speciosa,* Host.

Numerous forms of *S. viridis*[1] occur, some being close to *S. fragilis*, others nearer to *S. alba*; and no exact definition of this hybrid is possible; but it is easy to recognise when *S. alba*, *S. fragilis*, and *S. cærulea* are excluded. According to Buchanan White, *S. viridis* is widely distributed in Britain, occurring from Cornwall and Surrey to Perth; but it is less abundant and more local than the parents, *S. alba* and *S. fragilis*. The second-class bat willow is often *S. viridis*.

*S. viridis* doubtless grows to as large a size as either of the parents, with which it is generally confused. The largest[2] we know are growing beside a stream at Thornbury, Gloucestershire. One tree is 60 ft. high by 18 ft. in girth. Near it are two trees growing from the same root, one of which is 65 ft. by 11 ft. 6 in. and the other 60 ft. by 15 ft. 7 in., according to measurements which were kindly taken for us by Mr. Samuel Fudge in July 1912.

4. *Salix Russelliana*, Smith, *Fl. Brit.* iii. 1045 (1804), *Eng. Bot.* t. 1808 (1808), and *Eng. Flora* iv. 186 (1823); Forbes, *Sal. Woburn.* 55, t. 28 (1829); Loudon, *Arb. et Frut. Brit.* iii. 1517, fig. 1311 (1838).

*Salix fragilis*, var. *Russelliana*, Koch, *Syn. Fl. Germ.* 643 (1837).

A tree, remarkably vigorous in growth, with long, straight and slender branches, not angular in their insertion like those of *S. fragilis*. Leaves in the young state white and silky pubescent beneath; when fully grown, lanceolate, acuminate, $3\frac{1}{2}$ in. long, $\frac{5}{8}$ in. wide, finely serrate, with scattered appressed hairs on the under surface near the midrib. Pistillate trees only known, with flowers similar to those of *S. fragilis*, having pedicellate ovaries and bipartite styles, but with looser catkins.

This seems, judging from the type specimens in Smith's herbarium at the Linnean Society, to be a form of *S. viridis*, which is not clearly known to botanists[3] at the present time. The normal leaves are well figured by Loudon (fig. 1311); those depicted by Sowerby (*Eng. Bot.* t. 1808) from coppice shoots are larger than the normal leaves, and have the glands on the petiole often developed into slender leaflets.

*S. Russelliana* was sent to Woburn about 1800 by Mr. Bakewell from Leicestershire, and was called the Dishley or Leicester Willow. It seems to have been extraordinarily vigorous, as Lowe, *Agric. Survey of Notts*, p. 118, is quoted[4] for the fact that "this willow in a plantation yielded at eight years' growth poles which realised a net profit of £214 per acre." It was also remarkable for the large per-

[1] Scaling mentions in his pamphlet, *Salix or Willow*, i. *Cat.* p. 8 (1871) and ii. 19 (1872) two willows, which he sent out in 1871 from his nursery at Basford, Notts:—(1) *S. sanguinea*, the branches of which were brilliant red in winter. This was obtained by Scaling some years previously in the Ardennes, where it was known as the red willow. (2) *S. basfordiana*, with branches of a deep orange colour. This appeared as a seedling in the nursery, and grew with great vigour.

Salter, in *Gard. Chron.* xvii. 298, figs. 41, 42 (1882), apparently confused these two distinct trees, which he described as *S. basfordiana*. There are two trees at Kew under this name—a male, No. 58, which is a form of *S. viridis*, and a female, No. 80, which is identical with *S. vitellina*. Whether these correspond to Scaling's *S. sanguinea* and *S. basfordiana* is uncertain.

[2] These trees have pubescent leaves, simulating those of *S. alba*, but larger and thinner. They also differ from the latter species in having distinctly pedicellate fruit.

[3] *S. Russelliana* has been misunderstood by most botanists since Smith's time. White's view that it was true *S. fragilis*, while Smith's *S. fragilis* was *S. viridis*, is untenable.

[4] Duke of Bedford, in Forbes, *Sal. Woburn.*, p. v (1829). The frontispiece represents Johnson's Willow.

centage of tannin contained in the bark, which exceeded, according to Mr. Biggin's analysis,[1] even that of the oak.

The Rev. W. Dickenson assured the Duke of Bedford and Sir James Smith that the tree known as Johnson's Willow,[2] which grew between Lichfield and Stow Hall, was *S. Russelliana.* This tree, which was so called because Dr. Johnson frequently rested under its shade, was figured by Loudon, *Arb. et Frut. Brit.* iii. figs. 1312, 1313, who states that it was about 60 ft. high, with a great bole, about 20 ft. in length, which girthed,[3] in 1810, 21 ft. at six feet from the ground. It contained 130 cubic ft. of timber, and was perfectly sound. It was blown down in 1829, when it was supposed to have been 130 years old.

Mr. L. Fosbrooke, of Ravenstone Hall, Ashby de la Zouch, believes that *S. Russelliana* still occurs in Leicestershire in the Trent valley, where the trees are supposed to be female *S. fragilis.* They differ in foliage from male trees of true *S. fragilis,* according to Mr. Fosbrooke, who tells me that though they grow fast when young, they become round-topped as they approach maturity, and are certainly not nearly as vigorous as *S. cœrulea.* I have not been able to examine specimens of these trees; and cannot say whether they are identical with the original *S. Russelliana,* which may have been a solitary sport or hybrid of exceptional vigour, now lost to cultivation.

### Distribution

The distribution of *S. fragilis* in the wild state cannot be determined with certainty, as it has been largely planted outside of its original area; but it is supposed to be indigenous in the greater part of Europe and in western Asia. It is not wild in Norway, but occurs there as a planted tree as far north as lat. 64° 5′. In Sweden it is also only known in cultivation, the male tree being seen occasionally in Werm-land and Schone, and the female tree in the district round Kalmar. It is wild in Jutland, in the island of Oesel, and in Russia,[4] where it extends as far north as Esthonia, Livland, Kostroma, and Kazan; but is absent from the Crimea. It is widely spread in the Caucasus, Persia, and Asia Minor. Farther east it is cultivated[5] in the Kuram valley, in Gilgit, Ladak, and Lahaul, as well as at Quetta, where it was found by Lace at 5600 ft. elevation.

In Europe it extends southwards to Spain and Portugal, Sicily, and Greece. It is planted largely in northern and central Germany, where it is wild in many localities; but in southern Germany, Austria, and Hungary, it is only seen in river valleys, where it prefers a deep loamy soil; and ascends along the edges of streams in the mountains, as high as 1700 ft. altitude in the Bavarian Alps, and 2500 ft. in Tran-sylvania. In order to succeed it requires considerably more moisture in the soil than *S. alba,* and on this account is most often seen on the Continent generally, on the banks of rivers, streams, and lakes, being rare in the interior of the forests.

[1] Davy, *Agric. Chem.* 89 (1814) analysed the bark of a large tree of the "Leicester Willow" and found it to contain more tannin than any other British tree, having a little more than coppice oak, and three times as much as the "common willow."

[2] See note 4, p. 1757.

[3] Withering, *Arr. Brit. Plants,* ii. 68 (1818) gives these measurements on the authority of Rev. W. Dickenson. Loudon, *Derby Arboretum,* 55 (1840), states that a young tree, raised from the branches of Johnson's Willow, was growing at Lichfield in 1836, when it was 20 ft. high.

[4] Köppen, *Holzgew. Europ. Russlands,* ii. 257 (1889).

[5] Brandis, *Indian Trees,* 637 (1906).

The crack willow is probably wild in marshy ground in Britain from Perthshire southwards; but it is extremely difficult to decide in what stations it is really indigenous. It is supposed to have been introduced[1] into Ireland, where however it is often seen as a planted tree. (A. H.)

This species occasionally attains as great dimensions as *S. alba*. In *Gard. Chron.* i. 447 (1874), E. Lees figures some remarkable old willows. The largest of these (*S. fragilis*) grew in the Wye valley near Ross, and was over 70 ft. in height, with a girth of 24 ft. at two feet from the ground. Mr. H. Marshall can find no trace of this tree at the present time.

Jackson, *Syon House Trees and Shrubs*, 29 (1910), mentions an immense tree on the south side of the lake in a decaying condition. "Judging from the length of the bole, which is now prostrate, this may well have been the specimen mentioned by Loudon[2] (p. 1521) as being 89 ft. high and about 13 ft. in girth, and called by him *Salix Russelliana*."

In Messrs. Samson's nursery at Kilmarnock there was a crack willow, which in 1904 measured, according to Mr. Renwick, 80 ft. high, with a bole 22 ft. in length and girthing 16 ft. 1 in. at five feet from the ground. This tree was blown down in November 1911, when the trunk was found to be much decayed at the heart. No shoots have sprung from the root since. Mr. Renwick[3] records a tree at Bruntwood Mains near Galston, Ayrshire, which measured 13 ft. 1 in. in girth in 1902, when it was reputed to have been 62 years old. (H. J. E.)

## SALIX ALBA, White Willow

*Salix alba*, Linnæus, *Sp. Pl.* 1021 (1753); Smith, *Eng. Bot.* t. 2430 (1812), *Eng. Flora*, iv. 231 (1828); Forbes, *Sal. Woburn.* 271, t. 136 (1829); Loudon, *Arb. et Frut. Brit.* iii. 1522 (1838); Andersson, *Monog. Salic.* 47 (1863), and in De Candolle, *Prod.* xvi. 2, p. 211 (1868); Willkomm, *Forstliche Flora*, 469 (1887); Buchanan White, in *Journ. Linn. Soc.* (*Bot.*) xxvii. 370 (1890); Mathieu, *Flore Forestière*, 451 (1897); Camus, *Monog. des Saules*, 69 (1904).

A tree, attaining about 90 ft. in height and 20 ft. in girth, with more or less ascending branches, but with spreading or pendulous ultimate branchlets. Bark less deeply fissured than in *S. fragilis*, the depressions between the ridges being narrower and shallower than in that species. Young branchlets covered with whitish appressed pubescence, partly retained in winter and in the following year. Buds flattened, appressed against the twig, silky pubescent. Leaves lanceolate, 2 to 2½ in. long, rarely exceeding ½ in. in width, tapering to a long acuminate, straight or curved apex; upper surface greyish green, more or less covered with silky white appressed pubescence; lower surface whitish, densely covered with similar pubescence; margin densely ciliate, and with minute glandular serrations; petiole short, pubescent.

Catkins, appearing with the leaves, on short lateral leafy branchlets; axis densely tomentose; flowers crowded. Staminate catkins, about 1½ in. long;

[1] Praeger, in *Proc. Roy. Irish Acad.* vii. 283 (1901).

[2] Loudon's identifications of these large willows are very uncertain.

[3] *Trans. Nat. Hist. Soc. Glasgow*, vi. pt. iii. 353 (1902).

stamens two, united and pubescent at the base; glands two, anterior quadrate and entire, posterior small and ligulate; scale concave, ciliate, half the length of the stamens. Pistillate catkins, 1½ in. long; ovary sessile, glabrous, about $\frac{1}{10}$ in. long, ovoid, abruptly narrowed towards the apex; style short, the two arms apparently undivided (really slightly bilobed); scale, fringed with long hairs, and as long as the ovary; gland, one, posterior, quadrate. Fruiting capsules sessile, $\frac{1}{6}$ in. long.

*S. alba* appears to be very variable in the size and pubescence of the leaves; but doubtless some of the varieties attributed to *S. alba* are forms of the hybrid, *S. viridis*, in which the characters of *S. alba* are dominant. The following form is in cultivation:—

1. Var. *argentea*, Wimmer, *Sal. Europ.* 17 (1866).

Var. *splendens*, Andersson, in De Candolle, *Prod.* xvi. 2, p. 211 (1868).
Var. *leucophylla*, Simon-Louis, *Cat.* 1869, p. 85.
Var. *regalis*, Beissner, in *Gartenflora*, xxvi. 40 (1877).
*Salix splendens*, Bray, *ex* Opiz, *Boehm. Gew.* 110 (1823).
*Salix regalis*, Wesmael, in *Bull. Cong. Bot. Brux.* 1864, p. 280.

Young branchlets and leaves on both sides covered with dense silvery white tomentum. This is highly effective in beds in gardens when cultivated in masses, the single stems being kept about 12 ft. high by pruning. At Glasnevin there is a thriving tree about 35 ft. high.

### Distribution

*S. alba* is widely spread through central and southern Europe, extending southwards to Algeria and Morocco, and eastwards to western and northern Asia. It is impossible to define its northern limit in Europe, as its natural area has been much extended by planting; but, according to Schübeler, it is not wild in Scandinavia, where it is occasionally planted as far north as lat. 63° 52′ in western Norway. In Russia,[1] it is probably wild as far north as southern Livonia, Vitebsk, Smolensk, Moscow, Vladimir, Kostroma, Viatka, and Perm; and is common in the Crimea, where it attains a large size. It is widely spread in the Caucasus, north Persia, and Asia Minor; and occurs in Siberia as far east as Lake Baikal; but it is only cultivated[2] in the northern Himalayas and in western Tibet.

It is doubtfully wild in northern Germany, where it is, however, planted to a considerable extent. Farther south, it is undoubtedly indigenous in southern Germany, Austria, the Balkan peninsula, Italy, Spain, and the greater part of France.[3] In central and southern Europe, it is often the dominant tree in the forests of the alluvial plains and in the woods on the banks of rivers, as on the Danube and its great tributaries, where it either forms pure stands or grows in mixture with *Quercus pedunculata*, *Populus nigra*, and *Salix fragilis*. It grows in similar situations to the last species, but thrives in lighter and more sandy soils. It ascends the river valleys in the mountains to 1400 ft. in the Bavarian Forest, to 2700 ft. in the Bavarian Alps, to 2900 ft. in Transylvania, to 3200 ft. in the Caucasus, and to 5400 ft. in the Sierra Nevada. The comparatively low altitude

[1] Köppen, *Holzgew. Europ. Russlands*, ii. 259 (1889).
[2] J. D. Hooker, *Flora Brit. India*, v. 629 (1888).
[3] It is common on the banks of all the large rivers of western France from the Seine to the Adour; and appears to be the sole native large willow in these situations where I saw no *S. fragilis* in 1912.

which it attains in the mountains is not due to the fall in temperature, as it ripens seed and grows to be a fine tree in Livonia.

The white willow is supposed to be indigenous in Britain in marshy ground from Sutherland southwards; and is considered by Praeger[1] to be probably in Ireland an original tree of river banks, but now generally planted.

Witches' brooms[2] on this species are apparently formed by the irritation set up by a mite (*Eriophyes salicis*), which causes a shoot to branch repeatedly and produce small narrow soft leaves, the whole mass often measuring a foot in diameter, and turning bright red in autumn.

The white willow was early introduced into North America, where it is now planted both in Canada and the United States. Together with *S. fragilis*, it has proved very useful as a windbreak in the prairie regions, and where timber is scarce is valuable for fuel. It is also a useful tree for reclaiming and holding the soil along streams. The wood is fairly durable in contact with the soil, and has been employed for fence-posts in the north-western plains of the United States. Pinchot[3] recommends that it should be grown as coppice, when required for posts or fuel; and says that plantations should be tilled frequently till they are well shaded. This cultivation destroys weeds, and prevents excessive evaporation of moisture from the soil. (A. H.)

### Remarkable Trees

A tree at Bury St. Edmunds, figured by Strutt, *Sylva Britannica*, plate xxiii, as the Abbot's Willow, is one of the largest white willows of which we have record. It was measured by a surveyor, named Lenny, in 1822, when it was 72 ft. high by 18½ feet in girth, and was estimated to contain 440 feet of timber. Loudon was informed that it was almost dead in 1836.

The largest white willow known to us is at Haverholme, near Sleaford, in the park east of the priory, and measured[4] in 1907 about 80 ft. high by 25½ ft. in girth, with a spread of branches about 80 ft. in diameter. A photograph, for which I am indebted to Miss F. H. Woolward (Plate 380) shows it to be past its prime; but though partly decayed, it is said to be still increasing in girth.

At Water Hall farm, Bayfordbury, Herts, there is a tree, about seventy years old, which Mr. H. Clinton-Baker measured in 1912. It is about 65 ft. high; and has a short bole, 27 ft. in girth at one foot from the ground, and divided at four and a half feet up into two stems, the larger of which is 17 ft. in girth at six feet from the ground.

At Compton Wynyates, Kineton, a beautiful old mansion belonging to the Marquess of Northampton, there is a group of sound healthy trees from 80 to 90 ft. high, one of which girthed 10 ft. 4 in. in 1905.

At Highclere, there is a fine tree, a picture of which by Alfred Parsons, now in the possession of Lady Carnarvon, was reproduced by Robinson, *Wild Garden*, 258 (1895). It formerly had three stems, but the two largest were blown down some

[1] *Proc. Roy. Irish Acad.* vii. 283 (1901).

[2] *Proc. Roy. Hort. Soc.* xxxvi. pt. ii. p. cxvii. (1910).

[3] *U. S. Dept. Agric. Forest Circ.* No. 87 (1907).

[4] Measured by Mr. J. Cowance. In *Woods and Forests*, 1884, p. 642, this tree was said to have been, in 1881, 27 ft. 4 in. in girth at one foot from the ground, 20 ft. 5 in. at four feet, and 28 ft. at seven feet. The branches spread 40 ft. on one side, and 28 ft. on the other. Cf. *Gard. Chron.* xiv. 362 (1893).

years ago. Mr. Storie informs us that the stem remaining, which leans to one side, is over 50 ft. in height and girths 13 ft. at five feet from the ground.

At Fawsley Park, Daventry, a tree, of which a photograph was reproduced in the *Journal of the Northampton Natural History Society* for 1882, was said to have been no less than 102 ft. high and 9 ft. 2 in. in girth. This tree was blown down in February 1882; and apparently all that now survives is a small tree, grown from a cutting, specimens of which were kindly sent us by Lady Knightley of Fawsley.

In Scotland, Henry measured in 1904 two trees at Palnure, Kirkcudbright, which were 86 ft. by 10 ft. 8 in., and 82 ft. by 12 ft. 9 in.

At Coodham House, Kilmarnock, there are two great trees, the largest of which was 17 ft. 1 in. in girth in 1904, when it was measured by Mr. Renwick. Mr. J. M'Gran, the gardener, informs us that in 1910 it was about 60 ft. high, and girthed 17 ft. 8 in. at three feet from the ground, 19 ft. 8 in. at five feet, and 21 ft. at six feet.

At Moncreiffe I saw in 1907 a remarkable old willow, of which the original trunk, now broken, measured 19 ft. 4 in. in girth. The branches had become rooted in several places; and one of these, now severed from the trunk, is over 6 ft. in girth. The total circumference of the branches is 106 paces. The foliage of this tree, of which we have not seen flowers, is not so white as usual; and it may be a form of *S. viridis.*

In Ireland we have no records of any very large trees; but in Mucksna Wood, a mile from Kenmare, I saw by the road-side in 1910 some very fine willows, which the Marquess of Lansdowne believes to have been planted eighty or ninety years ago. I measured one of these 85 ft. by 9 ft. 10 in.

The white willow attains a large size in Germany. T. Schube[1] gives reproductions from photographs of two enormous trees which are growing in Silesia, one at the Primkenauer factory, 20 ft. in girth, and the other, with a taller stem not quite so thick on the road between Stronn and Korschlitz. The largest willow[2] in Berlin, which grew beside a canal and had a girth of 23 ft., fell to the ground in 1894. Beissner reports[3] a tree, 92 ft. high and 9 ft. in girth in 1907, which is growing in a park at Hohenmistorf in Mecklenburg. As the park was laid out in 1854, this willow probably dates from that year.

## Timber

The wood of this species is used for making cricket-bats of an inferior kind, those of the best class being mainly obtained from *S. cærulea.* Apart from this special use, the wood of the white willow is of considerable value, though it is not so much esteemed now as formerly. It is tough, and indents without splintering from blows or hard usage, and is used on that account for brakes on railway waggons, and for the sides and bottoms of carts; it is also used for the rims of riddles and milk-pails, and by turners for making toys. It is also valuable for hurdle-making. G. W. Newton states[4] that George Stephenson had a high opinion of willow, as forming durable blocks for paving. Gorrie states[5] that "in roofing it has been known to stand one hundred years as couples, and with the exception of about ½ in. on the

[1] *Mitt. Deut. Dend. Ges.* 1910, p. 52, figs. 8 and 9.
[2] *Ibid.* 1894, p. 29.
[3] *Ibid.* 1907, p. 56.
[4] *Timber Trees*, 34 (1859).
[5] Loudon, *Gard. Mag.* i. 45 (1826).

outside, the wood has been found so fresh at the end of that period as to be fit for boat-building." Boards of willow were laid for floors in 1700.[1] (H. J. E.)

## SALIX CŒRULEA, Cricket-Bat Willow

*Salix cœrulea*, Smith, *Eng. Bot.* t. 2431 (1812), and in Rees, *Cycl.* xxxi. 141 (1819); Aiton, *Hort. Kew.* v. 365 (1813) (not Forbes).

*Salix alba*, var. *cœrulea*, Smith, *Eng. Flora*, iv. 231, 232 (1828); Loudon, *Arb. et Frut. Brit.* iii. 1523 (1838); Bean, in *Kew Bulletin*, 1907, p. 312, plate, and 1912, p. 205.

*Salix viridis*, Pratt,[2] in *Journ. Roy. Agric. Soc.* lxvi. 22 (1905), and *Quart. Journ. Forestry*, i. 325, fig. No. 02 (1707) (not Fries).

A tree, attaining 100 ft. in height, with ascending branches, making a narrow angle with the stem, and forming a pyramidal crown.[3] Terminal branchlets erect, not spreading or drooping. Bark smoother than in *S. alba.* Young branchlets appressed pubescent, becoming reddish brown in winter and the following year. Leaves similar to those of *S. alba,* but thinner in texture, more translucent, and less densely pubescent; lower surface not white, but bluish grey; margin ciliate, with minute glandular serrations.

Pistillate catkins, differing from those of *S. alba* as follows: ovary[4] slightly stalked, more tapering at the apex, about $\frac{1}{6}$ in. in length; each of the two style-arms distinctly bilobed; scale more concave, about two-thirds the length of the ovary. Fruiting capsule, nearly $\frac{1}{4}$ in. long, on a distinct short pedicel.

This remarkable tree, which is best distinguished by its pyramidal habit, with stiff ascending branches and branchlets, is *par excellence* the true cricket-bat willow, as it exceeds the other kinds in rapidity of growth. Bean calls it the "best close-bark willow." In addition to its different habit, the leaves are readily distinguishable from those of *S. alba* by their different colour, and also by their translucency, as when viewed against the light with a lens, the tertiary venation is always plainly visible. In the forms[5] of *S. viridis,* which approach *S. alba* in foliage, the leaves simulate those of *S. cœrulea* in colour and translucency, but are considerably larger and are also more coarsely serrate in margin; and no form of *S. viridis,* known to me, has either the peculiar habit or rapid growth of *S. cœrulea.*

*S. cœrulea* is considered by most botanists to be a variety of *S. alba*; but it differs from the latter in the flowers, which approximate in their size and shape and stalked ovaries to those of *S. fragilis*; and it is possible that *S. cœrulea* may be the first cross between these two species, most of the characters of *S. alba,* if this hypothesis is correct, being dominant.

*S. cœrulea* was first distinguished by Smith, who was unable to find any botanical

[1] Ellis, *White Woods*, quoted by Mitchell, *Dendrologia*, 56 (1827).

[2] Mr. E. R. Pratt at first accepted, on Mr. Linton's identification of certain specimens, the name *S. viridis* for the true cricket-bat willow, though in *Journ. Roy. Agric. Soc.* lxvi. 22 (1905), he agrees that the East Anglia cricket-bat willow has leaves indistinguishable from those of *S. cœrulea.*

[3] Owing to the fastness of growth of *S. cœrulea*, the nodes are at greater intervals, and the crown of foliage in consequence is remarkably sparse. The bark is darker in colour than that of *S. alba.*

[4] Bean, *Kew Bull.* 1907, p. 313, states that the ovaries of *S. cœrulea* are identical with those of *S. alba.* A careful examination shows that the pyramidal tree with bluish foliage has always the distinct flowers described above.

[5] These forms of *S. viridis* are often referred to *S. alba*, var. *cœrulea*, especially on the Continent; but I restrict the name *S. cœrulea* to the pistillate pyramidal tree here described.

character to separate it from *S. alba*, except "that the under side of the leaves loses at an early period most of the silky hairs." However, he went on to say: "Its qualities are of the highest importance. The superior value of the wood and bark, the rapid growth as well as handsome aspect of the tree, its silvery blue colour, its easy propagation and culture, in dry as well as wet situations, all render it so superior to our common white willow, that a cultivator might justly think lightly of any one who should tell him that there was no difference between them." He states that a cutting planted in Norfolk "became in ten years a tree 35 ft. high and 5 ft. 2 in. in girth, which is a rapidity of growth beyond all comparison with the common white willow." Succeeding botanists have applied the name *cœrulea* to pendulous slow-growing trees, of which the foliage is bluish white beneath; but I consider that Smith distinctly described the quick-growing pyramidal tree which is now recognised by cultivators as the true cricket-bat willow; and his specimen of *S. cœrulea* in the herbarium of the Linnean Society, London, is undoubtedly this tree.

Apparently no staminate tree[1] of *S. cœrulea* exists; and Smith knew it only as a female tree.

The origin of *S. cœrulea* is obscure; but it appears to be confined to the eastern counties of England, where it has been known[2] since 1804 at least. Bean,[3] relying upon Shaw, states that it is only found at the present time in Essex, Hertford, and Suffolk; but it undoubtedly occurs also in Norfolk and Cambridgeshire; and is said to have been formerly a rare tree in Kent and Surrey. Dealers restrict their purchases to the eastern counties, and have not yet ever found any suitable willow for making bats in other parts of Britain or on the Continent. It is probable, however, that *S. cœrulea*, when planted in other parts of the British Isles, will prove satisfactory, as there is no reason for believing that the peculiar qualities of the wood of this tree are dependent upon the climatic conditions of the eastern counties.

Mr. J. A. Campbell, who planted in 1904 about 150 trees of *S. cœrulea* at Arduaine, Lochgilphead, Argyllshire, has received a favourable report[4] from Mr. D. J. Carter, willow-dealer at Waltham Cross. One of these trees, which was cut down in 1912, when it was 6 in. in diameter, would, if it had been large enough for making bats, have fetched the normal price. This tree was too small to give a certain result; but judging from Mr. Carter's report, *S. cœrulea*, grown in the west of Scotland, where the rainfall is 60 inches annually, apparently retains the qualities which render its timber so valuable in East Anglia, where the climate is dry and sunny, and the mean summer temperature much higher than in the west.[5]

Loudon identified *S. cœrulea* with the upland or red-twigged willow of Pontey[6];

[1] Forbes, *Sal. Woburn.* 273, t. 137 (1829), describes as *S. cærulea*, a male tree, which is doubtless a form of *S. viridis.*

[2] Smith, *Fl. Brit.* iii. 1072 (1804), refers to it as a bluish, quick-growing variety of *S. alba.*

[3] *Kew Bulletin*, 1907, p. 313.

[4] Cf. Bean, in *Kew Bulletin*, 1912, p. 205.

[5] Notwithstanding the above statement, I should hesitate to advise the planting of this willow in any part of Scotland or the west of England, until it has been proved by actual sale that the timber is of equal value to that grown in the eastern counties. Even supposing that it should prove to have the same qualities when large enough, it is not easy to convince leading manufacturers that their reputation is worth risking; and though eminent cricketers whom I have consulted do not seem to judge bats so much by the appearance of their wood, as by their balance, handle, and weight, they would rather pay a high price for a bat which is guaranteed by the maker than use a substitute whose durability and driving powers are more or less uncertain.—H. J. E.

[6] *Profitable Planter*, 72 (1814). Sang, Nicol's *Planter's Kalendar*, 68 (1812), says that the red-twigged willow "forms a large tree and has a fine silvery foliage; it is probably the same as the upland willow of Mr. Pontey."

but the latter's description is vague and uncertain; and I have been unable to find any accounts of this remarkable tree as early as the eighteenth century.

*S. cœrulea* is much faster in growth than either *S. alba* or any of the varieties of *S. viridis*. It is true that well-shaped trees of the latter kinds, grown with clean stems on proper soil, are frequently purchased at a fair price[1] by bat-makers; yet *S. cœrulea* should always be preferred for planting. It comes to market earlier, on account of its rapid growth; and produces a wood light in weight,[2] very elastic and tough, which is found to be the most suitable for making the best kinds of cricket bats.[3]

## Cultivation

The best account of the cultivation of this tree is given by Mr. E. R. Pratt, from whose article[4] I shall quote largely.

The choice of soil is most important, as many failures are due to the prevalent idea that willows will thrive in any wet or marshy situation. This is erroneous, and soils sodden with stagnant water or of a peaty nature should be avoided. The best ground is undoubtedly rich light alluvial land by the side of a running stream; but good willows are also seen growing in fertile loam, where there is a good supply of moisture. Clay and gravel soils are usually quite unsuitable. If willows are planted in grass land beside a stream, they must be protected against cattle, during the whole period of their growth; and, in all cases, rabbits must be excluded.

In peaty soil, except in very rare cases where the drainage is good and the properties of the peat modified in consequence, willows never thrive, and after a few years often die. In such cases the soil, when examined, proves to be very acid in reaction; and the willows are frequently attacked by a fungus, *Physalospora gregaria*, Saccardo, which produces cankerous spots on the young stems. The epidermis at first looks as if scorched, then dries up, turns brown, and becomes cracked by the protrusion of very small black spots (the fruit of the fungus); it ultimately peels off, exposing the inner part of the stem. Dr. T. Johnson, who has given a good account,[5] with figures, of this fungus, believes that its ravages are much favoured by raw peaty soil; and certainly, in some cases, it is the cause of extensive failures in willow plantations. In wet or marshy situations, where there is great growth of grass, the willow seems more liable also to the attacks of the beetle, *Saperda carcharias*, Linn.

Though rooted cuttings are frequently advertised by nurserymen, there seems to be no doubt that it is much more advantageous to plant large sets. Scaling,[6] who had great experience, says: "All varieties of tree willows grow better and more

[1] Two trees in Suffolk, growing in good soil, which I examined before they were cut down—both *S. alba*, but one male, and the other female—were sold at 6s. 8d. per cubic foot in 1910; but these trees, twenty years after planting, had only attained 14 in. in diameter, and had short stems which made only four bat lengths.

[2] Mr. L. Fosbrooke, Ravenstone Hall, Leicester, states in *Gard. Chron.* xxxix. 46 (1906), "The close-barked white willow is *Salix cœrulea.* It is the quickest grower of all the tree willows in a moist soil, reaching 18 in. in diameter in as many years. The quicker the growth, the lighter the timber, and the better the price."

[3] Mr. E. R. Pratt, in *Quart. Journ. Forestry*, i. 336 (1907), shows that the specific gravity of the wood of *S. alba* (male trees) exceeds that of *S. cœrulea* (female trees) by 14½ per cent.

[4] *Journ. Roy. Agric. Soc.* lxvi. 19-34 (1905).

[5] *Proc. Roy. Dublin Soc.* x. pt. ii. 153-166, plates 13-15 (1904).

[6] *Salix or Willow*, i. *Cat.* p. 8 (1871).

vigorously from cuttings than from rooted plants." Gorrie[1] found that shoots of the white willow "6 to 8 ft. long and about 2 in. in diameter succeed better than rooted plants; they require to be put from 18 in. to 2 ft. deep, in marshy soil, which should be drained." At the present time, it is the invariable custom amongst expert growers of cricket-bat willow to use large sets, up to 20 ft. long and 3 in. in diameter, and scarcely smaller than 6 to 10 ft. long by 2 in. in diameter. The sets are best obtained from young trees that have either been felled for sale or that have been specially pollarded for the purpose; and should never be taken from the tops of old trees, as these are seldom straight, and require much trimming. A young tree, about ten years old, gives the best sets when pollarded, and can be again pollarded every five years for six or seven times till decay sets in. The sets should be cut in early spring. Mr. Pratt, after trimming them for four-fifths of their length, has them tied up in bundles of ten, and kept in water for a month, after which they are planted out.

As the willow is a light-demanding tree, and the grower's object is to produce, as quickly as possible, a short stem clean of branches for about 12 to 15 ft., a good crown of foliage must be preserved from the start, and the trees should be planted so wide apart that they do not interfere with each other by lateral shade. If closely planted, they grow more slowly, and often develop an elliptical instead of a circular stem. The distance apart along the side of a stream should not be less than 10 yards. Close planting to kill undergrowth or grass is a mistake, as the latter, if necessary, can be removed by cultivation, though this is seldom done.

Holes may be made for the sets by driving in a stake two or three feet deep, which can afterwards be levered out. The sharpened end of the set is then dropped into the hole, and tightly rammed in position. It is very important to insert the set deeply and firmly, so that it may not be shaken by the wind. The after care consists in rubbing off in the first three years, with the gloved hand, as high up as possible, the buds which appear on the stem, so as to prevent the development of side shoots. Mr. Pratt advocates the pruning of the stem afterwards to a height of 25 ft., but this is seldom if ever done in Herts or Essex, where growers are content with a stem clear of branches to 12 or 15 ft.

Willow trees become saleable to bat-makers, when they are about 15 in. in diameter at six feet from the ground, or about 13 in. at twelve or fifteen feet up. The original set remains as a useless core in the centre of the stem. As the width of a bat is about $5\frac{1}{2}$ in. and the clefts are taken radially, the minimum diameter should be $5\frac{1}{2}$ in. + $5\frac{1}{2}$ in. + 2 or 3 in. (the diameter of the set) = 13 or 14 in. The trees may of course be allowed to grow for another period of years, until a second ring of clefts is formed around the first ring, or even for a further period; but it is usually most profitable to dispose of the trees when they are young.

The trees are generally sold standing, and are deemed of first quality when the stem is straight, clear of knots or branches, and covered with a smooth scaly bark, which is indicative of rapid growth. As the length of a bat is about 28 in., the trees

[1] Loudon, *Gard. Mag.* i. 46 (1826). J. Harrison, *New Method*, 20-24 (1766) recommended, in planting willows, the use of sets 16 ft. long, with all the side branches pruned off; to be preserved from cattle by three stakes two feet from the set, tied up with thorns; and an after treatment of disbudding.

as soon as felled, which is done in winter, are cross-cut into lengths of 28 to 30 in., and these in turn are split into clefts. The clefts are split up along the radii so that the annual rings run from the front to the back of the bat. The best clefts come from the lower part of the tree, which is far tougher than the upper portion. In the best bats I have counted from seven to nine annual growths.

The clefts are ultimately fashioned into blades, which are subjected to hydraulic pressure; and it is here that the value of *S. cœrulea* shows itself, as blades made of *S. fragilis* are unable to stand the requisite amount of pressure. The further process of manufacture is detailed in a pamphlet[1] written by W. E. Bussey, which should be consulted by those interested in the growth of willow.

The extraordinary value of the true cricket-bat willow is not exaggerated in the following statement,[2] made by Mr. John Barker of Pishiobury, Sawbridgeworth: "A good set costs 1s. to 1s. 6d., and, if planted in a suitable soil and does well, is worth from £5 to £8 in fifteen years." He instances a case where a piece of land was bought for £50, and planted with willows, which were sold, when sixteen years old, for £2000 in 1905.

The following figures have been given to us by a reliable grower of willows in Herts. In December 1910 he sold twenty-four trees which grew on the bank of a stream, and had been thirteen years planted, for an average of £5 per tree. In January 1912 he sold eleven willows, which had been planted fourteen years previously, for £81. These trees averaged 55 to 60 ft. in height, and were clear of branches for about 18 ft., their stems ranging from 42 to 46 in. in girth at five feet from the ground. The best tree contained 12 cubic feet of timber, available for making bats, and as it sold for £8, the price per cubic foot came to 13s. 4d. In 1906 fifty-three trees growing on the same property were sold for £190. These had short stems, averaging 13 in. in diameter, and yielding only three bat lengths.

Many trees of remarkable size, but comparatively young, have been felled for conversion into cricket bats. One of the largest on record[3] was a tree at Boreham, Essex, which was planted in 1835, and felled in 1888, when it was 101 ft. high and 5 ft. 9 in. in diameter. It weighed upwards of eleven tons, and was perfectly sound. It was felled by B. Warsop and Sons, who made from it no less than 1179 bats.[4]

Mr. H. Clinton-Baker tells us of another willow which grew in a field at Aspenden, near Buntingford, and was purchased by the same firm. It was 6 ft. in diameter, at three feet from the ground, and divided at five feet up into two stems, which were clear to a height of 50 ft., where they still measured 2 ft. in diameter.

Mr. Stuart Surridge purchased for £25 in 1910 a tree near St. Albans, which was about 80 ft. high and measured 5 ft. in diameter at three feet from the ground. Judging from a photograph, it had a clean stem of about 16 feet. He states that the largest tree known to him grew at Robertsbridge, in Sussex, and measured 21 ft. in girth. This was felled in 1902, and produced over 1000 cricket bats.

Sir Thomas B. Beevor made the following note in his copy of Evelyn's *Sylva*:

[1] Published in 1910 by Geo. G. Bussey and Co., Queen Victoria St., London, E.C.

[2] *Gard. Chron.* xxxix. 62 (1906).

[3] *Trans. Eng. Arbor. Soc.* iv. 122 (1899).

[4] Mr. Edwin Savill informs us that he sold a tree in 1911 for £70, but he has no measurements. The dealer told him that the number of feet utilized worked out at £1 per cubic foot.

"Mar. 12, 1808. Willow tree, planted about eighteen years in a meadow at Wortham, in Suffolk, in length about 22 ft., and measured, at four feet from the ground, 7 ft. 10 in., and in its bark appearing as green as when young. This willow is now found to be the cœrulean willow, and was taken down in 1816, and cut into boards which measured 22 in. broad; and increased (in the last) eight years in girth to 11 ft. 8 in."

There are many fine trees on the Copped Hall estate, near Epping, where the agent, Mr. P. W. Dashwood, informed me in 1907 that he had refused a short time previously an offer of £1500 for 100 trees to be selected by the purchaser. He sold one tree for £25 which was afterwards re-sold for £40 to another dealer. This tree had a stem 3½ ft. in diameter, and was clean of branches for about 25 ft.; it yielded eleven lengths of bats, twenty-eight inches being allowed for a bat-length.

In 1910 I measured at Ryston, Downham, Norfolk, a true *S. cœrulea* which was twenty years old and growing in good soil—alluvial silt over clay. It was 80 ft. high by 5 ft. 4 in. in girth. A male *S. alba* of the same age growing beside it was 65 ft. high by 3 ft. 11 in. in girth. Mr. E. R. Pratt sent me a photograph of a tree, growing on the bank of the river Wissey at Hilgay, south of Downham, which, in October 1912, measured 84 ft. in height, and 6 ft. 5 in. in girth. This tree was planted in 1889; and is growing very rapidly, as it was only 5 ft. 6 in. in girth in February 1910.

Elwes saw in 1907 a fine tree growing in a meadow near Spains Hall, Braintree, Essex, which was considered by Mr. A. W. Ruggles-Brise, to be a true cricket-bat willow. It was quite sound and healthy, and measured 90 ft. by 12½ ft.

The illustration (Plate 381) is reproduced from a negative for which we are indebted to Mr. M. C. Duchesne. It was taken from a tree in a meadow called Hartham, near Hertford, of which the grazing is annually let by the municipality. The soil and situation are considered ideal for growing willow. (A. H.)

## SALIX VITELLINA, Golden Willow

*Salix vitellina*, Linnæus, *Sp. Pl.* 1016 (1753); Smith, *Eng. Bot.* t. 1389 (1805), and *Eng. Flora*, iv. 182 (1828); Host, *Salix*, tt. 30, 31 (1828): Forbes, *Salic. Woburn.* 39, t. 20 (1829); Loudon, *Arb. et Frut. Brit.* iii. 1528 (1838).

*Salix alba*, Linnæus, var. *vitellina*, Stokes, *Bot. Mat. Med.* iv. 506 (1812); Seringe, *Ess. Saule Suisse*, 83 (1815); Andersson, in De Candolle, *Prod.* xvi. 1, p. 211 (1868); Buchanan White, in *Journ. Linn. Soc.* (*Bot.*) xxvii. 371 (1890); Camus, *Monog. Saules*, 75 (1904).

(?) *Salix basfordiana*,[1] Scaling, *Salix or Willow*, *Cat.* p. 8 (1871), and ii. 19 (1872).

A tree, attaining about 60 ft. in height, with spreading branches. Young branchlets, pubescent at the tips and near the nodes, becoming glabrous and bright yellow in winter, and in the following year. Leaves lanceolate, averaging 2½ in. long and ⅜ in. wide, gradually tapering to a slender acuminate caudate apex; shining green above, bluish white beneath, with scattered appressed silky hairs, usually slight

[1] Cf. page 1757, note 1.

on the upper surface, more abundant beneath; serrations minute with incurved glandular tips; margin either ciliate throughout the season, or with the cilia deciduous in summer; petiole slightly pubescent.

Catkins terminating short leafy branches and appearing with the leaves; axes white tomentose, densely flowered. Staminate catkins, 1 to 1¾ in. long, curved; stamens two, rarely three, base of filament slightly pilose; scales concave, ovate-lanceolate, almost glabrescent, slightly ciliate, nearly twice as long as the stamens; glands two, posterior quadrate and usually bilobed, anterior smaller and usually entire. Pistillate catkins, very slender, 1¾ in. long; ovary with a distinct pedicel, which is twice as long as the posterior bilobed quadrate gland, conic, glabrous, about ⅙ in. long, ending in a short style; stylar arms spreading and bilobed; scale as long as the ovary, pubescent on both surfaces.

*S. vitellina* is related to *S. alba*, but is distinct in the flowers, and has narrower and less pubescent leaves. The flowers are occasionally unstable, three stylar arms and three stamens being present instead of two, and the ovary is often peculiarly inflated towards the apex. *S. vitellina* is variable as regards the amount of the pubescence; in one form the leaves are scarcely pubescent on the upper surface, and the margin becomes non-ciliate; while in another form there is appressed pubescence on both surfaces, often dense beneath, and the cilia are retained on the margin till late in the season. *S. vitellina* is possibly of hybrid origin, and though long known in cultivation and naturalised in many parts of Europe is very doubtfully wild. Smith states[1] that Crowe found it wild in pastures at Ovington, near Watton, Norfolk.

The following varieties are known:—

1. Var. *pendula*, Späth, *Cat.* No. 69, p. 110 (1888).

Branchlets pendulous. Leaves narrow, non-ciliate. This is a beautiful weeping tree,[2] which is often sold under the erroneous name of *S. babylonica*, var. *aurea.*

2. Var. *britzensis*, Späth, *Cat.* No. 57, p. 67 (1883).

Young branchlets bright red, pubescent. Leaves appressed-pubescent on both surfaces, ciliate till autumn. This is a staminate tree. It is the finest of all the coloured willows, the twigs assuming a beautiful red colour in winter. A thriving specimen at Glasnevin is about 40 ft. high, and is narrowly pyramidal in habit.

The golden willow is planted for ornament both in England and the Continent, and is occasionally cultivated in osier beds.[3] It is very striking in winter, and seems to thrive in this country, though Smith states that the twigs are often killed by severe cold, like those of *S. babylonica.* We have seen no trees of great size, the finest being probably two trees at Glasnevin, which are about 65 ft. in height and 8 ft. in girth. These are probably of considerable age, and have rough bark like that of *S. fragilis.* It was early introduced into North America, where it is now very common in New England. (A. H.)

[1] *Fl. Brit.* iii. 1050 (1804).

[2] Dode, in *Bull. Soc. Bot. France*, lv. 655 (1910), considers this to be a hybrid between *S. vitellina* and *S. babylonica*, and names it *S. chrysocoma*, Dode.

[3] Cf. Ellmore and Okey, in *Journ. Board. Agric.* xviii. 914 (1912), who say that it is one of the toughest willows grown, if used with the bark on in a green state. Hence it mostly produces rods, which are used for tying purposes.

# POPULUS

*Populus*, Linnæus, *Gen. Pl.* 307 (1737) and 456 (1754), *Sp. Pl.* 1034 (1753); Wesmael, in De Candolle, *Prod.* xvi. 2, p. 323 (1868), in *Mém. Soc. Sci. Hainaut*, iii. 186-250 (1869), and in *Bull. Soc. Bot. Belg.* xxvi. pt. i. p. 371 (1887); Bentham et Hooker, *Gen. Pl.* iii. 412 (1880); Schneider, *Laubholzkunde*, i. 2-23 (1904), and ii. 869, 870 (1912); Dode, in *Mém. Soc. Hist. Nat. Autun.* xviii. 1-76 (1905); Ascherson and Graebner, *Syn. Mitteleurop. Flora*, iv. 14-54 (1908); Gombocz, in *Math. Termes. Közl.* xxx. 5-238, with two maps (1911).

Deciduous trees, belonging to the natural order Salicaceæ. Terminal buds large, with one or two outer pairs of opposite scales at the base, and several imbricated scales above; axillary buds smaller, with fewer scales, the lowest of which is short, broad, and open next the stem; scales accrescent, marking when they fall the base of the branchlet with ring-like scars. Branchlets terete or angled, with five-angled pith, and showing, when the leaves have fallen, three-dotted leaf-scars, on the sides of which are visible two minute scars left by the early deciduous stipules. Leaves simple, alternate, penninerved, usually long-stalked; entire, dentate, or lobed; often different in shape, pubescence, and margin on the long and on the short shoots.

Flowers without honey, fertilised by the wind, appearing before the leaves in early spring; diœcious, in pendulous stalked catkins, which arise from buds in the axils of the leaf-scars of the previous year or at the ends of short shoots. Catkins bearing on a slender axis numerous flowers, each of which is subtended by a caducous stalked lobed, dentate, or laciniate scale (bract). Perianth absent, replaced by a stalked disc. Male flowers densely crowded; stamens four to twelve, or twenty to sixty, with short white filaments, and purple or red two-celled anthers, arising from the oblique, flat or concave disc. Pistillate flowers not so dense in the catkin as the male flowers; ovary sessile in the oblique cup-shaped disc, one-celled, with two, three, or four placentæ; style short or obsolete, with as many entire or bifid stigmas as there are placentæ. Fruiting catkins elongated, ripening early before the leaves are fully grown; capsule one-celled, separating when ripe into two to four recurved valves, girt at the base by the persistent disc.[1] Seeds numerous, minute, without albumen, elliptic, compressed, acuminate at the apex, surrounded by tufts of long white silky hairs, attached to their short stalks and deciduous with them. Seedling,[2] with two stalked suborbicular cotyledons, sagittate at the base, thick and succulent; primary leaves either in opposite pairs as in *P. canescens*, or alternate as in the black poplars.

[1] In section *Turanga* the disc is deciduous, and does not persist on the fruit.

[2] An account of the germination of poplars, with figures of the seedlings, is given by Miss F. H. Woolward, in *Journ. Bot.* xlv. 417, t. 487 (1907). Cf. also Hickel, in *Bull. Soc. Dend. France*, No. 25, p. 88 (1912).

The poplars are typical light-demanding trees, incapable of bearing shade, their branchlets and leaves dying when not exposed to full light. This is well seen when two Lombardy poplars are planted close together, the shade of the taller of the two killing the branches on the adjacent side of the other. In connection with this demand for light, which necessitates sparse branches and foliage, the poplars normally shed many of their smaller branchlets in autumn. The process[1] by which this is effected is similar to that by which the leaves are cast off,—a zone of corky tissue being formed at the point where the rupture subsequently takes place,—the branchlet leaving when it falls a circular scar on the main branch to which it was attached.

The genus *Populus* comprises about twenty-five species, inhabiting the extra-tropical regions of the northern hemisphere from the Arctic Circle southwards; in North America extending to Lower California and northern Mexico; throughout Europe and northern Africa; and in Asia, extending as far south as Asia Minor, Syria, Mesopotamia, Persia, Afghanistan, the Himalayas, China, and Japan. Towards the extreme north certain species often form great forests; elsewhere poplars are most common in alluvial land bordering rivers, streams, and swamps; but occasionally they form a part of the deciduous forests.

The genus is divided into five sections; and the following key, based mainly on the characters of the leaves and buds, includes all the species in cultivation, with the more important hybrids; and in addition, three species, mainly of botanical interest, which we have not seen in England in the living state.

I. TURANGA, Bunge, *Beit. Kennt. Fl. Russ.* 498 (1848).

This section differs from the others in the remarkable polymorphic leaves, and in the deeply cleft disc of the flowers, which does not remain persistent at the base of the fruit.

1. *Populus euphratica*, Olivier.[2]

Northern and eastern Africa, Syria, Mesopotamia, Persia, Turkestan, Afghanistan, north-west India, Mongolia, north China.

A tree 50 ft. high. Leaves coriaceous, greyish green, of the same colour on both surfaces; on young trees, linear or oblong, entire, short-stalked, and willow-like; on old trees extremely diverse, ovate, oblong, rhombic, or orbicular, lobed or cut, long-stalked.

Not hardy in Great Britain.[3] This poplar, and not a willow,[4] is the *ʿarabim* of the Psalms, cxxxvii. 2, the trees growing by the rivers of Babylon, on which the Jews in captivity hanged their harps.[5]

1 This natural fall of branchlets, effected by a vital process, was termed *cladoptosis* by Berkeley, in *Gard. Chron.* 1855, p. 596. It has been observed in oaks and willows, as well as in poplars, and Shattock gave a complete account of it in the case of the aspen in *Journ. Bot.* xxi. 306 (1883). This phenomenon appears to have been first noticed by J. Main, *Hort. Register*, iv. 193, fig. A (1835), where a fallen branch of the black poplar is figured.

2 *Voy. Emp. Othom.* figs. 45, 46 (1807). This is a very variable species, which has been variously treated by botanists. Gombocz, in *Math. Termes. Közl.* xxx. 71, 72 (1911), recognises several varieties, and treats *P. pruinosa*, Schrenk, as a distinct species. *P. illicitana*, Dode, in *Bull. Soc. Dend. France*, 1908, p. 163, lately found near Elche in Spain, appears to be only naturalized there, and is identical with the ordinary form of the species in Morocco and Algeria.

3 Späth, *Catalogue*, No. 91, p. 51 (1893-1894), states that a plant sent to Lauche from Turkestan in 1881 soon died. It was reintroduced by General Korolkow, who sent it to Späth in 1892.

4 *Salix babylonica*, Linnæus, was so called, because it was erroneously supposed to be the tree of the Psalms. Cf. p. 1752.

5 Cf. Koch, *Dendrologie*, ii. pt. i. p. 507 (1872), and Ascherson, in *Sitzb. Ges. Nat. Fr. Berlin*, 1872, p. 92.

II. Leuce, Duby, in De Candolle, *Syn. Pl. Fl. Gall.* i. 427 (1828).

White poplars and aspens. — Bark smooth on young stems, ultimately breaking on the surface into rough rhombic cavities. Buds variable, tomentose and dry, or glabrous and viscid. Leaves variable, tomentose or glabrous; lobed, dentate, or serrate. Flower scales fringed with long hairs; stamens few, about ten; capsules slender, oblong.

A. White poplars.—Leaves densely tomentose on the long shoots; less tomentose or glabrescent, and different in shape on the short shoots.

2. *Populus alba*, Linnæus. Europe, North Africa, Asia Minor, Caucasus, Central Asia, Himalayas. See p. 1777.

Leaves on the long shoots palmately lobed, snowy white beneath; on the short shoots oval, dentate, greyish beneath.

3. *Populus tomentosa*, Carrière. North China. See p. 1786.

Leaves on the long shoots triangular-ovate, without lobes, biserrate, grey tomentose beneath; on the short shoots, with a few sinuate teeth, glabrescent beneath.

4. *Populus canescens*, Smith. Western Europe. See p. 1780.

Leaves on the long shoots ovate-deltoid, greyish white beneath, with irregular triangular serrated teeth; on the short shoots suborbicular or broadly ovate, glabrescent beneath, with a few sinuate teeth.

B. Aspens.—Leaves on the long and short shoots not markedly different, glabrous or glabrescent beneath.

* *Branchlets glabrous.*

5. *Populus tremula*, Linnæus. Europe, North Africa, Asia Minor, Caucasus, Siberia. See p. 1787.

Leaves suborbicular, 1½ to 2 in. in diameter, thin in texture, acute or rounded at the apex; margin with rounded or sinuate teeth.

6. *Populus tremuloides*, Michaux. North America. See p. 1791.

Leaves orbicular or ovate, 1½ to 2 in. in diameter, thin in texture, cuspidate at the apex; margin ciliate, finely and regularly serrate.

* * *Branchlets slightly tomentose.*

7. *Populus Sieboldii*, Miquel. Japan. See p. 1794.

Leaves ovate, 3 in. long, 2 in. wide, thick in texture, shortly acuminate at the apex; margin with minute sinuate teeth or glandular serrations.

8. *Populus grandidentata*, Michaux. North America. See p. 1792.

Leaves ovate-deltoid, 3 to 4 in. long, 2 to 3 in. wide, thin in texture, acuminate; margin with a few sinuate triangular large teeth.

III. Aigeiros, Duby, in De Candolle, *Syn. Pl. Fl. Gall.* i. 427 (1828).

Black poplars.—Trees with furrowed bark. Buds viscid, but not very odorous. Leaves green on both surfaces, with a clearly defined translucent border.

*A. Leaves never ciliate in margin.*

9. *Populus nigra*, Linnæus. Europe, Caucasus, Siberia. See p. 1795.

Leaves rhomboid, about 3 in. long, 2 in. broad, cuneate at the base, gradually tapering above into a long acuminate apex; glands never present at the base. Stigmas always two. Stamens fifteen to thirty, purple.

Branchlets and petioles glabrous in var. *typica*, pubescent in var. *betulifolia*. The Lombardy poplar (p. 1798) is the fastigiate form of var. *typica*; and var. *plantierensis* (p. 1802) is the fastigiate form of var. *betulifolia*.

*B. Leaves ciliate in margin.*

* *Glands absent at the base of the leaf.*

10. *Populus Fremontii*, Watson. North America. See p. 1794.

Leaves on young cultivated trees, reniform or rhombic, with a cuneate base; on old trees deltoid with a truncate base; about 2½ in. wide; apex cuspidate; serrations few, coarse, and incurved. Stigmas three. Stamens about sixty, with dark red anthers.

* * *Glands always present on the base of the leaf.*

11. *Populus monilifera*, Aiton. North America. See p. 1807.

Branchlets rounded. Leaves deltoid-ovate, about 3 in. wide, shallowly cordate or truncate at the base, cuspidate at the apex; serrations sinuate, with incurved tips, fewer and coarser than in the hybrids. Stigmas three or four. Stamens fifty to sixty.

12. *Populus angulata*, Aiton. North America. See p. 1810.

Branchlets angled. Leaves triangular-ovate, longer than broad, up to 7 in. long, and 5 in. broad, truncate or cordate at the base, acute or shortly acuminate at the apex. Stigmas four. Stamens fifty to sixty.

*** *Glands variable at the base of the leaf, absent or one or two in number.*

Hybrids between the European *P. nigra*, and one or other of the American *P. monilifera* and *P. angulata*. Leaves with irregular marginal cilia, which are often sparse, and usually deciduous in summer; serrations not coarse and sinuate.

* *Branchlets glabrous.*

13. *Populus serotina*, Hartig. See p. 1816.

Branches ascending, wide-spreading. Leaves unfolding latest of all the poplars and with a red tint, ovate-deltoid, about 3 in. wide, with a broad truncate base, and a short cuspidate or acuminate apex. A staminate tree; stamens twenty to twenty-five.

14. *Populus regenerata*, Schneider. See p. 1824.

Similar to *P. Eugenei* in habit, with foliage like that of *P. serotina*, but unfolding a fortnight earlier. A pistillate tree; stigmas usually only two.

15. *Populus Eugenei*, Simon-Louis. See p. 1826.

A narrow columnar tree, with ascending short branches. Leaves unfolding early with a reddish tint, smaller than those of *P. serotina*, broadly cuneate at the base, with a slender sharp-pointed non-serrated acuminate apex. A staminate tree; stamens about twenty.

16. *Populus marilandica*, Bosc. See p. 1828.

A tree with irregular branches. Leaves early in unfolding, rhomboid, cuneate at the base, tapering above into a long acuminate apex. A pistillate tree; stigmas variable, two, three, or four.

17. *Populus Henryana*, Dode. See p. 1829.

A tree with irregular branches. Leaves opening early, not tinged with red, ovate-triangular, cuneate at the broad base, and ending in a long acuminate apex. A staminate tree; stamens thirty to thirty-five. Flower buds pubescent.

* * *Branchlets pubescent.*

18. *Populus robusta*, Schneider. See p. 1829.

Branchlets grey. Leaves unfolding early with a deep red tint, variable in shape, ovate-deltoid or rhomboid; cuneate, rounded, or truncate at the base; acuminate at the apex; petioles pubescent. A staminate tree; stamens twenty.

19. *Populus Lloydii*, Henry. See p. 1830.

Branchlets yellow. Leaves similar to those of *P. robusta*, but smaller; petioles pubescent. A pistillate tree; stigmas usually two.

IV. TACAMAHACA, Spach, in *Ann. Sci. Nat.* xv. 32 (1841).

Balsam poplars.—Trees with rough furrowed bark, and fragrant foliage. Buds very viscid, exhaling a strong balsamic odour. Leaves whitish beneath, without a clearly defined translucent cartilaginous border; petioles rounded or quadrangular, channelled on their upper side.

*A. Branchlets rounded, without projecting ribs, except occasionally on vigorous shoots of young trees.*

* *Branchlets and petioles pubescent.*

20. *Populus candicans*, Aiton. North America. See p. 1834.

Leaves broad, ovate-deltoid, palminerved and cordate at the base, 5 to 6 in. long, 4 in. wide, ciliate in margin.

21. *Populus tristis*, Fischer. Himalayas (?). See p. 1840.

Leaves narrow, ovate, palminerved and cordate or rounded at the base, 4 in. long, 2 in. wide, ciliate in margin.

22. *Populus Maximowiczii*, Henry. Japan, Amurland, Manchuria, Korea. See p. 1838.

Leaves nearly orbicular, oval, or elliptic; palminerved and subcordate at the rounded base; 4 in. long, 3 to 3½ in. wide; densely ciliate in margin, and pubescent on the midrib, nerves, and veinlets of both surfaces.

23. *Populus suaveolens*, Fischer. Siberia, Mongolia. See p. 1841.

Leaves ovate or ovate-lanceolate, palminerved and rounded at the base, abruptly contracted into an acuminate apex, 3 to 3½ in. long, 1¼ to 2 in. broad; margin ciliate; glabrous on both surfaces.

* * *Branchlets and petioles glabrous.*

24. *Populus balsamifera*, Linnæus. North America. See p. 1832.

Leaves ovate, palminerved and rounded at the base, 4 to 5 in. long, 2 to 3 in. wide, sparsely and minutely ciliate in margin.

25. *Populus angustifolia*,[1] James. Rocky Mountains of North America. See p. 1831.

Leaves lanceolate, resembling those of *Salix fragilis* in shape, cuneate at the base, greenish beneath, 2 to 4 in. long, ¾ to 1 in. wide; lateral nerves, fifteen pairs, all pinnate.

*B. Branchlets with projecting linear ridges.*

* *Branchlets pubescent.*

26. *Populus laurifolia*, Ledebour. Altai Mountains. See p. 1842.

A wide-spreading tree. Leaves lanceolate, 3 to 5 in. long, 1 to 2 in. broad, rounded at the base, gradually tapering to an acuminate apex, finely and regularly serrate.

27. *Populus berolinensis*,[2] Dippel. A hybrid. See p. 1844.

A narrow columnar tree. Leaves ovate or ovate-rhombic, 3 in. long, 2 in. wide, rounded or cuneate at the base, contracted at the apex into an acuminate point, crenately and occasionally irregularly serrate.

* * *Branchlets glabrous.*

28. *Populus Wobstii*, Schrœder. A hybrid. See p. 1843.

Branchlets only slightly ribbed in the first year, the ribs more apparent in the second year. Leaves lanceolate, tapering to a narrow rounded base, 4 to 6 in long, 2 in. broad; ciliate in margin; nerves all pinnate.

29. *Populus Simonii*, Carrière. North China. See p. 1839.

Leaves rhombic-elliptic, cuneate and pinnately nerved at the base, contracted at the apex into a short cuspidate point, 3 in. long, 1¾ in. broad.

30. *Populus trichocarpa*, Torrey and Gray. Western North America. See p. 1836.

Young stems with bark peeling off in papery shreds. Leaves ovate or ovate-deltoid, palminerved and rounded or subcordate at the base; 5 in. long, 3 in. broad; whitest beneath of all the balsam poplars.

V. Leucoides, Spach, in *Ann. Sci. Nat.* xv. 30 (1841).

Trees with rough bark, breaking into loosely attached plates. Leaves very large, cordate, simply serrate, covered when unfolding with greyish tomentum, which speedily disappears except on the nerves beneath; petioles rounded, much shorter than the blades, which do not flutter in the wind. Buds pubescent, viscid. These poplars are not easily propagated by cuttings.

31. *Populus lasiocarpa*, Oliver. Central China. See p. 1846.

Leaves ovate-cordate, longer than wide, about 9 in. long, and 6 in. broad; green beneath, with uniform serrations, extending regularly to the sinus at the base.

32. *Populus heterophylla*, Linnæus. United States, from the Mississippi valley eastwards to the Atlantic coast. See p. 1846, note 2.

[1] This species is very distinct from the other balsam poplars, in the willow-like leaves, which are not whitish beneath.

[2] This hybrid differs from the balsam poplars in the very narrow translucent border on the leaves, which are less fragrant and greenish, or only slightly whitish beneath. *P. rasumowskyana* and *P. petrowskyana* are similar hybrids, of which I have not seen complete material. See pp. 1843, 1844.

Leaves ovate-cordate, nearly as broad as long, 4 to 7 in. long, 3 to 6 in. wide; pale beneath, with regular serrations except at the base, where they are few and wide apart. (A. H.)

### The Raising of Poplars from Seed

No good description or illustration of the germination of poplars seems to have been published in England before that of Miss Florence Woolward[1] in 1907.

I have never found in England a poplar grown from seed either naturally or by nurserymen; and though Grigor[2] describes the process, it seems doubtful whether he ever practised it; and all my own attempts to raise poplars from seed were fruitless, until I followed nature as closely as possible. Having observed that poplars only germinated freely on the sandy banks of rivers, I sowed seed as soon as ripe on a pot of sand, and placed the pot in a pan of water in the full sun. Germination was then extremely rapid, according to Miss Woolward, in ten hours to two days. I found that after four days no more seeds came up, and that in all cases the proportion of germinating seed was quite small. In five or six days the cotyledons are well developed, but the growth of the young plants is very slow and does not exceed 2 to 5 inches in height in the first season. Some one-year seedlings of *P. nigra*, sent me from the banks of the Allier in central France, were only 3 to 4 inches above ground, though their thick rather fleshy root was 6 to 10 inches long. Some seedlings of *P. monilifera* from the banks of the St. Lawrence in Canada, collected by Mr. Jack, were equally small; as were some of *P. nigra* raised by Mr. Hankins from seed collected at Bury St. Edmunds.

Miss Woolward and I also raised *P. canescens* from seed collected at Upcot near Colesborne, where both sexes of this tree grow on my own property; and she raised *P. marilandica* from seed of a tree in Kew Gardens. In both cases the seedlings were much less vigorous than cuttings from the same trees.

I also raised seeds collected under a tree, which I believe to be *P. monilifera*, of American origin, in the botanic garden at Padua, which was perhaps fertilized by a *P. alba*, the only poplar growing near it; but none of those seedlings have shown the least sign of hybrid origin, and have grown slowly and seem tender as compared with poplars raised from cuttings.

I am therefore convinced that, though we may succeed, by crossing different species, in obtaining new races of superior vigour, like *P. Eugenei* and *P. robusta*, the raising of poplars from seed is not a practice which can be recommended for general purposes. Some species of poplar are said to be difficult or impossible to strike from cuttings, among which Grigor includes *P. alba* and *P. canescens*, but I have found no difficulty in the case of the latter. In the case of species which will not strike, recourse must be had to layers, root-suckers, or grafting. (H. J. E.)

[1] *Journal of Botany*, xlv. 417, t. 487 (1907).

[2] *Arboriculture*, 328 (1881). Hickel, in *Bull. Soc. Dend. France*, No. 25, p. 88 (1912), gives directions for raising poplars from seed, which he has followed with success at Versailles.

## POPULUS ALBA, White Poplar

*Populus alba*, Linnæus, *Sp. Pl.* 1034 (1753); Loudon, *Arb. et Frut. Brit.* iii. 1638 (1838) (in part); Willkomm, *Forstl. Fl.* 516 (1887); Mathieu, *Flore Forestière*, 483 (1897); Schneider, *Laubholzkunde*, i. 21 (1904); Ascherson and Graebner, *Syn. Mitteleurop. Fl.* iv. 17 (1908); Boissier, *Flora Orient.* iv. 1193 (1879); Hooker, *Fl. Brit. India*, v. 638 (1888); Aigret, in *Ann. Trav. Publ. Belg.* x. 1213 (1905); Gombocz, in *Math. Termes. Közl.* xxx. 141 (1911).

*Populus major*, Miller, *Dict.* ed. viii. No. 4 (1768).

*Populus nivea*, Willdenow, *Berl. Baumz.* 227 (1796).

A tree, similar in size and bark to *P. canescens.* Young branchlets and buds densely white tomentose.[1] Leaves (Plate 408, Fig. 1), very variable in size and shape, dependent upon the position and vigour of the branchlets and on the age of the tree; on vigorous long shoots and suckers, large, up to 4 or 5 in. in diameter, ovate, palmately three- to five-lobed; lobes triangular, with deep sinuses; margin ciliate with minute glandular teeth; subcordate at the rounded or broad base, acute at the apex; dark green and slightly tomentose above; covered beneath with a dense snowy white tomentum; petiole rounded, tomentose. Leaves on short shoots (and occasionally at the base of the long shoots), elongated oval, rounded at the base, with a few sinuate teeth, usually covered beneath with a thin greyish tomentum; in some cases (var. *denudata,* Wesmael) glabrescent; in some forms, broad and suborbicular like the similarly placed leaves of *P. canescens*, but uniformly tomentose beneath, and with fewer teeth in the margin; petiole tomentose.

Pistillate catkins similar to those of *P. canescens*, but more silvery in colour; scales obovate to lanceolate, with minute teeth, and fringed with long hairs; stigmatic lobes four, green, erect, long and linear, each of the two pairs united at the base. Staminate catkins[2] (probably not perfectly developed), 1 inch long; axis tomentose; scales concave, spatulate, dentate, fringed with long hairs; disc pubescent; stamens six.

Spread over an immense area, *P. alba* is very variable in the form of the foliage, and might be divided into several distinct geographical races; but the wild specimens in herbaria are scanty and incomplete; and only a few important varieties will be here mentioned. *P. canescens*, a native English tree, and *P. tomentosa*, wild in North China, are easily distinguishable; and are treated here as distinct species.

1. Var. *nivea*, Aiton, *Hort. Kew.* iii. 405 (1789).

This is the typical form[3] of the species described above. It has been long in cultivation in western Europe, where it is known in gardens as var. *argentea*, var. *acerifolia*, var. *arembergiana*, etc. It appears to be indigenous in eastern and south-eastern Europe, in the Caucasus, in Persia, Turkestan, Afghanistan, and the Altai Mountains, where specimens have been gathered by travellers. In Kashmir, according to Mr. Lovegrove, of the Forest Department, from whom I have received specimens, the typical form is fairly common in the wild state, attaining on an average 90 ft. in height and 8 ft. in girth.

In cultivated trees, the leaves are extraordinarily variable in size and shape, but it

[1] The suckers of *P. alba* are described by Dubard, in *Ann. Sci. Nat.* xvii. 163 (1903).

[2] The only male flowers of *P. alba* which I have been able to examine were produced by a small tree at Kew, labelled var. *nivea*, and did not seem to be properly developed. Most, if not all, of the white poplars in Britain are female trees.

[3] Gombocz, *op. cit.* 148, 149, recognizes as forms of var. *nivea*, the slight variations, which are doubtfully constant, called by Dode, *op. cit.* 21, 22, *P. triloba*, *P. Treyviana*, *P. Paletskyana*, *P. heteroloba*, *P. Morisetiana*, and *P. palmata*.

is doubtful if these constitute distinct varieties; on one tree I have observed ordinary branches with small leaves, varying in shape from oval and slightly toothed to trilobed, and more vigorous shoots showing large palmately lobed leaves. Old trees put forth feeble long shoots, on which the terminal white leaves are often imperfectly developed.

2. Var. *subintegerrima*, Lange, in Willkomm and Lange, *Prod. Fl. Hisp.* i. 233 (1861).

Var. *integrifolia*, Ball, in *Journ. Linn. Soc.* (*Bot.*) xvi. 668 (1878).
*Populus subintegerrima*, Dode, *op. cit.* 20 (1905).
*Populus monticola*, Brandegee, in *Zoë*, i. 274 (1890); Sargent, in *Garden and Forest*, iv. 330, fig. 56 (1891), vi. 190 (1893), and vii. 313, fig. 51 (1894).
*Populus Brandegeei*, Schneider, *Laubholzkunde*, i. 23 note, and 803 (1906).

Leaves coriaceous, often sub-evergreen; on the long shoots ovate or orbicular, sub-cordate or cuneate at the base, sub-entire or irregularly and slightly toothed, white beneath; on the short shoots almost orbicular, slightly sinuate or quite entire, grey beneath.

A native of southern Spain, Algeria,[1] and Morocco, where it was gathered in the greater Atlas by Hooker. It occurs also in the Canaries and the Azores. It appears to have been introduced into Mexico by the Spaniards, and has been found, apparently naturalised, along streams in the high mountains of Lower California, where it is called *guerigo* by the inhabitants. According to Sargent,[2] who speaks of it as a distinct native American species, its wood in this locality is quite unlike that of the other poplars, being light red, satiny, and useful for making furniture.

Similar forms[3] occur near Askabad, east of the Caspian Sea, and in Kashmir.

3. Var. *pyramidalis*,[4] Bunge, in *Rel. Bot. Mém. Ac. St. Pétersb.* vii. 498 (1851).

Var. *Bolleana*, Lauche, *ex* Huttig, in *Deut. Gart.* 500 (1878); Masters, in *Gard. Chron.* x. 502 (1878) and xviii. 556 (1882).
*Populus Bolleana*, Carrière, in *Rev. Hort.* liii. 40, 123 (1881), and lxiii. 188, fig. 48 (1891); Masters, in *Gard. Chron.* xviii. fig. 96 (1882).

Resembling the Lombardy poplar in habit. Leaves on the long shoots palmately 3- to 5-lobed, very white tomentose beneath; on the short shoots orbicular, with coarse triangular teeth, green beneath with traces of tomentum.

The fastigiate form of the white poplar was first described by Bunge from specimens found by Lehmann in 1841, apparently wild[5] on the bank of a stream on the north side of the Karatau mountain, between Bokhara and Samarkand. It appears to have been introduced[6] in 1872 into Hoser's nursery at Warsaw, from a cutting sent by Col. Korolkow.[7] Lauche procured it for the Horticultural Society of Potsdam from the same source in 1875.

[1] Collected near Ronda in Spain, by M. P. Price, and near Affreville in Algeria by A. Henry. A specimen from Gibraltar, in the Cambridge Herbarium, has slender female catkins, 4 to 5 in. long, with a woolly axis and pedicellate flowers; scales long, concave, irregularly toothed or lacerate; styles 4, spreading.

[2] In *Garden and Forest*, vi. 190 (1893).

[3] Specimens in the Kew Herbarium.

[4] *P. alba*, var. *croatica*, Wesmael, in De Candolle, *Prod.* xvi. 2, p. 324 (1868) is an erroneous name founded on *P. croatica*, Waldstein and Kitaibel, in *Flora*, xv. 2 Beil. p. 14 (1832), which is a narrow form of *P. nigra*. Cf. Koch, *Dendrologie*, ii. pt. i. p. 489 (1872).

[5] Lehmann also saw it planted at Bokhara, where he gathered a leafless branch bearing staminate flowers, on 5th March 1842.

[6] Cf. *The Garden*, 10th Dec. 1887, p. 543.

[7] According to Carrière, *Rev. Hort.* lxiii. 188 (1891), it was introduced by Col. Korolkow, under the name of *P. alba pyramidalis* into various places in France in 1878, as Orleans, Segrez, Angers; and was first sold by Simon-Louis in 1879-1880. E. Morren, in *Belg. Hort.* 1879, p. 269, says that it was sent to Späth from Tiflis by Scharrer in 1879.

The oldest tree in England is probably a fine specimen at Kew, growing near the large Ginkgo tree, which measured 67 ft. by 4 ft. 8 in. in 1910. Another at Terling Place, Essex, planted in 1886, measured, in 1910, 60 ft. high by 3 ft. 3 in. in girth. A good specimen at Over Bridge, near Gloucester, was 55 ft. high in 1911. The largest tree we know of in Germany is one at Späth's nursery, which in 1908 was 66 ft. by 4 ft. It was first sent out from here in 1879.

4. Var. *globosa*, Späth, *Cat.*, No. 66, p. 3 (1886); Dippel, *Laubholz.* ii. 191 (1892).

A small tree or shrub, of dense habit, oval in outline, with small slightly lobed deltoid cordate leaves, grayish beneath, and said to be pink in the young state. This has been propagated by Späth of Berlin, since 1886, when it originated in their nursery from a cutting of *P. alba*; but is little known in England.

5. Var. *Richardi.* Leaves yellow on the upper surface, the colour lasting throughout the season. This was shown at the International Horticultural Exhibition of 1912 by Richard, nurseryman at Naarden-bussum, near Amsterdam.

6. Var. *pendula*, Loudon, *Arb. et Frut. Brit.* iii. 1640 (1838). Branchlets pendulous. This is mentioned by Loudon as a continental variety not introduced in 1838. It has lately been noticed[1] at Breslau; but we have seen no specimens.

### Distribution

Both the grey and the white poplars have been so widely planted in Europe that their exact distribution in the wild state is very uncertain. According to Willkomm, *P. alba* is certainly not a native of northern Germany, or of Scandinavia, though it thrives as a planted tree in Norway as far north as lat. 68°. Apparently *P. canescens* is a native of the British Isles, central and northern France, and of the Rhine valley in Baden and Alsace; while typical *P. alba* is more southerly and easterly in its distribution, occurring throughout the Mediterranean region in Spain and Portugal, Italy, the Balkan States, Greece, Algeria, Morocco, Syria, and Asia Minor, and extending northwards into the Alps and Carpathian ranges, being a characteristic tree of the alluvial flats of the Danube and all its tributaries from Bavaria, throughout Austria and Hungary, to its mouth in the Black Sea. Farther east, *P. alba* occurs in Russia, the Caucasus, western Siberia, the north-western Himalayas, and western Tibet. It is usually a native of low-lying moist woods, especially those fringing the banks of rivers; but it ascends in the valleys of the Alps, and in Spain and Italy to about 2000 feet elevation.

The white poplar is certainly not indigenous in England, and records of it as a wild tree all probably refer to *P. canescens*, which often shows a considerable amount of white foliage on the long shoots on young and vigorous trees. Turner,[2] in 1568, says, "As touching the whyte aspe, I remember not that I ever saw it in any place in England," and he appears to have known the tree, as he says it was plentiful in Germany and Italy. The date of the introduction of *P. alba* is uncertain.[3] The

[1] Behnsch, in *Mitt. Deut. Dend. Ges.* 1906, p. 212.

[2] *Herball*, 99 (1568).

[3] Hartlib, *Complete Husbandman* (1659), *ex* Loudon, *op. cit.* 1641, states that some years before the time of his writing, 10,000 abeles were sent into England from Flanders, and transplanted into many counties. These trees were probably *P. canescens*.

white poplar is often called *abele* in England, but this is the Dutch name, *abeel*,[1] of *P. canescens*. The white poplar is known in Dutch and Belgian nurseries as *peuplier blanc de Hollande*.

The true white poplar has never become a common tree in England. Most of the large trees known as white poplars that we have seen are *P. canescens*, and it is impossible to separate the two species in Loudon's account of them. There are three good trees of *P. alba* at Bayfordbury, Hertford, the largest of which was 95 ft. by 10 ft. 4 in. in 1911. Mr. J. E. Little of Crofton, Hitchin, who has kindly sent specimens, tells me that between Norton Mill and Radwell Mill, near Baldock, Herts, on the bank of the river Ivel, there are four large white poplars, about 80 feet in height and 12 to 13 feet in girth. At Syston Park, Grantham, Miss F. H. Woolward, in 1905, measured a tree 105 ft. by 10 ft. 10 in. In Ireland the finest are at Adare Manor, where Mr. R. A. Phillips measured a tree 80 ft. by 9 ft. in 1910. Another at Nenagh was 60 ft. by 5 ft. in 1911. All these trees bear female flowers.

Pynaert[2] saw in 1882, at Troyes in France, a true white poplar which measured 140 ft. in height and 21 ft. 3 in. in girth, at six and a half feet above the ground. It divided at thirty-one feet up into three main stems, the largest of which girthed 14 ft. 9 in.; while the total spread of branches was 260 ft. in circumference. This magnificent tree, which was supposed to be about 400 years old, was destroyed[3] by a storm on 1st February, 1902. I saw, in 1912, a fine tree in the botanic garden at Toulouse, 90 ft. high by 8 ft. in girth. (A. H.)

## POPULUS CANESCENS, Grey Poplar

*Populus canescens*, Smith, *Fl. Brit.* iii. 1080 (1805), and *Eng. Flora*, 245 (1828); Smith and Sowerby, *Eng. Bot.* xxiii. t. 1619 (1806), and vii. 114, t. 1392 (1840); Loudon, *Arb. et Frut. Brit.* iii. 1639 (1838); Bromfield, in *Phytologist*, iii. 841 (1850); Schneider, *Laubholzkunde*, i. 23 (1904); Dode, in *Mém. Soc. Hist. Nat. Autun*, xviii. 26 (1905); Aigret, in *Ann. Trav. Publ. Belg.* x. 1214 (1905).

*Populus alba*, Linnæus, var. *β*, *foliis minoribus*, Lamarck, *Fl. Franc.* ii. 235 (1778).

*Populus alba*, Linnæus, var. *canescens*, Aiton, *Hort. Kew.* iii. 405 (1789).

*Populus alba*, Willdenow, *Berl. Baumz.* 227 (1796) (not Linnæus); Hunter, in Evelyn, *Silva*, i. 208, plate (1801); Smith, *Eng. Bot.* t. 1618 (excl. description) (1806).

*Populus alba*, Linnæus, var. *genuina*, Wesmael, in De Candolle, *Prod.* xvi. 2, p. 324 (1864).

*Populus alba*, Linnæus, var. *typica*, Gombocz, in *Math. Termes. Közl.* xxx. 151 (1911).

*Populus megaleuce* and *alba*, Dode, *op. cit.* 24, 25 (1905).

*Populus hybrida*, Reichenbach, *Icon. Fl. Germ.* xi. 29, t. 615 (1849) (not Bieberstein[4]).

*Populus Steiniana*, Bornmüller, in *Gartenfl.* xxxvii. 173, figs. 37, 38 (1888).

*Populus Bachofenii*, Reichenbach, *Icon. Fl. Germ.* xi. 29, t. 616 (1849) (not Wierzbicki[5]).

---

[1] Murray, *New Eng. Dict.* i. 15 (1888), says that the Dutch name corresponds to an old French word, *abel*, derived from late Latin, *albellus*, a name given to the white poplar in the twelfth century. Worlidge, *Syst. Agric.* 96 (1681), speaks of the abele tree as "a finer kind of white poplar."

[2] *Bull. Arbor. Belg.* 1882, p. 190.

[3] *Ibid.* 1902, p. 72.

[4] Bieberstein, *Fl. Taur. Cauc.* ii. 422 (1808), describes *P. hybrida*, a doubtful plant which, *op. cit.* iii. 633 (1819), he abandons, as not being different from *P. alba*. A specimen, however, in the Cambridge Herbarium, labelled "*P. hybrida*, M. B., Caucasus," is quite distinct from *P. alba*, and appears to be a form of *P. canescens* with orbicular leaves, sinuately toothed, and glabrous beneath. It bears fruiting catkins 4 in. long.

[5] *P. Bachofenii*, Wierzbicki, in Rochel, *Banat. Reise*, 77 (1838), is said by Gombocz, *op. cit.* 148, to be a form of *P. alba*, var. *nivea*, and identical with *P. heteroloba*, Dode, authentic specimens of which I cannot distinguish from typical *P. alba*, var. *nivea*.

A large tree, attaining 100 ft. or more in height and 15 ft. in girth. Bark on young stems thin, smooth, grey or whitish; on older stems breaking on the surface into small roughened dark-coloured rhombic cavities, which ultimately unite together, making the bark towards the base of old trunks deep and longitudinally furrowed. Young branchlets, towards the top of the long shoots, covered with a dense whitish tomentum, which towards their base and on the short shoots diminishes in quantity, the twigs becoming dull grey or shining and glabrescent. Buds ovoid, more or less tomentose, according to their position on the branchlet. Leaves (Plate 408, Fig. 3) of two kinds—on long shoots and on suckers ovate-deltoid, cordate, acute, dark shining green and slightly tomentose above, covered beneath with a thick greyish tomentum; margin ciliate, with a few triangular teeth, variable in size, and like the rest of the margin irregularly glandular-serrate; petiole rounded, tomentose. Leaves on the short shoots, suborbicular or broadly ovate, subcordate, obtuse; margin non-ciliate, with a narrow translucent border, and a few sinuate non-serrate teeth; dark green, shining and glabrescent above; lower surface light green, with traces of scattered grey tomentum; petiole laterally compressed, glabrescent or slightly tomentose.

Staminate catkins,[1] 2½ to 4 in. long; axis tomentose; scales obovate, toothed, yellowish brown, fringed with hairs at the apex; pedicels pilose; disc oblique, entire in margin, glabrous; stamens eight to fifteen. Pistillate catkins about 1 in. long; with similar scales; disc pubescent; ovary glabrous, with four spreading sub-sessile yellowish simple stigmatic lobes.[2] Fruiting catkins, 3 to 4 in. long; capsules glabrous, two-valved. The stigmatic lobes like those of *P. alba*, are simple and undivided; whereas in *P. tremula*, each lobe is deeply scolloped and waved.

*Populus Bogueana*, Dode, *op. cit.* 24 (1905), is a vigorous form of *P. canescens*, in which the leaves on the long shoots are very large, 5 in. or more in length and breadth; and appears to be now sold by some nurserymen as *P. tomentosa*, the white poplar from Peking. A tree of this at Kew, obtained from Simon-Louis in 1904, is now about 25 ft. high. Another at Grayswood, obtained from Barbier in 1906, is 15 ft. high and very thriving. There are also small trees with similar large foliage in the Botanic Gardens at Edinburgh and Glasnevin. I found this form wild in the forest of Orleans, where it was growing evidently from vigorous suckers of the typical grey poplar beside it.

By *P. canescens*,[3] Smith meant the common grey poplar in England, which I have described above. It differs mainly from typical *P. alba* in the lesser amount of

[1] A tree in the Cambridge Botanic Garden bore monœcious catkins in 1907, the staminate flowers at the base, the pistillate towards the apex. Similar cases are recorded by Penzig, *Pfl. Teratologie*, ii. 321 (1894), and by Baillon, in *Bull. Soc. Linn. Paris*, 1897, p. 658.

[2] The eight stigmas, figured by Smith, *Eng. Bot.* xxiii. t. 1619 (1806), have not been observed by any other botanist, and were probably abnormal. In *Eng. Bot.* vii. 114, t. 1392 (1840), the number of stigmas is said to be inconstant, but I have found them invariably four.

[3] The grey and white poplars were first distinguished by Lobelius, *Icon. Pl. seu Stirp.* ii. 193 (1591). Their synonymy is much involved, but has been clearly elucidated in an interesting article by Tidestrom, in the *American Midland Naturalist*, i. 113 (1909), to which the reader interested in such matters is referred. I am not at all certain that Smith clearly understood the distinctions between the two species, as Sowerby's plate of *P. alba*, t. 1618 (1806), is the leaf of a vigorous shoot of *P. canescens*; and Smith's statement that *P. alba* is not uncommon in moist woods (in England) is erroneous, as the latter is never seen in England, except as a planted tree. Most of his description refers to *P. alba*; and his specimen in the herbarium of the Linnean Society, London, certainly belongs to this species.

the tomentum on the branchlets and leaves, the latter having a different shape. It is adapted to the humid and mild climate of western Europe; while *P. alba*, with its dense protective covering against evaporation of water, is a native of drier and more continental regions.

In the grey poplar, as in the white poplar, the leaves are fundamentally of two kinds, different in shape, margin, tomentum, and petiole; but conveniently distinguished as "white leaves" and "green leaves." Some shoots bear only white leaves, some only green leaves; whilst others have green leaves towards the base and white leaves towards the tip. The green leaves may be greyish beneath or nearly devoid of tomentum. The size of the white leaves is inconstant, and depends on the vigour of the branchlets. In old trees the green leaves are preponderant, especially on the lower branches; and such trees are popularly known as *P. canescens.* In younger trees, and on the upper branches of older trees, the white leaves are conspicuous; and such trees are often erroneously called *P. alba.* In addition to this variation, occurring on the same tree in its different stages of growth, there are differences in the foliage of different trees, which require further study.

A peculiar form, in which the leaves are thinner in texture, orbicular and not deltoid in outline, and nearly glabrous beneath, is considered by Dr. C. E. Moss to be possibly the hybrid *P. canescens* × *P. tremula.* Two pistillate trees[1] with this foliage, one at Hitchin, and another growing beside a stream between Caverham and Icklingham, Suffolk, have pink stigmas, the colour of which suggests the influence of *P. tremula.* These two trees flower a fortnight earlier than ordinary *P. canescens.* Staminate trees with similar foliage have been found both in England and Ireland. Mr. R. A. Phillips, who sends specimens[2] from Lorrha, Tipperary, points out that the catkins are shorter than in the typical grey poplar, about 1½ in. long, with fewer (4 to 7) stamens, and narrower scales. There is also a difference in habit, the round-leaved trees have thick stiff ascending branches and short erect twigs; while the ordinary form has spreading branches and pendulous twigs.

In *P. tremula*[3] the foliage is all of one kind, similar in form to the green leaves of *P. canescens*, but devoid of any trace of tomentum. The theory that typical *P. canescens* is a hybrid[4] between *P. alba* and *P. tremula* rests on this apparent similarity of foliage, and is not supported by evidence from the floral organs. Female trees of the grey poplar are, however, rare, and the pistillate flowers have only been examined in a few cases. So far as I can judge, *P. canescens* is a good species, being the

[1] The pistillate tree with pink stigmas is *P. canescens*, Dode, in *Mém. Soc. Hist. Nat. Autun.* xviii. 26 (1908) (not Smith).

[2] Mr. Phillips recently sent me another form with leaves intermediate in shape and with flowers having eight stamens.

[3] The suckers of *P. tremula* bear leaves which, in their shape and in the presence of a slight tomentum, simulate the white leaves of *P. canescens.*

[4] The older continental botanists, as Reichenbach, *Icon. Fl. Germ.* xi. 30, t. 617 (1849), and Hartig, *Naturgesch. Forst. Culturpfl.* 434 (1851), considered *P. canescens*, Smith, to be identical with the pubescent form of the aspen, *P. tremula*, var. *villosa.* This has silky hairs on the leaves and branchlets, very different from the tomentum of *P. canescens.* The occurrence of true hybrids between *P. tremula* and *P. alba* is possible on the Continent. Radde, *Pflangenverb. Kaukas.* 153 (1899), saw a group of trees near Sotschi on the eastern shore of the Black Sea, which he considered to be of this origin, as the trees strongly resembled *P. tremula.* Rechinger, in *Verh. Zool. Bot. Ges. Wien*, xlix. 284 (1899), discusses the hybrids in Austria. Adamovic, *Veg. Balkanländ.* 145 (1909), reports similar hybrids in eastern Roumelia. Cf. also *P. albo-tremula*, Krause, in *Jahrb. Schles. Ges.* 1848, p. 130.

western representative in Europe of the white poplar. It appears to be common in France, extending eastward to the Rhine valley; and is known in Belgium and France as *franc-picard* or *grisard*. It often attains an immense size in France, the finest that I have seen being one in the Botanic Garden at Toulouse, which was 100 ft. in height and 12 ft. in girth in 1912. In Holland, where it is wild on the dunes near Haarlem,[1] it is always called *abeel*, a name which is often erroneously applied to the white poplar by English writers. The grey poplar is undoubtedly a native[2] of England, and is the tree referred to by old authors as the white poplar; but has been probably always better known to woodmen and peasants as the aspe.

(A. H.)

Both the grey and white poplars were known to Evelyn,[3] who spoke of the first as the white "to be raised in abundance by every set or slip. Fence the ground as far as any old poplar roots extend and they will furnish you with suckers innumerable, to be slipped from their mothers and transplanted the very first year, but if you cut down an old tree you shall need no other nursery." Later on, he says: "There is something a finer sort of white poplar, which the Dutch call abele, and we have late much of it transported out of Holland. They are also best propagated of slips from the roots, the least of which will take, and may in March, at three or four years' growth, be transplanted." The latter was still an uncommon tree in Plot's time, for he says:[4] "Of unusual trees now cultivated in Oxfordshire is the abele tree, advantageously propagated by Sir J. Croke of Waterstock, by cutting stakes out of the more substantial parts of the wood; which put into moist ground grew more freely than willows, coming in three or four years' time to an incredible height."

French and English authors agree that the white poplar will not bear lopping like the black poplars; and though I have no experience in this matter, yet as I have never seen a pollarded tree I presume that the grey poplar is equally liable to injury when large branches are cut. It seems able to attain its largest dimensions in poor stiff soil, and in cold situations, provided that there is sufficient moisture in summer; and though I cannot say that the tree is equal to the black Italian poplar from an economic point of view, or equal to *P. alba* as an ornamental tree, yet as the illustration shows, it is a stately tree when well grown.

The range of this tree in Great Britain is obscure, because it has been planted for a long period.[5] Watson[6] says, "It is given as an unquestioned native in the floras of Surrey, Essex, Herts, Suffolk, Norfolk, Cambridge. In the floras of Tyne and North Yorkshire it is reduced to the grade of denizen. And if I rightly know the distinction between *alba* and *canescens*, I should now deem the latter native, and the former planted, as seen in Surrey and elsewhere."

Though the male tree is found of large size in many counties, the female is in

[1] Mr. L. Springer, a distinguished landscape gardener of Haarlem, says that the true white poplar will not endure the sea wind on the coast of Holland.

[2] Clement Reid, *Origin of Brit. Flora*, 150 (1899), states that leaves collected by Prestwich in interglacial beds at Greys, Essex, suggest *P. canescens*, though they may belong to *P. tremula*. The latter species is only recorded from one locality, Caerwys, Flintshire, in neolithic deposits; so that the geological evidence as to the existence of poplars in England is very scanty.

[3] *Sylva*, 78 (1679).

[4] *Natural History of Oxfordshire*, 175 (1705).

[5] Loudon quotes M'Culloch, that it is the only tree found in the island of Lewis, but probably he mistook the aspen for it.

[6] *Topographical Botany*, i. p. 357 (1873).

my experience very rare, the only ones I have seen being two rather stunted specimens of considerable age, which grow on my land by the side of a watercourse in a meadow on Upcot Farm, near Withington, Gloucestershire. There are male specimens close to them, and in some seasons fertile seed is produced, from which, in 1907, a few seedlings[1] were raised by Miss F. Woolward. Two of these were planted at Colesborne, but one was weakly from the first and the other was killed by a fungoid disease soon after I planted it out. I have never seen a natural seedling of this or of the white poplar in England.

## Remarkable Trees

The finest grey poplars that we have seen are in a group called the Grove in the park at Longleat. Mitchell,[2] writing in 1827, mentions these as being then 100 ft. high, with trunks 3 to 4 ft. in diameter, and 40 to 60 ft. of clear bole. In *Trans. Eng. Arbor. Soc.* v. 391 (1903), an illustration of one of these trees from a photograph by A. C. Forbes was given. He calls it an abele poplar, and gives the measurement of the largest as 120 ft. by 15 ft. 3 in., and the cubic contents 450 ft. Ten trees in this grove, all with straight clean boles of 50 to 80 ft., then averaged 115 ft. by 13 ft., with an average cubic content of 240 ft. When I last visited Longleat in 1909 I found the majority of them still healthy, and measured one which was 125 ft. by 11 ft.

There are several very fine trees at Colesborne, one of which (Plate 382) has two stems rising from the base, and measuring 115 ft. by 10 ft. and 10 ft. 3 in. respectively. These trees sucker freely, and some young trees grown from the suckers, which have been pruned but not transplanted, are growing very fast —about 40 ft. in ten years.

In a field about a mile from Overbury Court, near Tewkesbury, Mr. F. R. S. Balfour discovered a tree 86 ft. high, with a trunk 18 ft. long by 18 ft. 10 in. in girth and with very large spreading branches. At Kingston Lacy, Dorsetshire, a tree with many suckers round it measured, in 1906, 118 ft. by 13 ft. 9 in. At Strathfieldsaye I measured in 1905 a tree 108 ft. by 16 ft., which had a sucker coming up 80 yards from its base. In the park at Syon House there is a well-shaped tree 101 ft. by 12 ft. 5 in.[3] In the home park of Windsor Castle there is a short avenue of grey poplars close to the bank of the Thames, below Victoria Bridge. Mr. Mackellar informs us that these were planted between 1840 and 1850, and measured in January, 1913, 90 to 100 ft. in height and 9 to 12 ft. in girth. At Gilbert White's old home at Selborne, the tallest tree is a grey poplar, which Mr. H. B. Watt[4] measured in 1912 as 109 ft. high by 9 ft. 5 in. in girth. At Youngsbury, Ware, Hertford, a tree measured 95 ft. by 10 ft. 4 in. in 1911.

In Wales I have seen none of any great size; and I have not heard of any remarkable ones in the eastern or northern counties.

In Scotland, where the tree is common, Mr. J. Renwick[5] tells us of two very

[1] *Journ. Bot.* xlv. 417, t. 487 (1907). [2] *Dendrologia*, 51 (1827). [3] A. B. Jackson, *Cat. Trees Syon*, 22 (1910). [4] *Selborne Magazine*, xxiii. 122 (1912). [5] *Glasgow Naturalist*, iii. 119, pl. ii. (1911).

large trees at Mauldslie Castle, Lanarkshire, the seat of Lord Newlands, which in 1911 measured 100 ft. by 21 ft. 3 in. at two feet from the ground, and 117 ft. by 16 ft. 5 in. at two feet nine inches from the ground. Both these trees suffered severely in the great gale of October 5, 1911, the larger tree losing one of its heavy limbs. Walker, *Essays of Natural History*, 49 (1812), states: "In the year 1769 there was a row of abeles at Stevenston, East Lothian, which was soon after cut down. It contained 122 trees, all about 80 ft. high and from 20 to 30 ft. of clear trunk, without a branch. The trunks were 5 to 7 ft. in circumference, and yet they stood only 7 ft. distant from each other. They were planted in a deep moist soil, were then eighty years old, and afforded a great quantity of timber. It is doubted whether or not the abele is a native of England. It certainly has the appearance of being an indigenous tree in several parts of Scotland. It was planted in many places about the end of the last and in the beginning of the present century, but it has since been neglected." The tallest recorded in the *Old and Remarkable Trees of Scotland* was at Glenarbuck in Dumbartonshire, and was said to be 110 ft. by 12 ft. in 1867.

In Ireland the grey poplar is found on the banks of the Suir, Nore, Barrow, and other rivers in the south, where it may very possibly be wild; and Mr. R. A. Phillips has sent us specimens of several trees, one of which, growing near Birr, was 90 ft. by 12 ft. in 1910. I measured a tree at Markree Castle, Sligo, which in 1909 was 120 ft. by 13 ft. There are two fine trees at Abbeyleix, one of which was 100 ft. by 13 ft. 4 in. in 1910.

In Belgium and France this tree is known as *grisard*, and is often mistaken for the white poplar. I have not noticed any of remarkable size.[1]

### Timber

I am inclined to think that the timber is of superior quality to that of the black Italian poplar; and from what I heard in Belgium and France,[2] that opinion is also held there. But this may depend on the age of the trees and the rate of growth, which, according to Crowe,[3] who paid much attention to this tree in Norfolk, is slower than that of any other British poplar. It is rarely distinguished by the timber merchants in England. Messrs. Howes and Sons, Norwich, however, inform me that "it is the only poplar found to be of any service in the coach-building trade. It is light in weight, exceedingly strong, and can be cleaned up into a nice finish to receive the paint." They pay for the best quality of the wood of this tree, which they know under the name of abele, about 2s. 6d. per foot, as compared with about 1s. 3d. for the wood of the black Italian poplar, which is much less durable and not so firm in texture. I prefer to use it at home for cart bottoms and linings, and for flooring cottage bedrooms, for which its toughness and non-inflammable nature make it valuable. (H. J. E.)

[1] In *Mitt. D. Dendr. Ges.* 1904, p. 18, a tree was recorded at Schloss Dyck, near Dusseldorf, which, at about 95 years old, was 48 metres high by 3½ metres in girth. The height is evidently much exaggerated.

[2] Mouillefert, *Essences Forestières*, 301 (1903), says that its fine-grained wood is one of the best of the poplars.

[3] Smith, *Eng. Flora*, iv. 244 (1828), who obtained his information from Crowe, states that "the wood is much finer than that of any other British poplar, making as good floors as the best Norway fir in appearance, and having moreover the valuable property, that it will not, like any resinous wood, readily take fire." Cf. *Gard. Chron.* 1848, p. 172.

## POPULUS TOMENTOSA

*Populus tomentosa*, Carrière, in *Rev. Hort.* x. 340 (1867); Wesmael, in De Candolle, *Prod.* xvi. 2, p. 325 (1868), and *Mém. Soc. Sc. Hainaut*, 228, t. 17 (1869); Schneider, *Laubholzkunde*, i. 21 (1904); Dode, in *Mém. Soc. Hist. Nat. Autun*, xviii. 25 (1905); Gombocz, in *Math. Term. Közl.* 140 (1911).

*Populus alba*, Burkill, in *Journ. Linn. Soc. (Bot.)* xxvi. 535 (1899) (not Linnæus).

*Populus alba*, Linnæus, var. *tomentosa*, Wesmael, in *Bull. Soc. Roy. Bot. Belg.* xxvi. 373 (1887); Burkill, in *Journ. Linn. Soc. (Bot.)* xxvi. 535 (1899).

*Populus alba denudata*, Maximowicz, in *Bull. Soc. Nat. Mosc.* i. 49 (1879) (not Hartig).

*Populus pekinensis*, L. Henry, in *Rev. Hort.* lxxv. 355, fig. 142 (1903).

A large tree, similar in size and bark to *P. alba*. Branchlets grey tomentose. Buds ovoid, slightly tomentose, chestnut brown. Leaves (Plate 408, Fig. 2) on the long shoots of old trees, 4 to 6 in. long, 3 to 5 in. broad, triangular-ovate, without lobes or lobules, subcordate or truncate at the broad base, acuminate at the apex; margin with a few (not exceeding ten on each side) sinuate teeth; dark shining green above, glabrescent beneath with traces of grey tomentum. Leaves on the long shoots of vigorous young trees, similar in shape, but the margin biserrate with acute glandular teeth, and the lower surface covered with a grey tomentum. Leaves on the short shoots, small, ovate or triangular, cuneate at the base, sinuately toothed, glabrous beneath. Flowers not seen.

This fine poplar, which attains an enormous size in north China,[1] was discovered by Simon at Siwan, north-west of Peking. His specimen, described by Carrière, is identical with those in the Kew Herbarium, collected near Peking by Sir Rutherford Alcock and by Prof. Sargent, and with another gathered in Shantung by the Rev. A. Williamson, which is preserved in the Edinburgh Herbarium. Gombocz records it also for the mountains of Shensi, and Kiaochow in Shantung. Elwes, in 1912, saw old trees in the grounds of the Summer Palace, Peking, which were about 75 ft. high and 10 ft. in girth. This poplar is called *pai-yang* by the Chinese.[2]

A young living plant sent from Peking in 1897 by Père Provost to the Museum at Paris, where it was propagated, has grown vigorously, and was in 1912 about 35 ft. high by 2 ft. 3 in. in girth. We have seen no trees in England of this species.[3] Jack[4] introduced in 1905 into the Arnold Arboretum cuttings from Peking, which have produced thriving and hardy young trees. He states, however, that it is more readily propagated by grafting. (A. H.)

[1] According to Schneider, it was collected also by Père Giraldi, farther south in Shensi; but the latter's specimen is identified by Diels, *Flora von Central China*, 274 (1901), with *P. tremula*, and is possibly *P. wutaica*, Mayr, *Fremdländ. Wald- u. Parkbäume*, 494, fig. 215 (1906).

[2] Cf. Bretschneider, *Bot. Sinic*, ii. 359 (1892), who refers to this tree as *P. alba*.

[3] The trees often sold by French nurserymen as *P. tomentosa* appear to be *P. Bogueana*, Dode. Cf. *ante*, p. 1781.

[4] *Mitt. Deut. Dend. Ges.* 1909, p. 281.

## POPULUS TREMULA, Aspen

*Populus tremula*, Linnæus, *Sp. Pl.* 1043 (1753); Loudon, *Arb. et Frut. Brit.* iii. 1645 (1838); Wesmael, in *Mém. Soc. Sc. Hainaut*, iii. 229 (1869), and De Candolle, *Prod.* xvi. 2, p. 325 (1864); Willkomm, *Forstl. Flora*, 521 (1887); Mathieu, *Flore Forestière*, 486 (1897); Schneider, *Laubholzkunde*, i. 19 (1904); Dode, in *Mém. Soc. Hist. Nat. Autun*, xviii. 30 (1905); Ascherson and Graebner, *Syn. Mitteleurop. Flora*, iv. 24 (1908); Gombocz, in *Math. Termes. Közl.* xxx. 123 (1911).

*Populus australis*, Tenore, *Ind. Sem. Hort. Neap.* 1830, p. 15.

*Populus græca*, Grisebach, *Spic. Fl. Rum.* ii. 345 (1844) (not Aiton, *Hort. Kew.*).

A tree, occasionally attaining in Scandinavia, Russia, France, and Germany 100 ft. in height and 6 to 8 ft. in girth, but usually much smaller, especially in the British Isles. Bark of young trees smooth, thin, greenish or whitish; on old trunks thick, with small rhomboidal fissures, as in *P. alba*, and ultimately deeply furrowed. Young branchlets glabrous, rounded, shining, with orange lenticels. Buds ovoid, acute, shining brown, slightly viscid, with ciliate scales, the uppermost of which are slightly pubescent. Leaves (Plate 408, Fig. 4) suborbicular, variable in size, averaging 2 in. in diameter, thin in texture, truncate or subcordate at the base, rounded or acute at the apex; margin with a narrow translucent border, and a few rounded or sinuate small teeth; tomentose when young, speedily becoming glabrous on both surfaces, pale or glaucous beneath; venation pseudo-five-palmate; glands[1] at the base two, cup-shaped, well-developed on the terminal leaves of long vigorous shoots, absent on the basal leaves and on those of the short shoots; petioles slender, glabrous, laterally compressed, often as long as the blade. Leaves on young plants and on sucker shoots, and in rare cases on sporadic branches of adult trees, different in shape and much larger, 4 to 5 in. long, 3 to 4 in. broad; ovate, acuminate at the apex, truncate or cordate at the base, greyish and slightly woolly beneath, glandular-serrate, with short pubescent terete petioles.

Catkins sub-sessile, densely and greyish tomentose; axis pubescent; scales long persistent, obovate, deeply lobed and fringed with long white hairs; flowers dense, numerous, on very short pilose pedicels. Stamens about 10, with short filaments and purple anthers, on an oblique disc with an entire and incurved margin. Ovary glabrous; stigmas two, reddish, each divided, forming four widely dilated curving arms; disc funnel-shaped, oblique, glabrous; capsule two-valved.

This species in the wild state displays a considerable amount of variation in the shape, size, and colour of the leaf, and in the amount of pubescence on the branchlets and leaves. The most noteworthy[2] are :—

1. Var. *Freyni*, Hervier, in *Bull. Herb. Boiss.* iv. app. i. 18 (1896), and *Rev. Gen. Bot.* viii. 177 (1896). Leaves rhombic, cuneate at the base, ciliate, pubescent beneath when young. Central France and Prussia.

[1] According to Kerner, *Nat. Hist. Plants*, Eng. Trans. i. 238, fig. 55 (1898), these glands exude resin and serve for absorbing water in rainy weather; but Trelease, in *Bot. Gaz.* vi. 284 (1881), states that they contain honey at the beginning of the season, and are visited by bees and other insects.

[2] Var. *purpurea*, Simon-Louis, ex Späth, *Cat.* No. 102, p. 108 (1898-1899), with purplish young leaves, does not seem to differ from the type, as seen in cultivation at Kew.

2. Var. *villosa*, Wesmael, in De Candolle, *Prod.* xvi. 2, p. 325 (1868).

*Populus villosa*, Lange, *Syll. Soc. Ratisb.* i. 185 (1824), and *ex* Reichenbach, *Fl. Germ. Excur.* 173 (1830).

*Populus canescens*, Reichenbach, *Icon. Fl. Germ.* xi. 30 t. 617 (1849) (not Smith); Hartig, *Forstl. Kulturpfl.* 434 (1851).

Branchlets and leaves at first densely pubescent with long silky hairs, more or less persistent in summer. This variety[1] appears to be quite as common in the wild state as the typical glabrous form.

The origin of the following cultivated form is unknown :—

3. Var. *pendula*, Loudon, *Arb. et Frut. Brit.* iii. 1646 (1838).

*Populus pendula*, Burgsdorf, *Anl. Anpfl. Holzart.* ii. 175 (1787).

Branchlets pendulous. Usually seen as a small grafted tree.

The common aspen is widely spread through Europe, northern Africa, Asia Minor, the Caucasus, and Siberia, being replaced by closely allied species in China, Japan, and the Himalayas. It occurs in every European country; but is absent from the south of Spain, Sicily, and the islands in the western Mediterranean. It is much more common in the north, where it reaches the Arctic Circle both in Europe and Asia; and either in pure woods or mixed with birch covers extensive tracts in Scandinavia, Russia, and Siberia. Towards the south, it only occurs as a scattered tree in mixed woods; and ascends in the Pyrenees to 6000 ft., and in the Alps to 4000 ft. In England it is not uncommon in coppiced woods; but it is of more frequent occurrence in the Highlands of Scotland, ascending to tree limit.

The aspen is a short-lived tree, rarely attaining an age of over 100 years. When cut down, it produces coppice shoots of no great vigour, and usually reproduces itself in such cases by abundant suckers,[2] which are given off to a considerable distance by its widely-spreading superficial roots. (A. H.)

As an ornamental tree the aspen is in northern countries one of the most beautiful, on account of the splendid red and yellow tints which the leaves assume in autumn; but in England these colours are seldom seen in the same degree, and though the bright pale green of its trembling leaves gives it a certain beauty, it is hardly worth growing in any quantity. It is not particular about soil, and may serve to clothe waste places such as old pit banks. It is one of the latest trees to come into leaf in spring.

The largest trees I have noticed in England are two at Little Sodbury Manor, in Gloucestershire, which do not much exceed 60 ft. in height. Sir Hugh Beevor tells me that he has seen none larger in the eastern counties. There are four trees forming a handsome group near the river Gade at The Grove, Watford, the largest being 54 feet by 3 ft. 3 in. in 1904. These were photographed[3] by Mr. Henry Irving.

In Wales the tree does not seem common, but apparently attains a greater size than it usually does in England. In May 1911 I saw some large trees at an

[1] M. Dode informed me that in the Forêt de Bondy this variety attains 35 metres in height.—H. J. E.

[2] The suckers of *P. tremula* are described by Dubard, in *Ann. Sc. Nat.* xvii. 160 (1903).

[3] Reproduced by Groom, *Trees and their Life Histories*, figs. 192, 193 (1907).

elevation of about 800 ft. at Abergwessin in North Breconshire. The Welsh name for it in this district is *aethnen*; but at Hafodunos, in North Wales, W. Jones, the head gardener, told me that the local name was *tafoden merched*, which means "women's tongues." In the Isle of Man a name of the same meaning, *chengey-ny-mraane*, was used.

In the Highlands, where it ascends to the upper limit of trees—in Braemar up to at least 1600 ft.[1]—it seems commoner, and attains larger dimensions. I have never seen any, however, which in size equal those in the north of Norway, the largest I know of being a tree on the shores of Loch Garry, which Captain Ellice of Invergarry found in 1910 to be about 60 ft. by 7 ft.[2] In the birch woods of Strathglass, Glenaffric, and Guisachan the aspen occurs in clumps which appear to have grown from suckers, but old trees are scarce. The largest I saw was a fallen tree above the falls in Glenaffric, which, when standing, was about 50 ft. high, with a trunk 7 ft. in girth. The belief[3] of the Highlanders, who call the aspen *crithean* or *critheac*, that the Cross of Christ was made of this tree still exists both among Catholics and Protestants in this district. I was assured by reliable persons that it is looked on as an accursed tree, and that no Highlander will use the wood for any purpose, even for fuel.[4] Notwithstanding this belief, I saw on the croft of Peter Macdonald at Balnaith, near the head of Glen Urquhart, a group of well-shaped aspen about 40 ft. high, which, as he told me, he had trained up from suckers, and were about forty years old.

In Ireland, the aspen is recorded from almost every county, but is by no means a common tree. Mr. R. A. Phillips informs us that it is native on mountain cliffs, rocky lake shores, the banks of rivers, and in old hedgerows in uncultivated bogland districts, and also on the islands off the west coast. In the mountains of Kerry and Antrim it is a mere bush, but in the lowlands is a small tree, rarely over 40 ft. in height. The finest which he has seen grows on the edge of the river Nore near Durrow, and measured 65 ft. by 5 ft. in 1908.

According to Schübeler it extends in Scandinavia as far north as Alten (lat. 70°), where it attains 60 ft. in height, and in the south ascends to 3500 ft. The tallest aspens that he mentions in Norway grew some miles east of the farm of Viken, in Niderven, and were 90 to 100 ft. high; whilst at Femrade, in Sogndal (lat. 61°), there was a very old aspen 58 ft. high, with a girth of 16 ft. at four feet from the ground. I saw myself in Junkersdal, in lat. 67°, trees of 80 feet high and 6 to 7 ft. in girth, which were finer than any I know in Great Britain.

In France, Mouillefert says, that though common in the damper parts of the forest on sandy or gravelly soils, it is rare on dry or calcareous formations, and that owing to the freedom with which it produces suckers it tends to supersede other trees in suitable places.

1 White, *Flora of Perthshire*, p. 268, says that it ascends to 2100 ft. in Athole, and 1400 ft. in Rannoch; and that both the forms—*villosa*, Lange, and *glabra*, Syme,—occur in the county.

2 Loudon mentions trees of much greater size at various places in England and Scotland, but there is little doubt that they were *P. canescens*.

3 Cf. Loudon, *op. cit.* 1648, and Cameron, *Gaelic Names of Plants*, 70 (1883).

4 Cf. Carmichael, *Carmina Gadelica*, 104 (1900), who states that in Uist the hateful aspen is banned. If it still exists in Uist, it is now an uncommon tree, as I saw none in North or South Uist in 1910.

### TIMBER

Though usually looked on as a forest weed, it is possible that when grown naturally from suckers thickly enough to clean the stems from branches, it may have some value for making matches. In Sweden it is largely employed for this purpose, and according to Schübeler 1,400,000 cubic ft. were thus used in 1882, and valued at 70 ore (about 8d.) per cubic foot. Cargoes of poplar timber, which I believe to be mainly aspen, are now imported from the Baltic for making matches, and cost delivered in Gloucester from 38s. to 48s. per load of 50 cubic ft. caliper measure.

It is also largely used for pulp-making, for which it is very suitable; but it could scarcely be produced here in sufficient quantities at a price that would compete with the produce of northern Europe and America.

On this subject, however, a valuable paper[1] by Weigle and Frothingham on the American aspens should be consulted, as these species resemble our native aspen in life-history, characteristics, and uses. The conclusions of these authors may be summarised as follows:

1. No other trees have so wide a distribution in Europe and Asia as the aspens, *P. tremula* covering 140° of longitude and 35° of latitude, whilst *P. tremuloides* ranges over 112° and 41° respectively.

2. They are both pre-eminently cold- and moisture-loving trees, requiring a very short season of growth and thriving—as *P. tremula* does at Colesborne—where frosts may occur during every month of the year.

3. For their best growth they require deep fresh or moist, porous and well-drained soils; but they will grow on thin dry soil and in poorly drained situations.

4. They are strikingly intolerant of shade; and this applies even more strongly to the suckers by which they are commonly reproduced, and which are often mistaken for seedlings. The latter are rarely seen in England.

5. Their growth is rapid during the first twenty to thirty years, and though they may attain considerable size, yet they are short-lived, usually decaying before 100 years of age, and often much sooner; and root-suckers do not produce such large or well-shaped trees as seedlings.

6. They are best managed as a pure crop under a short rotation; and on account of their extreme light-demanding character require timely thinning.

7. The wood produces the best and whitest pulp, which can be produced and manufactured, more cheaply than other species, into paper which is peculiarly suitable for books and magazines. As the fibre is too short to make good paper alone, it is mixed with a proportion (usually about 40 per cent) of sulphite spruce pulp which adds strength. The finished paper is tough, white, and easily sized, and though inferior to rag paper for the finest uses, is much cheaper.

Waste land suitable for profitable planting of *P. tremula* (and possibly also *P. canescens*) might be found in some parts of Scotland and Ireland; and experiments in this direction are advisable.

In France, according to Mouillefert, the wood is valued for charcoal, but as firewood it burns out very quickly. (H. J. E.)

[1] *U.S. Forest Service, Bull.* No. 93, *The Aspens: their Growth and Management* (1911).

## POPULUS TREMULOIDES, American Aspen

*Populus tremuloides*, Michaux, *Fl. Bor. Amer.* ii. 243 (1803); Sargent, *Silva N. Amer.* ix. 158, t. 487 (1896), and *Trees N. Amer.* 154 (1905); Schneider, *Laubholzkunde*, i. 19 (1904); Dode, in *Mém. Soc. Hist. Nat. Autun*, xviii. 33 (1905).

*Populus trepida*, Willdenow, *Sp. Pl.* ii. 803 (1805); Loudon, *Arb. et Frut. Brit.* iii. 1649 (1838).

*Populus tremuliformis*, Emerson, *Trees Massachusetts*, 243 (1846).

*Populus atheniensis*,[1] Ludwig, *Neue Wilde Baumz.* 35 (1783), *ex* Koch, *Dendrologie*, ii. 486 (1872); Koehne, *Deut. Dend.* 80 (1893).

*Populus græca*,[1] Loudon, *Arb. et Frut. Brit.* iii. 1651 (1838) (not Aiton); Lauche, *Deut. Dend.* 316 (1883).

A tree, attaining in America 100 ft. in height and 9 ft. in girth. Bark like that of *P. tremula.* Young branchlets slender, glabrous, shining reddish brown, with orange lenticels. Buds ovoid, sharp-pointed, shining brownish, slightly viscid, with glabrous scales. Leaves (Plate 408, Fig. 5) ovate to nearly orbicular, 1½ to 2 in. in diameter, thin in texture; truncate, rounded or cuneate at the base; shortly cuspidate at the apex; glabrous on both surfaces, pale beneath; margin with a narrow translucent border, ciliate especially on the leaves of the short shoots, finely glandular-serrate; pseudo-three- to five-palminerved at the base, where the glands in occurrence and appearance are like those of *P. tremula*; petiole slender, glabrous, laterally compressed, variable in length, often as long as the blade. Leaves on sucker shoots, similar to those of *P. tremula*, but glabrous on both surfaces, with ciliated margins.

Flowers scarcely distinguishable from those of *P. tremula*, but with more slender catkins, smaller in all the parts of the flowers; disc of the pistillate flower crenate.

1. Var. *pendula.* A weeping form, with pendulous branches, always grafted.

This is generally known as the *Parasol de St. Julien*, which is said[2] to have been first propagated by Messrs. Baltet, who found in 1865, on the bank of a canal at St. Julien, near Troyes, a tree with weeping branches, which they crown-grafted on the white poplar, and considered to be a weeping form of *P. tremula.*

It is said by Koch[3] to have been much more common in commerce in 1872 than the weeping variety of the common aspen, and it is possible that the preceding history is applicable rather to a weeping variety of *P. tremula.*

2. *P. cercidiphylla*, Britton, *N. Amer. Trees*, 180 (1908), seems to be a form with small entire or undulate leaves, which was found in Wyoming by Dr. C. C. Curtis in 1900. (A. H.)

[1] Loudon describes this poplar under both the names *P. trepida* and *P. græca*, and states in *Gard. Mag.* 1840, p. 231, and *Trees and Shrubs*, 823 (1842), that *P. græca* was "named after the village called Athens, on the banks of the Mississippi, where the tree grows abundantly." *P. atheniensis* is said by Koehne to derive its name from the town of Athens in New York State, whence it was introduced. *P. græca*, Aiton, *Hort. Kew.* iii. 407 (1789), was insufficiently described, and said to be a native of the Greek archipelago.

[2] Cayeux, in *The Garden*, 1886, p. 2. The pendulous variety of *P. tremuloides* is called *Parasol de St. Julien*, in Simon-Louis's catalogue, 1899-1900, and appears to be now always sold in France under this name. Späth, *Catalogue*, No. 57, p. 61 (1883), identifies the *Parasol de St. Julien* with *P. canescens pendula*; and the latter name, now no longer employed, would seem to show that Späth's tree was rather a weeping *P. tremula* than a pendulous form of *P. tremuloides.*

[3] *Dendrologie*, ii. pt. i. 487 (1872).

The American aspen is similar in its habits to the common species, and is widely spread throughout North America, from southern Labrador, the eastern shores of Hudson's Bay, the mouth of the Mackenzie river, and the Yukon valley in Alaska, southwards to Pennsylvania and Nebraska, and through all the mountain regions of the west to central California, the San Pedro mountain in Lower California, northern Arizona, New Mexico, and the state of Chihuahua in Mexico. It is common in the east on moist sandy soil, and often borders the western prairies with a wide belt; and to the northward is often mixed with spruce and birch. In the mountain regions of the western and Pacific states, it ascends to 10,000 ft. above the sea. Sargent says that on account of its remarkable power of germinating on burnt soil, and rapidly covering mountain sides which have been devastated by fire, it has had a greater influence than any other tree on the composition and distribution of the subalpine and boreal forests of North America. Macoun says that in the North-west, Athabasca, and Mackenzie districts it is everywhere common on dry soil, but not on alluvial flats, and that it reproduces freely after a forest fire by root-suckers, but not from seed.

It is said to have been introduced[1] in 1779 by Hugh, Duke of Northumberland; but we have seen no trees of considerable size, and the weeping variety appears to be now the only kind usually cultivated.

Loudon says that a tree in the Chiswick Garden was 12 ft. high eight years after planting; and on account probably of its northern habitat it produced leaves so early that on the 20th April 1835 they were cut by frost. It is so like the aspen of the old world in appearance that even if it would grow, it is hardly worth cultivation except in botanic gardens. (H. J. E.)

## POPULUS GRANDIDENTATA

*Populus grandidentata*, Michaux, *Fl. Bor. Am.* ii. 243 (1803); Loudon, *Arb. et Frut. Brit.* iii. 1650 (1838); Sargent, *Silva N. Amer.* ix. 161, t. 488 (1896), and *Trees N. Amer.* 155 (1905); Schneider, *Laubholzkunde*, i. 17 (1904); Dode, in *Mém. Soc. Hist. Nat. Autun*, xviii. 28 (1905). Gombocz, in *Math. Termes. Közl.* xxx. 138 (1911).

A tree, attaining in America 70 ft. in height and 6 ft. in girth; bark like that of the common aspen. Young branchlets covered at first with a greyish tomentum, persistent more or less during summer. Buds ovoid, acute, grey tomentose. Leaves (Plate 408, Fig. 7) on the long shoots, 3 to 4 in. long, 2 to 3 in. broad, ovate-deltoid; truncate, rounded, or cuneate at the base; acuminate at the apex; glabrescent and dark green above; lower surface pale or glaucous green, glabrescent or with traces of the grey tomentum, which is dense at the time of unfolding of the leaves; margin with a narrow translucent border, and with a few sinuate triangular teeth, but entire near the base; petioles slender, laterally compressed, glabrescent. Leaves on the short shoots, oval, with sharper teeth, and often with two glands at the summit of the petiole.

[1] Loudon, under *P. græca*, p. 1651.

Catkins with deep narrowly lobed scales, fringed with long hairs; differing chiefly from the other aspens in the pubescent disc and ovary; stamens six to twelve; style divided into four long filiform lobes; capsule two-valved.

This species is much less widely spread in North America than the other aspen (*P. tremuloides*), apparently requiring a moister soil, and mainly growing in deep sand on the banks of rivers and swamps. It occurs in Ontario, southern Quebec, Nova Scotia, and New Brunswick, extending southward in the United States to north Delaware on the coast, and along the Alleghany Mountains to North Carolina, and westwards to Minnesota, Wisconsin, Illinois, Indiana, Kentucky, and Tennessee.

According to Dame and Brooks,[1] it is best distinguished from *P. tremuloides* in early spring, by the colour of the unfolding leaves, which are cottony white, whilst those of *P. tremuloides* appear yellowish green. The leaves when open are much larger and more coarsely toothed, and the buds divergent, dull, and dusty-looking; whilst those of *P. tremuloides* are mostly appressed and highly polished with a resinous lustre. In Canada it generally grows on sandy soil, mixed with pines, and is often mistaken for aspen. It is never a large tree, though usually larger than *P. tremuloides*, and as Elwes saw it, near Ottawa, is a straggling, ill-shaped tree of 40 to 50 ft. high, liable to be broken by the wind, and of little or no value either for use or ornament.

Though introduced, according to Loudon,[2] in 1772, it has always been a scarce tree in England, and the only specimen at Kew died about a year ago. At Grayswood, Haslemere, a tree obtained from Meehan in 1887 is only 16 ft. high, and apparently this species does not thrive in our climate.

The weeping grafted tree, commonly cultivated under the name *P. grandidentata*, var. *pendula*, differs from that species in flowers and other characters,[3] and may be distinguished as follows:—

*Populus pseudo-grandidentata*, Dode, in *Mém. Soc. Hist. Nat. Autun*, xviii. 31 (1905).

*Populus tremula*, Linnæus, var. *pseudo-grandidentata*, Ascherson and Graebner, *Syn. Mitteleurop. Flora*, iv. 26 (1908).

Young branchlets stout, dark reddish, with orange lenticels, covered with whitish tomentum in spring, which persists in summer at the base of the shoots. Buds viscid, tomentose near the top of the branchlet, glabrescent elsewhere. Leaves (Plate 408, Fig. 8) similar in shape and dentation to *P. tremula*, but larger, 3 to 4 in. in diameter, and thicker in texture; margin with a translucent border, ciliate in spring. Staminate catkins, 2 in. long, with a slender pubescent axis; pedicels glabrous; stamens five, on an oblique glabrous shallow spatulate disc, which is entire in margin; filaments slender, white; anthers red.

The origin of this plant is unknown, but it is probably a hybrid; and if it came

[1] *Trees of New England*, 32 (1902).

[2] *Trees and Shrubs*, 823 (1842). It was introduced earlier into France, as it is well figured as *P. tremula, ampliori folio*, by Duhamel, *Traité des Arbres*, ii. 178, pl. 38, fig. 8 (1755).

[3] In *P. grandidentata* the leaves are ovate, long acuminate, with fewer and larger teeth than in the weeping tree; stamens more numerous, with short filaments; pedicels and disc pubescent.

from America may be a cross between *P. tremuloides* and *P. grandidentata.*[1] It appears to have been first mentioned by Simon-Louis[2] in 1869. Koch[3] speaks of it as a pendulous tree existing in England in 1872. There are good specimens at Abbotsbury and at Glasnevin. (A. H.)

## POPULUS SIEBOLDII

*Populus Sieboldii*, Miquel, in *Ann. Mus. Bot. Lugd. Bat.* (*excl.* pl. masc.[4]) iii. 29 (1867); Wesmael, in De Candolle, *Prod.* xvi. 2, p. 327 (in part) (1864); Schneider, *Laubholzkunde*, i. 17 (1904); Dode, in *Mém. Soc. Hist. Nat. Autun*, xviii. 32 (1905); Gombocz, in *Math. Termes. Közl.* xxx. 131 (1911).

*Populus tremula*, Linnæus, var. *villosa*, Franchet and Savatier, *Enum. Pl. Jap.* i. 465 (1875) (not Wesmael); Maximowicz,[4] in *Bull. Soc. Nat. Mosc.* liv. 49 (1879); Shirasawa, *Icon. Ess. For Japon*, i. text 37, t. 18, figs. 1-10 (1900).

*Populus rotundifolia*, Simon-Louis, *ex* Dippel, *Laubholzkunde*, ii. 192 (1892).

A tree, attaining in Japan 60 ft. in height. Young branchlets stout, covered in spring with a dense white tomentum, persistent in part during summer. Buds more or less tomentose, not viscid. Leaves (Plate 408, Fig. 6) thicker in texture than those of *P. tremula*, densely tomentose and ciliate when young, glabrescent in summer, dark shining green above and yellowish or pale beneath, about 3 in. long and 2 in. broad, ovate, rounded or cuneate at the base, abruptly contracted into a glandular short acuminate apex; margin with a translucent border, minutely (varying even on the same leaf) sinuately toothed or glandular serrate; basal glands usually well developed; petiole slender, pubescent, laterally compressed. Flowers similar to those of *P. tremula*, but with the disc slightly pubescent.

This species, which is very distinct in appearance from the common aspen, appears to be confined to Japan, where it was collected at Aomori by Elwes.

It appears to have been introduced by Simon-Louis about 1887, but the only tree which I have seen is a grafted one at Glasnevin, planted in that year, which has only attained a height of 15 ft. and is not very thriving. (A. H.)

## POPULUS FREMONTII

*Populus Fremontii*, Watson, in *Proc. Am. Acad.* x. 350 (1875); Sargent, *Silva N. Amer.* ix. 183, t. 496 (1896), and *Trees N. Amer.* 164 (1905); Dode, in *Mém. Soc. Hist. Nat. Autun*, xviii. 40 (1905); Gombocz, in *Math. Termes. Közl.* xxx. 76 (1911).

*Populus monilifera*, Torrey, in *Sitgreave's Rep.* 172 (1853) (not Aiton).

A tree, attaining 100 ft. in height and 15 ft. in girth; bark at first smooth and thin, ultimately becoming on old trunks deeply fissured into broad rounded scaly

[1] Nuttall, *Gen. Pl.* ii. 239 (1818), describes as *P. grandidentata*, var. *pendula*, a tree "with pendulous branches, as in the weeping ash, on the Alleghany Ridge, Pennsylvania, rare." Loudon, *op. cit.* 1651, states that there was a tree bearing this name in the Horticultural Society's garden in 1838; but that its branchlets were not pendulous. There is, therefore, no evidence that Nuttall's tree, seen only in the wild state, was ever introduced into cultivation.

[2] *Cat. Général*, 1869, p. 73, where it is called *P. grandidentata*, var. *pendula*, and is described as a weeping form with large teeth to the leaves.

[3] *Dendrologie*, ii. pt. i. 488 (1872).

[4] The staminate specimen described by Miquel, which is preserved in the Leyden Herbarium, is a species of *Carpinus*, according to Maximowicz.

ridges. Young branchlets glabrous. Buds small, viscid. Leaves on old trees deltoid, about 2½ in. broad, truncate at the base, and abruptly contracted at the apex into broad short entire points: on young cultivated trees (Plate 409, Fig. 13), reniform or rhombic, with a cuneate base and a similar apex; serrations few, coarse, with incurved points; margin with dense minute cilia, discernible with a good lens; glands absent at the base; petiole glabrous.

Staminate catkins, 2 in. long; axis glabrous; disc broad, oblique, entire in margin; stamens sixty, with dark red anthers. Pistillate catkins, 2 in. long; pedicels short; disc crenate, cup-shaped; stigmas three, irregularly and crenately lobed. Fruiting catkins, 4 to 5 in. long; capsule thick-walled, three- to four-valved.

This species grows on the banks of streams in California, Lower California, Nevada, southern Utah, southern Colorado, and western Texas.

It[1] has only lately been introduced into cultivation, and small specimens may be seen at Kew and Glasnevin.

Var. *Wislizeni*, Watson, in *Amer. Journ. Sci.* xv. 136 (1878).

*Populus Wislizeni*, Sargent, *Silva N. Amer.* xiv. 71, t. 732 (1902), and *Trees N. Amer.* 165 (1905); Dode, in *Mém. Soc. Hist. Nat. Autun*, xviii. 39 (1905); Gombocz, in *Math. Termes. Közl.* xxx. 78 (1911).

This appears to have similar foliage, and is mainly distinguishable by the long pedicels of the flowers. It is the common poplar in the Rio Grande valley of western Texas and New Mexico, and the adjacent parts of Mexico. (A. H.)

## POPULUS NIGRA, Black Poplar

*Populus nigra*, Linnæus,[2] *Sp. Pl.* 1034 (1753); Loudon, *Arb. et Frut. Brit.* iii. 1652 (1838); Wesmael, in De Candolle, *Prod.* xvi. 2, p. 327 (1868), and in *Mém. Soc. Sc. Hainaut*, iii. 258 (1869); Willkomm, *Forstliche Flora*, 527 (1887); Mathieu, *Flore Forestière*, 491 (1897); Schneider, *Laubholzkunde*, i. 5 (1904); Dode, in *Mém. Soc. Hist. Nat. Autun*, xviii. ("groupe *nigra*") 37 (1905); Ascherson and Graebner, *Syn. Mitteleurop. Fl.* iv. 36 (1908); Gombocz, in *Math. Termes. Közl.* xxx. 85 (1911).

A tree, attaining above 100 ft. in height and 20 ft. in girth, usually with a straight and single stem, but occasionally dividing near the base into several limbs; with wide-spreading stout and irregular branches, not slender and regularly ascending as in many of the hybrids. Bark deeply furrowed on old trunks, and often covered with large burrs. In all its forms this species is readily distinguishable from the American species and the hybrids by the leaves, non-ciliate on the margin, without glands at the base, and when well-developed gradually tapering from the middle of the blade to a long acuminate apex.

The black poplar, and apparently all the poplars of the same section, rarely if ever produce suckers while the trees are living, but if one is cut down suckers are

[1] According to Dode, the trees introduced are var. *Wislizeni*; but until they flower their identification is uncertain. However, Späth, *Cat.* No. 95, p. 89 (1895-1896), states that the young plants first introduced in 1894 were obtained from Colorado, where only the typical form of the species exists.

[2] By this name Linnæus meant the black poplar inhabiting temperate Europe; though he quotes a Virginian poplar *ex Herb. Gronov.* It is most convenient to assume, as the typical form of the species, the tree planted by Linnæus at Upsala, which is still living, and from which I gathered specimens in 1908.

produced abundantly from the roots. Root cuttings are as readily propagated as ordinary cuttings. Dubard, in *Ann. Sc. Nat.* xvii. 147 (1903), gives an elaborate account of the peculiarities of the suckers of this tree.

This species comprises two distinct forms, one glabrous in all its parts, the other more or less pubescent, described in detail as follows :—

1. Var. *typica*, Schneider, *op. cit.* 5 ; Ascherson and Graebner, *op. cit.* 39. Continental Black Poplar.

Young branchlets rounded, glabrous, ashy grey in the second year. Buds reddish, viscid, glabrous, closely appressed to the twig at their base, with a sharp apex curving outwards. Leaves on the long shoots (Plate 409, Fig. 11), about 3 in. long, and 2 in. broad, cuneate at the base, gradually tapering from about the lower third, where they are widest, towards the long acuminate apex ; glabrous ; dark green above, light green below ; margin with a narrow translucent border, non-ciliate, finely and crenately glandular-serrate ; petiole glabrous, laterally compressed. On the short shoots the leaves are smaller, broader at the base, which is often less cuneate, and truncate or rounded ; and similar leaves often occur on old trees even on the long shoots.

Catkins about 1½ in. long, with early deciduous scales, which are broadly obovate, and divided into numerous irregular, linear entire or lanceolate toothed lobes ; axis glabrous. Staminate flowers sub-sessile ; stamens twenty to thirty on an oblique concave non-ciliate glabrous disc, which is slightly waved and upturned in margin ; filaments white, thread-like, as long as the deep red anthers. Pistillate flowers shortly stalked ; ovary glabrous, globose, in a cup-like glabrous disc ; stigmas two, dilated, crenate in margin, closely appressed at first to the sides of the upper part of the ovary. Capsules two-valved, glabrous, on long pedicels.

2. Var. *betulifolia*, Torrey, *Fl. New York*, ii. 216 (1843) ; Skan, in *Bot. Mag.* t. 8298 (in part[1]) (1910) ; Schneider, *Laubholzkunde*, ii. 870 (1912). English Black Poplar.

Var. *viridis*,[2] Lindley, *ex* Loudon, *Arb. et Frut. Brit.* iii. 1652 (1838).
Var. *betulæfolia*, Wesmael, in De Candolle, *Prod.* xvi. 2, p. 328 (1868).
Var. *hudsonica*, Schneider, *Laubholzkunde*, i. 5 (1904); Ascherson and Graebner, *Syn. Mitteleurop. Fl.* iv. 39 (1908).
*Populus nigra*, Michaux, *Fl. Bor. Am.* ii. 244 (1803) (not Linnæus).
*Populus hudsonica*, Michaux f., *Hist. Arb. Amer.* iii. 293, t. 10, fig. 1 (1813), and *N. Amer. Sylva*, ii. 114, t. 96, fig. 1 (1819).
*Populus betulifolia*, Pursh, *Fl. Amer. Sept.* ii. 619 (1814) ; Loudon, *Arb. et Frut. Brit.* iii. 1656 (1838) ; Dode, in *Mém. Soc. Hist. Nat. Autun*, xviii. 48 (1905).
*Populus nigra*, *Vaillantiana*, and *Muelleriana*, Dode, in *Mém. Soc. Hist. Nat. Autun*, xviii. 48 (1905).

Young branchlets rounded, covered with a dense short pubescence, orange or yellow in the second year. Buds greenish, tinged with brown, viscid, otherwise as

[1] The female catkins figured show ovaries with three stigmas, and were taken from a hybrid tree (*P. Lloydii*) growing near Turnham Green station. Cf. p. 1831, note 1.

[2] Lindley's specimen in the Cambridge Herbarium is a vigorous branch from a young tree of *P. nigra*, var. *betulifolia*. This is confirmed by a tree labelled var. *viridis* in the Cambridge Botanic Garden. Loudon's description of this variety, "leaves of a brighter green than the species," is inadequate ; and the name var. *viridis*, though older than that of var. *betulifolia*, cannot be used. Mackie, in Loudon, *Gard. Mag.* xiii. 230 (1837), says that var. *viridis* was discovered at Bealings, near Woodbridge, and had been grown in his nursery at Norwich for twenty years.

in var. *typica*. Leaves[1] (Plate 409, Fig. 12) similar in shape, colour, size, and margin to those of var. *typica*, but slightly pubescent when young; petioles pubescent.

Catkins $1\frac{1}{2}$ to 2 in. long, as in the typical variety, but with a pubescent axis; stamens in the specimens examined, fewer, about twelve to fifteen; scales, ovary, and stigmas, identical. Fruiting catkins—on the Bury St. Edmunds tree, which was probably fertilised by staminate trees of the same variety close beside it—about 4 in. long, with ovoid capsules about $\frac{1}{4}$ in. in length, glabrous and tuberculate on the outer surface; seed oblanceolate, yellowish, about $\frac{1}{8}$ in. long, covered with dense cottony hairs enveloping the whole catkin after the dehiscence of the capsules.

Forms, in which glabrous catkins are associated with pubescent leaves and branchlets, occur; and on this account I have refrained from making var. *betulifolia* a distinct species.[2]

The pubescent variety of the black poplar is the only form occurring in England in the wild state; and it is also a native of the greater part of France, from Normandy and Picardy to the foot of the Pyrenees. In 1912 I saw it apparently wild in many places, as in hedges on hills not far from Argentan, where it grows in a small and stunted form. It is most common, however, as a fine tree, often with a burry trunk, along the banks of the great rivers, as on the Seine at Mantes, on the Loire near Tours, on the Garonne in the vicinity both of Bordeaux and of Toulouse, on the Adour between Bayonne and Dax, and on the Gave de Pau, where there are two good trees in a meadow opposite the shrine at Lourdes. This poplar is also frequently planted in botanic gardens, as at Le Mans, Tours, and Montauban, a fine tree in the latter place measuring 90 ft. by 13 ft. There is a specimen in the Montpellier Herbarium, gathered at Ganges on an island in the river Herault; but I have seen no specimens from Provence.[3] It is remarkable how this variety has escaped the notice of British botanists, though it has been collected from early times, as there are specimens in the British Museum[4] gathered by Plukenet and Buddle towards the end of the seventeenth century.

This tree was first distinguished by the younger Michaux, who found it growing on the banks of the Hudson river above Albany, and mentions large specimens planted in New York city; but adds that he never saw it in the forest. Sargent[5] in 1896 stated that it was growing then on an island in the Delaware river near Easton, Pennsylvania; but in a letter to Kew, dated 31st July 1902, he mentions only a single specimen known to him, an old tree near Boston;[6] and adds that it is

[1] On very old trees the leaves are smaller, truncate or occasionally subcordate at the base, and with a shorter acumen at the apex. These appear to be *P. Muelleriana*, Dode.

[2] *P. nigra*, var. *pubescens*, Parlatore, *Fl. Ital.* iv. 289 (1867), was described from trees growing in moist valleys at S. Martino, Palermo; and is recorded in Thessaly by Halacsy, *Consp. Fl. Græc.* iii. 136 (1904). A specimen in the Cambridge Herbarium has branchlets, leaves, petioles, and female catkins covered with long white hairs; and is much more pubescent than trees growing in England. *P. hispida*, Haussknecht, which I have not seen, is probably the same as var. *pubescens*.

[3] Pardé, in *Bull. Soc. Dend. France*, 1911, p. 255, states that *P. nigra* is pretty common in Provence; but this is probably var. *typica*.

[4] *Herb. Sloan*, 83, fol. 8, and 126, fol. 6.

[5] *Silva N. Amer.* ix. 153, note (1896). In the Montpellier Herbarium there is a specimen of this tree labelled "*Populus*, New York, growing planted opposite Dr. Hosack's door in Broadway, May 7, 1807." Another specimen named *P. hudsonica*, Michaux, was taken from a tree growing at Versailles in 1808.

[6] This is no doubt the tree which I saw growing on the shore of Jamaica Pond, when staying with Prof. Sargent in 1904, and recognised at once as the English black poplar by its burry trunk and foliage.—H. J. E.

sold in the United States by Ellwanger and Barry of Rochester under the name of *P. elegans.*[1] There is little doubt that this poplar was introduced into the United States in the eighteenth century from England.

3. Var. *italica*, Du Roi, *Harbk. Baumz.* ii. 141 (1772). Lombardy Poplar.

Var. *pyramidalis*, Spach, in *Ann. Sc. Nat.* xv. 31 (1841).
*Populus italica*, Moench, *Bäume Weissenstein*, 79 (1785).
*Populus dilatata*, Aiton, *Hort. Kew.* iii. 406 (1789).
*Populus pyramidata*, Moench, *Meth.* 339 (1794).
*Populus pyramidalis*, Borkhausen, *Forstbot.* i. 541 (1800).
*Populus fastigiata*, Poiret, in Lamarck, *Encycl.* v. 235 (1804); Loudon, *Arb. et Frut. Brit.* iii. 1660 (1838).

Branches directed nearly vertically upwards, forming a narrow fastigiate tree. A sport of the typical glabrous variety of *P. nigra*, differing in no respect except in habit.[2] The leaves are variable, many being the same as those of the ordinary form; but others are often broader than long, truncate or subcordate at the base, with a short acuminate apex, due to increased vigour, as is usual in this species.

The common Lombardy poplar is a staminate tree, always reproduced by cuttings; and for aught we know, all the numerous individuals planted throughout the world may have originated from a single tree, as happened, without any doubt, in the case of the upright form of the common yew. No instances of a second origin have been recorded.

A few trees[3] of similar habit, though with branches not quite so vertically inclined, have been observed bearing pistillate flowers.[4] Plate 383, reproduced from a photograph sent to us by the late Prof. W. Blasius, shows a remarkable female tree at the village of Greene, near Kreiensen, in the Duchy of Brunswick, which has ascending and not erect branches, and differs considerably in habit from the ordinary Lombardy poplar. There is a good specimen at Kew with nearly erect branches, about 50 feet high, which was covered with woolly catkins in 1908. It produced flowers in the spring of 1910, which did not, however, ripen into fruit

[1] This is referred to as a variety of *P. nigra*, commonly sold by nurserymen in the United States, by L. H. Bailey, in *Cornell Univ. Bull. Agric.* No. 68, p. 227 (1894).

[2] I carefully compared in 1908, in Servia, the branchlets, foliage, and buds of a Lombardy poplar with those of some wild common black poplars growing near it, and did not detect the slightest difference. The bark of some trees in this region, and also in Algeria, is remarkably whitish; while the colour of their third year and older twigs is peculiarly greyish. This form has been named *P. thevestina*, Dode, in *Mém. Soc. Hist. Nat. Autun*, xviii. 52 (1905). Siehe, in *Mitt. Deut. Dend. Ges.* 1912, p. 123, describes the remarkable pale bark of the Lombardy poplar in Asia Minor. Cuttings were obtained in Algeria by Mr. A. W. Hill in 1910, which are now growing in the nursery at Kew. Vigorous shoots from near the base of old Lombardy poplars at Cambridge show the same coloration.

[3] Spenner, *Fl. Friburg*, i. 274 (1825), mentions a female tree near the Carthusian monastery at Friburg in Germany. Another was reported to have been noticed in the University Botanic Garden at Göttingen in 1828 (cf. Denson, in Loudon, *Gard. Mag.* vi. 419 (1830)). Loudon, *Derby Arboretum*, 57 (1840), states that the female tree was introduced in 1840 into the Horticultural Society's garden from Monza near Milan. There is a specimen in the Kew Herbarium, sent from Carlsruhe by A. Braun in 1845, and others undated, which are labelled Frankfort-on-Oder and Switzerland. Mr. W. L. Wood also noticed, in 1910, two smaller trees with pistillate catkins growing in the Walpole Road, Twickenham.

[4] *P. pannonica*, Kitaibel, *ex* Besser, *Pl. Enum. Volhyniæ*, 38 (1822), and Reichenbach, *Icon. Fl. Germ.* xi. 30, t. 619 (1845) (figured with rhombic acuminate leaves), is possibly the correct name of the female Lombardy poplar. Besser, in *Flora*, 1832, ii. Suppl. 14, states that *P. croatica*, which was published at the same time as *P. pannonica*, is the name that should be applied to the supposed cross between typical *P. nigra* and the Lombardy poplar, which grew in the Theresa garden at Vienna, the sex of which is not mentioned. It was supposed to occur wild on the Dnieper. Zawadski, *Enum. Pl. Galic.* 117 (1835), saw fine specimens of this on the Dniester in Podolia. Petzold and Kirchner, *Arb. Musc.* 593 (1864), state that they received *P. fastigiata*, var. *pannonica*, from many sources, but never were able to discover any distinguishing characters. Cf. also *P. nigra*, var. *pannonica*, Dippel, *Laubholzkunde*, ii. 198 (1892).

in this season, though there are staminate trees at no great distance. It opens its flowers at the same time as the Lombardy poplar. The latter is about three weeks earlier than the native black poplar, an indication of its southern origin.[1] The history of these pistillate trees is quite unknown; but they may have arisen as the result of hybridisation between the staminate Lombardy and the ordinary poplars.

The staminate Lombardy poplar appears to have originated on the banks of the river Po in northern Italy, probably in the beginning of the eighteenth century, as it was unknown to classical writers [2] and is not mentioned by mediæval Italian authors.[3] Moreover, it was not noticed by Ray and other English travellers in Italy in the seventeenth century. Séguier,[4] an old writer, states that it was known anciently in Lombardy, and mentions a superb avenue, which he saw in 1763 at Colorno, the residence of the Duke of Parma. It was apparently carried [5] by the Genoese to the Levant; and there are no grounds for supposing that it originated in Asia Minor or Afghanistan, as Royle,[6] who first made this statement, simply relied on the fact that it bore a native Persian name.[7] W. G. Browne,[8] who travelled in Asia Minor in 1798, makes the first reference to its occurrence in western Asia, where he states that it abounds all over the plain of Damascus, and when old becomes rugged and uncouth, as usual in other regions.

It was introduced from Lombardy into France in 1749; and is usually stated to have been brought into England and planted at St. Osyth's in Essex, in 1758, by the Earl of Rochford, who was ambassador in Turin at the time. It was possibly, however, first planted at Whitton some years earlier by Archibald, Duke of Argyll, who died in 1761, as the tree still growing there in 1838 was much larger than any of the others recorded by Loudon, being 115 ft. high and 19 ft. 8 in. in girth at 2 feet from the ground.[9]

[1] Sargent, *Silva N. Amer.* ix. 154, note (1896), says that the fact that the Lombardy poplar does not suffer in the cold of the Canadian winter, shows that it originated in a climate much more severe than that of northern Italy. (The winter in the plain of the Po, it may be stated, is very cold, the mean temperature being below 35° Fahr.; and in Milan the thermometer sometimes sinks below zero.) Prof. Budd, quoted by L. H. Bailey, *Cornell Univ. Bull. Agric.* No. 68, p. 228 (1894), however, explains that the Lombardy poplar, grown in Canada, was imported from Voronej in central Russia, where it has become acclimatised, and is perfectly hardy. The Russian botanists assured him that its hardiness depended on the region from whence it was obtained. Bailey, *Survival of the Unlike*, 297 (1896), in an interesting chapter on acclimatisation of trees, states that cuttings of the white poplar, taken from trees at Montpellier and at Geneva, which were planted at the latter place, differed as much as twenty-five days in their time of coming into leaf; and similar results were obtained at Ithaca (New York) with cuttings of the Lombardy poplar.

[2] The often quoted lines of Ovid, *Met.* ii. 345-360, and of Virgil, *Æneid*, x. 190, do not refer to this tree, as has been supposed. The poplars depicted by Perugino (1446–1524), as in a picture in the London National Gallery, are slender, but not with vertical branches, and are probably *Populus alba*. Cf. Rosen, *Die Natur in der Kunst*, 293 (1903).

[3] Rostafinski, in *Verhandl. K. K. Zool. Bot. Gesell. Wien*, xxii. 170 (1872), states that it was introduced from Italy into Poland by King Sobieski, who reigned from 1624 to 1696, and that the original trees are still standing in the garden of the Wilanow Castle, near Warsaw. Miss Ivanovska, at my request, examined the old poplars there, which proved to be all of the ordinary wide-spreading form; and Prof. Rostafinski in a letter acknowledges that he made a mistake.

[4] *Hist. Plant. Nat. Envir. Vérone*, ii. 267 (1745).

[5] Fougeroux de Bondaroy, in *Mém. d'Agric.*, *Paris*, 1786, p. 84.

[6] *Illust. Bot. Him.* i. 344 (1839). Griffith's statement that it is wild near Kabul, at 7000 ft. altitude, is not confirmed. Aitchison, in *Trans. Bot. Soc. Edin.* xviii. 162 (1891), says: "I only met with this tree cultivated in orchards or near houses in Afghanistan and north-east Persia."

[7] Boissier, *Fl. Orient.* iv. 1194 (1879), doubts its existence in the wild state in western Asia; and his reference to it being perhaps wild in the Karatau mountain in Turkestan is an error, as the wild fastigiate poplar in this locality is *P. alba*, var. *pyramidalis*. Cf. p. 1778.

[8] *Travels in Africa, Egypt, and Syria*, 397, 408 (1799). Siehe, in *Mitt. Deut. Dend. Ges.* 1912, p. 123, states that in Asia Minor the Lombardy poplar is extensively cultivated and is a most useful tree, producing after twenty or thirty years' growth, long slender but tough beams, which are much used in house-building.

[9] Loudon, *Arb. et Frut. Brit.* i. 58 (1838).

According to Sargent,[1] it was brought to America in 1784 by W. Hamilton, who planted it in his garden at Woodlands near Philadelphia.

The Lombardy poplar appears to be a short-lived tree, and is said to be dying out in Germany. As it does not now apparently attain the immense size recorded in former years, there may be some truth in the opinion advanced by Focke [2] that as all the trees have been raised by cuttings since the origin of the first sport, they may now be dying of old age. (A. H.)

The most reliable account we have of the introduction of this tree is that given by Aiton, *Hortus Kewensis*, iii. 406 (1789), who states that it was brought by Lord Rochford from Turin, where he was ambassador about 1758. He planted cuttings at St. Osyth's Priory in Essex, where two trees now much decayed still survive. I am indebted to Mr. J. Edge, the gardener there, for photographs and measurements of the larger of these trees, which show a large hollow stump divided into two trunks about 20 feet high and measuring 18 ft. in girth at 3 feet from the ground. Living branches have sprung from different places in the trunk, two of which attain a height of about 50 feet.

Loudon records a large tree of the same age as the last, which was blown down at Canterbury in 1836; but the tallest tree mentioned by him was at Great Tew in Oxfordshire, said to have been 125 feet high when only fifty years old. A tree was recorded by Thomas Hogg, forester at Hampton Court, Herefordshire,[3] as growing at Wharton Court farm near that place, which in 1879 was said to measure no less than 160 feet. I visited this place in 1905 and found no trace of its remains; but if the height was correct, which from the other measurements given of trees on the estate seems probable, it was much taller than any that I have measured in England or France.

Sir Hugh Beevor tells me of a tree at Pitchford, Shropshire, which in 1907 was 120 ft. by 14 ft. 8 in.; and I have measured many of 100 to 115 feet, but none which can be said to stand out from the average of mature trees. Henry measured a tree at Shiplake House, near Henley, which was 105 ft. by 10 ft. 10 in. in 1905; and another at Alderbourn Manor, Gerard's Cross, which was 100 ft. by 12 ft. in 1912, and visible for many miles around. An old tree in Lensfield Road, Cambridge, 90 ft. high in 1904, of which a photograph was sent me by Mr. Lynch, was removed in 1912.

J. Smith [4] recorded a tree growing at Fox Mills near Romsey as 125 ft. by 13 ft. 2 in., but when I was there in 1900 I could not find it; and another at Greatbridge House, near Romsey, which was 130 ft. by 13 ft. 9 in. These died in 1881, no doubt from the effects of the inclemency of the weather in 1879-80.

The seasons of 1879-81 appear to have killed a very large number of Lombardy poplars in the eastern and midland counties, not perhaps so much by their excessive

[1] Sargent, *op. cit.* 154, note (1896).

[2] In *Gart. Zeit.* September 1883, quoted in *Gard. Chron.* xx. 571 (1883). Cf. also *Rev. des Eaux et Forêts*, xxiv. 277 (1885). Manetti's letter quoted by Loudon, *Gard. Mag.* xii. 450 (1836), is rather obscure, and his statement that plants were raised in Italy from seed, which preserved the characters of their parents, is extremely doubtful. In the Cambridge Herbarium, however, there is a specimen, with female flowers and fruit, of a poplar sent by Manetti, which he considered to be the female Lombardy poplar.

[3] *Trans. Scot. Arb. Soc.* ix. 151 (1879).

[4] *Ibid.* xi. 534 (1887).

cold, as because of the two unusually cold and wet summers. A paper on the subject by Mr. H. D. Geldart[1] gives many interesting particulars; and all his correspondents seem to agree that the damage was much greater than in the colder winter of 1860-61, when the thermometer at Audley End near Saffron Walden went down on Christmas morning to − 11°. Mr. Geldart quotes the replies received by Mr. Southwell to his inquiries as to the death of Lombardy poplars in other parts of England as follows. Twenty miles round York, death or severe injury was almost universal; in Wilts, they had suffered very much; in west Dorset, most of them were killed or seriously injured; near Doncaster, all were more or less killed; but at Oxford among the College gardens it was the exception to find a damaged tree; and Mr. C. B. Plowright wrote that in the west of England the Lombardy poplars did not seem to be injured at all.

In Scotland it was, according to Dr. Walker, introduced at New Posso in Tweeddale as early as 1765 from cuttings sent by the Earl of Hertford, and was extensively distributed some years later by Lord Gardenstone; but the climate of most parts of Scotland is evidently too cold or too wet to suit this tree, as none of those mentioned by Loudon are still alive so far as I can learn, except one at Brahan Castle, Ross-shire, that was 70 ft. by 6 ft. in 1838. When I visited this place in 1907 I found a fine tree 98 ft. by about 9 ft., which may be the same. Six miles north of Inverness, on the high-road to Beauly, I also saw four well-shaped trees, of which the largest was 90 ft. by 11½ ft., showing the excellent climate of that district. Mr. Renwick tells us of one near Braidwood, Lanarkshire, which was 93 ft. by 10½ ft. in 1910; this was blown down on 5th November 1911.

In France the Lombardy poplar is common, though now on account of its inferior growth often replaced by the hybrid poplars. It commonly attains 110 ft. to 120 ft. in height, but I have seen none approaching the trees near Rouen mentioned by Loudon[2] which, according to M. Dubreuil, were then 150 ft. high. A tree[3] at the Trianon, Versailles, was 17½ ft. in girth at four feet from the ground in 1888.

The Lombardy poplar has been planted largely in the irrigated districts of Utah, and, according to F. C. Sears,[4] rows of tall Lombardy poplars, marking the irrigation canals, are a feature in the landscape. (H. J. E.)

In Chile,[5] especially about Valparaiso, the Lombardy poplar is largely planted, both in gardens and on the margins of the irrigation canals, where it grows so rapidly as to be ready for felling in fifteen years. Its timber is used for indoor work in houses. Dode states[6] that in Chile and Argentina there is a form of the Lombardy poplar which keeps its leaves evergreen.

Mr. Lovegrove has sent us specimens of the Lombardy poplar from Kashmir, where it is planted along roads, and often attains 100 ft. in height and 7 ft. in girth.

[1] *Trans. Norf. and Norw. Nat. Soc.* iii. 354-366 (1880-1884). Cf. *Gard. Chron.* xv. 764, 798, and xvi. 246 (1881), where instances are given of the death of many trees also in the north of France and in Belgium.

[2] *Arb. et Frut. Brit.* iii. 1670 (1838).

[3] *Garden and Forest*, i. 174 (1888).

[4] *Ibid.* x. 357 (1897).

[5] Dr. W. Balfour Gourlay, in *Trans. Bot. Soc. Edin.* xxiv. 74, plate 7 (1910). It is much attacked in Chile by the quintral, *Loranthus tetrandus*.

[6] *Bull. Soc. Dend. France*, 1908, p. 29, and 1909, p. 152, where this form is named *P. pyramidalis*, var. *Thaysiana*, Dode.

Its timber is used in house-building, and lasts well when protected from rain. A tree 5 ft. in girth sells for 12 to 26 shillings.

4. Var. *plantierensis*, Schneider, *Laubholzkunde*, i. 803 (1906).

*Populus fastigiata plantierensis*, Simon-Louis, *Cat.* 1884-1885, p. 51.
*Populus plantierensis*, Dode, in *Mém. Soc. Hist. Nat. Autun*, xviii. 43 (1905).

A fastigiate form of var. *betulifolia*, similar to the Lombardy poplar in habit and foliage, but with reddish pubescent petioles, and shortly pubescent branchlets. It originated in Simon-Louis's nursery at Plantières, near Metz, whence it derives its name, and is said to occur in both sexes, and to be the result of a cross between the Lombardy poplar and var. *betulifolia*; but this origin is unlikely. The original tree is a male, and when measured by Elwes in 1908 was 74 ft. by 5 ft. It is claimed for it that it is more vigorous than the ordinary fastigiate poplar, and not liable to die off at the top, as is frequent in the latter. There are specimens at Kew about 25 ft. high.

5. Var. *viadri*, Ascherson and Graebner, *Syn. Mitteleurop. Flora*, iv. 40 (1908).

*Populus viadri*, Rüdiger, in *Abhand. Naturw. Ver. Reg. Bez. Frankfurt*, viii. *Mon. Mitt.* 12 (1891); Koehne, in *Verh. Bot. Ver. Brandenb.* xxxvii. p. xxviii (1895), and in *Gartenflora*, xxxix. 447 (1890).

A narrow pyramidal tree, with ascending branches, which in branchlets, buds, and foliage resembles the typical glabrous form of the species.[1] It occurs along the banks of the Oder, near Frankfort, whence it derives its name (*Viadrus* being the Latin name for this river). It is said to produce pistillate flowers identical with those of *P. nigra*; but a tree[2] at Kew, about 25 ft. high, produced staminate flowers in 1910. These differ slightly from the type in occasionally having peculiar scales deeply bilobed at the summit. (A. H.)

## Distribution

The distribution of the black poplar is very wide in Europe, but difficult to define accurately, as it has been much planted in former times. In Norway and Sweden, Schübeler only knew it as a planted tree, and figured a very large one, which from its burry trunk has the appearance of the English tree. This grew on the banks of a river at Ronneby in Sweden; and when measured in 1882 by Prof. Wittrock, was, at four feet from the base, 34 ft. 8 in. in girth, dividing into two main trunks a little higher up; and this is the largest girth of which I have any record.

In the Botanic Garden at St. Petersburg a tree, supposed to have been planted by Peter the Great, measured in 1908 about 90 ft. high by 17 ft. in girth, forking low down, and with a large burr on its trunk.

In north Russia it extends to 57° N., according to Von Herder; but in the St. Petersburg Herbarium I found a note by Kusnetsov stating that it was found at

[1] Koehne, who mentions trees of both sexes in *Deut. Dend.* 84 (1893), describes this peculiar poplar as a hybrid, *P. candicans* × *nigra*, but he afterwards withdrew this very unlikely hypothesis. The leaves are more cuspidate at the apex than in ordinary *P. nigra*, and it is possible that *P. viadri* is a hybrid, but I have seen no pistillate flowers. *P. viadri* appears to have been introduced into cultivation by Späth, as it is mentioned as a novelty in his *Catalogue*, No. 91, p. 96 (1893-1894).

[2] Another tree at Kew, labelled *P. viadri*, also obtained from Rüdiger, is different in habit, having spreading and not ascending branches. It is pistillate, and appears to differ in no respect from *P. nigra*, var. *typica*.

Emetskoie on the Dwina in lat. 63° 30'. It is found on the Volga[1] as far south as Astrachan, and on the Ufa river, where, according to Loffiewsky, it attains 100 ft. high. I also found specimens in the herbarium under the name of *P. nigra* from Zlataoust on the southern Ural, from Tobolsk and Barnaoul, and from the Yenesei river in lat. 66° N., collected by Brenner, with very small leaves. In France I saw a large female tree at Chenonceaux, near Tours, which measured 100 ft. by 10 ft. 10 in. in 1908. At Dijon there is in the Botanic Garden a very large tree[2] of this species, probably the oldest in Europe. It is said from historical documents to be over five centuries old, and bears an inscription to the effect that in 1866 it was at the ground 12 metres, and at two metres high 8 metres in girth, and contained about 40 cubic metres—about 1400 cubic feet of timber. It is quite hollow at the base, the shell being only a foot or so in thickness, and has deeply ridged bark, 4 to 6 in. in thickness. As nearly as I could estimate, its present height is about 125 ft., but having lost several branches has probably been taller. Its girth at 5 ft. is 26 ft. 7 in. The leaves though small are typical in shape, but the trunk is not as burry as in English trees. I could not learn whether it is male or female. M. Mathey, inspector of forests, told me that the black poplar which was formerly common around Dijon is now growing scarce, being superseded, as in England, by *P. serotina.*

In Spain and Portugal I only saw small and stunted trees; and in Italy I have seen none which looked like true *P. nigra.* In Austria it is common in the valley of the Danube; and in the Prater at Vienna there are many good-sized trees. In Greece, Heldreich records it from Mt. Pelion. In Morocco, Maw and Ball collected specimens in the Atlas at 3000 to 4000 feet.

In Great Britain it has been occasionally confused by local botanists with *P. serotina*; and I can find nothing in the older works, such as those of Evelyn, Miller, and Boutcher, to show that these authors knew the tree from personal observation. Though rare in most parts of England it seems to be a native of the counties on the Welsh border, where it is still fairly common; and it is probably indigenous in Norfolk, Suffolk, Cambridgeshire, and Essex. It is so very distinct in trunk, foliage, and time of leafing from *P. serotina,* that it is extraordinary that so little notice has been taken of it either by botanists or foresters, and as the trees given below have in almost every case been visited and identified by ourselves, it is very likely that it will be found in other districts now that attention has been called to it.

The English tree may be recognised in mature specimens by its trunk being usually covered more or less with large burry excrescences, which are formed by a mass of abortive buds, and which do not seem to be found to the same extent in Continental specimens; secondly, by its flowering later and leafing earlier than the much commoner black Italian poplar; and, thirdly, by the shape of the leaves, and also by the green and not reddish tint of the young foliage. As this tree does not seem to have been propagated by nurserymen for many years past, young trees

[1] Loudon, *Trees and Shrubs,* 824 (1842), states that it abounds on the banks of the Vistula, whence the mottled wood of knotty trunks was brought to Berlin and made into ladies' workboxes. Concerning *P. vistulensis,* Hort., see Jacquin, *Ann. de Flor.* ii. 96 (1833), and *Rev. Hort.* 1865, pp. 305, 346, and 405.

[2] Figured in *Gard. Chron.* xxi. 641, fig. 123 (1884), and in *Rev. Hort.* iii. 184, fig. 11 (1854).

are rarely seen, and the species appears to be dying out in all parts of England, except on the Welsh borders and in East Anglia. The glabrous Continental form is very rarely planted in Britain.

### Remarkable Trees

In the south-western counties I have seen none except at Dunster in Somerset. where, in a meadow below the castle there is an old hollow stump 24 feet in girth, from which a large limb has extended and taken root some way off, and thrown up two stems 5 and 5½ ft. in girth, but only about 40 ft. high. There are three other large trees near the park gate south-east of the castle, the first 92 ft. by 17 ft measured over a large burr, the second about 90 ft. by 16½ ft., also very burry.

In Hants I have seen none, but am told by Mr. J. Smith of Romsey that an old pollard exists near that town 15 ft. by 7 ft. 9 in. In Sussex there is a tree in the park at Beauport about 100 ft. by 8 ft. In Kent in Penshurst Park there are some small, old, and very stunted trees.

In Middlesex by far the finest are two at Syon. These are rather shut in by other trees, and in 1905 were[1] about 110 ft. by 17 ft. 8 in. In the MS. catalogue at Syon a tree of this species was recorded as 120 ft. by 15 ft. in 1849. There are two trees on Wandsworth Common, and Miss Woolward tells me of a young female tree near the Serpentine Bridge. In the Green Park there are three trees in a group about 100 yards from Piccadilly opposite Down Street.

In Suffolk on the banks of the Larke in the Abbey grounds at Bury St. Edmunds there are several old trees, one of which was figured by Strutt,[2] and was said by him to be 90 ft. by 15 ft., and to contain 551 cubic ft. of timber. I could not identify this particular tree in 1907, when on 7th April I found them covered with bright red male flowers, which gave a beautiful effect in the sun. A female tree here was only just showing flower at the same date, but produced fertile seeds later, from which Mr. Hankins, forester at Culford Park, raised numerous seedlings, thirty of which planted out in a plot measured 5 to 7 ft. high in November 1911.

A tree at West Stow, near Bury St. Edmunds, which measured 92 ft. high by 19 ft. in girth, was felled in March 1912. It had immense spreading superficial roots; and the original set, about 4 in. in diameter, was recognisable in the centre of the butt, being separated from the older wood by a ring-shake. Near the base of the trunk Mr. Hankins counted 225 annual rings. The timber was quite sound, the first length measuring 39 ft. by 42½ in. quarter-girth or 489 cubic ft., the total contents being 748 cubit ft.; a plank of it is now in the Cambridge Forestry Museum.

At Islington Hall, King's Lynn, there is a magnificent tree, which was measured by Mr. A. P. Long in October 1912, as follows: total height, 108 ft.; girth at five feet from the ground, 17 ft. 3 in.; volume of the bole (36 ft. in length to the point where the first branch is given off), 361 cubic feet; total volume, 620 cubic feet.

[1] A. B. Jackson, *Cat. Trees Syon.* 22 (1910), gives the height of these trees as 128 ft.

[2] *Sylva Britannica*, t. 24.

At Weeting Park near Thetford there are two large trees in the park about 90 ft. high and 17 ft. 4 in. in girth, whose trunks are covered with large burrs. Mr. Howell, the gardener here, told me that a much larger one had been blown down some time before my visit, the trunk of which was no less than 20 ft. in girth at eight feet from the ground. The rings in the wood showed this tree to be over 200 years old. At St. Helens, Norwich, in Mr. Skelton's garden on the banks of the Wensum, there are two fine trees, one of which in 1908 was 95 ft. by 15½ ft.

In Essex it is rare, but at Audley End, in a withy bed below the park, there are three trees, one of which measured 115 ft. by 15 ft. on 18th March, 1907, when it was not yet in flower. At Stanstead Bury, Hertford, Mr. H. Clinton Baker found a tree bearing pistillate catkins, and 85 ft. by 12 ft. 9 in., in April 1911. At Bishop Stortford Henry has seen a large female tree.

In Gloucestershire the tree is quite rare, the only ones I know of being a small tree by the roadside on Crickley Hill, and another on the banks of an old canal at Coombe Hill, half a mile west of the Gloucester and Tewkesbury Road. Miss F. Woolward discovered a female tree about 60 ft. high, near Bourton on the Water; and at Forthampton Court, near Tewkesbury, there are four trees about 90 ft. by 15 to 16 ft., by a pond north of Mr. Yorke's house. In Herefordshire the tree, I am told by Mr. T. E. Groom of Hereford, is not uncommon in the Wye valley, while Mr. Openshaw of Wofferton Court farm informs me that it was more abundant fifty years ago. In Worcestershire, Lees,[1] who seems to have known the tree better than recent botanists of the county, says, "A few scraggy native black poplars appear in various localities by brooksides, but this tree appears to be dying out." At Arley Castle there is a fine tree planted in the park, which is about 100 ft. by 15 ft.

In Shropshire I have seen more than in any other county, the finest being a tree at Oakley Park, Ludlow, which in 1908 was about 100 ft. by 15 ft. 10 in. Between Craven Arms and Lydbury North there are several comparatively young trees planted by the roadside about 90 ft. by 6 to 7 ft. In a meadow near Walcot Park a line of trees in which *P. nigra* and *P. serotina* are mixed, shows the difference of size, habit, and period of shedding their leaves very plainly. On 28th October 1908, *P. nigra* had shed all its upper leaves, while those of *P. serotina* were still green. *P. nigra* here averaged about 70 ft. by 6 ft., whilst *P. serotina* were about 100 ft. by 9 ft., all being apparently planted at the same time. Between Oswestry and Whittington on the main road are several large but not very old trees. Strutt says that the black poplar is oftener found in Suffolk and in Cheshire than in any other counties, but I can hear of none in the latter county.

The largest I have seen in Wales are at Gwernyffed Park, Breconshire, where there is a tree in a belt west of the lodge gate, overgrown with ivy, which is about 90 ft. by 17 ft. At Maesllwych Castle, Radnorshire, there are two fine trees in the park, with burry trunks, measuring about 85 ft. by 15½ ft. in 1907.

In the north I have seen none except at Alnwick, where in a field by the road between the town and the castle there is an old tree. Mr. A. C. Forbes tells me of a female tree near Hexham.

[1] *Botany of Worcestershire*, xl. (1867).

In Scotland it seems quite rare, and whether it is truly native seems doubtful. The finest I have seen is in the park at Gordon Castle, which in 1908 was about 100 ft. by 15½ ft. At Smeaton Hepburn I measured in 1911 a tree 90 ft. by 8 ft. 7 in. Mr. Renwick tells me of two trees at Auchentorlie, Dumbartonshire, of which the best measured 94 ft. by 12 ft. 8 in. in 1909; and of two trees at The Ross, Lanarkshire, one of which, bearing pistillate flowers, and 87 ft. by 14½ ft., was greatly injured by the storm of 5th November, 1911; the other measured 80 ft. by 12 ft. 5 in. in 1912. At the foot of the former tree, *Lathrea squamaria*, a parasitic plant, is growing. Mr. Renwick measured, in 1909, a fine tree at Cambusnethan House, Lanarkshire, 102 ft. by 11 ft. 8 in.; four trees at Dalzell House in the same county, the largest of which was 99 ft. by 17 ft. 7 in.; and a tree at Kilkerran, Ayrshire, 88 ft. by 15 ft. 3 in.

Selby[1] mentions an immense tree growing at Maxwellheugh near Kelso, which had a trunk 16½ ft. high and 31 ft. at the base, 21 ft. at two feet and 18 ft. at ten feet, respectively in girth, which was computed to contain over 900 ft. of timber, and, supposing it to be the black Italian poplar, says that it could not be above sixty years old.[2] But Sir George Douglas of Springwood Park, Kelso, informs me that this tree, which had to be cut down in 1902 as it overhung the road, and was decaying and dangerous, was, according to reliable oral tradition, nearly as large one hundred and twenty years previously; and in any case must have been a great deal older than Selby supposed. It was 21 ft. in girth at five feet from the ground when felled, and was said to have been 92 ft. high in 1859. Sir George Douglas further tells me that Andrew Brotherstone, a botanist of Kelso, called it the black Italian poplar. Particulars of it were sent to the Edinburgh Botanical Society, but Professor Balfour informs me that no trace can now be found of this in the records of the Society. Mr. G. Leven, forester at Bowmont Forest, informs us that the gardener who cut it down and a local nurseryman state that it was a black Italian poplar.

In Ireland it seems to be doubtfully native. I have seen none in positions where they seem really wild. Stunted trees which look like *P. nigra* are common by the roadsides in several counties, but the only fine trees I have seen are as follows:—At New Ross, Wexford, by the road about 500 yards south of the town, an old tree with a hollow trunk about 90 ft. by 18 ft. This is a female, and produced good seed in 1907, from which Miss Woolward raised seedlings.[3] At Mallow Park, close to the bridge out of the town, a very large tree[4] with a burry trunk, which in 1909 measured 90 ft. by 19½ ft. At Muckross, on the green near the hotel, an old tree about 70 ft. by 13 ft., which Mr. Greany, resident agent for the Muckross estate, told me was the only one he knew in the district. (H. J. E.)

[1] *British Forest Trees*, 202 (1842).

[2] M'Kay and Renwick, in *Trans. Nat. Hist. Soc. Glasgow*, iv. pp. 251, 261 (1891), found this tree to be 19 ft. 9 in. in girth at six feet three inches above the ground, in 1893.

[3] Cf. *Journ. Bot.* xlv. 417, t. 487 (1907).

[4] There are two photographs of this tree in the Timber Museum at Kew, which were sent in 1878, with a note that the girth in that year was 17 ft. 2 in. at 4½ ft. and 18 ft. 1 in. at 6 ft. above the ground.

## POPULUS MONILIFERA, Canadian Black Poplar

*Populus monilifera*,[1] Aiton, *Hort. Kew.* iii. 406 (1789); Willdenow, *Sp. Pl.* iv. 805 (1805); Watson,[2] *Dendr. Brit.* ii. t. 102 (1825) (excl. the staminate flowers); Hartig, *Naturges. Forstlich. Culturpfl.* 436 (1851).[3]

*Populus virginiana*,[4] Fougeroux de Bondaroy, in *Mém. d'Agric. Paris, 1786*, i. p. 87 (1787).

*Populus canadensis*,[5] Michaux, *Hist. Arb. Amer.* iii. 297, t. 11 (1813); Loudon, *Arb. et Frut. Brit.* iii. 1655 (1838) (not Moench, Hartig, or Koehne).

*Populus deltoides*,[6] Sudworth, in *Bull. Torrey Bot. Club*, xx. 44 (1893) (in part).

*Populus deltoidea*,[6] Sargent, *Silva N. Amer.* ix. 179, t. 494 (1896), and *Trees N. Amer.* 163 (1905) (in part).

A tree, attaining in America 100 ft. in height and 20 ft. in girth. Bark deeply divided into broad rounded scaly ridges. It is readily distinguishable in cultivation from *P. angulata* by its different foliage and the less angular branchlets, which are rarely marked with projecting ribs, but are similar in their greenish colour, with white lenticels. Buds brownish, viscid. Leaves (Plate 409, Fig. 14) smaller than in *P. angulata*, and broader in proportion to their length, averaging 3 in. in width, deltoid-ovate, abruptly cuspidate-acuminate at the apex; base wide, shallowly cordate, occasionally truncate; glabrous, marked at the base of the blade by two glands; margin with a narrow translucent border, densely ciliate, with sinuate serrations fewer and coarser than in *P. angulata*, absent on the base and at the apex, their glandular tips being incurved; petiole laterally compressed.

Catkins about 3 in. long; axes and pedicels glabrous; scales large, dilated at their apex, and irregularly divided into filiform lobes. Stamens 40 to 60; disc broad, orbicular or quadrate, oblique. Pistillate flowers; disc broad, cup-shaped; ovary sub-globose, 3- or 4-celled with 3 or 4 stalked dilated lobed stigmas.

[1] This is the oldest certain name for the northern form of the American black poplar, which is represented in the British Museum by a specimen from Fothergill's garden bearing staminate flowers, labelled *P. monilifera* in Solander's hand-writing. Fothergill lived 1712-1780, and his garden was at Upton, Westham.

[2] Watson figures a pistillate tree, growing at Kew in 1822, which was certainly the American species; but is no longer living, most of the poplars planted at Kew in early days having died or been removed. Watson mentions also a pistillate tree at Cottingham, near Hull, the only other specimen which he had seen. The staminate trees which he refers to as being plentiful were *P. serotina*.

[3] Cf. *P. carolinensis*, Moench, p. 1810, note.

[4] *Populus virginiana*, according to Fougeroux, was a pistillate tree, usually named in collections "Peuplier de Canada," and very sensitive to frost when young. He probably meant the true northern species. Usually in France, the "Peuplier de Virginie" was a synonym of the "Peuplier Suisse," the hybrid *P. serotina*. Fougeroux's name is wrongly used in *Journ. Bot.* l. 132 (1912).

[5] Michaux's name, *P. canadensis*, is accompanied by an accurate description. *P. canadensis*, Moench, *Bäume Weiss.* 81 (1785), identical with *P. latifolia*, Moench, *Meth.* 338 (1794), is possibly one of the female hybrids, *P. regenerata* or *P. marilandica*, but must remain a doubtful name. These names of Moench are supposed by Dode, *op. cit.* 65 (1905), to indicate *P. candicans*, but this is improbable, as Moench clearly describes the latter species as a variety of *P. balsamifera* with cordate leaves. *P. canadensis* is usually erroneously applied by Continental botanists to the various hybrid poplars which are commonly met with in cultivation. See pp. 1816, 1824, 1828. It is wrongly used for the Black Italian Poplar in *Journ. Bot.* l. 132 (1912).

[6] Both these names, *deltoides* and *deltoidea*, are modifications of *Populus deltoide*, Marshall, *Arb. Amer.* 106 (1785). This is an incomplete description of a poplar growing on the banks of large rivers in Carolina and Florida, and is perhaps meant for *P. angulata*. Dode, *op. cit.* 40 (1895), says: "Quant à la notation *deltoidea*, il semble absolument impossible de lui donner un sens quelconque, Marshall ne l'ayant fait suivre que de quelques mots équivoques."

Fruiting catkins, 6 to 8 in. long; capsules 3- to 4-valved; seed with long white hairs, enclosing the catkins, when the capsules dehisce, in a dense mass of cotton.

1. Var. *occidentalis*, Rydberg, in *Mem. New York Bot. Garden*, i. 115 (1900).

*Populus occidentalis*, Britton, *ex* Rydberg, *Fl. Colorado*, 91 (1906); Gombocz, in *Math. Termes. Közl.* xxx. 79 (1911).

*Populus Sargentii*, Dode, in *Mem. Soc. Hist. Nat. Autun*, xviii. 40 (1905); Britton, *N. Amer. Trees*, 178 (1908).

Leaves smaller, deltoid, truncate at the base, abruptly contracted at the apex into a long acuminate point, with few and coarse serrations.

This is the common black poplar in western North America, east[1] of the Rocky Mountains from Saskatchewan and Alberta southwards to New Mexico and western Texas; and is the characteristic tree on the river flats of the western prairies.

*P. monilifera* is distinguishable from all the hybrids by the dense persistent cilia on the margin of the leaf, the peculiar few hooked serrations, the abrupt cuspidate apex, and the constancy of the glands at the base; and when flowers are obtainable by the large number of stamens (about sixty), and the shape of the stigmas. It has been confused by most dendrologists with the hybrids, which both on the Continent and in England have entirely supplanted it in cultivation. It has become an extremely rare tree; and the only specimens which we know of in England are a tree apparently past its prime, though probably not over sixty years old, at Bradwell Grove, Burford (Oxon), the seat of W. H. Fox, Esq., which Elwes measured in 1910 and found to be 91 ft. by 9 ft. 3 in.; and another at Penrice Castle, Glamorganshire, about 80 ft. high and 13 ft. 9 in. in girth. The stem of the latter has deeply furrowed bark, and gives off the first branch at twenty-seven feet up; above are numerous wide-spreading branches.

An old tree in the Cambridge Botanic Garden was cut down two years ago, but a cutting from it is now making vigorous growth. Elwes has also obtained cuttings from America, which are thriving at Colesborne.

### Distribution of P. Monilifera and P. Angulata

*P. monilifera* and *P. angulata* are considered by modern American botanists and foresters to constitute a single species, which they term *P. deltoidea*, Marshall, with the following distribution, according to Sargent:—Province of Quebec and the shores of Lake Champlain, through western New England and New York, Pennsylvania west of the Alleghanies, and the Atlantic states south of the Potomac river to western Florida; and westwards as var. *occidentalis* to the base of the Rocky Mountains from southern Alberta to northern New Mexico; comparatively rare and of smaller size in the east and in the coast region of the south Atlantic states and east Gulf states; a large and abundant tree along the streams between the Alleghany range and the Rocky Mountains.

*P. monilifera* and *P. angulata* are undoubtedly connected by intermediate forms; and the view that they constitute one species is possibly true in a wide sense, but

[1] It is reported to occur in Idaho by M. E. Jones, *Montana Botany Notes*, 24 (1910).

they are very different in cultivation, and herbarium specimens justify them being kept separate. Var. *occidentalis*, Rydberg, more closely allied to *P. monilifera* than to *P. angulata*, may also rank, when further studied in the field, as a third species.

I am unable to limit correctly the distribution of the two species, but *P. monilifera*[1] is undoubtedly the poplar, wild in Ontario, Quebec, New England, New York, and Pennsylvania. *P. angulata*[2] appears to be that common in the basin of the Mississippi, and in the southern Atlantic states and the Gulf states.

Michaux states that *P. angulata* attains its most northerly point in lower Virginia, but is more common in the two Carolinas, Georgia, and Louisiana, growing in the marshy basins of the great rivers, ascending the Mississippi from its source to its junction with the Missouri, and continuing along the latter for a hundred miles. He had only seen *P. monilifera*[3] along the river Genesee in New York, in some parts of Virginia, and on several islands of the Ohio river; but believed from reports that it occurred in the Mississippi valley as far south as the Arkansas river. (A. H.)

*P. monilifera* is not generally distributed or common either in Canada or New England so far as I saw. Macoun states[4] that along the Grand Trunk Railway in Ontario there are many young trees which had grown from western seed carried by the railway cars, and he speaks of trees "over 50 ft. high, and some at least 2 ft. in diameter," at Big Stick Lake, north of the Cypress hills in Manitoba.

In New England it seems to be commonest in the Connecticut valley, where Dame and Brooks[5] speak of it as a stately tree 70 to 100 ft. high; and Emerson mentions a tree which he found at New Ashford in 1838, not over sixty years old, which was 20 ft. 5 in. in girth. Mr. Foxworthy sent us a photograph of a tree with a wide-spreading crown growing near Ithaca, New York. This measured 80 ft. by 13 ft., with the trunk dividing into three main stems at about twenty feet from the ground. Self-sown seedlings found by Mr. Jack on the sandy shore of the St. Lawrence, near Chateaugay, on 15th August 1895, where the fruit was ripe on 30th May, were only 4 to 6 in. long, with 4 to 8 small crenate ovate-lanceolate leaves.

S. C. Mason, of the Agricultural College, Kansas, gives a good account of *P. angulata* in that state in *Garden and Forest*, iv. 182, and figures a large tree on the banks of the Kansas river which shows its habit in the west. This tree was 24 ft. in girth near the ground, and was 80 ft. high, with a spread of over 80 feet. Another was 104 ft. high, and as much in the spread of its branches. A tree cut on the Saline river had a stump 8 ft. across, and furnished ninety-six loads of wood. Though most of the large old trees have been cut by the early settlers to build their houses and stockades, the tree is now largely planted for fuel.

[1] Herbarium specimens from Ontario, Vermont, New York, and Ohio are *P. monilifera*.

[2] Herbarium specimens from the banks of the river Ohio, "near North Bend," collected by C. W. Short in 1833, and from Missouri and New Orleans, collected by Drummond, are *P. angulata*, var. *missouriensis*. Specimens gathered by Sargent at Augusta, Georgia, have pubescent leaves and petioles, and may be another variety.

[3] Michaux describes this species under the name *P. canadensis*, and lays great stress on the difference in hardiness between it and *P. angulata*. He considered the latter to be a native only of the southern states, as its shoots were always injured by a few degrees of frost.

[4] *Cat. Canadian Plants*, i. 457 (1883).

[5] *Trees of New England*, 34 (1902).

Ridgway, in *Proc. U.S. Museum*, 86 (1882), speaking of *P. angulata*, states that it is a very common tree on rich bottom lands and along alluvial banks of streams, where it occasionally attains immense size. Trunks of 5 to 6 ft. diameter are not uncommon, while trunks of 7 to 8 ft. are sometimes found, the stem being usually more than 50 feet clear. The largest measurements were as follows:—

| | Girth. | Bole. | Height. | Measured by |
|---|---|---|---|---|
| In Posey Co., Ind. . . | 18 | 70 | 165 | C. Schneck. |
| In Wabash Co., Ill. . . | 18¾ | 75 | 175 | Dr. J. Schneck. |
| In Gibson Co., Ind. . . | 24 | — | — | R. Ridgway. |

### Timber

The timber varies very much in quality, a variety known as yellow cottonwood being much the best. Mason says that he has known strong and good houses built from it whose joists and sheathing boards were straight and sound thirty years after they were put up. But the timber of some trees is soft, spongy, and worthless. Sargent says that the wood is very difficult to season, and is apt to warp badly in drying, but of late years has been used in the Mississippi valley for packing-cases and other coarse work. (H. J. E.)

## POPULUS ANGULATA, Carolina Poplar

*Populus angulata*, Aiton, *Hort. Kew.* iii. 407 (1789); Michaux f., *Hist. Arb. Amer.* iii. 302, pl. 12 (1813); Loudon, *Arb. et Frut. Brit.* iii. 1670 (1838); Schneider, *Laubholzkunde*, i. 9 (1904); Dode, in *Mém. Soc. Hist. Nat. Autun*, xviii. 38 (1905); Gombocz, in *Math. Termes. Közl.* xxx. 81 (1911).

*Populus carolinensis*,[1] Fougeroux de Bondaroy, in *Mem. d'Agric. Paris, 1786*, i. 90 (1787); Dode, *op. cit.* 37 (1905).

*Populus heterophylla*, Du Roi, *Harbk.* ii. 150 (1772) (not Linnæus).

*Populus balsamifera*, Miller, *Gard. Dict.* ed. 8, No. 5 (1759) (not Linnæus).

*Populus angulosa*, Michaux, *Fl. Bor. Amer.* ii. 243 (1803).

*Populus macrophylla*, Loddiges, *Cat.* (1836), *ex* Loudon, *op. cit.* 1671 (1838).

*Populus Besseyana*, Dode, *op. cit.* 38 (1905).

A tree, attaining over 100 ft. in height and 15 ft. in girth. Bark deeply and regularly furrowed on old stems. Young branchlets angled, glabrous, greenish with white lenticels; on vigorous shoots, with projecting ribs, persistent two or three years. Buds greenish, glabrous, only slightly viscid. Leaves (Plate 409, Fig. 15) always longer than broad, averaging, when well-developed, 7 in. long and 5 in. wide, triangular-ovate; base broad, truncate, subcordate or deeply and narrowly cordate; apex acute or shortly acuminate; glabrous and firm in texture when

[1] *P. carolinensis*, Moench, *Bäume Weissenstein*, 81 (1785), is said by Willdenow, *Sp. Pl.* iv. 805 (1805) and by Hartig, *Naturges. Forst. Culturpfl*, 436 (1851) to be identical with *P. monilifera*, Aiton; but this is uncertain, and Moench's plant may have been the female tree of *P. angulata*.

mature, slightly pubescent when young; with two to six glands at the base of the blade, irregular in shape and size, either on the petiole or on the adjoining margin; margin, with a narrow translucent border, ciliate, crenately glandular-serrate, the teeth close together, except near the apex, where they are wanting, and on the base, where they are coarser and wide apart; petiole laterally compressed, greenish, pubescent at first, glabrous later.

Catkins, about 2 to 3 in. long; scales small, cucullate or concave, dentate and without long linear lobes; axis and short pedicels glabrous. Stamens, thirty to forty, on an oblique glabrous disc; filaments slender, white; anthers red. Ovary sessile, globose, glabrous, in a deep five-lobed glabrous cup-shaped disc; stigmas three or four, yellowish, widely dilated, crenulate, often subdivided each into two lobes, spreading upwards or horizontally outwards from the apex of the ovary. Fruit not seen.

The female tree of typical *P. angulata*, which is in cultivation in Europe, has branchlets less winged than those of the staminate tree, and is often distinguished as *P. angulata cordata*, though its leaves are as often truncate as cordate. At Plantières, Metz, it is said[1] to be very hardy, as young trees bore without injury a temperature of $-23°$ Cent. in December 1859; whereas the staminate tree is killed by a temperature of $-12°$ to $-15°$ Cent. The pistillate tree is represented at Kew by a specimen about 35 ft. high, which was obtained from Simon-Louis in 1885. A similar tree at Glasnevin regularly bears flowers, which have been drawn by Miss F. Woolward. There is an old tree of this at Plantières, which produced[2] fertile seed in 1860, and was 7 ft. in girth[2] in 1905.

1. Var. *missouriensis*, Henry. This variety has branchlets and leaves similar to those of typical *P. angulata*, but differs in the flowers, which are like those of *P. monilifera*. The scales of the flowers of both sexes are large, laciniate and fimbriate; while the stamens are 40 to 60, as in the latter species.

Var. *missouriensis*, judging from specimens in the Kew Herbarium, which agree with those sent to me by Rehder from North Bend on the Ohio river, appears to be the ordinary wild form of *P. angulata* in the Mississippi basin. This variety is in cultivation[3] in the south of France and in Italy; but requires further study.

Typical *P. angulata* has not yet been identified, so far as I know, in the United States; and is possibly a form with peculiar flowers, in which the scales have become modified, that has arisen in cultivation in the cooler climate of western Europe. It may, however, yet be found wild in the Carolinas or in Virginia, whence *P. angulata* is reputed to have been originally introduced into Europe. (A. H.)

[1] Thomas, in *Rev. Hort.* 1861, p. 75. It is called *P. cordata*, by Simon-Louis, *Cat. Gen.* 1869, p. 72, where it is said to have been a long time in their nurseries, and to be of unknown origin.

[2] Letter to Kew from Jouin. Its leaves appear earlier than those of the European poplars, and were opening on 15th April 1911, and were fully out on 6th May 1911; whilst those of *P. serotina* do not come out at Metz till late in May, as M. Jouin told me.—H. J. E.

[3] Specimens with staminate flowers collected by Elwes in the Borilly Park at Marseilles, and others of both sexes sent to me from Turin by Prof. Voglino, belong to var. *missouriensis*. Possibly all the trees in cultivation in the south of France and in Italy belong to this variety; but I see no means of distinction in the absence of flowers. The habit of these southern trees is scarcely distinct.

REMARKABLE TREES

The most remarkable tree in Europe is one at Danny Park, Sussex, the seat of W. H. Campion, Esq., of which Plate 384, reproduced from a photograph taken by Mrs. F. D. Godman in 1910, gives a good idea. When this wonderful tree was planted it is impossible to say, but Mr. Campion's grandfather told him that it was an old tree when he came to Danny about 1815; and judging from a drawing of the house made in 1787, in which a tulip-tree probably of the same age is shown as a small tree, it may have been planted about 1760. The branches extend over a space 150 yards round, as measured by Mr. Campion, and the spread on the longer axis is about 60 yards. The main stem was about 80 ft. in length and 12 ft. 10 in. in girth in 1909. It is a male tree, and flowers in April, holding its leaves till late in November. Cuttings have failed to root, but two trees were propagated by grafting on stocks of *P. serotina*, and these are growing at Conyboro, Lewes, the seat of Lord Monkbretton. An illustration of this tree was published in *The Garden*, xxv. 189 (1898), but the ground plan which I am able to give of it, from accurate measurements made by Mr. J. P. Williams, is the only way by which the extraordinary ramifications and rooting of its branches can be understood.

Another large male tree is at Syon Park, which in 1905 was 92 ft.[1] by 9 ft.; but this may not be the same as the one figured by Loudon, *Arb.* vii. 277, as 83 ft. high in 1838. I showed this tree to Prof. Sargent, who admitted it to be distinct from *P. monilifera*, but said that he had never seen any trees like it in America.

There are two male trees on the bank of the Thames in the playing fields at Eton, of considerable size, which keep their leaves till the end of November. Another at Beauport, in the lower part of the park, was 80 ft. by 6 ft. 4 in. in 1909, and, as Sir A. Lamb tells me, retains its leaves till the middle of December unless there has been a severe frost, which seems a good proof of its southern origin.

In Scotland I have seen none; but in Queen's County, Ireland, at Abbeyleix, the seat of Viscount de Vesci, there is a spreading tree of no great age, which in 1910 measured 63 ft. by 4 ft. It is a female tree and was obtained from Smith of Newry.

In France there are many old trees, especially around Paris. A large one at Baleine, near Moulins-sur-Allier, was blown down shortly before my visit in May 1909; and the log, which measured 14 metres long by 3.75 metres in girth, was sold for 140 francs. In the Parc Borilly at Marseilles there were a number of fine trees, all males, in full flower on 17th March 1910, whilst the trees which M. Coste, the head gardener, called *peuplier de Virginie* (*P. serotina*), growing close by, were not yet in bloom. (H. J. E.)

In the south of France in many places this species replaces the hybrid poplars, which are commonly planted farther north, as an avenue and roadside tree. There is a remarkable avenue, over a mile long, in the forest of Thétieu, near Dax, the trees averaging 90 to 110 ft. in height, and 7 to 9 ft. in girth, with stems clear of

[1] In Jackson's *Catalogue of Trees at Syon* the height is given as 111 ft.

branches to 30 or 40 ft. Near Puyoo I noticed, in 1912, some very large trees bearing mistletoe. In the park at Toulouse, this tree is largely planted in walks and

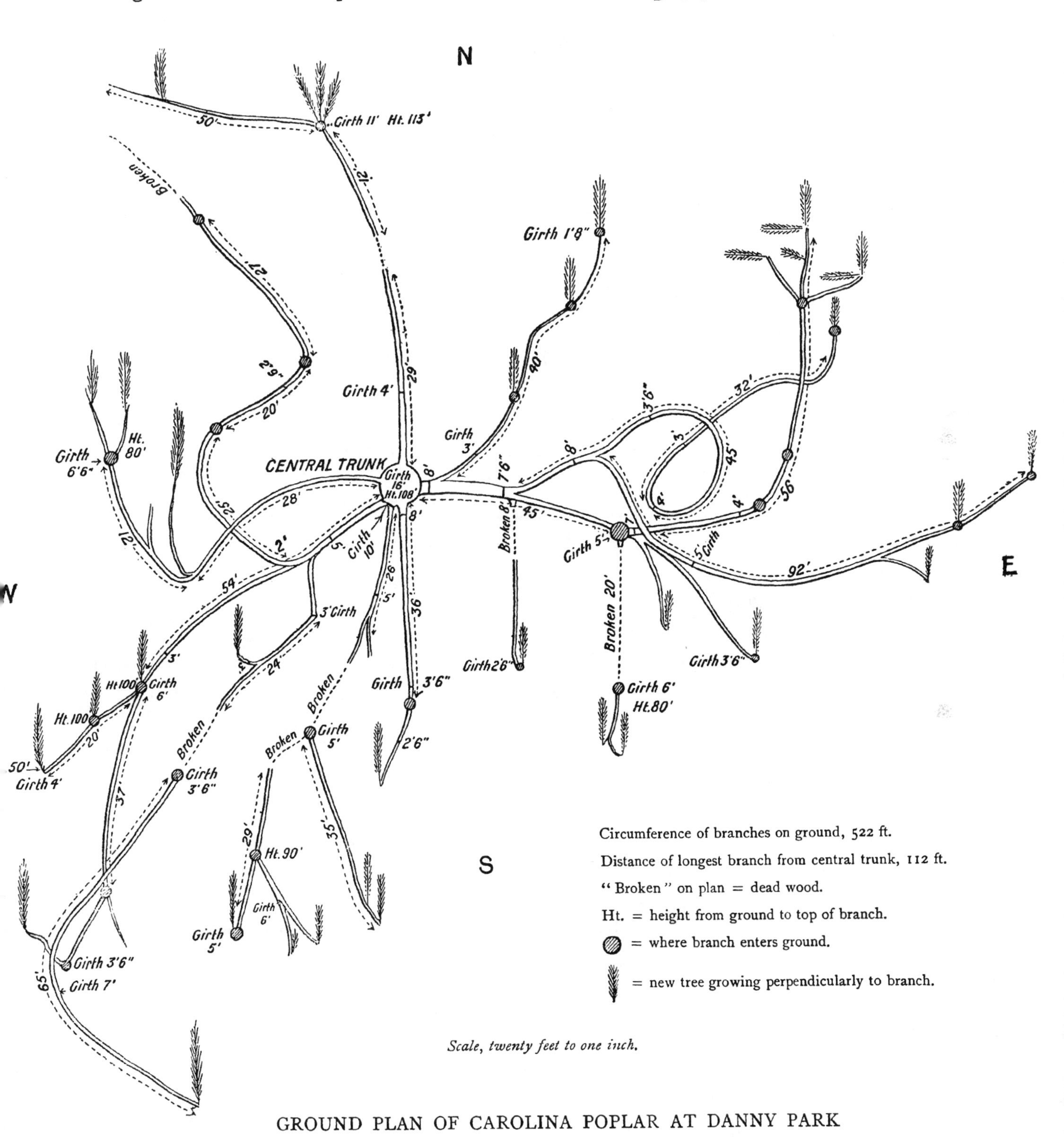


GROUND PLAN OF CAROLINA POPLAR AT DANNY PARK

avenues, and averages over 100 ft. in height, and 7 to 8 ft. in girth, much surpassing in size *P. nigra* which has been planted beside it. No tree in this region appears to have such a great demand for light; and in consequence it readily loses its lower branches, and quickly produces clean stems with useful timber. Prof. Voglino,[1] who has made a careful study of the poplars in cultivation in the park of Santena, near Turin, states that this species grows more slowly, but attains a greater height than *P. serotina.* The wood, though slightly darker in colour, is much stronger, and is always very free from knots. Young trees, however, have brittle branches, which suffer severely even from slight falls of snow. (A. H.)

## History of the American Black Poplars and their Hybrids in Europe

*P. angulata* was early introduced into France, as it is mentioned as a well-known tree by Duhamel du Monceau,[2] who described it in 1755 as the "peuplier noir de Virginie à très grandes feuilles, et dont les jeunes pousses sont relevées d'arêtes qui les font paraître quarrées." This species was in cultivation[3] in England in 1738.

The second American species of black poplar, *P. monilifera,* Aiton, appears also to have been early introduced into France, where, according to Michaux, it was probably imported from Canada. He accurately describes it under the name *P. canadensis,* and states that in cultivation it did not succeed in compact and clay soil and frequently branched into two near the base of the trunk. It was mentioned[4] by Fougeroux de Bondaroy in 1786 as the *Peuplier de Canada,* though he also called it *P. virginiana,* and states that he had only seen the pistillate tree. It is said[5] to have been introduced into England in 1769.

Poplars hybridise very freely, especially in France, where seedlings are much more frequently found[6] than in England; and probably the first hybrid to be selected was the tree now known as *P. serotina,* which, according to tradition, has been known in France as the *peuplier suisse* since the middle of the eighteenth century.[7] Its botanical characters show that it is a cross between one of the two American black poplars, presumably *P. monilifera,* and the glabrous form of the European *P. nigra.* It is quite unknown in America.[8] Michaux, who erroneously named it *P. monilifera* in 1813, is clear on the subject, as he states that "neither my father, myself, nor any of the educated English who traversed the Atlantic states, and a great part of the states of the West, ever found this species of poplar." He adds that it was commonly called *peuplier suisse* because it was more cultivated in Switzerland than elsewhere; and that only the male tree existed in France, which was always propagated by cuttings.

[1] *I Nemici del Pioppo Canadense di Santena,* 8 (1910). This valuable work, pp. 130, figs. 1-16, appeared in *Ann. R. Acad. Agric. Torino,* liii. (1910), and gives an account of the fungoid and insect enemies of the cultivated black poplars.

[2] *Traité des Arbres,* ii. 178, pl. 38, fig. 8 (1755). The figure represents exactly *P. angulata.*

[3] A branch, gathered by Miller in the Chelsea garden in 1738, is preserved in the British Museum.

[4] In *Mém. d'Agric. Paris,* 1786, p. 87.

[5] Loudon, *Trees and Shrubs,* 825 (1842), under *P. canadensis.*

[6] Loudon, *Trees and Shrubs,* 827 (1842), stated that many thousands of seedlings came up annually in the walks of Fontainebleau, most of which were destroyed, but some varieties had been selected from them.

[7] It seems to have been the poplar described and figured by Duhamel, *Traité des Arbres,* ii. 178, pl. 39, fig. 5 (1755), as "*P. nigra, foliis acuminatis, dentatis, ad marginem undulatis,* ou mal à propos *osier blanc.*" It was supposed by Duhamel to be a variety of *P. nigra,* and was planted in vineyards, instead of willow, where it was treated as coppice, the shoots being cut annually. Fougeroux, in *Mém. d'Agric. Paris,* 1786, i. 80, states that it grew with great vigour even in dry soils, and was called *osier blanc* in the districts south of Paris, and *alain* in other places.

[8] Prof. Sargent assured us that he had seen no tree in America which resembled *P. serotina.*

He quotes from Foucault that its branches are less spreading than the Canadian poplar, and that on account of its more rapid growth and greater aptitude to thrive on dry soils, it had been very largely planted in all parts of France.

Poiret's account[1] of the black poplars existing in France in 1804 is instructive. He enumerated five kinds :—

1. *P. nigra*, European species.

2. *P. angulata*, American species.

3. *P. monilifera*, American species, accurately described.

4. *Peuplier suisse*, which he regarded as of European origin, and therefore gave it the botanical name of *P. nigra helvetica* ; but he added that it resembled in many ways the next kind.

5. *P. canadensis*, apparently a hybrid from the description.

Mirbel,[2] also in 1804, distinguished five black poplars in cultivation :—

1. *P. nigra*.

2. *P. angulata*.

3. "*P. monilifera*, Aiton, with ciliate leaves," the true American species.

4. *Peuplier suisse*, also known as "*P. virginiana*, Hort. Paris."

5. "*P. monilifera*, Hort. Paris, with non-ciliate leaves," quite distinct from *P. monilifera*, Aiton, and commonly known as *peuplier de Canada*.

I am unable to identify with certainty the second hybrid, *P. canadensis*, Poiret, or *P. monilifera*, Hort. Paris ; but it was evidently distinct from the *peuplier suisse*, and was not the true American species ; and possibly may have been identical with *P. regenerata*, which originated independently in 1814, near Paris.

Poiret,[3] in 1816, added another poplar to his list of 1804, *P. marilandica*, Bosc, which on account of its rhombic leaves, with long points, similar to those of *P. nigra*, appears to be a second cross, probably the glabrous form of *P. nigra*, pollinated by *P. serotina*.

*P. angulata*, on account of its remarkable large leaves and conspicuously winged branchlets, was never confused with the other species and hybrids, and has continued to be cultivated both in England and on the Continent. The other American species, which was difficult to grow except in favourable moist soils, gradually ceased to be cultivated, and is now a rare tree in Europe. As the existence of hybrids was not suspected, both botanists and cultivators assumed the hybrids to be true species from America ; and in course of time there were practically only four black poplars generally recognised in France :—

1. *P. nigra*.

2. *P. angulata*, known generally as *peuplier de Caroline*.

3. *P. serotina*, known either as *peuplier suisse* or *peuplier de Virginie* ; always a staminate tree.

4. *P. marilandica* and *P. regenerata*, known commonly as *peuplier de Canada*, both pistillate trees.

The origin of the other hybrids is mentioned in their descriptions, which follow ;

[1] In Lamarck, *Encycl. Meth.* v. 235-239 (1804).
[2] In *Nouveau Duhamel*, ii. 186 (1804).
[3] In Lamarck, *Encycl. Meth. Suppl.* iv. 378 (1816).

and none appears to have arisen in England, except possibly *P. Lloydii*, one of the parents of which is the English or pubescent form of the black poplar.

*P. serotina* was first accurately and completely described by Hartig; and this name is preferable to *P. nigra helvetica* or *P. helvetica*. It has always been known in England as the black Italian poplar, and appears to have been imported from France, sometime before 1787, when Messrs. Dickson of Hassendeanburn, Scotland, sold some stock to Pontey.[1] Mr. A. Dickson, much later, in 1813, informed Pontey[1] that this poplar was obtained by his firm from a gentleman in Scotland, who received it from his son in America; but this account is unreliable, as there is no evidence that this poplar originated in America, and it fails to explain the name "black Italian poplar," by which it was always sold. As *P. serotina* and the Lombardy poplar came into England about the same period, both poplars, arriving from the Continent, were supposed to be from Italy, and hence the name.

(A. H.)

## POPULUS SEROTINA, Black Italian Poplar

*Populus serotina*, Hartig, *Naturges. Forst. Culturpfl. Deutschl.* 437 (1851); Schneider, *Laubholzkunde*, i. 11 (1904); Dode, in *Mém Soc. Hist. Nat. Autun*, xviii. 44 (1905); Ascherson and Graebner, *Syn. Mitteleurop. Fl.* iv. 44 (1908).

*Populus helvetica*, Poederlé, *Man. de l'Arbor*, ii. 148 (1792).

*Populus monilifera*, Michaux f., *Hist. Arb. Amer. Sept.* iii. 295, pl. 10, fig. 2 (1813) (not Aiton, *Hort. Kew.*); Loudon,[2] *Arb. et Frut. Brit.* iii. 1657 (1838) (excluding the pistillate tree).

*Populus nigra helvetica*, Poiret, in Lamarck, *Encycl. Méth.* v. 234 (1804).

*Populus virginiana*, Mirbel, in *Nouveau Duhamel*, ii. 186 (1804) (not Fougeroux).

*Populus canadensis*, Mathieu, *Flore Forestière*, 495 (excluding the pistillate tree) (1897) (not Michaux).

A large tree of hybrid origin (cf. p. 1814), attaining 130 ft. or more in height and 18 to 20 ft. in girth, with a single undivided straight stem, free from burrs, and slender wide-spreading ascending branches; bark regularly furrowed. Young branchlets at first green, turning brownish yellow in summer, glabrous, with white lenticels, slightly angled, becoming greyish and terete in the second year. Buds brownish, viscid. Leaves (Plate 409, Fig. 16), opening latest of all the poplars, with a reddish bronze tinge, glabrous, averaging on adult trees 3 in. in breadth and length, ovate-deltoid, with a broad truncate base and a short cuspidate or acuminate apex; margin crenate-serrate, with the serrations few and wide apart at the base of the blade; cilia short, at first continuous on the two sides of the blade and sparse on the base, deciduous in summer; glands near the insertion of the petiole, variable, one, two, or none being present; petiole reddish. On young plants and vigorous shoots, the branchlets have projecting ribs like those of *P. angulata*, and the leaves are much larger, up to 5 or 6 in. in length or more.

Staminate catkins about 3 in. long; axis glabrous; flowers on very short glabrous pedicels; scales early deciduous, obovate, with short irregular lobes

[1] *Profitable Planting*, 218 (1814).

[2] Loudon, in *Gard. Mag.* xiii. 536 (1837), says that this "is a very doubtful native of America, and much more likely, in our opinion, to be an improved European tree."

terminating in long filaments; stamens about twenty to twenty-five, with long slender white filaments and deep red anthers; disc oblique, shallow, concave, glabrous, crenate or slightly lobed in margin.

1. Var. *erecta*, Henry.

*Populus monilifera erecta*, Selys-Longchamps, in *Bull. Soc. Bot. Belg.* iii. 11, 13 (1864).

A fastigiate form [1] of *P. serotina*, with which it agrees in sex, branchlets, buds, and leaves, the latter being late in unfolding.[2] (A. H.)

More remarkable for its habit than its size is this poplar, which was described by the late Baron de Selys-Longchamps, a distinguished Belgian naturalist. I visited the Château de Longchamps near Waremme, about twenty miles west of Liège in 1908, on purpose to see these trees, and am indebted to the baron for a photograph (Plate 385), which shows their peculiar habit very well. The original tree was planted in a meadow close to the village of Willines in 1818, and was procured by chance from a neighbouring nursery. It is somewhat past its prime, and measures about 120 ft. by 8 ft. 8 in., having a somewhat less fastigiate habit than Lombardy poplars growing in the same field. It is a male tree with reddish petioles. The trees shown in the photograph were grown from cuttings of it, which were planted in 1862. The tree which has been pruned measures 120 ft. by 7 ft. 4 in.; the unpruned one 120 ft. by 8 ft. 2 in. The Lombardy poplar planted next to it at the same time is only 100 ft. by 6 ft. 4 in. The earlier leafing of the Lombardy poplars is well brought out in the photograph. On the other side of the same meadow there is a line of poplars, which though they seemed identical in foliage with *P. serotina*, had whitish bark; and I was informed by M. Edmond de Selys, who has now succeeded his father, that these were liable to be injured by a canker in the branches, from which the dark-barked form planted in the same field was free.

I may add that as an ornamental tree, the fastigiate form seems to be at least as good as the Lombardy poplar, whilst its timber is more valuable, its growth more rapid, and its hardiness superior. Cuttings from the fastigiate variety were kindly sent me by the baron, and are growing vigorously at Colesborne. Henry saw in 1912 a specimen about 25 ft. high in the Calmpthout Nursery near Antwerp.

(H. J. E.)

2. Var. *aurea*,[3] Henry.

*Populus canadensis aurea van Geerti*, André, in *Illust. Hort.* xxiii. 26, t. 232 (1876); Dippel, *Laubholzkunde*, ii. 200 (1892).

A sport, with yellow foliage, probably referable to *P. serotina*. This form was produced spontaneously on a single branch of a large tree in 1871; and was propagated by Ch. Van Geert in his nursery, which is now the Société Horticole de Calmpthout. We have seen no flowers or large trees of this variety; and the yellow colour of the foliage at Kew does not last throughout the season. (A. H.)

[1] This tree has been erroneously identified by Koch and other German dendrologists with the *peuplier régénéré* of Carrière, which is a pistillate tree, of quite independent origin. See p. 1824.

[2] Specimens obtained by Elwes from the trees planted in 1862. Baron Selys-Longchamps, in *Belg. Horticole*, 1864, p. 257, states that there were two original trees of the fastigiate variety, which were planted in 1818 with 60 ordinary *peupliers suisses*. They resembled the latter in all respects, except in habit; but proved to be less vigorous, as they had not attained, 46 years after planting, as great a girth.

[3] Simon-Louis, *Catalogue*, 1869, p. 73, mentions a variety with variegated leaves, which we have not seen.

### Cultivation

Though no English writer has as yet fully realised the economic importance of the black Italian poplar, and though consumers have hardly recognised the value of its timber for many purposes; yet in France and Belgium it is more generally planted than any other tree, and, as I shall be able to show, will produce a quicker and larger return than any other, if properly grown and converted into timber. None of the poplars are particular as to soil, provided it is moist in summer; and though, like all other trees, they grow faster and larger on good than on bad soils, yet they succeed in cold wet and undrained valleys and meadows, and are resistant to frost at all seasons, and are little liable to fungoid and insect attacks. From an ornamental point of view the black Italian poplar is inferior to the grey and the white species, but its large red clusters of flowers appearing in spring on the bare branches are very beautiful, and its extremely rapid growth makes it suitable for situations where no other tree will attain a large size in a lifetime.

Before planting poplars it is important to have a variety which has proved hardy and vigorous on a similar soil and situation; and though nothing seems likely to surpass *P. robusta* and *P. Eugenei* where the climate suits them, I shall not give up the propagation of the red-petioled *P. serotina* until the others have proved their ability to endure the worst vicissitudes of our climate. The green-petioled variety, though it seems equally vigorous and hardy in any soil, is not so erect, and seems to have more spreading branches.

Owing to the general confusion between this tree and the true black poplar which has prevailed among English botanists and foresters, few of whom seem to have distinguished the two trees, I must point out that many of the statements[1] which have been made about the latter species really apply to the former.

Grigor states[2] that on a sandy soil (probably in Morayshire) the black Italian poplar in a mixed plantation, twenty-four years planted, measured 60 ft. by 3½ ft., when larch was 48 ft. by 3 ft 4. in., beech 40 ft. by 2½ ft., sycamore 34 ft. by 2 ft. 1 in., and Scotch elm 33 ft. by 3 ft.

Thirty trees planted by my father in a cold clay soil not worth 5s. per acre, in a situation remarkably subject to late and early frosts, in forty-eight years averaged 120 cubic feet and realised £3 each; and two trees planted by myself in a clay soil close to the stream at Colesborne attained in fifteen years 56 ft. by 5 ft. 3 in. and 50 ft. by 3 ft. 2 in. respectively. Fifteen trees felled recently in a more shaded position and closer together in the same valley were at eighty-five to ninety years old of considerably smaller girth, and had increased but little during the last thirty years, some of them being more or less decayed and hollow at the base. This lot, though much older, only averaged 70 cubic feet, and proved to me that a sunny situation and plenty of room are essential. In another part of the same valley where the land is

[1] In *Trans. Surveyors' Institution*, 1904, p. 226, my advocacy of *P. serotina* as a valuable timber tree was erroneously printed under the heading Black Poplar, instead of Black Italian Poplar; and my remarks have been quoted by Nisbet and Sir Herbert Maxwell (in Green, *Encyclopædia of Agric.* iii. 308) as referring to *P. nigra*.

[2] *Arboriculture*, 326 (1868).

densely covered with meadowsweet and rushes, and where water often stands on the surface in winter, poplars of the ordinary red-petioled kind ten years after planting are about 35 feet high, whilst one with green petioles, distinguished by its greyer bark and more spreading habit, is 44 ft. by 2 ft. 1 in.

Forbes[1] agrees with me that about fifty years is the most profitable age at which to cut this tree; but he recommends that it should be planted at 6 ft. apart unmixed with other trees, in which I disagree with him—first, because at least three-quarters of the trees planted would be worthless as thinnings, and, secondly, because if it is desired to make a plantation close enough to suppress side branches the common alder seems to me the most suitable tree. I should not plant poplars nearer than 15 or 20 feet apart, and would fill up the intervals with alder, which could be cut out for clog soles at about thirty years, when it had attained 6 to 8 in. diameter, leaving the poplars to stand at the rate of about 50 to the acre. Assuming that the alder would pay the rent of the land, which it ought to do, 50 poplars, averaging 80 ft. at £2 a-piece, would give a very handsome return for the interest of the outlay on planting and pruning them.

In France, where the cultivation of poplars is well understood, the general practice is to plant them in lines at about 20 ft. apart along the sides of ditches, and leave the intervening spaces in pasture; and in this way the trees attain a profitable size much quicker than if planted as closely as advised by Forbes.

As the tree can be very quickly and easily struck from cuttings, the only question to consider is whether they shall be planted without roots where they are intended to grow, as is done with willows, or kept in a nursery till larger. I have tried both methods, and think that the latter is best, as the tree gets a start sooner, and is not so liable to be choked by the rank vegetation which is always found in places suitable to it. The best way to procure strong cuttings is to take one- or two-year shoots from stools; side branches may be used, but they are not so erect in habit, and require more pruning. These cuttings are best from 4 to 6 ft. long, put in during early spring, and in good soil will make strong plants in one season. If left longer than a year they do not transplant so well. After the tree has begun to grow freely, it will make 3 to 5 ft. of annual growth; and it is important to prune the side branches before they become too thick. This pruning should be carried on with a pruning chisel up to 30 or 40 ft. high, in order to avoid knots and make clean timber, and is best done in summer.

In cold and shady situations the branches are apt to die off, and a canker produced by *Didymosphæria populina* sometimes affects the trees.[2]

In a brochure which is published by M. Marion,[3] he recommends planting either large cuttings, put into the ground at a depth of 1 to 2 ft., or rooted plants in pits; and in both cases advises that the grass shall not be allowed to grow over the roots till the trees are well established. He recommends a width of 6 to 10 yards for avenues, or 4 yards apart in lines, and that the trees shall never be mixed with or crowded by other trees, as they must have plenty of room to grow well—150 to

[1] *English Estate Forestry*, 82 (1904).
[2] *Cf.* Hartig, *Diseases of Trees*, Eng. Trans., 104 (1894).
[3] *Petit Manuel du Propriétaire Sylviculteur* (Libraire Horticole, Paris (1909)).

200 to the hectare ($2\frac{1}{2}$ acres) being thick enough on the ground. He finds that though they grow best on damp soil, yet it should be drained first if the water stagnates; and that successive crops may be taken, at periods of thirty years, without diminishing the growth, provided the old stumps are grubbed. Pruning is considered essential, and this must be frequently attended to, up to a height of 20 to 30 ft. Though poplars are often pruned nearly up to the top in France, it is evidently very prejudicial to their good growth. The mistletoe, which infests these trees to such an extent in most parts of France, is not allowed to remain on the branches at Pontvallain, but when it appears on the trunk it cannot be eradicated, and usually ends by killing the tree.

### Remarkable Trees

One of the largest poplars in England was cut down in March 1907, near Cassio Bridge, Herts, and was sold to Messrs. East and Son, of Berkhamsted. It was described and figured in the *Timber Trades Journal* of 13th April 1907, and when measured by Sir Hugh Beevor in 1902 was 130 ft. high and 16 ft. 11 in. in girth, and the contents of the butt alone were 56 ft. by $42\frac{1}{2}$ in. quarter girth, making 701 cubic feet. With the top and branches it was said to contain upwards of 1000 ft. of timber. Messrs. East inform me that at 15 ft. from the butt they counted only 97 rings, which would make the age of this tree little over 100 years. Probably no other tree on record in England has attained so great a size in so short a time. Sir Hugh Beevor tells me that another tree near the same place, though not quite so tall, measures 18 ft. in girth at 6 ft. from the ground.[1]

A very large tree was blown down early in January 1908 in the meadow of Christ Church, Oxford. It was mentioned in the *Oxford Times* of 18th January 1908 as a Lombardy poplar, but leaves sent me by the late Prof. Fisher show that it was *P. serotina.* The approximate measurements of the trunk were given as follows: length 55 ft., girth 16 ft. at twelve feet from base, timber contents 1056 cubic feet, weight about 20 tons. The removal of this tree by a traction engine, aided by several horses, took several days.

Mr. A. B. Jackson measured in 1912 a tree growing in a dell by the stream near the old church at Albury, Sussex, which he could not make less than 150 ft. high, though on account of its situation it is difficult to measure accurately; its girth was 15 ft. 3 in.

At Shalford House, near Guildford, the property of Col. H. H. Godwin-Austen, F.R.S., there stands in a damp meadow an immense tree of this species, which in 1911 measured about 110 ft. high by 23 ft. in girth. Its upper branches were much damaged by a violent storm some years ago; and I am informed by its owner, who has a good photograph of it, that it was formerly at least twenty-five feet higher. It forks first at about ten feet up, and gives off very large branches at about twenty feet, which spread to a width of thirty-seven paces. This tree is known to have been

[1] *Cf.* D. Hill, in *Trans. Herts. Nat. Hist. Soc.* xiv. 133 (1911).

planted by Charles X. of France, who lived at Shalford House during his exile from France, sometime between 1789 and 1814.

Among notable trees of this species which I have measured are two at Belton, near Grantham—one in a meadow by the bridge, which, though difficult to measure, I believe to be at least 130 ft. high by 16 ft. in girth. Another by a pond in the private grounds is 125 ft. by 15 ft. (Plate 386). Most of the large trees of *P. serotina* at Belton were planted about 1818, and began to fall to pieces in 1907, from natural decay. After the death or cutting of the tree (but never before) suckers are produced.

Another even larger grows on the banks of a pond near the approach to Woburn Abbey; and though it has been injured by the wind, measured, in 1905, 125 ft. by 19 ft. 3 in.[1] On the banks of the Thames there may be larger trees than any I have seen; but at Fawley Court, near Henley, I saw one about 105 ft. high by 16 ft., with a clean bole 50 ft. long; and at Mapledurham, near the mill, there is a group of very tall slender trees, one of which measured, in 1907, 135 ft. by 6 ft. 7 in.

At Bicton, just outside the gardens, there is a very fine tree of this species, which I believe to be about 130 ft. high, though I could not measure it exactly, by 17½ ft. in girth. Sir Hugh Beevor measured a tree at Petworth in 1904 which was 114 ft. by 18 ft. 10 in., with a bole 20 ft. long. At Mote Park, Maidstone, there is a group of fine trees, one of which in 1902 was 120 ft. by 11½ ft. At Godinton, near Ashford, in Kent, I counted 100 rings on the stump of a tree 15 ft. in girth. At Hawstead, near Bury St. Edmunds, I measured in 1905 one of a row of young trees in a rich meadow, which, though only 6 ft. in girth, was 116 ft. high.

At Plas Machynlleth, Montgomeryshire, I saw in 1912 a black Italian poplar, about 100 ft. high, and 19½ ft. in girth at 3 ft. up under a large limb, and 17½ ft. in girth at 6 ft. from the ground, above the limb; a label attached to the tree stated that it was planted in 1794 by Sir John Edwards.

In Scotland the black Italian poplar is not so common as in England, but seems to grow equally well. At Scone Palace, Henry measured in 1904 one no less than 132 ft. high by 15 ft. 4 in. in girth, which eleven years previously was 14 ft. 9 in. It carried its full girth nearly to the first branch, over forty feet from the ground. Another, also measured by Henry, at Monzie, Perthshire, was 125 ft. by 9 ft. 2 in., with an absolutely straight trunk drawn up by beech trees, and clean to 71 ft. There are two good trees near Achnacarry, the seat of Cameron of Lochiel, one of which in 1910 measured 105 ft. by 10½ ft. Renwick in 1909 measured two trees at Cambusnethan House, Lanarkshire, 112 ft. by 11 ft. 1 in. and 119 ft. by 11 ft. 5 in. respectively.

In Ireland it has not been planted so generally as in England. The best I have seen are at Adare, Co. Limerick, near the river, a tree 95 ft. by 15 ft., in 1909; at Abbeyleix, Queen's County, several fine old trees, one of which was 120 ft. by 13 ft. 4 in. in 1910

On the Continent there are many large trees, of which the most interesting are those at Brunswick which formed the types described by Hartig. I visited Brunswick on purpose to see them in June 1910, and am much indebted to the late Dr. Blasius and to Mr. A. Hollmer for their guidance. The largest tree stands in the

[1] In 1911 this tree had increased in girth to 20 ft. 6 in.

Theatre Park, close to a pond, and is about 110 ft. by 18 ft. 9 in., with a bole about 30 ft. long. The habit was more spreading than usual in England, and the young leaves seemed less bronzy in tint. The second is a double-stemmed tree from a common base, in front of the Grand-ducal residence; it is about 110 ft. high, the two stems measuring 14 and 15 ft. respectively. The third, said to be the actual type described by Hartig, stands in the Railway Park, and has the trunk much covered with ivy. It was severely lopped about ten years ago, and now measures about 105 ft. by 16 ft. 4 in., but has been taller. All these trees are males and appear to be of the same age, probably about a hundred years.

In Denmark the black Italian poplar is rather commonly planted by the road as a shade tree; and I measured a fine old tree at Gisselfeld, the seat of Count Danneskjold-Samsö, 120 ft. by 14 ft. 9 in., in 1910. How far east and south this poplar has extended we are unable to say, as Hartig's description has been apparently overlooked by foresters, and many of the trees in Germany[1] named *P. canadensis* are *P. serotina.*

### Timber

The timber of the various species of poplar is, or has been in the past, so little valued by merchants that "Acorn," in his work on English timber, speaks of it as hardly worth hauling for any great distance from the place where it grows. For many years it was supposed to be the best material for making railway brake-blocks, which are now commonly made of iron, and was also used for beds of wagons and wheel-barrows, for second-class spade handles, and to some extent for other purposes,[2] but its use for packing-case making has been entirely ignored in England.

In France, however, this poplar is one of the most common and abundant timbers, and is almost the only material used in making wine-cases and packing-boxes of all descriptions; and there is no reason why it should not be so used in England, except the very low price of foreign deal. From fivepence to eightpence per foot is the price which I have been able to realise for standing trees of black Italian poplar containing 100 ft. and upwards, growing seven to nine miles from a station; but I have no doubt that, for well-grown trees whose lower branches have been pruned when young, and which can be converted with little waste, eightpence to a shilling per foot could be obtained when near the place of conversion.

"Acorn" states that in buying poplar standing, the merchant is almost sure to gain a considerable advantage if the trees are measured or estimated in one length, as they carry their girth higher and taper less than most trees. I have had good evidence of this myself; as a tree, which was estimated standing to contain 110 to 125 ft., and sold on the higher estimate, actually contained 130 ft. as measured in one length after felling, and 165 ft. when measured in three lengths, as it would have been cross-cut for sawing.

[1] In *Mitt. Deut. Dend. Ges.* 1904, p. 19, a tree at Schloss Dyck, near Düsseldorf, named *P. monilifera*, which is probably *P. serotina*, is said to be 57 metres high by 5.25 metres in girth; but the height measurement is probably much exaggerated.

[2] A large quantity of black Italian poplar is used as blocks for polishing plate glass in the course of its manufacture at St. Helens and other places. Cf. also *Quart. Journ. Forestry*, vi. 264 (1912).

The conversion of poplar timber into boards for various purposes requires considerable experience, which English sawmills do not seem to possess. A number of specimens were shown in the Franco-British Exhibition of 1908 which displayed great ingenuity in conversion; and I am assured by reliable authorities that, when properly cut and seasoned, poplar boards are considered a perfectly suitable wood on which to lay veneers, and are used in France for all but the highest qualities of furniture. The increasing price of mahogany and of white wood (*Liriodendron*), on which veneers in England are laid, seems to point to an increased demand for poplar. Mr. A. Howard gave me some handsomely figured veneers showing a good deal of satiny lustre, cut from a flitch of poplar wood which he bought in France. These I have used for the front of a large bookcase, made from the burry wood of an old black poplar (*P. nigra*) grown in Herefordshire. From an ornamental, as well as from a structural point of view, the result is satisfactory.

M. Breton-Bonnard,[1] President of the French Timber Merchants' Society, classifies the timber of the various poplars as follows:—

*White Poplar, Grey Poplar.*—Considered the best poplar wood; wood white, light, tender and soft, capable of taking a good polish. Used by furniture makers, largely by railway companies for wagon-building, by coach-builders for the panels and bodies of carriages and carts, and in mines and for boat-building. It must be thoroughly dry, as it warps when green.

*Aspen.*—Used when sound for the same purposes, also for matches, toys, turning, carving, broom handles, and for packing-cases and paper pulp.

*Black Poplar* (*P. nigra*).—This wood, which is very stiff, warps much if cut soon after it is felled, and ought to lie a year in the log before converting. It is valued by trunk makers, and for wagon beds, and other objects subject to rough usage.

*All the Hybrids.*—All kinds of cheap furniture and painted work, packing-cases, for laying veneers, etc.

He states[1] that poplar sawdust is the best litter for stables and pig-sties, and the cheapest litter used by the Paris Omnibus Company, and gives figures to prove that the manure produced from it was equal or superior in effect, on beetroot and wheat, to that produced by peat-moss litter or straw. He gives[1] the following table, representing the returns of *peupliers régénérés* on the assumption that the increase in diameter is 10 millimetres, or $\frac{2}{5}$ inch per annum:—

| Age. | Diameter at 1.33 m. from the Ground. | Girth at the same height. | Length of Timber. | Cubic Contents, ¼ Girth Measure. | Value at 17 francs per cubic metre = about 5d. per foot. |
|---|---|---|---|---|---|
| | | | | Cubic Metres. | Francs. |
| 5 years. | 0.08 m. | 0.25 m. | 5 m. = 16 ft. | ... | ... |
| 10 „ | 0.18 m. | 0.56 m. | 8 m. = 25 ft. | 0.09 | 1.53 |
| 15 „ | 0.28 m. | 0.86 m. | 10 m. = 33 ft. | 0.30 | 5.10 |
| 20 „ | 0.38 m. | 1.20 m. | 12 m. = 39 ft. | 0.70 | 12.00 |
| 25 „ | 0.48 m. | 1.50 m. | 14 m. = 45 ft. | 1.26 | 21.42 |
| 30 „ | 0.58 m. | 1.82 m. | 16 m. = 54 ft. | 2.10 | 35.70 |
| 35 „ | 0.68 m. | 2.05 m. | 18 m. = 60 ft. | 3.30 | 56.10 |
| 40 „ | 0.78 m. | 2.48 m. | 20 m. = 66 ft. | 4.80 | 81.60 |

[1] *Le Peuplier*, 176, 179, 182 (1902).

These returns, which are possible when the trees have been well cultivated, justify the common saying in France that poplar ought to produce one franc per tree annually. The price given in the table is often exceeded where the trees are of good size and in a favourable situation. Twenty-five to thirty francs per cubic metre is a common price, as I was informed by M. Marion at Pontvallain. M. Breton-Bonnard says that in 1900 the poplars of the valley of the Ourcq, near Paris, were sold at fifty francs per cubic metre and upwards; and he knew a case where a landowner took a meadow on a thirty years' lease, planted it with poplars, and, after paying the rent with the hay[1] and grazing, was able, by cutting down the trees before the end of his lease, to buy the land with the profit.

I certainly know of land in England, which could be bought at £10 per acre or less, that would produce in forty to fifty years poplars containing an average of 100 cubic ft., and if there were only thirty trees to the acre, and the price sixpence per foot, this would amount to 3000 cubic ft., worth £75 per acre.

According to Mathey,[2] the timbers of the different poplars may be distinguished as follows:—

A. Species in which the heart- and sap-wood are confused.
*Black and Lombardy Poplars.*—Heart-wood with black veins in old trees.
*Black, Italian, and other Hybrid Poplars.*—Heart-wood uniformly white or slightly reddish.

B. Species in which the heart- and sap-wood are distinct.
*White Poplar.*—Sap-wood white; heart-wood distinctly reddish.
*Grey Poplar.*—Sap-wood reddish; heart-wood reddish-brown.

C. *Aspen.*—Heart- and sap-wood confused in trees growing in the forests of the plain; heart-wood distinct and vinous red in trees growing on the hills and mountains.

(H. J. E.)

## POPULUS REGENERATA

*Populus regenerata*, Schneider, *Laubholzkunde*, i. 7 (1904).
*Peuplier régénéré*, Carrière, in *Rev. Hort.* 1865, pp. 58 and 276; Lambin, in *Rev. Hort.* 1873, p. 47.
*Populus canadensis*, var. *grandifolia*, Dieck, *Nacht. Haupt. Verz. Zöschen*, 1887, p. 16.

A tree of hybrid origin (cf. p. 1815), resembling *P. serotina* in branchlets and foliage, but bearing pistillate flowers, and unfolding its leaves about a fortnight earlier. Pistillate catkins similar to those of *P. marilandica*, but with usually only two stigmas.

This hybrid, according to Carrière, originated in 1814 in the nursery of M. Michie at Arcueil near Paris, and was apparently a seedling of unknown origin,

[1] Balzac, *Vie de Province*, i. 255 (1855) gives calculations by Eugénie Grandet, showing that the loss of hay, due to the shade of the poplars which he had planted in lines on good grass land on the banks of the Loire, near Saumur, was not made up by the proceeds of the sale of their timber, if compound interest was allowed. Grandet drew the conclusion, not generally accepted in France, that poplars could only be grown at a profit on poor soil. A correspondent in the *Gard. Chron.* 1855, p. 102, states that near Diss in Norfolk, 336 black Italian poplars, which were planted at intervals between 1819 and 1822, were sold in 1854 for £124. They grew on an area of 4⅛ acres; and whilst the trees were standing, the pasture beneath them was let at five shillings an acre. This tree grows in low, marshy, boggy land, where almost every other species ceases to thrive.—A. H.

[2] *Traité d'Exploitation Commerciale des Bois*, i. 23 (1906).

though one of the parents was supposed to be *P. serotina* (*peuplier suisse*). Carrière believed it to be simply the pistillate form of the latter, distinguishable by its greater vigour and more conical stem, swollen at the base and not so cylindrical as *P. serotina*. M. Romanet of Montmirail obtained some plants, which he multiplied under the name of *Peuplier régénéré*; and M. Bujot, of Chiary near Château Thiery (Aisne), sold others as *Peuplier Bujot*, a name no longer in use. In 1865 this poplar had become very common in the valley of the Ourcq, and was then sold by M. Terré, at Lizy-sur-Ourcq (Seine et Marne).

The *Peuplier suisse rouge*, one of the two so-called "Eucalyptus" poplars, now propagated by M. Marion in his nurseries at Pontvallain, Sarthe, is identical in all its characters with *P. regenerata*, being a female tree with similar foliage and flowers. I examined flowers which were kindly sent to me in the spring of 1911; and I obtained good specimens of branches with leaves, when I visited Pontvallain in August 1912. This tree was chosen as the fastest-growing variety of poplar in 1880, by M. Sarcé, grandfather of the present proprietor. It had been obtained by a continuous selection for many years, as sets for planting, of the uppermost and straightest branches of the most vigorous trees. The variety is said to have been called *rouge* on account of the reddish colour of the petioles, which is supposed [1] to distinguish it from the other kind of "Eucalyptus" poplar in the same nursery. The latter, known as *Peuplier suisse blanc*, because it is reputed to have whitish petioles, is a less vigorous tree, which is propagated only to a slight extent, its sole merit being that it forms a straighter stem. This variety opens its leaves late in the season, fifteen days after the *Peuplier suisse rouge*, and is reputed to be a male tree.[2] It is apparently a form of *P. serotina*, with which it agrees in leaf, but I have had no opportunity of examining the flowers.

The cultivation of poplars at Pontvallain is extensive, covering about 500 acres. The land, which is flat and intersected by a stream, is laid out in grass meadows of small size, around which are single or double lines of poplars, pruned up to 30 ft., and at least 12 ft. apart. Experience here shows that this is the best method of obtaining a quick and large yield of timber. Poplars, especially the quick-growing kinds, being light-demanding trees, do not bear well any lateral shade. Attempts to grow them crowded in large plantations never succeed, as all the trees, except those on the outside, remain very slender, and yield timber too small in diameter to be of any value for planking. The soil here is a sandy loam, and is rich in humus, as it was formerly a coppice of oak and hornbeam. The growth of the red "Eucalyptus" poplar is astonishing on this soil, almost double that of the other kind in diameter, and somewhat more in height. The stems are invariably curved near the base, which is attributable to its rapid increase in height while in the young stage. At thirty-five years old they average at least 100 ft. in height, and 7 ft. in girth, producing over 100 cubic feet of timber, which is worth about 100 francs. Elwes, who visited Pontvallain in May 1908, measured two trees in an avenue planted twenty-two years; one was 114 ft. by 6 ft. 2 in.; the other was 118 ft. by 6 ft. 10 in.

[1] No difference in the colour of the petioles was discernible in August.

[2] Cf. Breton-Bonnard, *Le Peuplier*, 62 (1902).

Whether either of the "Eucalyptus" poplars will equal in vigour the black Italian poplar, if planted in this country, remains to be proved; but the trial is worth making, as they are evidently most profitable to grow in France.

Other reputed forms of *P. regenerata* or of *P. serotina*, which are classed under the term *Peupliers régénérés*, are mentioned by M. Breton-Bonnard; but none of these have as great a reputation as those which are propagated at Pontvallain.

Specimens sent by the late Dr. Blasius from trees growing in the garden of Cramer von Klausbruch at Brunswick, which he considered to be the pistillate form of *P. serotina*, are similar in foliage to *P. regenerata.*

This hybrid has lately come into cultivation in England, the only large tree which I have seen being one in the Queen's Cottage grounds, Kew, which displays no vigour of growth. A tree at Kew, obtained under the name *P. deltoidea erecta* from Dieck in 1889, and now about 35 ft. high, produced female flowers in 1911, which showed it to be identical with *P. regenerata*, of which it has the foliage and habit. A female tree at Glasnevin, cultivated under the name *P. régénéré*, and about 60 ft. in height and 4 ft. 11 in. in girth in 1912, resembles *P. Eugenei* in habit, and is very thriving. It was obtained from Simon-Louis in 1892.

Mention may be made here of a peculiar poplar[1] at Beauport, Sussex, 60 ft. by 3 ft. in 1909, which was blown down in August 1912, the day before I last visited this remarkable collection of trees of all kinds. It differed from *P. regenerata* in having smaller leaves, 2 to 2¼ in. long and broad, similar to those of *P. monilifera*, var. *occidentalis*, in shape, size, and few coarse serrations; but like the hybrids in the sparse ciliation of the margin and the irregular number (0, 1, or 2) of glands at the base. It bore female catkins in April 1912; and the flowers with only two stigmas showed also its hybrid origin. A similar tree, judging from a specimen in the Kew Herbarium, grew at Carlsruhe in 1845. (A. H.)

## POPULUS EUGENEI

*Populus Eugenei*, Simon-Louis, *ex* Koch, *Dendrologie*, ii. pt. i. p. 493 (1872); Schneider, *Laubholzkunde*, i. 9 (1904); Dode, in *Mém. Soc. Hist. Nat. Autun*, xviii. 46 (1905); Ascherson and Graebner, *Syn. Mitteleurop. Fl.* iv. 45 (1908).

*Peuplier Eugène*, Carrière, in *Rev. Hort.* 1865, p. 58.

*Populus pyramidalis meetensis*, Mathieu, in *Gartenflora*, xxxvi. 674 (1887), translated in *Gard. Chron.* ii. 818 (1887).

A narrow pyramidal tree of hybrid origin, with a straight undivided stem and numerous short branches, mostly ascending at an angle of 30° to 45° with the stem. Young branchlets glabrous, slightly angled. Buds small, reddish-brown, viscid. Leaves (Plate 409, Fig. 17) unfolding early in the season, with a reddish tint, smaller than those of *P. serotina*, averaging 2½ in. in width, usually broadly cuneate, rarely

[1] This peculiar hybrid appears to be identical with *P. incrassata*, Dode, *op. cit.* 41 (1905), described as a pistillate tree. having flowers with two stigmas; leaves thick in texture, moderate in size, deeply cordate at the base, acuminate at the apex. The leaves of the Beauport tree, often cordate, occasionally truncate at the base, agree with those of a specimen kindly given me by Dr. Dode as the type of his *P. incrassata.*

truncate at the base, and terminating in a slender sharp-pointed non-serrated acuminate apex; margin with coarse crenate incurved serrations, few and wide apart on the base, and sparsely ciliate except near the insertion of the petiole, which with the midrib and veins is usually of a reddish tint; glands at the base variable in number.

Staminate catkins, 1½ to 2 in. long; axis slender, glabrous; scales obovate, concave, with irregular lobes ending in long filaments; disc shallow, cup-shaped, glabrous, entire in margin; stamens about fifteen to twenty, with white thread-like filaments and red anthers.

Messrs. Simon-Louis informed Carrière in 1865 that this tree, which bears staminate flowers, originated in their nursery at Plantières in 1832. It appeared in a seed-bed of silver firs, so that it was impossible to know from what poplar the seed had come. It is supposed to be a seedling of *P. marilandica* (*P. canadensis*, Hartig), which had been pollinated by a Lombardy poplar. The shape of the leaves, intermediate between those of the supposed parents, and the narrow pyramidal habit confirm to some degree this explanation of its origin.

(A. H.)

So far as I know, no attempt has ever been made by English nurserymen to improve poplars by selection, or to raise them from seed; but on the Continent, where they seed much more freely, some varieties have been selected by the firm of Simon-Louis frères at Metz. On this rich deep calcareous loam, poplars succeed to perfection; and the original tree of *P. Eugenei* is a marvellous instance of the size which a planted tree may attain in a man's lifetime. This was planted in 1834, and measured about 140 ft. high in August 1908. I was unable to see the topmost branch from the measuring point. At the ground it was 39 ft. round, at five feet 22½ ft. The main trunk divides at about 30 ft. into several immense limbs, one of which, broken off by the wind, was about 85 ft. long. A younger tree, planted in 1870 in rather better soil, measured, in 1908, 128 ft. by 14 ft., with a clean bole 56 ft. long, and contains 500 to 600 cubic feet, which at thirty-eight years old must I think, be a record for any species of planted tree. As M. Jouin told me that the rapidity of growth increases after the first twenty years, this tree is likely to surpass its parent.

I have planted this variety at Colesborne, where its growth, though not so rapid as in France, is very satisfactory, and have found it to resist late and early frosts without injury.

There are eight examples of this tree at Kew, which were procured from Metz in 1888. The two largest measured[1] in June 1912, 90 ft. by 5 ft. 1 in. and 84 ft. by 4 ft. 5 in.; whilst the others ranged from 50 to 60 feet in height and 2 ft. 4 in. to 3 ft. 5 in. in girth. All preserve the narrow pyramidal form, and are growing vigorously in sandy soil. The bark is slightly fissured into narrow longitudinal ridges. Another tree at Glasnevin in poor soil, planted in the same year, was about 55 ft. high and 4 ft. 5 in. in girth in April 1913. (H. J. E.)

[1] Cf. *Kew Bull.* 1911, p. 310, where the height of the largest tree is given in excess.

## POPULUS MARILANDICA

*Populus marilandica,*[1] Bosc, *ex* Poiret, in Lamarck, *Encycl. Suppl.* iv. 378 (1816).
*Populus canadensis,*[2] Hartig, *Naturges. Forstl. Culturpfl.* 436 (1851) (not Michaux); Koehne, *Deut. Dendr.* 81 (1893); Schneider, *Laubholzkunde,* i. 7 (1904) (in part); Ascherson and Graebner, *Syn. Mitteleurop. Fl.* iv. 33 (1908) (in part).
*Populus euxylon,* Dode, in *Mém. Soc. Hist. Nat. Autun,* xviii. 41, 69 (1905).

A tree of hybrid origin (cf. p. 1815) less vigorous in growth and not attaining so great a height as *P. serotina,* with branches wider apart and not regularly ascending. Young branchlets glabrous, rounded. Buds small, viscid. Leaves (Plate 409, Fig. 19) considerably earlier in unfolding than *P. serotina,* resembling in shape more those of *P. nigra* than those of the American parent; when well developed about 4 in. long and 3 in. broad, rhomboid, cuneate at the base, tapering above into a long acuminate apex; glabrous; serrations crenate, with incurved points, few and wide apart on the cuneate base; margin with scattered deciduous minute cilia; glands absent or one or two in number at the base; petiole greenish.

Pistillate catkins about 2½ in. long; axis slender, glabrous; pedicels stout, glabrous; scales obovate, with irregular lobes ending in long filaments; disc cup-shaped, glabrous, undulate in margin; stigmas sessile, yellowish, variable in number —two, three, or four—each appressed at the base to the summit of the glabrous globose ovary, and dividing above into two erect arms. Fruiting catkins, fertilised by other kinds of poplars, about 4 or 5 in. long; capsules usually three-valved on slender pedicels, ultimately immersed in the dense white silky wool of the seeds.

This tree, which is always female, is not uncommon on the Continent, where it is usually considered to be the pistillate form of the Canadian species. It is occasionally met with in nurseries; and should not be selected for planting either for ornament or profit, as its masses of cottony catkins in late spring are disagreeable, and its vigour of growth[3] is considerably inferior to that of *P. serotina.*

The largest tree of this hybrid which we know in England is growing on the lawn near the Palm House at Kew, and measures 90 ft. by 8 ft. 9 in. Its history is unknown. On 5th July 1907 it was covered with downy seeds, a few of which proved fertile; and there is a young plant at Colesborne which was raised from them by Miss F. Woolward. (A. H.)

[1] The description by Bosc is imperfect, and there is no authentic specimen in the Paris Herbarium. A specimen in the Herbarium at Montpellier, from a tree cultivated in the garden there in 1833 under the name *P. marilandica,* and another in the Kew Herbarium labelled *P. marilandica,* from a tree cultivated at Carlsruhe in 1845, were probably correctly named, and may be accepted as the species meant by Bosc. The name *P. marilandica* may be objected to, as implying that the tree is a native of Maryland; but I prefer it to the later name of *P. euxylon,* Dode.

[2] Koehne's specimen of *P. canadensis* in the British Museum is identical with the Carlsruhe specimen. So far as I can judge, the hybrid female poplar, which I consider to be *P. marilandica,* is the *P. canadensis* of German dendrologists. Hartig's description is unmistakable, as he refers to its rhombic and scarcely ciliate leaves, with variable glands at the base, and to its peculiar stigmas. Schneider states that *P. canadensis* is nearly always a female tree in cultivation, and figures its rhombic leaves cuneate at the base.

[3] Simon-Louis, *Catalogue,* 1869, p. 72, speaking of this tree as *P. canadensis,* states that it is much less vigorous than the *peuplier de Virginie,* meaning by the latter name *P. serotina,* Hartig.

## POPULUS HENRYANA

*Populus Henryana*, Dode, in *Mém. Soc. Hist. Nat. Autun*, xviii. 39 (1905).

A large tree of hybrid origin, similar in bark to *P. serotina*, but differing in habit, forming a wide-branching round-headed crown of foliage. Young branchlets glabrous, becoming dull grey in the second year, with whitish raised lenticels. Buds viscid, with slightly pubescent scales. Leaves (Plate 409, Fig. 18) opening earlier than those of *P. serotina*, and not tinged with a bronzy red tint, ovate to ovate-triangular, about 3 in. long and 2¼ in. wide, truncate or slightly cuneate at the broad base, ending above in a long acuminate apex; serrations small, crenate, numerous, with incurved points; margin sparsely ciliate; glands minute, usually two present, occasionally absent; petiole reddish, glabrous. Leaves on the short shoots smaller, usually as broad as long, ovate, cuneate at the base.

Flower buds with densely pubescent ciliate scales, very viscid. Staminate catkins about 1½ in. long; axis glabrous; scales deeply divided into irregular lobes ending in long or short filaments; disc glabrous; stamens, thirty to thirty-five, with white filaments and dull red anthers.

This hybrid, which has peculiar pubescent buds, is of unknown origin and is staminate. The only tree which we have seen is in the park in front of the house at The Wilderness, White Knights, Reading. It measured in 1907 about 100 ft. by 14 ft., and was probably obtained as a young plant from France, where M. Dode[1] has observed trees of a similar kind. (A. H.)

## POPULUS ROBUSTA

*Populus robusta*, Schneider, *Laubholzkunde*, i. 11 (1904); Dode, in *Mém. Soc. Hist. Autun*, xviii. 45 (1905); Ascherson and Graebner, *Syn. Mitteleurop. Fl.* iv. 45 (1908).
*Populus angulata cordata robusta*, Simon-Louis, *Catal.* 61 (1899).

A tree of hybrid origin remarkably vigorous in youth, with ascending branches. Young branchlets angled, minutely pubescent; on strong shoots, stout, grey with white lenticels, and with projecting ribs as in *P. angulata*. Buds reddish brown, viscid. Leaves (Plate 409, Fig. 20) unfolding early in the season with a deep red tint,[2] variable in shape, about 3 in. in length, ovate-deltoid or rhombic; cuneate, rounded, or truncate and sub-cordate at the base; acuminate at the apex; serrations coarse, with incurved tips, few and wide apart on the base; margin with scattered cilia; glands one, two, or absent; petiole reddish, with scattered short hairs.

Staminate catkins, borne on a tree at Glasnevin, 2½ in. long; flowers numerous, crowded; axis glabrous; scales deeply and irregularly lobed, ending in long

[1] M. Dode tells me that there is a fine tree at Paris in the Champs de Mars, near the Eiffel Tower, which was planted about 1888. There was an old tree, no longer living, in the Jardin de Luxembourg. He adds that the propagation by cuttings of this poplar fails in the open air, but succeeds under glass.

[2] As seen at Kew, it is the most beautiful of all the poplars when coming into leaf.

filaments; disc small, glabrous, oblique; stamens about twenty, with white filaments.

*P. robusta* is readily distinguished from most of the other hybrid poplars by the minute pubescence on the branchlets and petioles. It is said to have been raised about 1895 at Metz from seed of a female *P. angulata*, supposed to have been fertilised by the pollen of the large tree of *P. Eugenei* which grows near it; but this is improbable, as the branchlets of the latter tree, as well as those of *P. angulata*, are quite glabrous. In all probability the male parent was the common pubescent *P. nigra*, from which *P. plantierensis*, its fastigiate variety, was also derived at Metz. (A. H.)

I saw in the nursery at Metz a large bed of plants from cuttings put in in 1909, which were 10 to 12 ft. high in 1910, and was told that its stem is straighter and more erect than that of the *peuplier régénéré*. The extraordinary vigour of the young trees justifies M. Jouin in believing that it will, in warm districts at least, surpass *P. Eugenei* in size. Whether it will succeed in England remains to be proved; but it has been largely planted for several years by Baron Aheré in Belgium, and in low damp ground near Metz, which is specially liable to spring frost. I have now had it planted for two years in the same situation in which I grow *P. serotina* at Colesborne. It seems perfectly hardy, but has not yet had time to show whether it is superior in vigour to *P. Eugenei* or *P. serotina*. A tree planted at Glasnevin in 1900 measured 42 ft. by 1 ft. 1 in. in April 1913. It is very narrow, almost columnar in habit, with short ascending branches. (H. J. E.)

## POPULUS LLOYDII

*Populus Lloydii*, A. Henry (*hybrida nova*).

A tall tree of hybrid origin, with bark similar to that of *P. serotina*. Young branchlets slender, covered with a minute erect pubescence; glabrous and yellowish brown in the second year. Buds small, brownish, viscid. Leaves (Plate 409, Fig. 21) about 2½ in. wide and long; truncate, rounded, or cuneate at the base; tapering above into a short non-serrated acuminate or cuspidate apex; margin with crenate serrations, ending in incurved points, ciliate till late in summer; glands at the base minute, variable, often absent; petiole reddish, with a scattered minute erect pubescence.

Pistillate catkins, 2 to 2½ in. long; axis glabrous; pedicels short, glabrous; ovary globose, glabrous, in an oblique cup-shaped entire glabrous disc, crowned by two, rarely three, widely dilated spreading yellow stigmas. Fruiting catkins, about 4 in. long; capsules two-valved.

This remarkable hybrid, of which the parents are probably the common English black poplar (*P. nigra betulifolia*) and *P. serotina*, resembles the former in the pubescent branchlets and petioles, and the latter in the shape, ciliation, and glands of the leaves, which are bronze-coloured when opening. The leaves, being borne on old trees, are probably smaller than normal; and this tree, which is pistillate, is scarcely to be distinguished in technical characters from *P. robusta*, which is staminate and bears considerably larger leaves, known from young trees only. *P. robusta* appears to differ, however, in the grey and not yellowish colour of the branchlets in their second year. (A. H.)

The only large trees which we have seen of this are at Leaton Knolls, Shrewsbury, the residence of Major Lloyd, who has no record of their origin or date of planting. The largest, which is growing on a hill at the edge of an old pit-hole, now dry, was 120 ft. by 13½ ft. in July 1910. The second tree is 110 ft. by 11 ft. 2 in., and the third is 95 ft. by 10½ ft., as measured by myself in July and by Major Lloyd in September 1910. None of the down lying on the ground in July contained good seed as far as I could see. The trees are clearly of the same age —possibly sixty or seventy years—and being all females, perhaps originated from cuttings of a tree which may have existed in the neighbourhood, where *P. nigra* is not uncommon.

This hybrid may possibly in some cases have been hitherto confused by botanists with *P. nigra*, var. *betulifolia*; but it is apparently rare. The only other specimen which we have seen is a tree[1] in a garden near Turnham Green Station, which is about 35 feet high. It produced numerous natural seedlings in 1907, ten of which were transplanted into the nursery at Kew, and were in 1912 vigorous plants 2 to 3½ ft. high. (H. J. E.)

## POPULUS ANGUSTIFOLIA

*Populus angustifolia*, James, *Long's Expedition*, i. 497 (1823); Sargent, *Silva N. Amer.* ix. 171, t. 492 (1896), and *Trees N. Amer.* 159 (1905); Schneider, *Laubholzkunde*, i. 14 (1904); Dode, in *Mém. Soc. Hist. Nat. Autun*, xviii. 58 (1905); Gombocz, in *Math. Termes. Közl.* xxx. 105 (1911).

*Populus salicifolia*, Rafinesque, *Alsograph. Amer.* 43 (1838) (not Loudon).

*Populus canadensis*, Desfontaines, var. *angustifolia*, Wesmael, in De Candolle, *Prod.* xvi. 2, 329 (1868).

*Populus balsamifera*, Linnæus, var. *angustifolia*, Watson, *King's Rep.* v. 327 (1871).

*Populus coloradensis*, Dode, in *Mém. Soc. Hist. Nat. Autun*, xviii. 58 (1905).

A tree, attaining in America 60 ft. in height and 5 ft. in girth. Bark smooth, yellowish green, becoming fissured at the base of old trunks. Young branchlets glabrous, rounded, yellowish grey. Buds minute, viscid, sharp-pointed. Leaves (Plate 410, Fig. 26), lanceolate on long shoots, resembling those of *Salix fragilis* in shape, about 2 to 4 in. long, ¾ to 1 in. wide, cuneate at the base, gradually tapering to a gland-tipped acute or rounded apex, glabrous, pale green and not whitish beneath; margin revolute, with close fine glandular serrations; lateral nerves, about fifteen pairs, all pinnate; petiole short, glabrous, flattened above. On the short shoots, the leaves become shorter and broader, almost rhombic in outline.

Catkins densely flowered, glabrous; scales obovate, with irregularly cut dark brown filiform lobes. Stamens twelve to twenty in a cup-shaped slightly oblique disc, with a thickened reflexed margin. Ovary two-lobed, with two stigmas, enclosed

[1] The pistillate catkins of this tree were figured in *Bot. Mag.* t. 8298 (1910), and show three-styled flowers, which never occur in true *P. nigra*. Cf. p. 1796, note 1.

in a shallow cup-shaped crenate disc. Fruiting catkins, 4 in. long; capsule two-valved, glabrous, on a $\frac{1}{8}$ in. long pedicel.

This species is a native of the Rocky Mountain region of North America, usually growing on the banks of streams between 5000 and 10,000 ft. It occurs as far north as south-western Assiniboia, extending southward through the Black Hills of Dakota, Montana, eastern Idaho, Wyoming, Utah, Colorado,[1] to central Nevada, Arizona, and New Mexico.

It is readily distinguishable amongst the balsam poplars by its willow-like leaves, which scarcely show any whitish tint beneath.

It was introduced into cultivation by Späth[2] of Berlin, who received young plants from Colorado in 1893. It forms at Kew small trees of spreading irregular habit, and may be looked upon as rather a shrub than a tree in this country. (A. H.)

## POPULUS BALSAMIFERA, Balsam Poplar

*Populus balsamifera*, Linnæus, *Sp. Pl.* 1034 (*excl. syn.* Catesby et Gmelin) (1753); Loudon, *Arb. et Frut. Brit.* iii. 1673 (in part) (1838); Sargent, *Silva N. Amer.* ix. 167, t. 490 (1896), and *Trees N. Amer.* 157 (1905); Schneider, *Laubholzkunde*, i. 14 (1904); Dode, in *Mém. Soc. Hist. Nat. Autun*, xviii. 62 (1905); Gombocz, in *Math. Termes. Közl.* xxx. 108 (1911).

*Populus Michauxi*, Dode, *op. cit.* 62 (1905).

A tree, attaining in America 100 ft. in height and 20 ft. in girth. Bark at first smooth, light reddish brown; on old trunks deeply divided into broad rounded ridges, Young branchlets terete, without projecting ridges, glabrous. Buds elongated, sharp-pointed, exuding a yellowish strong-smelling resin. Leaves (Plate 410, Fig. 27) on long shoots averaging 4 in. long and 2 in. broad, ovate, rounded at the base, narrowing towards the apex, which is often abruptly acuminate, glabrous on both surfaces, whitish and often tinged with rusty red beneath; margin minutely and sparsely ciliate, with crenate serrations, ending in short incurved glandular points; lateral nerves about eight pairs, each of the lowest pair giving off at its origin usually one secondary nerve, making with the midrib the base of the blade pseudo-five-palminerved; petiole quadrangular, channelled above, with a minute scattered pubescence. Leaves on short shoots smaller, broader in proportion to their length.

Staminate catkins about 3 in. long; axis with a few scattered hairs; pedicels long and similarly pubescent; scales broadly obovate, often irregularly three-lobed at the apex, with numerous short thread-like divisions; stamens about twenty on an oblique crenate deep saucer-shaped glabrous disc. Pistillate catkins: disc cup-shaped; ovary ovoid, two-lobed, with two nearly sessile large oblique dilated crenulate stigmas. Fruiting catkins 5 in. long; capsule ovoid, curved at the apex, two-valved, on a slender pedicel about $\frac{1}{12}$ in. long.

This species, extending over a wide area in North America, is probably variable; and may hybridise with *P. candicans*. The form distinguished by Dode as

[1] F. von Holdt, of Arvada, Colorado, in *Mitt. Deut. Dend. Ges.*, 1912, pp. 118, 119, describes this poplar in its native home, and gives a fine photograph of it growing on the edge of a mountain lake.

[2] *Catalogue*, No. 91, p. 49 (1893-1894).

*P. Michauxi* has slightly pubescent branchlets and leaves which have occasionally a subcordate base, and is intermediate between the two species; but it has the narrow leaves of *P. balsamifera*, and cannot be confused with *P. candicans*, which has broadly ovate deeply cordate leaves. (A. H.)

*P. balsamifera* is confined to North America, where it is known as Balsam or Tacamahac, and ranges from far north in Alaska and Canada southwards to northern New England, New York, central Michigan and Minnesota, the Black Hills of Dakota, north-western Nebraska, northern Montana, Idaho, Oregon, and Nevada. It is the largest of sub-arctic American trees,[1] attaining its greatest size on the Peace river and other tributaries of the Mackenzie river, where, according to Macoun, it is often nearly 150 ft. high, with a trunk occasionally over 7 ft. in diameter and free of branches from 60 to 100 ft. up. It is the characteristic tree along the streams in the prairie regions of British America, and is common throughout the northern border of the United States, growing on alluvial lands liable to floods and on the borders of swamps.

According to Aiton it was cultivated at Hampton Court in 1692, and was again introduced in 1731, when a tree given by Queen Caroline to Sir Hans Sloane was planted in the Chelsea Botanic Garden. It was introduced[2] into Scotland in 1768 by seeds sent from Canada. We have seen no pistillate trees of this species, which has become exceedingly scarce in cultivation, being now almost entirely supplanted by *P. candicans*, the balsam poplar usually grown by nurserymen in England. It differs from *P. candicans* in habit, being a narrower tree with ascending branches, and seems to be short-lived in our climate, which is perhaps too warm[3] for it.

The largest tree which we have seen in England is one (Plate 387) at Bute House, Petersham, which in 1905 was 71 ft. high by 7 ft. 10 in., and surrounded by a great number of suckers. The leaves of the suckers, as in other species of poplars, attain a large size, occasionally 6 to 8 inches in length. At Kew there are two small trees, obtained from Späth in 1905. At Syston Park, Lincolnshire, there was a tree of considerable size, growing on an island in the lake, which was cut down thirty years ago; but numerous suckers remain, about 18 ft. high, which produce flowers abundantly. A younger tree at Belton, 20 ft. high, planted fifteen years ago, produced staminate flowers, which were drawn by Miss F. H. Woolward. A tree at New Humberstone, Leicester, of which specimens in flower were collected by Mr. H. Burbank, was 50 ft. by 5 ft. 3 in. in 1905. It has since been topped.

In Scotland there is a large tree in the park at Castle Menzies, which looks older than any other I have seen, and in 1907 measured 68 ft. by 11 ft. 10 in. Another very old-looking tree grows by the schoolhouse at Achnacarry, and has bark more like that of an ash than a poplar. It measured in June 1910 about 60 ft. by 7 ft. I also saw several good-sized trees in Glen Urquhart.

[1] E. T. Seton, *Arctic Prairies*, 330 (1912), says: "The balsam poplar attains a large size on the lower Athabasca and the Slave rivers, at least 100 ft. We observed it as far as the eastern extremity of Great Slave Lake, but there it is scarcely more than a shrub. The leaves had partly turned colour near Caribou island on September 22."

[2] Walker, *Essays Nat. Hist.* 65 (1812).

[3] Britton and Shafer, *N. Amer. Trees*, 172 (1908), state that this poplar is not much planted south of its natural range in America, as it does not well endure hot summers.

There are small trees doing badly at Glasnevin; but Mr. R. A. Phillips has seen four thriving trees at Staffordstown, Co. Antrim, the largest of which measured 40 ft. by 4 ft. in 1910.

According to Schübeler, it thrives in Norway as far north as Tromsö, lat. 69° 40′, and succeeds in the Gudbrandsdal at an elevation of 1350 ft. (H. J. E.)

## POPULUS CANDICANS, Ontario Poplar

*Populus candicans*, Aiton, *Hort. Kew.* iii. 406 (1789); Michaux, *Hist. Arb. Am.* iii. 308 (1813); Loudon, *Arb. et Frut. Brit.* iii. 1676 (1838); Schneider, *Laubholzkunde*, i. 13 (1904); Dode, in *Mém. Soc. Hist. Nat. Autun*, xviii. 65 (1905); Gombocz, in *Math. Termes. Közl*, xxx. 115 (1911).

*Populus Tacamahaca*,[1] Miller, *Gard. Dict.* ed. 8, No. 6 (1768).

*Populus macrophylla*, Lindley, *ex* Loudon, *Encyc. Plants*, 840 (1829).

*Populus ontariensis*, Desfontaines, *Cat. Hort. Paris* (1829).

*Populus balsamifera*, Linnæus, var. *candicans*, Gray, *Manual*, 419 (1856); Sargent, *Silva N. Amer.* ix. 169, t. 491 (1896), and *Trees N. Amer.* 159 (1905).

A tree, similar in size and bark to *P. balsamifera*, but different in habit, with more spreading branches, forming a broad crown of foliage. Young branchlets terete or slightly angled, but without projecting ridges, pubescent. Buds large, reddish brown, sharp-pointed, parallel with the twig, exuding resin; scales ciliate. Leaves (Plate 410, Fig. 22) variable in size, averaging 5 to 6 in. in length, and 4 in. in width, broadly ovate-deltoid; widest near the base, which is cordate, sub-cordate, or rarely truncate; cuspidate at the apex; margin conspicuously ciliate, with coarse serrations, ending in incurved glandular points; upper surface green, with scattered pubescence, glandular at the junction of the petiole; lower surface whitish, but usually less so than in *P. balsamifera* and *P. trichocarpa*, and with scattered hairs, dense on the midrib and nerves; nervation as in *P. balsamifera*; petiole terete, grooved above, densely pubescent.

Staminate flowers not seen. Pistillate catkins; axis covered with stiff hairs; flowers numerous on short glabrous pedicels; scales broadly obovate, fimbriated with long thread-like lobes; ovary glabrous, half enclosed in a deep cup-shaped glabrous crenate disc, and surmounted by two sub-sessile large dilated crenulate yellowish or orange stigmas. Fruiting catkins,[2] 6 in. long; capsule, on a glabrous $\frac{1}{8}$ in. long pedicel, with two glabrous tuberculate valves; the whole catkin covered with the down of the seed after the capsules have opened.

A variegated form of the species is known, in which the leaves are blotched with yellow. Specimens of this may be seen at Woburn, and in the Glasnevin and Edinburgh Botanic Gardens.

This species is very distinct from *P. balsamifera*, and is closely allied to *P. ciliata* of the Himalayas, from which it differs in flowers, though the shape of the leaf is very similar.

[1] Miller's diagnosis applies plainly to this species; but his detailed description includes also *P. balsamifera*. I have, therefore, not thought it desirable to resuscitate Miller's name, though the oldest, for the Ontario poplar.

[2] Described from a tree at Belton, perhaps fertilised by some other species of poplar.

*P. candicans* appears to be unquestionably[1] a native of North America, though Sargent states that he has seen no wild specimens, and that it "does not appear to be indigenous in New England or eastern Canada, where the pistillate plant[2] has been used as a shade tree from very early times, as it has been in the Middle States and in Europe." L. H. Barclay,[3] however, states that there was a grove of this species, "with many large trees, at South Haven, Michigan, when the first pioneers visited the place, and these appeared to be coeval with the surrounding forest, with which they were interspersed for some distance back from the lake shore." Gates states[4] that on the west coast of Michigan, north of Waukegan, the sand dunes in some places are surmounted by narrow groves of *P. candicans*, consisting of trees of moderate size, which are associated with *Prunus pumila*. Seedlings of this poplar are rarely found on the dunes themselves, but are common on the adjoining heaths. A specimen in the Kew Herbarium, collected by Fernald along the St. John River in Aristook County, Maine, has a note attached stating that it is common north of lat. 47° along river banks and in low-lying woods. No further information of a positive kind is obtainable concerning its distribution; but Dame and Brooks[5] state that trees of both sexes are found by collectors in New Hampshire and Vermont; while in central and southern New England the staminate tree is rarely, if ever seen. These authors conclude that the evidence points to the habitat of the wild tree being a narrow belt extending through northern New Hampshire, Vermont, the southern section of Ontario, and Michigan. Macoun says that it is the prevalent balsam poplar in Ontario, and that it is apparently wild in the neighbourhood of Picton, Nova Scotia.

(A. H.)

The Ontario poplar was introduced into England in 1772, and has been widely planted ever since, owing its popularity, no doubt, to the fact that it is more easily propagated by cuttings than the true balsam poplar, which it surpasses also in beauty of foliage. The Ontario poplar has no economic value; and as it is always female,[6] producing downy fruit in quantity, and also produces suckers freely, it is objectionable in ornamental plantations.

Nearly all the balsam poplars in England belong to this species, and the finest specimen is probably a tree at Syon, which measured, in 1906, 85 ft. by 9 ft. Another at Bayfordbury was 70 ft. by 6½ ft. in 1910. Of the numerous trees which we have seen in other places, none are noteworthy as regards size, 60 to 70 ft. being apparently the average in England for full-grown trees.[7]

(H. J. E.)

[1] Michaux, however, states that it was common in Rhode Island, Massachusetts, and New Hampshire, but always planted. He had never seen it in forests. It is mentioned by Duhamel, *Traité des Arbres*, ii. 181 (1755), as a poplar found in the neighbourhood of Quebec, where it was known as *liard*, with a leaf like a maple, white beneath, and exhaling a very odorous balsam.

[2] Sargent, however, figures staminate as well as pistillate flowers.

[3] In *Bot. Gaz.* v. 91 (1880).

[4] *Bull. Illinois State Lab. Nat. Science, Urbana*, ix. 287, plate xlviii. fig. 2 (1912).

[5] *Trees of New England*, 37 (1902). Britton and Shafer, *N. Amer. Trees*, 169 (1908), apparently do not agree with these authors, and simply state: "Evidence that it is wild in Michigan has been adduced, and it is probably indigenous farther to the north-west."

[6] It is evident from Fougeroux, in *Mém. d'Agric., Paris*, 1786, i. pp. 91, 94, that in France at that early period, *P. candicans* was always female, and *P. balsamifera* always male, as is now the case in England. His account is however confused, as he transfers the name *liard* to *P. balsamifera*, and *beaumier du Pérou* to *P. candicans*, the converse being correct.

[7] In *Mitt. Deut. Dend. Ges.* 1904, p. 19, a tree named *Populus balsamea*, growing at Schloss Dyck near Dusseldorf, was reported to be 38 metres high by 4.25 metres in girth. It is unlikely that this tree was any species of balsam poplar.

## POPULUS TRICHOCARPA, Western Balsam Poplar

*Populus trichocarpa*, Torrey et Grey, *ex* Hooker, *Icon. Plant.* ix. t. 878 (1852); Sargent, *Silva N. Amer.* ix. 175, t. 493 (1896), and *Trees N. Amer.* 161 (1905); Schneider, *Laubholzkunde*, i. 16 (1904); Dode, in *Mém. Soc. Hist. Nat. Autun*, xviii. 64 (1905); Jepson, *Flora California*, 346 (1909), and *Silva California*, 188 (1910); Gombocz, in *Math. Termes. Közl.* xxx. 112 (1911).
*Populus balsamifera*, Linnæus, var. γ, Hooker, *Fl. Bor. Am.* ii. 154 (1839).
*Populus balsamifera*, Lyall, in *Journ. Linn. Soc.* (*Bot.*) vii. 134 (1864) (not Linnæus).
*Populus angustifolia*, Newberry, *Pacific R. R. Rep.* vi. pt. iii. 89 (1857) (not James).
*Populus hastata*, Dode, in *Mém. Soc. Hist. Nat. Autun*, xviii. 64 (1905).

A tree, attaining on the Pacific coast of North America 200 ft. in height and 20 ft. in girth. Bark of young stems peeling off in papery scales; on old stems grey and deeply divided into broad rounded scaly ridges. Young branchlets glabrous, shining, brown, marked with white linear lenticels, angled and with five prominent ridges on vigorous shoots, often retained in the second year. Buds brownish, very fragrant, resinous, glabrous, elongated, sharp-pointed, parallel with, but not appressed to the twig. Leaves (Plate 410, Fig. 31) variable in size, very large on upper vigorous branches; on lower terminal shoots averaging 5 in. long and 3 in. broad; ovate or ovate-deltoid; slightly cordate, truncate, or rounded at the base; broadest near the base, gradually narrowing towards the gland-tipped acuminate apex; upper surface light green, glabrescent; lower surface whiter than in the other balsam poplars, with a few short scattered hairs; lateral nerves about seven or eight pairs, each of the lowest pair giving off at its origin two secondary nerves, making with the midrib the base of the blade pseudo-seven-palminerved; margin ciliate, with incurved crenate glandular shallow serrations, often absent on large leaves towards the apex; petiole reddish, minutely pubescent, terete, channelled above, about 1 in. in length.

Staminate catkins (described from a living tree at Kew), about 2½ in. long; axis green, pubescent, crowded with numerous sessile flowers; scales with numerous filiform ciliated divisions, quickly deciduous; stamens about fifty, on an oblique orbicular flat glabrous disc, which has an entire ciliate margin; filaments slender, white; anthers deep red. Pistillate catkins, according to Sargent, with a loosely-flowered tomentose axis; ovary densely tomentose, with three nearly sessile broadly dilated, deeply lobed stigmas, enclosed in a crenate or entire deep cup-shaped disc. Fruiting catkin, 4 in. to 5 in. long; capsule sub-sessile, globose, thick-walled, three-valved.

This species, which is readily distinguished from the other balsam poplars by its very white leaves, winged twigs, and bark peeling off in thin papery shreds, was founded by Hooker on a specimen collected by C. C. Parry in 1850, on the Santa Clara river near Buenaventura, California. This specimen and others from southern California in the Kew Herbarium, have small deltoid leaves scarcely acuminate at the apex, and with considerable pubescence on the midrib and veins beneath and on the petiole; the capsules are also very pubescent. Specimens from Oregon, Washington, and British Columbia have usually larger leaves, much longer than broad, distinctly acuminate, and with less pubescence; and their capsules are sometimes nearly glabrous. The latter form, which is the one in cultivation, is considered by Dode (*op. cit.* 64) to be a distinct species, *P. hastata*; and may, on further investigation in the field, turn out to be worth ranking as a distinct variety, *P. trichocarpa*, var. *hastata*. Jepson also mentions a form *ingrata* from San Bernardino county, which has small lanceolate leaves.[1]

This magnificent poplar, the largest of the genus, is a native of the Pacific coast region of North America from southern Alaska to San Diego county in California, extending inland in British Columbia as far as the valley of the Columbia river, and in California to the western slope of the Sierra Nevada. It is reported to occur in eastern Washington[2] and in western Montana;[3] but the balsam poplar in these regions is possibly *P. balsamifera*. *P. trichocarpa* grows mainly in open groves in river valleys, attaining its largest size near sea-level in the coast region north of California; southward, it is a small tree rarely more than 40 ft. high,[4] ascending to 6000 ft. altitude in the Sierra Nevada in central California. It is most abundant[5] in Oregon and Washington, where its timber is used for making staves and wooden ware. Jepson states that the wood is light, soft, and fairly close-grained, but not strong. (A. H.)

It is the largest deciduous tree[6] of the Pacific coast region, attaining its greatest size on Puget Sound, where it is sometimes 200 ft. in height and 6 to 8 ft. in diameter. I saw it in perfection on a farm in Vancouver Island called Swallowfield, some miles north of Duncans, and measured the tree here figured (Plate 388), which was growing in a meadow on rich alluvial soil, and was at least 140 ft. high with a bole 70 ft. long and 28 ft. 3 in. in girth. Another tree, with the top broken, had a bole 80 ft. by 21 ft. 3 in. which I estimated to contain 1300 to 1400 ft. of timber; and on 8th June it was covered with seed capsules which I gathered, but being unable to sow them till I came home, failed to raise any trees. A beautiful picture of a group of these trees on the banks of the Merced river in the Yosemite valley is given in *Garden and Forest*, v. 281.

This species is of recent introduction[7] in England, the oldest specimen known to us being a tree in the Edinburgh Botanic Garden, which was planted in 1892, and measured in 1906, 39 ft. by 2 ft. 6 in. Another at Grayswood, planted in 1898, measured in 1906, 32 ft. by 2 ft. 1 in. The largest specimen at Kew was obtained from Späth in 1896, and measured 55 ft. by 3 ft. 10 in. in 1911.

[1] Var. *cupulata*, Watson, in *Amer. Journ. Science*, cxv. 136 (1878), was a name given to a specimen from Plumas County, California, with flowers, in which the disc was campanulate and pubescent.

[2] Piper in *Cont. U.S. Nat. Herb.* xi. 217 (1906).

[3] M. E. Jones, *Montana Botany Notes*, 24 (1910).

[4] Hilyard, *Soils*, 480 (1906), says that this species tolerates white alkaline soil, containing sodium sulphate and chloride; but remains dwarf and stunted on black alkaline soil, containing sodium carbonate.

[5] Jepson, *op. cit.*

[6] Dawson, *Cat. Can. Plants*, i. 457 (1884), states that there is some difficulty in separating this tree from *P. balsamifera* in the northern and north-eastern part of British Columbia, and perhaps the tree of the Yukon valley, which he refers to *P. balsamifera*, is the same which Sargent considers *P. trichocarpa*.

[7] Dieck, *Neuh. Offert. Zöschen*, 1889-1890, p. 13, introduced it on the Continent from British Columbia in 1889.

It seems likely to succeed as well and to grow larger in this climate than the common balsam poplar. Those which I have planted in a cold heavy soil and a situation where severe spring and autumn frosts occur regularly, grow very fast and are perfectly healthy. I received them from Messrs. Meehan of Philadelphia in 1903, and one of those measured in 1911 no less than 32 ft. high by 1 ft. 5 in. in girth, though its branches have been cut several times to propagate. On account of its large handsome leaves, which in hot weather diffuse a most fragrant smell for some distance, this is one of the best of all poplars for ornamental planting.

The largest tree on the continent, appears to be one in the Dresden Botanic Garden, which Mr. Bean[1] in 1908 estimated to be 70 to 80 ft. high, with a trunk 5 ft. 10 in. in girth. In the Copenhagen Botanic Garden it is 30 ft. high, and very thriving. A small tree in the Christiania Botanic Garden was also doing well in 1908. (H. J. E.)

## POPULUS MAXIMOWICZII

*Populus Maximowiczii*, A. Henry, in *Gard. Chron.* liii. 198, fig. 89 (1913).

*Populus suaveolens*, Regel, *Tent. Fl. Ussur.* 132 (1861); Maximowicz, in *Bull. Soc. Nat. Mosc.* liv. 51 (1879); Komarov, *Fl. Manshuriæ*, ii. pt. i. 17 (1903) (not Fischer).

*Populus balsamifera*, Linnæus, var. *suaveolens*, Burkill, in *Journ. Linn. Soc.* (*Bot.*) xxvi. 535 (1899) (not Loudon); Shirasawa, *Icon. Ess. Forest. Japon*, i. text 37, t. 18, figs. 11-24 (1900).

A tree, attaining 100 ft. in height and 10 ft. in girth. Branchlets densely pubescent. Leaves (Plate 410, Fig. 24), about 4 in. long, and 3 to 3½ in. broad, nearly orbicular, oval, or broadly elliptic; subcordate at the rounded base; cuspidate at the apex; pubescent on the midrib, nerves, and veinlets of both surfaces, whitish or tinged with rusty red beneath; densely ciliate and sharply serrate in margin; petiole densely pubescent. Fruiting catkins 7 to 10 in. long; capsules glabrous, sub-sessile, 3- to 4-valved.

This species, owing to the peculiar shape and pubescence of the leaf, is remarkably distinct from *P. suaveolens*, with which it has been confounded, and is most closely allied to the Himalayan *P. ciliata*. It is the common balsam poplar in eastern Asia, extending from Kamtschatka southwards through Saghalien to Japan, Amurland,[2] Manchuria, and Korea. (A. H.)

This is the only poplar which attains a large size in Japan, where it is common in Hokkaido and in the north-eastern district of Honshu, and is called *doro-noki*. I saw it in the central parts of Hokkaido, the largest near Lake Shikotsu being about 100 ft. by 12 ft., growing on river banks and also mixed with other trees in deciduous forests;[3] but in some parts of the country there

[1] Cf. *Kew Bull.* 1908, p. 397.

[2] Elwes found what he believes to be this tree planted at the railway stations on the Siberian Railway near Harbin in May 1912, and brought home cuttings which have rooted at Colesborne. It is recorded for Saghalien by Koidzumi in *Journ. Coll. Sci. Tokyo*, xxvii., art. 13, p. 44 (1910).

[3] Such a forest is figured in *Forestry of Japan*, 96, issued in 1910 by the Bureau of Forestry, Department of Agriculture and Commerce, Tokyo. Jack, in *Mitt. Deut. Dend. Ges.* 1909, p. 282, fig. 285, gives an illustration of an old tree in Hokkaido with deeply furrowed bark.

are pure woods of this and *P. Sieboldii*. Both of these poplars are valued for their timber, which is considered the best for matches, and are being planted extensively by private persons. The seed was not ripe till the middle of July, but none of that which I brought home germinated, and I am not aware that any trees of Japanese origin have yet been introduced. (H. J. E.)

## POPULUS SIMONII

*Populus Simonii*, Carrière, in *Rev. Hort.* 1867, p. 360; Wesmael, in *Mém. Soc. Sc. Hainaut*, iii. 247 (1869); Schneider, *Laubholzkunde*, i. 16 (1904); Dode, in *Mém. Soc. Hist. Nat. Autun.*, xviii. 58 (1905); Gombocz, in *Math. Termes. Közl.* xxx. 105 (1911).

*Populus balsamifera*, Linnæus, var. *Simonii*, Wesmael, in *Bull. Soc. Roy. Bot. Belg.* xxvi. 378 (1887); Burkill, in *Journ. Linn. Soc.* (*Bot.*) xxvi. 536 (1899).

A tree, the dimensions of which in the wild state are unknown. Young branchlets glabrous, usually reddish brown, with five projecting ribs. Buds viscid, glabrous, sharp-pointed, parallel with the twig. Leaves[1] (Plate 410, Fig. 28), on young trees rhombic-elliptic, averaging 3 in. long and 1¾ in. broad, cuneate at the base, contracted at the apex into a short cuspidate point, glabrous, dull whitish beneath; margin with deciduous cilia and crenulate serrations, ending in minute glandular incurved points; lateral nerves seven to nine pairs, all pinnate; petiole short, often not ½ in. long, channelled above, glabrous, reddish, the red colour being continued along the midrib[2] on the upper surface of the blade. Leaves on short shoots and on older trees, smaller and on long petioles, which are often 1½ in. in length. Leaves on vigorous shoots and at the summit of the tree, very large, often 5 in. long and 4 in. broad, ovate-elliptical, cuspidate, on petioles ½ to 1 in. long.

Staminate catkins, according to Schneider, 1 in. long; stamens about eight in a cup-shaped disc. Pistillate flowers and fruit unknown.

This species occurs in north China, where it was collected by Simon in 1862, at Si-wan, north-east of Kalgan, and by Bushell[3] in 1868 in the neighbourhood of Peking. Simon sent living plants about 1862 to the Museum at Paris, where the original tree, described by Carrière, is still living, and to Simon-Louis at Plantières, Metz.

This poplar is little known in England, where we have only seen small living trees at Kew. Elwes received some short truncheons of this from Pekin in 1907, of which, though three months in transit, two grew, and seem likely to succeed at Colesborne. A thriving tree at Grignon, near Paris, which measured 58 ft. by 2½ ft. in 1906, is narrow in habit, with short branches and pendulous branchlets; and young specimens show a similar weeping habit.

Prof. Craig, of the Central Experimental Farm, Ottawa, states[4] that this tree

[1] This species is remarkable for the variation in the length of the petioles, which appear to remain short only in young trees. The large leaves are characteristic of the summit of adult trees, but may appear on vigorous lateral branches also.

[2] The midrib is yellow throughout on old trees.

[3] Erroneously identified as *P. laurifolia* in *Journ. Linn. Soc.* (*Bot.*) xxvi. 536 (1899).

[4] Quoted by L. H. Bailey, in *Cornell Univ. Agric. Station, Bull.* 68, p. 221 (1894).

grows very rapidly in Canada, making a yearly growth of 6 to 10 ft. It is not killed by the severe winters of Manitoba, and is useful for planting where wind-breaks are desired quickly.

*P. yunnanensis*, Dode, *op. cit.* 63 (1905), agrees in technical characters with *P. Simonii*, from which it mainly differs in the brilliant red colour of the branchlets, petioles, and midrib and nerves on the upper surface of the leaves. As in that species, the glabrous branches have five projecting ribs, and the leaves are variable, both in size and shape, and in the length of their petioles. The leaves on the long shoots of young trees are narrowly elliptic or rhombic-elliptic, 3 to 4 in. long, 1½ to 2 in. wide, glabrous, white beneath, cuspidate at the apex, cuneate at the base, closely crenulate in margin, and with very short petioles, about ¼ in. long. On vigorous shoots they are broadly ovate, 5 in. long, 3½ in. wide, with the cuspidate point directed to one side; petioles about 1 in. long.

This poplar, which is very ornamental, is probably a sport or a geographical form of *P. Simonii*; and was introduced from Yunnan, China, in 1906 by a cutting, sent to Dr. Dode, which has been propagated by Chenault of Orleans. There are young and thriving plants at Glasnevin, Casewick, Borde Hill, and Colesborne, the leaves of which were still green on 15th November 1912. (A. H.)

## POPULUS TRISTIS

*Populus tristis*, Fischer, in *Allg. Gartenzeit.* ix. 402 (1841), and *Bull. Sc. Acad. Imp. Pétersb.* ix. 343 (1842); Koehne, *Deut. Dend.* 82 (1893); Schneider, *Laubholzkunde*, i. 13 (1904); Dode, in *Mém. Soc. Hist. Nat. Autun*, xviii. 62 (1905); Ascherson and Graebner, *Syn. Mitteleurop. Fl.* iv. 49 (1908); Gombocz, in *Math. Termes. Közl.* xxx. 98 (1911).

(?) *Populus balsamifera*,[1] J. D. Hooker, *Fl. Brit. India*, v. 638 (1888) (not Linnæus); Gamble, *Indian Timbers*, 691 (1902).

A small tree. Branchlets as in *P. candicans*. Buds viscid, pubescent, ciliate, often subtended by persistent ovate acuminate pubescent stipules. Leaves (Plate 410, Fig. 23), similar to those of *P. candicans* in colour, pubescence, and ciliated margin; but narrowly ovate, about 4 in. long and 2 in. broad, acuminate at the apex, subcordate or rounded at the base.[2] Flowers and fruit of the cultivated plant unknown.

This species was described by Fischer from a cultivated tree at St. Petersburg, supposed to have been introduced[3] from central Asia. Koehne identifies with *P. tristis*, the poplar which has been found by various travellers[4] in the north-west

[1] *P. balsamifera* is confined to North America, and is very distinct from any of the Asiatic species.

[2] Glands are only present at the base of large, well-developed leaves.

[3] *P. tristis* was mentioned as newly introduced into Germany, by Späth, *Catalogue*, No. 91, p. 96 (1893-1894).

[4] Thomson, who collected this poplar in the Zanskar district and other places near Leh in Ladak, describes it in *W. Himalaya and Tibet*, 180 (1852), as "a spreading tree, with large cordate leaves, which was first seen in Upper Kunawar, and is common in all the Tibetan villages, up to the highest limit of tree cultivation." It was also collected "in N.-W. India" by Dr. J. S. Stewart (specimens in the Edinburgh Herbarium); and in the Nubra valley in Tibet by Schlagintweit in 1856. It appears to be a pubescent form of *P. ciliata*, modified by a high and arid situation. The latter species, which, so far as we know, is not in cultivation, has much larger, broad ovate-cordate densely ciliate leaves; glabrous branchlets, buds, and pistillate catkins; disc of the flower lobed; and grows at lower elevations in the Himalayas, from 4000 to 10,000 feet.

Himalayas, occurring both wild and cultivated, at high elevations, 8000 to 14,000 ft. Herbarium specimens of this poplar agree with *P. tristis* in the pubescent buds, with occasional persistent stipules, hairy branchlets, and narrow leaves usually rounded but occasionally subcordate at the base. They bear pistillate catkins, 4 to 6 in. long, with densely pubescent axes and pedicels; disc crenate; ovary pubescent, globose, crowned by three two-lobed stigmas; capsule, pubescent with scattered long hairs, and splitting into three valves, each with a deep longitudinal furrow on the outer surface, not seen in any other species. Koehne's identification is probably correct, but the material for comparison is not sufficient.

I am indebted to Späth of Berlin for fresh branches of this poplar which agree perfectly with Fischer's type specimen of *P. tristis*, preserved in the Kew Herbarium. So far as I know, it is rare in cultivation in England. At Grayswood, Haslemere, it is slow in growth, a plant obtained from Späth in 1896 being only about 4 ft. high, but quite healthy. At Kew there are several small specimens, which retain their foliage hanging withered on the branches during winter. There is also a specimen in the Edinburgh Botanic Garden, about 6 ft. high, and not in a thriving state.

(A. H.)

## POPULUS SUAVEOLENS

*Populus suaveolens*,[1] Fischer, in *Allgem. Gartenzeit.* ix. 404 (1841), and *Bull. Sc. Ac. Imp. Pétersb.* ix. 348 (1842); A. Henry, in *Gard. Chron.* liii. 198, fig. 88 (1913).

*Populus balsamifera*, Pallas, *Fl. Ross.* i. 67, t. xli. fig. 1, A and C (1784) (not Linnæus).

*Populus balsamifera*, Linnæus, vars. *intermedia* and *suaveolens*, Loudon, *Arb. et Frut. Brit.* iii. 1674 (1838).

A tree, attaining in central Asia 50 ft. in height. Young branchlets terete, slightly pubescent above the nodes. Leaves (Plate 410, Fig. 25) ovate-lanceolate or ovate, 3 to 3½ in. long, 1¼ to 2 in. wide, rounded at the base, usually abruptly narrowed towards the acuminate apex; margin ciliate, finely crenate-serrate, with incurved glandular points; nerves running to the margin about eight pairs, the lower two or three pairs arising close together at the base, making it palmately five-nerved; glabrous on both surfaces, whitish beneath.

Staminate catkins not seen. Pistillate catkins about 3 in. long; axis densely pubescent, with white stiff hairs, which are also present on the three-valved globose ovary, and on the extremely short pedicels; disc orbicular, slightly concave, pubescent beneath, with a wavy and densely ciliate margin; stigma two-lobed.

This species in the wild state appears to be variable in the width of the leaf, and in the amount of pubescence on the branchlets and petioles. It has smaller and narrower leaves than any of the balsam poplars, which have whitish leaves beneath.

*P. Przewalskii*,[2] Maximowicz, in *Mél. Biol.* xi. 321 (1881), founded on

[1] Fischer's description is founded on Pallas's figure, and agrees with a specimen collected in Soongaria by Schrenk in 1840. Cf. my article in *Gard. Chron.* cited above.

[2] Dode, in *Mém. Soc. Hist. Nat. Autun*, xviii. 55 (1905), identifies with this species *P. rasumowskyana* and *P. petrowskyana*, hybrid poplars originating in Europe, which resemble in no respect the type specimen of *P. Przewalskii*, preserved in the Kew Herbarium. Cf. pp. 1843, 1844.

specimens collected by Przewalsky in the Ordos territory in Mongolia, seems to be a variety with quite glabrous branchlets and petioles.

*P. suaveolens* appears to be confined[1] to western and northern Siberia, and to Mongolia. It has sweet-scented foliage, resembling in this respect *P. trichocarpa*. Introduced in 1834, it has apparently never thriven in this country, where we have seen no living specimens, except small stunted trees at Kew. It appears to thrive better in Späth's nursery at Berlin, from which we have received specimens, showing healthy foliage and vigorous shoots. Elwes found a fine tree of it in the rich arboretum at Gisselfeld in Denmark, the property of Count Danneskjold-Samsö. It measured 65 ft. by 4 ft., and bore the name of *P. Simonii*. (A. H.)

## POPULUS LAURIFOLIA

*Populus laurifolia*, Ledebour, *Fl. Alt.* iv. 297 (1833), and *Icon. Fl. Ross*, v. t. 479 (1834); Fischer, in *Allgem. Gartenzeit.* ix. 404 (1841); Schneider, *Laubholzkunde*, i. 16 (1904); Dode, in *Mém. Soc. Hist. Nat. Autun*, xviii. 59 (1905); Gombocz, in *Math. Termes. Közl.* xxx. 102 (1911).
*Populus balsamifera*, Linnæus, var. *viminalis*, Loudon, *Arb. et Frut. Brit.* iii. 1673 (1838).
*Populus balsamifera*, Linnæus, var. *laurifolia*, Wesmael, in De Candolle, *Prod.* xvi. 2, p. 330 (1868).
*Populus Lindleyana*,[2] Carrière, in *Rev. Hort.* 1867, p. 380 (not Booth, *ex* Loudon, *op. cit.* 1657); Dode, *op. cit.* 59.

A tree, attaining 40 ft. in height. Young branchlets greyish yellow, pubescent especially near the apex, angled with five ridges, which are very prominent in the second year; lenticels few, scattered, lanceolate. Buds elongated, acute, parallel but not appressed to the twig, exuding a brownish strong-smelling resin. Leaves (Plate 410, Fig. 30) on long shoots, 3 to 5 in. long and 1 to 2 in. broad, lanceolate, rounded at the base, gradually tapering to an acuminate apex; margin finely and regularly glandular-serrate and ciliate; upper surface green, pubescent on the midrib; lower surface whitish, with scattered pubescence, most marked on the midrib; petiole short, terete, but channelled above, pubescent. Leaves on short shoots, smaller, oval, abruptly acuminate at the apex, rounded at the base, on long petioles, which are often two-thirds the length of the blade.

Staminate catkins (described from a tree at Kew) about 2 in. long, with a greenish white densely pubescent axis, bearing about thirty flowers; scales large, with about ten densely fringed lobes, and covered with long hairs on both surfaces; stamens about sixty, with short white filaments and red anthers, on a shallow oblique disc, which is entire and ciliate in margin, glabrous on both surfaces, nearly orbicular, and with a projecting point on one side; pedicel $\frac{1}{12}$ in., with a few scattered hairs. Pistillate flowers not seen. Fruit two- to three-valved, slightly pubescent.

[1] *P. suaveolens* is the poplar referred to by Gmelin, *Fl. Sibirica*, i. 152, fig. 33 (1747), who says that it grows in most places on the rivers emptying into Lake Baikal, being a little less in size than the tree willow; while in the upper regions of the Lena, Yenesei, and other northern rivers, it assumes a dwarf form, not exceeding 3 ft. in height.

[2] Dode keeps this distinct, though it is identical in foliage, on account of its pubescent capsules. Wild specimens of *P. laurifolia*, which I have seen, appear, however, to have pubescent capsules.

*P. laurifolia* appears to be confined[1] to the Altai mountains; and little is known of its habit in the wild state. It is said to be often planted as a street tree in northern Russia, and apparently thrives in Sweden, where I saw two fine trees in the public park at Gefle, about 70 ft. by 5 ft., in 1908.

Though introduced in Loudon's time, it has always been a scarce tree in England; and the only specimen which we know of considerable size is a tree at Kew, about 30 ft. high, with a furrowed bark, a crooked stem, and pendulous branches and branchlets. There is also a small stunted tree at Beauport, Sussex. This poplar is only suitable for cultivation in this country as a curiosity in botanic gardens; but it is said[2] to bear pruning well, and on the continent is often trimmed to form pyramidal and globose shrubs.

*P. laurifolia* is supposed to be one of the parents of the four hybrids which follow.

I. Populus Wobstii, Schroeder, *ex* Dippel, *Laubholzkunde*, ii. 207 (1892).

This poplar, of which I have only seen branches sent me by Späth, has glabrous branchlets, slightly ribbed and marked with orange lenticels. Buds very viscid. Leaves (Plate 410, Fig. 32), 4 to 6 in. long, about 2 in. broad, lanceolate, widest about the middle, rounded but narrow at the base, gradually tapering towards the blunt acuminate apex, slightly pubescent on the midrib and veins, very white beneath, ciliate and crenately serrate in margin; lateral nerves pinnate at the base; petioles with a few scattered hairs.

Its origin is unknown;[3] but it is possibly a hybrid of *P. laurifolia* with *P. tristis*, as surmised by Schneider, *Laubholzkunde*, i. 16 (1904). According to Späth it is slow in growth, and apparently has little to recommend it, except as a curiosity in botanical gardens. I have seen no living specimens in England.

The remaining three hybrid poplars show *P. laurifolia* parentage in the peculiar greyish yellow colour of the branchlets, which are slightly ribbed. The other parent is one of the black poplars, from which the leaves derive the translucent border to their margin, which is, however, very narrow, and can only be made out on careful examination. The foliage and buds have merely a feeble balsamic odour, and the under surface of the leaves is only slightly whitish, being in these respects intermediate between the balsam and the black poplars.

II. Populus rasumowskyana, Schroeder, in Regel, *Russ. Dend.* 133 (1889).

Young branchlets glabrous, angled, with projecting ridges and numerous white lenticels. Buds viscid, sharp-pointed. Leaves on young trees on vigorous shoots, 4 in. long, 3 in. wide, orbicular-ovate, rounded or subcordate at the base, contracted above into a gland-tipped acuminate apex, glabrous except for slight pubescence at

[1] The Chinese specimens referred to this species in *Journ. Linn. Soc.* (*Bot.*) xxvi. 536 (1899), appear to be *P. Simonii.*

[2] Cf. Ascherson and Graebner, *Syn. Mitteleurop. Fl.* iv. 47 (1908).

[3] *P. Wobstii* appears as a novelty in Späth's *Catalogue*, No. 76, p. 108 (1889-1890); and in his *Catalogue*, No. 95, p. 100 (1894-1895), is said not to be a hybrid, but a narrow-leaved form of *P. suaveolens*; but Schneider controverts this, while admitting the possibility of its being a distinct species.

the base of the midrib on the upper surface, pale beneath ; margin glandular-crenate, non-ciliate, with a very narrow translucent border; petiole terete, grooved above. Leaves on older trees smaller, about 3 in. long and 2½ in. wide, elliptic-rhomboidal, with an acute or a short cuspidate apex.

III. Populus petrowskyana, Schroeder, *ex* Dippel, *Laubholzkunde*, ii. 200 (1892).

Very similar to the last, but with minute pubescence on the branchlets and the petioles. Leaves on vigorous shoots about 5 in. long and 4 in. wide, ovate, cordate at the base, where there are usually one or two glands on the upper surface; contracted above into a long acuminate apex; pale beneath; serrations deeper than in the preceding hybrid.

Both these hybrid poplars[1] originated in the garden of the Imperial Agricultural Institute at Petrowskoje-Rasumowskoje, near Moscow, and were exhibited by Schroeder at the Moscow Exhibition in 1882, with three other hybrid poplars, which do not appear to have been propagated. *P. rasumowskyana* was reported to have originated from a *P. nigra*, pollinated by *P. suaveolens*; and *P. petrowskyana* from a *P. canadensis*,[2] pollinated by the same species; but in all probability the balsam poplar concerned was *P. laurifolia.*

Both kinds are said by Späth to be vigorous in growth, but apparently are much less known than *P. berolinensis*, and have only lately been tried at Kew. I have no information concerning the habit of adult trees, and am doubtful as to whether these two hybrids are really distinct, the material which I have examined being very scanty.

IV. Populus berolinensis.

*Populus berolinensis*, Dippel, *Laubholzkunde*, ii. 210 (1892); Schneider, *Laubholzkunde*, i. 11 (1904).
*Populus hybrida berolinensis*, Koch, in *Wochenschr. Gärtn. Pflanzenkunde*, viii. 225 (1865), and *Dendrologie*, ii. pt. i. p. 497 (1872).
*Populus certinensis*, Dieck, *Hauptcatalog. Baumschul. Zözchen* (1885).
*Populus nigra*, var. *italica* × *laurifolia*, Koehne, *Deut. Dendr.* 85 (1893).
(?) *Populus pseudobalsamifera*,[3] Fischer, in *Allgemein. Gartenzeit.* ix. 402 (1841), *ex* Dode, in *Mém. Soc. Hist. Nat. Autun*, xviii. 55 (1905).

A tree, columnar in habit, with short ascending branches, and bark similar to that of *P. serotina.* Young branchlets slightly winged, densely pubescent; older branchlets also pubescent, rounded, yellowish grey. Buds greenish, viscid, sharp-pointed. Leaves (Plate 410, Fig. 29) on long shoots, ovate or ovate-rhombic, 3 to

[1] First mentioned with a brief description as *P. rasumovskoe* and *P. petrovskoe*, Schroeder, in *Gard. Chron.* xviii. 108 (1882). These ill-spelled names were perhaps misprints for the correct names given above, which appear to have been first published with marks of interrogation by Dieck, *Haupt-Catalog Zöschen*, 1886, p. 56.

[2] By *P. canadensis*, possibly *P. marilandica* is meant and not *P. monilifera.* The glands at the base of the leaf confirm the correctness of this parentage. Schneider, *Laubholzkunde*, i. 11, gives the different opinions that have been advanced concerning these hybrids. He considers *P. Rasumowskyana* to be a cross between the Lombardy poplar and *P. suaveolens.*

[3] Fischer describes here a balsam poplar, commonly cultivated in Russia, the earliest in leaf at St. Petersburg, and on that account probably not a native of Russia. There is no specimen of this poplar available for comparison; and it is doubtful if it is the same as *P. berolinensis.* In any case the name is invalid, as *P. pseudobalsamifera*, Fischer, *ex* Turczaninow, in *Bull. Soc. Mosc.* i. 101 (1838) is the name given earlier to a specimen, collected by Turczaninow near Lake Baikal, which is preserved in the Kew Herbarium, and is *P. suaveolens.*

4 in. long, about 2 in. broad, rounded or occasionally cuneate at the base, contracted at the apex into a long glandular acuminate point; glabrous on both surfaces, greenish or slightly whitish beneath, but never so plainly white as in the ordinary balsam poplars; margin non-ciliate, with a very narrow translucent border, only visible with a strong lens, and usually with regular crenate serrations ending in incurved glandular points, occasionally irregular with shallow lobes; nerves seven to eight pairs, those at the base arising together and making the leaf pseudo-five- to seven-palminerved; glands at the base of the leaf on the upper surface variable, none, one, or two being present; petiole variable in length, with scattered pubescence, terete, with the groove on the upper surface well-marked or not apparent. Leaves on short shoots, small, rhombic.

Staminate catkins, 2½ to 3 in. long; scales pale green with a brownish edge, ending in about fifteen long filaments; axis glabrous, crowded with flowers; stamens about sixty, with salmon-red anthers and very short slender white filaments; disc glabrous, oblique, slightly concave. Pistillate catkins and fruit not seen; according to Koehne the ovary, fruit, and disc are glabrous.

This poplar occurs in both sexes, the staminate tree being the one commonly known as *P. certinensis*, which is a finer tree at Kew than those labelled *P. berolinensis*, which are probably pistillate, though they have not yet flowered; the latter differ in the looser longer pubescence of the branchlets.

This beautiful tree is stated by Koch to have originated in the old Botanic Garden at Berlin, through the pollination of a tree of *P. laurifolia* by the pollen of either an adjacent black Italian poplar or of a Lombardy poplar. It is a remarkable hybrid, between two species belonging to different sections of the genus (a black poplar and a balsam poplar), and shows intermediate characters, the thin translucent border to the leaf being a character of the black poplar, while the viscid buds and slightly whitish leaves show the influence of the balsam poplar parent. The original tree is no longer living, but dried specimens of it in the Berlin Herbarium show that it was pistillate. The origin of the staminate tree is unknown, but the late Herr Späth informed us that he received it under the name of *P. certinensis* from Dieck of Zöschen in 1885, and from Transon of Orleans in 1886; and, as Dieck's catalogue of 1885 mentions it as "*P. certinensis* (?) *h. Gall.*," it is likely that it came originally from France.

There are two fine examples of the staminate tree at Kew, which were obtained from Dieck in 1889; these measured, in June 1912, 57 ft. by 3 ft. 5 in. and 50 ft. by 2 ft. 8 in. There is also a good specimen, nearly as tall, at Grayswood, near Haslemere; and another at Glasnevin, about 30 ft. high in 1913.

This species was introduced into America by the Arnold Arboretum, and, according to Professor S. B. Green,[1] is perhaps the best poplar for planting on the prairies of the north-west, as it is perfectly hardy in even the most exposed situations, and is rarely if ever affected with leaf rust, which so often checks the growth of the native cottonwood. Its timber is useful for buildings and floors. It grows readily from cuttings, and bears close planting well. (A. H.)

[1] Cf. L. H. Bailey, *Cornell Univ. Agric. Exp. Station, Bull.* 68, p. 213 (1894).

## POPULUS LASIOCARPA

*Populus lasiocarpa*, Oliver, in Hooker, *Icon. Plant.* xx. t. 1943 (1891); Burkill, in *Journ. Linn. Soc.* (*Bot.*) xxvi. 536 (1899); J. H. Veitch, in *Journ. R. Hort. Soc.* xxviii. 65, fig. 27 (1903); Schneider, *Laubholzkunde*, i. 17 (1904); Dode, in *Mém. Soc. Hist. Nat. Autun*, xviii. 66 (1905); Gambocz, in *Math. Termes. Közl.* xxx. 120 (1911).

*Populus Fargesii*, Franchet, in *Bull. Mus. Hist. Nat. Paris*, ii. 280 (1896).

A tree, attaining in China about 60 ft. high. Young branchlets angled, more or less covered with a loose pubescence, very short and yellowish on old trees. Buds large, slightly viscid, with pubescent basal scales. Leaves (Plate 408, Fig. 9) larger than in any other species, about 9 in. long and 6 in. broad, ovate, deeply cordate at the base, with a gland-tipped acuminate apex; margin revolute, glandular and crenately serrate, the serrations uniform and regular throughout; upper surface with a dense pubescent tuft[1] at the base, elsewhere glabrescent; lower surface pale green, with a scattered tomentum, dense on the midrib and nerves, pseudo-five- or seven-palminerved at the base; petiole slightly tomentose, about one-third as long as the blade, rounded with a groove above.

Staminate catkins with numerous flowers on a short pubescent axis; stamens thirty to forty, on a thick slightly concave disc, which has a thin and lobed margin. Fruiting catkins, 5 to 8 in. long, axis pubescent; capsules two- to three-valved, densely tomentose, shortly stalked, with a lobed glabrous disc.

This remarkable species, which is closely allied to the North American *P. heterophylla*,[2] was discovered by me in 1888, in the mountains of central China, where it occurs in the provinces of Hupeh and Szechwan at 4000 to 6000 ft. elevation.

It was introduced by E. H. Wilson in 1904 into Veitch's nursery at Coombe Wood, where it has proved perfectly hardy.[3] It is worth cultivating on account of its large handsome foliage. (A. H.)

[1] On old trees this pubescent tuft covers two large glands, which appear to be absent on the leaves of young trees.

[2] *Populus heterophylla*, Linnæus, *Sp. Pl.* 1034 (1753). This species (Plate 408, Fig. 10), which grows in swamps in the United States, along the Atlantic coast and in the Mississippi valley, does not thrive in England, where we have seen no living specimens. Loudon, *Arb. et Frut. Brit.* iii. 1672 (1838), mentions two plants at Syon and in the Mile End Nursery, which, though over fifty years old, were only 5 or 6 ft. in height. According to Späth, *Catalogue*, No. 91, p. 51 (1893-1894), it was not in cultivation on the Continent in 1893, but in that year he reintroduced it. Ascherson and Graebner, *Syn. Mitteleurop. Flora*, iv. 52 (1908), state that it is grown in school gardens for the study of the flowers, which are borne on quite small plants.

[3] It was awarded a first-class certificate by the Royal Horticultural Society in 1908. Cf. *Proc. R.H.S.* xxxiv., p. ccxxi, fig. 111 (1909).

# ULMUS

*Ulmus*, Linnæus, *Gen. Pl.* 68 (1737) and *Sp. Pl.* 225 (1753); Lindley, in Rees, *Cyclopædia*, xxxvii. Nos. 1 to 13 (1818); Planchon, in De Candolle, *Prod.* xvii. 154 (1873); Bentham et Hooker, *Gen. Pl.* iii. 351 (1880); Schneider, *Laubholzkunde*, i. 212 (1904); Ascherson and Graebner, *Syn. Mitteleurop. Flora*, iv. 546 (1911).

*Microptelea*, Spach, in *Ann. Sc. Nat.* xv. 358 (1841).

*Chætoptelea*, Liebmann, in *Vidensk. Medd. Kjöbenh.* 1850, p. 76.

Deciduous or rarely sub-evergreen trees, with furrowed bark, and zigzag branchlets, which are often provided with corky wings. Terminal leaf-buds not formed, the tip of the branchlet dying and dropping off early in the season, leaving a small circular scar close to the uppermost axillary bud, the latter in the following season prolonging the branch. Buds composed of numerous ovate rounded scales, imbricated in two ranks, those of the inner rows accrescent, and marking, when they fall, the base of the branchlet with ring-like scars.

Leaves simple, alternate, placed on the branchlet in two ranks, stalked, simply or doubly serrate, penninerved, asymmetrical at the base,[1] the inner half of the blade being the larger. Stipules lateral, entire, free or connate at the base, usually early deciduous.

Flowers minute, perfect, appearing either in early spring before the leaves in the axils of the leaf-scars of the previous year, or in autumn in the axils of the leaves of the current year; in stalked or sub-sessile fascicles or cymes; articulated on slender two-bracteolate pedicels: calyx, campanulate or funnel-shaped, with four to nine short or deeply divided lobes: stamens three to eight, inserted under the ovary, with thread-like filaments, and two-celled dorsifixed extrorse anthers, which dehisce by two longitudinal openings; ovary usually one-celled by abortion, sessile or stalked; style with two spreading lobes, stigmatic on the inner surface; ovule solitary, suspended from the apex of the cell.

Fruit, a samara, ripening in two or three months after flowering, surrounded at the base by the remains of the calyx, notched at the apex, the notch open or closed by the incurved persistent stigmas and ciliate within; seed-cavity in the centre or above it, compressed, slightly thickened in margin, and produced into a thin peripheral reticulate membranous wing. Seed solitary, suspended from the apex of the cavity, without albumen; cotyledons flat, raised above the ground in germination.

[1] Cf. Van Tieghem in *Ann. Sci. Nat.* (*Bot.*), iii. 377 (1906), on the peculiar asymmetry in the leaves and stipules of the elm.

The genus Ulmus comprises about twenty species, natives of the extra-tropical parts of the northern hemisphere, extending in North America to the mountains of southern Mexico, but not occurring in the Pacific coast region; widely distributed in the Old World, throughout Europe, and in Algeria and Morocco, and spread through Asia in Siberia, Asia Minor, Persia, Turkestan, Afghanistan, the Himalayas, China, Tongking, Japan, and Formosa.

The genus is divided into three sections:—

I. Blepharocarpus, Dumortier, *Fl. Belg.* 25 (1827).

*Oreoptelea*, Spach, in *Ann. Sc. Nat.* xv. 363 (1841).

Leaves, deciduous in autumn. Flowers opening early in spring before the leaves, on elongated unequal pedicels; calyx oblique, with five to eight unequal short lobes. Samaræ densely ciliate in margin.

Of the species in cultivation, *U. americana*, *U. racemosa*, *U. alata*, and *U. pedunculata* belong to this section.

II. Madocarpus, Dumortier, *Fl. Belg.* 25 (1827).

*Dryoptelea*, Spach, in *Ann. Sc. Nat.* xv. 363 (1841).

Leaves deciduous in autumn. Flowers opening early in spring before the leaves, on very short pedicels; calyx with four to seven equal short lobes. Samaræ non-ciliate.

Of the species in cultivation, *U. montana*, *U. nitens*, *U. campestris*, *U. major*, *U. minor*, *U. pumila*, *U. japonica*, and *U. fulva* belong to this section.

III. Microptelea, Planchon, in *Ann. Sc. Nat.* x. 260 (1848).

Leaves sub-persistent or tardily deciduous. Flowers opening in autumn, on short pedicels; calyx deeply divided into four to eight equal long lobes. Samaræ ciliate or non-ciliate.

Of the species in cultivation, *U. parvifolia* and *U. crassifolia* belong to this section.

About fifteen species of Ulmus are in cultivation, and may be arranged as follows:—

I. *Leaves with sixteen to twenty pairs of lateral nerves.*

* *Leaves, with lateral nerves rarely forked; axil-tufts inconspicuous or absent.*

(a) *Leaves ciliate in margin. Branchlets without corky ridges.*

1. *Ulmus pedunculata*, Fougeroux. Europe. See p. 1851.

Leaves usually obovate and widest above the middle, 2 to 4 in. long, biserrate with incurved serrations, smooth above, densely pubescent beneath. Buds elongated, fusiform, sharp-pointed.

2. *Ulmus americana*, Linnæus. North America. See p. 1855.

Leaves oval, widest about the middle, 3 to 5 in. long, biserrate with incurved serrations, scabrous or smooth above, more or less pubescent beneath. Buds ovoid, obtuse.

(b) *Leaves non-ciliate in margin. Branchlets developing corky ridges in the second or third year.*

3. *Ulmus racemosa*, Thomas. North America. See p. 1860.

Leaves oval or elliptic, about 3 in. long, usually sub-cordate at the base, biserrate with incurved points, glabrous and smooth above, slightly pubescent beneath without any trace of axil-tufts.

** *Leaves, with lateral nerves often forked, and with conspicuous axil-tufts.*

(a) *Young branchlets scabrous with numerous minute tubercles. Leaves ciliate in margin.*

4. *Ulmus fulva*, Michaux. North America. See p. 1862.

Leaves oval or obovate, 5 to 7 in. long, scabrous above with minute tubercles and short bristles, densely pubescent beneath, coarsely biserrate.

(b) *Young branchlets smooth. Leaves non-ciliate.*

5. *Ulmus montana*, Stokes. Europe, Asia Minor, Caucasus, Amurland, Manchuria, Japan. See p. 1864.

Branchlets stout, pubescent with stiff hairs. Leaves obovate or oval, 3 to 5 in. long, with short stout petioles, not exceeding ⅛ in. in length; scabrous above, densely pubescent beneath.

6. *Ulmus vegeta*, Lindley. A hybrid.[1] See p. 1879.

Branches long, straight, and ascending. Branchlets slender, glabrous or with a few scattered hairs. Leaves oval, 3½ to 5 in. long; smooth or nearly so and glabrous above; glabrous beneath except for axil-tufts; petiole ¼ to ⅜ in. long.

II. *Leaves, with eight to fourteen pairs of lateral nerves; very unequal at the base; plainly biserrate; and with conspicuous axil-tufts beneath.*

* *Young branchlets glabrous, or with only a few scattered hairs.*

7. *Ulmus major*, Smith. Europe. See p. 1883.

Leaves broadly oval, 3 to 5 in. long, nearly smooth above, with a scattered minute pubescence on both surfaces, and dense axil-tufts beneath; petiole ¼ to ⅜ in. long, pubescent. Epicormic branches with large corky ridges are usually present on the stem.

8. *Ulmus nitens*,[2] Moench. Europe, Algeria, Asia Minor, Caucasus, Persia, Turkestan. See p. 1887.

Leaves oval or obovate, 2 to 3½ in. long, shining and smooth above; with scattered minute pubescence on both surfaces in spring, disappearing in summer; glandular beneath; petiole ¼ to ½ in. long, pubescent.

9. *Ulmus minor*, Miller. Europe. See p. 1901.

Leaves elliptic, 1½ to 2½ in. long, acute or acuminate at the apex; dull and

[1] The other less common hybrids, having *U. montana* as one of the parents, are described, pp. 1868-1874. Most of these hybrids have leaves with numerous lateral nerves; but differ in habit and other characters from *U. montana* and *U. vegeta*.

[2] This species is very variable, and only the typical form is here indicated.

slightly scabrous above; glandular and ultimately glabrescent beneath, except for conspicuous axil-tufts; lateral nerves few, 8 to 10 pairs; petiole, $\frac{1}{5}$ in. long.

** *Young branchlets densely pubescent.*

10. *Ulmus campestris*, Linnæus. Southern England, Spain (?). See p. 1903.

Leaves broadly oval or ovate, 2 to 3 in. long; scabrous and minutely pubescent above; covered beneath with a dense soft pubescence: lateral nerves, 10 to 12 pairs; petiole $\frac{1}{5}$ in. long, densely pubescent. Branchlets without corky wings.

11. *Ulmus japonica*, Sargent. Japan, Manchuria. See p. 1923.

Young branchlets light brown, often roughened with minute tubercles or ridges. Leaves obovate or elliptic, 3 to 4 in. long; scabrous above with minute tubercles and short bristles; densely pubescent beneath; lateral nerves 12 to 16 pairs. Branchlets often with corky wings.

III. *Leaves, with eight to twelve pairs of lateral nerves; often nearly equal at the base; often simply serrate; axil-tufts inconspicuous or absent.*

* *Leaves deciduous in autumn.*

(a) *Branchlets with corky wings.*

12. *Ulmus alata*, Michaux. North America. See p. 1924.

Leaves light green, thin in texture, oblong-lanceolate, about 2 in. long, acute or acuminate at the apex, smooth above, with axil-tufts beneath, biserrate; nerves rarely forked.

13. *Ulmus crassifolia*, Nuttall. North America. See p. 1925.

Leaves light green, coriaceous, oval, 1 to 2 in. long, acute or rounded at the apex, often subcordate at the base, scabrous above, without axil-tufts beneath; often biserrate; nerves usually forked; stipules persistent till May.

(b) *Branchlets without corky wings.*

14. *Ulmus pumila*, Linnæus. Turkestan, Eastern Siberia, Manchuria, Korea, North China. See p. 1926.

Leaves thin and flexible in texture, ovate to ovate-lanceolate, 1 to 2 in. long, acute or acuminate at the apex, scabrous or smooth above, with axil-tufts and scattered minute pubescence beneath.

** *Leaves deciduous in January.*

15. *Ulmus parvifolia*, Jacquin. China, Tongking, Formosa, Japan. See p. 1928.

Leaves coriaceous, obovate or ovate-lanceolate, 1 to $1\frac{3}{4}$ in. long; shining, dark green, smooth and glabrous above; glabrous beneath, but occasionally with axil-tufts near the base. (A. H.)

## ULMUS PEDUNCULATA, European White Elm

*Ulmus pedunculata*,[1] Fougeroux, in *Mem. Acad. Roy. Sc.* 1784, p. 215, t. 2.
*Ulmus lævis*, Pallas, *Fl. Ross.* i. 75, t. 48, fig. F. (1784); Ascherson and Graebner,[2] *Syn. Mitteleurop. Flora*, iv. 548 (1911).
*Ulmus effusa*, Willdenow, *Fl. Berol. Prod.* 97 (1787) and *Berl. Baumz.* 393 (1796); Loudon, *Arb. et Frut. Brit.* iii. 1397 (1838); Willkomm, *Forstl. Flora*, 559 (1887); Fliche, *Flore Forestière*, 304 (1897).
*Ulmus ciliata*, Ehrhart, *Beitr.* vi. 88 (1791).
*Ulmus octandra*, Schkuhr, *Handb.* i. 178, t. 57 (1791).
*Ulmus racemosa*, Borkhausen, *Forstbot.* i. 851 (1800) (not Thomas).

A tree, attaining about 100 ft. in height and 12 to 20 ft. in girth. Bark similar to that of *U. americana*, smooth at first, then exfoliating in broad thin scales, and ultimately deeply fissured as in the other elms. Young branchlets densely clothed with white short wavy pubescence, partly retained on the branchlets of the second year which are slightly fissured but not finely striate. Buds longer and more sharply pointed than those of *U. americana*; with glabrous scales, which are minutely ciliate in margin. Leaves (Plate 411, Fig. 7), obovate and widest above the middle, or oval, 2 to 4 in. long, and 1½ to 2½ in. wide; very oblique and unequal at the base, the upper side rounded, the lower side rounded or straight; suddenly contracted at the apex into a serrated point; upper surface smooth to the touch, pubescent on the midrib and veins, elsewhere glabrous or with a few scattered hairs; lower surface pale green, covered with a dense short white pubescence; margin coarsely biserrate, with sharp incurved points, ciliate; nerves about sixteen pairs, running straight and parallel to the margin, with one or two rarely forked; petiole ¼ in. long, densely pubescent.

Flowers, twenty to twenty-five in a cluster, on long slender pedicels (¼ to ¾ in. in length); calyx campanulate, oblique, with five to seven short pink lobes; stamens five to seven, with white filaments and red anthers; ovary green, pubescent on the margins, with white stigmas. Samaræ, on long slender stalks, oval or ovate, ⅜ to ½ in. long, conspicuously reticulate and glabrous on the surface, densely ciliate in margin with long white hairs; apex with a deep cleft, usually closed by the incurved stigmas; seed situated towards the base of the samara, with its apex close to the base of the notch.

Seedlings, raised in 1909 at Cambridge, were very uniform, all bearing in the first year six to eight pairs of opposite leaves. In 1911, these seedlings still preserved their uniform character, and were readily distinguishable by their ciliate leaves, and the absence of corkiness on the branchlets.

This species, like *U. americana*, is often remarkable in old age for the sharp

[1] This name was published by Fougeroux in a paper, read at Paris on 1st September 1784. Pallas's name was published later in the same year, according to a note by Fougeroux, appended to his paper when it appeared in the volume of the *Mém. Acad. Roy. Sc.*, for 1784, which was issued in a complete form in 1787.

[2] *U. alba*, Kitaibel, in Willdenow, *Berl. Baumz.* 318 (1796), judging from the description, and also from a tree so named in the Leyden Botanic Garden, is a variety of *U. nitens*, and not a synonym of this species, as stated by Ascherson and Graebner. *U. alba*, Besser, *Enum. Pl. Volhynia*, pp. 43, 92 (1822), is *U. pedunculata*.

protruding ribs at the base of the trunk, with deep concave recesses between them. The stem, as in the American species, usually produces an abundance of pendulous epicormic shoots. Fliche states that it suckers freely from the roots, and gives abundant coppice shoots.[1] It does not appear ever to form corky wings on the twigs. The leaves usually turn a yellowish colour in autumn.

Klotzsch,[2] who was curator of the Berlin Herbarium, crossed *U. pedunculata* with "*U. campestris*" (probably *U. nitens*) in 1845; and raised seedlings from both the cross-fertilised seeds and from the seeds of the parent trees; and after eight years' growth under similar conditions, the hybrid seedlings were one-third taller than the others.

## Varieties

*U. pedunculata* does not differ from *U. americana* in the characters of the flowers or fruits; and can only be kept distinct[3] from that species, on account of its usually smaller and more oblique leaves, which are smooth above to the touch. *U. pedunculata* as described above has usually leaves densely pubescent beneath; but trees with almost glabrous leaves are occasionally seen in cultivation. The most remarkable variety in the wild state[4] is the following:—

1. Var. *celtidea*, Rogowicz, *Fl. Kief*, 229 (1869); Köppen, *Holzgewächse Europ. Russlands*, ii. 33 (1889); Chitrovo, in *Bull. Soc. Nat. Orel*, i. 50, t. 1 (1907).

Var. *glabra*, Trautvetter, in *Bull. Phys. Math. Acad. Imp. Sc. St. Petersb.* xv. 349 (1857).
*Ulmus celtidea*, Litwinow, *Schedæ Herb. Fl. Ross.* vi. 167 (1908).

Leaves oblong-lanceolate, about 1 in. in length, long acuminate at the apex, coarsely and sharply serrate, cuneate and subequal at the base, nearly quite glabrous beneath. Stamens five or six. Samaræ smaller than in the type.

This peculiar elm was found by Rogowicz in 1856 near Chernigof. Lately, another remarkable tree has been discovered near Briansk in the Orel province, which has similarly shaped trees, but much larger in size, 4 to 5 in. long, and pubescent beneath. Seedlings of this tree raised at Kief have broad leaves as in the typical form of the species.

The following horticultural varieties,[4] none of which are in cultivation in England, have been described.

2. Var. *punctata*, Schelle, *Laubholz-Benennung*, 87 (1903).

Leaves variegated with white.

[1] The leaves on coppice shoots, which I gathered in 1912 near Rochefort in Belgium, are very large, 6 to 9 in. long and 4 to 5 in. broad, oval, with a single long cuspidate point at the apex, nearly equal at the rounded base, scabrous with short bristles on the upper surface, sparsely pubescent beneath.

[2] In *Monatsbericht K. Preuss. Akad. Wiss. Berlin*, 1854, pp. 535-562, abstracted in *Bull. Soc. Bot. France*, ii. 327 (1855). Klotzsch, whose article is of great interest, seems to have been the first botanist to make experiments in crossing forest trees. In 1845 he also crossed *Pinus sylvestris* and *P. austriaca*, *Quercus sessiliflora* and *Q. pedunculata*, *Alnus incana* and *A. glutinosa*; and in each case raised hybrid seedlings of great vigour; and claimed that by hybridisation, both the rapidity of growth and durability of the timber of forest trees could be augmented considerably. Cf. Darwin, *Animals and Plants under Domestication*, ii. 130 (1868).

[3] Jouin showed Elwes in Simon-Louis's nursery at Metz young trees of both species. Those of *U. pedunculata* had much smoother bark; but it is doubtful if this character is a constant one.

[4] Hayne, *Arz. Gew.* iii. t. 17 (1813), described four varieties, which are unworthy of retention, being based on trifling variations in the flowers and leaves. Zapalowicz, *Consp. Fl. Galic.* ii. 96 (1908), has lately described two varieties, which I have not seen.

3. Var. *aureo-variegata*, Schelle, *loc. cit.*

Leaves spotted with yellow.

4. Var. *erubescens*, Von Schwerin, in *Mitt. Deut. Dend. Ges.* 1911, p. 423. Leaves turning a beautiful red in autumn.

5. Var. *urticæfolia*, Jacques, *ex* De Vries, *Plant Breeding*, 614 (1906).

A chance seedling, with laciniate leaves, which was raised by Jacques in 1830, and subsequently was multiplied by grafting. This variety appears to be lost, as it is not now known in cultivation.

6. Bolle mentions a pyramidal form, which is growing in the cemetery at Fredericksfelde near Berlin.

## Distribution

*U. pedunculata* is a native of central Europe, being distributed from eastern France to Russia. It is not a native of the British Isles or of Scandinavia,[1] and is nearly absent from the Mediterranean region, being unknown in Spain and Portugal, only recorded from two stations in northern Italy, and very rare in Greece.

It appears to be most widely spread in Russia,[2] where it is prevalent except in the extreme north and the region of the steppes, its northern limit extending from southern Finland across Lake Onega, the valley of the Dwina as far north as lat. 63°, and Perm, to the western side of the Ural range. It usually occurs scattered in the broad-leaved forests, but is more common than *U. montana* in the governments of Tula and Moscow. In Russia[3] it is planted along the railways as a protection against snowdrifts in winter. It also occurs in the Crimea and the Caucasus.

It is a rare tree in Belgium, Holland, and Denmark; but is met with in central and northern Germany, especially in Brandenburg,[4] where the finest trees occur on the banks of the Havel and Spree, some being 100 feet in height and 20 ft. in girth. In Germany it usually grows in damp deciduous woods, and on the swampy banks of streams and lakes. It is recorded for a few stations in Switzerland,[5] Montenegro, Roumania, Bulgaria, and Servia, where I saw in 1909 a tree about 50 ft. high on the banks of the Drina, near Zvornik. (A. H.)

In France it is only found in the north-east, in a few stations, though according to Fliche[6] it has probably been more abundant, as it occurs mostly on rich and fertile soils in the plains which are now cultivated. It is most numerous in the south-east of the department of the Ardennes in the forest of Mondieu and at Stenay, both in the valley of the Meuse. It never grows in quantity, but is scattered in mixture with *Quercus pedunculata* in damp places, associated with ash, willow, aspen, birch, and alder. It is also found near Gray in the upper Saône valley, and

[1] Schübeler says it grows well on the south and west coast of Scandinavia, and Elwes saw it planted near Stockholm.

[2] Köppen, *Holzgewächse Europ. Russlands*, ii. 26 (1889). It occurs in Finland on the shores of the Gulf, but Elwes did not see it in the forest of Raivola.

[3] *Garden and Forest*, 1890, p. 475.

[4] Bolle, in *Garden and Forest*, 1888, p. 381.

[5] Dr. Christ says it occurs wild only at Schaffhausen; but I have a specimen from Mr. A. B. Jackson, collected at the foot of the Harderberg, near Interlaken.

[6] In *Bull. Soc. Bot. France*, xlviii. 381 (1901).

near Luneville, in the Meurthe valley. M. Guinier, of the Forestry School at Nancy, to whom I am indebted for the above information, was good enough to show me this tree on the banks of the Moselle, just below the Château de la Fli, near Liverdun, on 5th May 1911 when it was in flower, and the leaves about half expanded. The trees were not large, and their bark seemed more scaly than that of *U. montana.* Though the habit seemed more pendulous than that of the wych elm, I do not think the tree could be certainly distinguished in winter except by the buds, and some trees looked as though they might be hybrids. Reuss states[1] that a few trees of *U. pedunculata* occur in the forest of Fontainebleau. M. Jouin at Plantières, near Metz, informed me that, as this species is much less subject to the attacks of the elm-leaf beetle[2] than *U. nitens*, it is likely to be much more generally planted in that part of France, but the wood is considered inferior to that of the other elms by those who distinguish it.[3] Loudon[4] mentions in 1841 an enormous tree of this species, with three great trunks, in the nursery ground at Neuilly, near Paris; but we do not know if it is still living. I saw a log under this name in the Hungarian Exhibition at London in 1908, which was very similar in appearance to *U. montana.*

In Belgium, according to Huberty,[5] this tree is only found in the district about Rochefort, where it grows on calcareous soil in fertile valleys on the edge of woods, but is rather local, and of no economic importance. The largest tree I heard of grows at the mouth of the Gouffre de Belvaux, where a stream disappears in the limestone rock and comes up again some way off at Han-sur-Lesse, celebrated for its extensive caverns. This tree was pollarded many years ago, and has a sound trunk about 15 ft. high by 16 ft. in girth, its total height being about 60 ft. It has been figured by Huberty.[5] I could find no other trees of the sort near it except some small bushy ones apparently produced from suckers.

Hempel and Wilhelm[6] give an illustration of a tree, about 80 ft. high, in the Prater at Vienna, which has wide-spreading branches. I saw this elm in June 1910 in west Slavonia, in the virgin forest of Subanja, which is mainly composed of old oak trees, with a mixture of hornbeam, lime, maple, and ash.

Loudon states that the date of the introduction of this species is unknown; and

[1] *Compte Rendu Congrès Internat. Sylviculture, Paris, 1900*, p. 683.

[2] *Galerucella luteola*, Müll. (more widely known as *G. xanthomelæna*, Schr.), a small beetle belonging to the family Chrysomelidæ, 6 to 8 mm. long, pale sordid yellow or yellowish brown above, with a stripe along the outer margin of the elytra, a short line on each side of the scutellum, a central and two lateral spots on the thorax, and a spot on the head, black. This insect is very destructive to the foliage of young elms both in its larval and perfect states; and, as I was assured by M. Jouin, is a serious hindrance to the raising of elms in his nursery. It is not, however, a native of Britain, and in the opinion of Mr. G. C. Champion, a competent coleopterist who has observed its ravages on the Continent, the climate of this country is unsuitable for its establishment, even if it were introduced with plants. In America, where it was introduced from Europe in 1837, its depredations have of late years attracted considerable attention. Pardé, in *Bull. Soc. Dend. France*, 1909, p. 100, agrees with Jouin as to the comparative immunity from this pest of *U. pedunculata*.

[3] Fliche, in *Bull. Soc. Bot. France*, xlviii. 381 (1901), states that the wood is little esteemed, partly on account of its pale yellowish colour, whence it is called *orme blanc*; but mainly because it fails in strength and elasticity, and is unfit for many purposes for which the other elms are valuable. It even makes poor fuel; and is always classed in commerce with soft woods, like poplar, etc.

[4] *Gard. Mag.* xvii. 389 (1841).

[5] In *Bull. Soc. Cent. Forest. Belg.* xi. 634, pl. xii. (1904).

[6] *Baüme u. Straücher des Waldes*, iii. 9, figs. 234, 235 (1889). The botanical details are well illustrated in t. 39 of this work.

the only large tree which he mentions, one at White Knights, which was 63 ft. high in 1838, is no longer living.

The finest specimen in England is a tree (Plate 389) at Syon,[1] which is 90 ft. in height and 12 ft. 8 in. in girth. This was long supposed to be *U. americana*; but it agrees better in the characters of the foliage with *U. pedunculata*, and like the latter species at Kew and elsewhere, produces flowers regularly every year, which is not the case with trees of *U. americana* in this country. There are two good specimens at Kew, one about 25 ft. high, which was obtained from Späth in 1895. It produced good seed in 1909, from which seedlings were raised at Cambridge. Another tree, obtained from Booth in 1872, is about 35 ft. high, and very thriving.

At Ugbrooke Park, Devonshire, the seat of Lord Clifford of Chudleigh, I found in April 1908 a row of seven large trees, in the park near the house, which I at once recognised by their flowers, though the leaves were not yet out. I saw these again on 28th November of the same year, when the leaves had fallen about three weeks previously, though those of the English elm were still of a golden colour. Their buds were more swollen than those of the Dutch elms near, which also had some leaves on, and the bark and habit differs from that of English and wych elms. I could find no suckers near any of the trees,[2] one of which is figured in Plate 390. The average height was 80 to 85 ft., and the girth of ten trees, three of which are on the other side of the avenue above the oaks, varied from 11 ft. to 14 ft. 8 in. Neither Lord Clifford nor the gardener, Mr. Abraham, could tell me anything of the age or history of these trees, which must have been brought from abroad over a century ago. (H. J. E.)

## ULMUS AMERICANA, American White Elm

*Ulmus americana*, Linnæus, *Sp. Pl.* 226 (1753); Loudon, *Arb. et Frut. Brit.* iii. 1406 (1838); Sargent, *Silva N. Amer.* vii. 43, t. 311 (1895), and *Trees N. Amer.* 289 (1905).
*Ulmus mollifolia*, Marshall, *Arb. Amer.* 156 (1783).
*Ulmus pendula*, Willdenow, *Berl. Baumz.* 519 (1811).
*Ulmus alba*, Rafinesque, *Fl. Ludovic.* 115 (1817), and *New Fl.* iii. 39 (1836), (not Kitaibel).[3]
*Ulmus floridana*, Chapman, *Flora*, 416 (1865).

A tree, attaining in America 120 ft. in height and 30 ft. in girth. Bark grey, divided by deep fissures into broad scaly ridges. Young branchlets occasionally glabrous but usually clothed with a soft white pubescence, more or less retained on the second year's branchlets, which are fissured but not finely striate on the surface. Buds ovoid, obtuse, with glabrous ciliate scales. Leaves (Plate 411, Fig. 11) oval, usually widest about the middle or below it, 3 to 5 in. long, $1\frac{1}{2}$ to $2\frac{1}{2}$ in. wide; oblique and unequal at the base, with the upper side rounded and the lower side straight;

[1] In the Syon catalogue of 1849, this tree is noted as having then measured 85 ft. by 10 ft.

[2] These trees were found to be infested with the elm Psylla (*Psylla ulmi*, L.), but the circumstance is not of economic importance, since there is no reason to believe that this insect is more detrimental to the elm than is its congener *Psylla alni* to the common alder. The nymphs of the elm Psylla live in the axils of the leaves on the ends of the twigs, and enter the imago state in June. The imagines do not hibernate, but lay their eggs in the autumn; a circumstance which would point to the introduction of the species with young trees. It is remarkable that the elm Psylla has never been found in this country on any other species of elm.

[3] Cf. p. 1851, note 2.

abruptly narrowed at the apex into a serrated point; upper surface dark green, smooth or scabrous to the touch, with scattered stiff short hairs arising from minute tubercles; lower surface pale, covered with a soft dense pubescence, conspicuous on the midrib and nerves, and occasionally forming minute axil-tufts at their junctions;[1] nerves sixteen to twenty pairs, running straight and parallel to the margin, only one or two being forked; margin coarsely biserrate with sharp incurved points, ciliate; petiole $\frac{1}{4}$ in. long, densely pubescent.

Flowers, about twenty in a cluster, on long slender pedicels (about $\frac{1}{2}$ to $\frac{3}{4}$ in. in length); calyx broadly campanulate, oblique, with six to eight short pinkish lobes; stamens six, seven or eight, exserted, with white filaments and red anthers; ovary green, with ciliated margins, and white stigmas. Samaræ, on long slender stalks, ovate or elliptic, $\frac{3}{8}$ to $\frac{1}{2}$ in. long, with prominent reticulations, glabrous on the surface, densely ciliate on the margin with long white hairs; with a deep notch at the apex, closed by the incurved stigmas. Seed towards the base of the samara, with its apex close to the base of the notch.

This tree, which never develops corky ridges on the branchlets, usually produces short pendulous epicormic shoots on the stem. It is very variable in habit, as is well seen in the trees at Hargham in Norfolk. In America, when growing in the forest, the stem is usually undivided for a great height, and surmounted by a compact crown. When growing in the open the tree shows well-marked differences in habit, which are described by Dame and Brooks,[2] as follows:—

"1. In the vase-shaped tree,[3] usually regarded as the type, the trunk separates into several large branches, which rise, slowly diverging 40 to 50 ft., and then sweep outward in wide arches, the smaller branches and spray becoming pendent.

2. In the umbrella form, the trunk remains entire nearly to the top of the tree, when the branches spread out abruptly, forming a broad shallow arch, fringed with long drooping branchlets.

3. The slender trunk of the plume elm[4] rises, usually undivided, a considerable height, begins to curve midway, and is capped with a one-sided tuft of branches and delicate elongated branchlets.

4. The drooping elm[5] differs from the type in the height of the arch, and greater droop of the branches, which sometimes sweep the ground.

5. In the oak form the limbs are more or less tortuous, and less arching, forming a wide-spreading rounded head."

Only a few horticultural varieties are known:—

1. Var. *pendula*, Aiton, *Hort. Kew.* i. 320 (1789).

Branches wide-spreading and arching downwards, with pendulous branchlets. This is said by Aiton to have been cultivated by Mr. James Gordon in 1752.

[1] Loudon says that *U. americana* is readily distinguished by the peculiar membrane in the axils of the veins. This membrane, uniting the base of the nerve with the midrib, is usually present only on the nerves of the outer half of the leaf.

[2] *Trees of New England*, 95 (1902).

[3] Figured by Sargent, in *Garden and Forest*, iii. 287 (1890).

[4] A tree of this kind, called the "feathered" form by Sargent, growing at Sandwich, New Hampshire, is figured in *Garden and Forest*, iii. 467 (1890).

[5] The Clark elm at Lexington, with pendulous branches, which sweep the ground, is figured in *Garden and Forest*, iii. 443 (1890).

There are two specimens at Kew, about 20 ft. high, which were obtained from Späth in 1896. These occasionally develop stray branchlets, bearing enormous leaves, 7 in. long by $4\frac{1}{2}$ in. broad.

This is probably not quite the same as Beebe's weeping elm, which was distributed by Meehan as *U. fulva*, var. *pendula*, but which, according to Sargent[1] is really a form of *U. americana*. It was propagated from a tree, resembling in habit a weeping willow, which was found growing wild near Galena, Illinois.

2. Var. *aurea*, Temple, *ex* Rehder, in Bailey, *Cycl. Amer. Hort.* 1880 (1902). Leaves yellow. This was found in Vermont by F. L. Temple.

*U. americana*, which is known, on account of the light colour of its bark, as "white elm" in America, is widely distributed, occurring from southern Newfoundland through Canada to the northern shores of Lake Superior and the eastern base of the Rocky Mountains, ascending the Saskatchewan river to lat. 54° 30′; and extending through the United States southwards to Florida, and westwards to the Black Hills of Dakota, western Nebraska, western Kansas, Indian Territory, and the valley of the Rio Concho in Texas.

The tree does not occur in pure stands, but sparingly in mixture with oak, ash, plane, tulip tree, and other hard woods; and attains its best development on deep fertile alluvial soil. It, however, readily adapts itself to less favourable soils, and is a hardy tree enduring great extremes of temperature and moisture. It is somewhat intolerant of shade.[2]

This species produces coppice shoots, when the tree is cut down; and at Hargham, numerous suckers, some of which are now trees 30 ft. in height, were produced after a tree had been felled. No suckers are produced, so far as we know, by living trees either in this country or in America. (A. H.)

No tree has attracted more attention among American writers, or is more dear to the natives of New England, than the American elm, which is a conspicuous ornament and the favourite shade tree in the older cities, and has quite a literature of its own.

Though some of the historic trees mentioned by Emerson and other writers are now dead or decayed, there are still many splended survivors of the original forest. Among these none is larger and more symmetrical than the Lancaster Elm in Massachusetts, which Prof. Sargent showed me in May 1904. It grows on deep sandy soil in the rich valley of the Nashua river, and measured 105 ft. by 24 ft. at five feet from the ground. The roots spread so widely that at ground level they are 45 ft. round. The trunk forks at 10 ft. into three tall stems; and though some holes and cracks show that the tree is past maturity, these wounds have been so carefully stopped with cement, by its owner, Mr. J. E. Thayer, that it may live for many years. Plate 391, for which I am indebted to Mr. A. H. French, of Brooklime, Mass., gives an excellent picture of it.

A monster elm[3] on the Avery Durfee farm in Wayne County, New York,

[1] *Garden and Forest*, i. 286 (1889).

[2] Cf. Pinchot, *U.S. Forest Circ.* No. 66 (1907), which gives directions for planting this tree in the United States.

[3] *Garden and Forest*, iii. 60 (1890).

between Palmyra and Marion, measures, at two feet above the ground, 33 ft. 10 in. in circumference, and at five feet above the ground 20 ft. 10 in. It is 60 ft. to the first limb; and the total amount of lumber in the body of the tree is 16,250 board ft. Eighty years ago, when the farm was cleared, this tree was left as a landmark. It was then a giant amongst the surrounding forest trees.

A very fine elm, presumably of this species, which differs much in habit from any *U. americana* that I have seen, is figured and described[1] by Mr. T. H. Hoskins, who, however, does not say what the species is. The tree is evidently a relic of the original virgin forest, and grows near the highway at Derby Line close to the Canadian boundary in Orleans County, Vermont. It has a clean bole for more than half its height with a very small head, and measures 102 ft. in height with a girth of 18 ft. 2 in. at five feet from the ground.

To give an idea of the rapidity with which this species grows I may cite Emerson, who says,[2] quoting from the *N. E. Farmer*, vii. 299, "An elm tree nearly opposite the house of Heman Day, Esq., in West Springfield, was planted by him on 8th January 1775, when it was a sapling carried in the hand. In 1829 the trunk was 18 ft. in circumference to the height of 12 ft. above the ground, where it divides into branches which overhang a circle of more than 300 ft. in circuit covering 7500 square ft. of area." It had thus grown 216 in. in girth in fifty-four years, or at a rate of 4 in. a year. In 1845 this tree was carefully measured by a gentleman at Springfield, who found it 7 ft. in diameter at three feet from the ground, and 7 ft. 4 in. at eleven feet. The spread of the top was 134 ft.

## Cultivation

According to Loudon this tree was introduced to England by James Gordon in 1752, though unnoticed in Miller's Dictionary published sixteen years later. It appears to have been short lived, as none were to be found in Loudon's time either at Kew or Syon, and the only specimens he mentions as then living in England were young trees 15 to 30 ft. high in the Horticultural Society's Garden at Chiswick. We have been able to recognise no old trees of it now in the country, and very few young ones. Of these the best are those at Hargham, Norfolk, where they were planted by Sir Thomas Beevor about 1854. Here there are now thirty trees varying somewhat in habit. The largest of these is one of a group of three by a sunk fence south-west of the house, and measured 85 ft. by 7½ ft. in 1911. Another in 1905 measured 68 ft. by 8 ft. 4 in. (Plate 392). A third near the farm buildings has more spreading and drooping branches, some of which have been broken by wind; its leaves are thicker and more glabrous than those of the others. I showed fresh specimens of these to Dr. N. L. Britton of New York, who thought that they all belonged to *U. americana.* I could find no suberous branchlets on any of these trees, but noticed on 24th August 1909, that their leaves were much eaten by caterpillars, which had not

[1] *Garden and Forest,* v. 303, fig. 55 (1892).

[2] *Trees and Shrubs of Massachussets,* ii. 332 (1875).

injured the native elms. They seldom flower, but one of these trees has borne seed from which plants were raised in 1909. The soil on which they grow is a light sandy loam of one or two feet depth on the chalky boulder clay. At one spot only, and in a hawthorn thicket, two young trees about 30 feet seem to have sprung from suckers. The parents of these, a thickly-planted cluster, died about seventeen years ago. Sir Hugh Beevor states that there are six trees of the same age at Wilby Hall, about a mile and a half from Hargham; and says that this species has a distinctive charm and moreover very rarely suckers, and that he will plant it for ornament in future at Hargham, in preference to other elms.

Another tree, growing at Hildenley, Yorkshire, was raised from seed by the late Sir Charles Strickland about 1870 and measured in 1905 44 ft. by 3 ft. There are also two small trees by the garden road at Tortworth; a small tree in Silkwood, Weston Birt; small trees planted by myself in Congham Wood, Norfolk, which are growing fairly well in sandy soil; and several at Colesborne which at present do not seem to suffer from the lime in the soil.

In Scotland Henry found a tree at Methven, Perthshire, which in 1904 was about 60 ft. by 5½ ft. Loudon says that in 1828 there was a tree in the Botanic Garden at Edinburgh which we cannot now find.

In Ireland Henry saw in 1903 the stump of a tree[1] blown down in that year which grew in Trinity College Botanic Garden, Dublin, and was said by the late F. W. Burbidge to have been 80 ft. high.

In France it seems to succeed no better than in England; and M. Jouin informs me that many specimens in the east of France and in Alsace grown as *U. americana* are in reality *U. pedunculata.*

### Timber

Sargent says of this wood that it is heavy, hard, strong, tough, difficult to split, and rather coarse-grained, light brown in colour, with thick sap wood which is paler. It is largely used in the States for wheel hubs, saddle trees, flooring, and cooperage, and in boat- and ship-building. It is now imported to London in the form of boards which are used as a cheap substitute for coffin boards. I am informed by Mr. A. H. Ross of the Toronto University that a great deal of what is now known in the trade as rock elm, a name properly used for *U. racemosa*, is really taken from *U. americana*; and judging from the specimens I have seen, and from those in Hough's *American Woods*, the two are difficult to distinguish. (H. J. E.)

[1] The lower part of the stem of this tree, which is 21 in. in diameter and shows eighty-two annual rings, is preserved in the Forestry Museum at Avondale.

## ULMUS RACEMOSA, Rock Elm

*Ulmus racemosa*, Thomas, in *Amer. Journ. Sci.* xix. 170 (1831) (not Borkhausen[1]); Sargent, *Silva N. Amer.* vii. 47, t. 312 (1895), and in *Bot. Gaz.* xliv. 225 (1907).
*Ulmus Thomasi*, Sargent, *Silva N. Amer.* xiv. 102 (1902), and *Trees N. Amer.* 290 (1905).

A tree, attaining in America 100 ft. in height and 10 ft. in girth. Bark deeply divided by wide irregular interrupted fissures into broad flat scaly ridges. Young branchlets densely clothed with soft white pubescence, more or less persistent in the second year, when the branches are smooth, brown, and not finely striated; usually in the third year furnished with three or four irregular corky wings. Buds conic, sharp-pointed, with chestnut-brown scales, covered with appressed hairs and ciliate in margin. Leaves (Plate 411, Fig. 10) elliptic or oval, averaging 3 in. long and 1¾ in. broad, unequal and usually subcordate at the base, shortly acuminate at the apex; upper surface glabrous, smooth, shining; lower surface with a scattered minute white pubescence throughout, conspicuous on the sides of the midrib and on the nerves, not forming axil-tufts; lateral nerves fourteen to eighteen pairs, running parallel and straight to the margin, occasionally forked; margin biserrate, non-ciliate; petiole ¼ in. long, glabrescent.

Flowers in a racemose inflorescence, about 1½ in. long, composed of two or three cymes, each with about three flowers; axis pubescent; pedicels slender, up to ⅜ in. long, pubescent; calyx funnel-shaped, with six to eight rounded red lobes; stamens six to eight, with slender filaments and red anthers; stigmas pale green. Samara obovate or oval, ½ to ¾ in. long, with a slight notch at the apex, pubescent on both surfaces, and densely ciliate with long white hairs on the slightly thickened margin.

The rock elm is distributed in Canada from the province of Quebec westward through Ontario, where it mainly grows on limestone, and extends southwards in the United States through northern New Hampshire to southern Vermont and northern New Jersey, and westward through northern New York, southern Michigan, and central Wisconsin to north-eastern Nebraska and western Missouri. It is rare in the east and towards the extreme western and southern limits of its range; and is most abundant and of its largest size in Ontario and the southern peninsula of Michigan. It grows mainly on dry gravelly uplands, on low heavy clay soils, and on rocky slopes and cliffs, in company with the sugar maple, butternut, lime, white ash, beech, and other trees, but is less abundant everywhere than *U. americana.*

*U. racemosa* is one of the rarest of American trees in this country, the only specimens which we have seen being a tree at Kew, about 25 ft. high and doing very badly, which was obtained from Sargent in 1875; a smaller tree in the Edinburgh Botanic Garden, which is not thriving; and another about 20 ft. high, growing slowly at Tortworth. The tree at Kew has never produced flowers or fruit, and scarcely ever develops foliage of a normal size or appearance, as it is hurt by both late and early frosts. Judging from these specimens, this species, like certain

[1] *U. racemosa*, Borkhausen, *Forstbot.* i. 851 (1800), is *U. pedunculata*, Fougeroux.

other Eastern American trees, as white oak and black ash, is totally unsuitable for our climate. (A. H.)

Plants at Colesborne, received from Meehan in 1903, have succeeded better, though this may be only for a time. These are now about 14 ft. high, and show the characteristic corky branches, but bear leaves much smaller in size and with fewer nerves than those on adult trees in America. This small foliage is probably characteristic of the juvenile stage in the life of the tree, and matches a branch which I gathered at Ottawa. In France the tree seems to thrive no better, the only specimen which I have seen being a poor tree at Segrez, and it does not exist in the National Arboretum at Les Barres.

## TIMBER

Macoun[1] says:—"The rock elm grows in southern Quebec and westward to Lake Superior, being best developed in south Ontario, to which part of Canada it is, as a commercial wood, now confined. It is much superior to the other elms, and for many purposes unequalled by any other wood. It is tough, strong, elastic, and very heavy. Its chief use is in the manufacture of agricultural implements, bicycle rims, and wheel stock. It is largely used in bridge- and ship-building and for heavy furniture. When highly polished the wood is very beautiful, and repays a greater expenditure of time in polishing than is usually given to elm."

Known in the trade as Canada rock elm, this wood long had a high reputation among ship- and boat-builders, perhaps, because it was grown very slowly in dense forests, and free from knots and defects. A large quantity of elm still comes under this name to the Liverpool and London importers, but it is difficult to say what proportion of it is genuine rock elm.

Laslett[2] gives a good account of its properties and resistance to transverse, tensile, and crushing strains from experiments made by him in the Royal Dockyards, where in his time 600 to 700 loads were annually used for garboards and planking on account of its durability under water, and for ladder-steps and gratings on account of its clean whitish appearance. Laslett said that it was one of the slowest grown woods he knew of, making only 1 in. in diameter in fourteen years. I have seen in the yard of Messrs. White and Company, the well-known boat- and yacht-builders of Cowes, a square log 59 ft. long and only 9½ in. square at 56 ft. from the butt, in which with a lens upwards of 250 annual growths could be counted. This was being cut into gunwales for boats. The wood is also used by makers of agricultural implements in this country, who complain of the difficulty of getting it of reliable quality. As the supply of the genuine slow-grown rock elm seems to be rapidly diminishing, whilst logs of good size are now worth 3s. 6d. per foot and upwards, it seems as though a substitute must be sought for, and may be found in home-grown wych elm if this was closely crowded on suitable land. (H. J. E.)

[1] *Forest Wealth of Canada*, 24 (1900).

[2] *Timber and Timber Trees*, 225 (1875), where the rock elm of Canada is erroneously named *U. americana*.

## ULMUS FULVA, Slippery Elm

*Ulmus fulva*, A. Michaux, *Fl. Bor. Amer.* i. 172 (1803); Loudon, *Arb. et Frut. Brit.* iii. 1407 (1838); Bentley and Trimen, *Medicinal Plants*, t. 233 (1880); Sargent, *Silva N. Amer.* vii. 53, t. 314 (1895), and *Trees N. Amer.* 293 (1905).

*Ulmus americana*, Linnæus, var. *rubra*, Aiton, *Hort. Kew.* i. 319 (1789).

*Ulmus rubra*, F. A. Michaux, *Hist. Arb. Amer.* iii. 278, t. 6 (1813).

(?) *Ulmus pubescens*,[1] Walter, *Fl. Carol.* 112 (1788); Sudworth, *U.S. Forest. Bull.* No. 17, p. 60 (1898); Pinchot, *U.S. Forest. Circ.* No. 85 (1907).

A tree, attaining in America 70 ft. in height and 6 ft. in girth. Bark with shallow fissures and covered with large thick appressed scales. Young branchlets, fawn-coloured, with numerous minute tubercles, and densely clothed with a short white pubescence; those of the second year tuberculate, fissured but not finely striate on the surface; remaining smooth and not developing corky ridges in the third and fourth years. Buds ovoid, with reddish brown scales, covered with long silky appressed hairs. Leaves (Plate 411, Fig. 8) oval to obovate-oblong, about 5 to 7 in. long and 2 to 3½ in. broad, very oblique and unequal at the base, acuminate at the apex; upper surface scabrous with minute sharp-pointed tubercles, and short stiff hairs; lower surface densely clothed with white soft pubescence, conspicuous on the midrib and veins, and forming axil-tufts at their junctions; lateral nerves about sixteen pairs, often forking before reaching the margin, which is ciliate and coarsely biserrate; petioles stout, ⅕ to ¼ in. long, glandular, densely pubescent.

Flowers in crowded fascicles on short pedicels; calyx campanulate, with a narrow tubular part below; sepals five or six, faintly pink, often irregular in size; stamens five, six, or seven, with white filaments and dark red anthers; stigmas tinged with pink on their inner side. Fruit nearly orbicular or obovate, ½ in. in diameter, rounded or slightly emarginate at the apex, with a minute notch closed by the incurved stigmas, pubescent over the seed-cavity, but elsewhere glabrous and non-ciliate; seed in the centre of the samara.

The slippery elm is distributed from the valley of the lower St. Lawrence southwards to western Florida, central Alabama and Mississippi, and the valley of the San Antonio river in Texas, and westwards through Ontario to north Dakota, eastern Nebraska, and northern and western Kansas. Throughout its entire range it is less common than *U. americana*, often occurring as a solitary tree in open woods, or less frequently on the moist banks of streams in almost pure stands. It is always a small tree, the largest measured by Emerson being 6 ft. 10 in. in girth. It thrives best in rich alluvial soil, but grows fairly well on rocky hill sides, and is often found on dry limestone ridges. It is called *orme gras* by the French Canadians, and is more common than *U. americana* in some localities near Montreal.[2]

The inner bark is fragrant and mucilaginous, and is officinal in the United States

[1] It is not certain whether the description by Walter, which is unsatisfactory, refers to *U. fulva* or another species, and his name, though older than that of Michaux, cannot be adopted.

[2] *Garden and Forest*, 1894, p. 413.

Pharmacopœia. It is frequently chewed by children, and when macerated in water yields a thick and abundant mucilage, which was formerly used as a refreshing drink for colds. The powdered inner bark is used for poultices, and is said to preserve butter and lard from rancidity if the latter are melted with it.[1]

The date of introduction of *U. fulva* into England was unknown to Loudon, who mentions no large trees. It is probably short-lived in our climate, as we have seen no specimens except young trees, of which there are good examples, 15 to 20 ft. high, at Kew. It is rare except in botanic gardens; but there is a tree at Hildenley of no great size, and another at Colesborne, procured from Simon-Louis, which seems healthy at present.

The timber is not of much importance commercially, and is not found anywhere in great quantity. Macoun[2] says that it is more durable than that of the other elms, and is better suited for railway ties, fence-posts, and rails. Pinchot[3] recommends the planting of the slippery elm in the Mississippi valley, as it grows fast in youth, and can be utilised for fence-posts when quite young, since the sapwood, if thoroughly dried, is quite as durable as the heartwood.

*U. fulva* has been much confused with the following species :—

Ulmus elliptica, Koch, in *Linnæa*, xxii. 599 (1849), and *Dendrologie*, ii. pt. i. 420 (1872) (not[4] Koehne, Schneider, or Ascherson and Graebner).

A tree,[5] similar in size and habit to *U. montana*. Young branchlets pubescent, smooth and not tuberculate as in *U. fulva*. Leaves similar to those of *U. montana*, in being nearly sessile, with the inner side of the base overlapping the branchlet; elliptic, $3\frac{1}{2}$ to 6 in. long, $1\frac{1}{2}$ to $2\frac{1}{4}$ in. wide, scarcely scabrous above, sparingly pubescent and with inconspicuous axil-tufts beneath; lateral nerves 18 to 20 pairs, often forked; biserrate in margin. Samaræ, obovate or oval, $\frac{7}{8}$ in. long, $\frac{5}{8}$ in. wide, emarginate at the apex, with a minute aperture formed by the incurved stigmas; seed-cavity in the centre, and pubescent on both surfaces, the rest of the samara being glabrous and non-ciliate.

This species is closely allied to *U. montana*, differing mainly in the pubescence on the centre of the samara, in which respect it resembles *U. fulva*; but the samara in the latter species is much smaller. *U. elliptica* is imperfectly known,[6] but is said by Koch to form extensive woods in the Caucasus, which are either pure or mixed with other broad-leaved trees. It is not in cultivation, so far as I know.

The trees cultivated by Späth, under the names *U. elliptica* or *U. Heyderi*,[7] are

[1] Flückiger and Hanbury, *Pharmacographia*, 557 (1879). Cf. also Loudon, *Gard. Mag.* xv. 574 (1839), and xvi. 231 (1840).

[2] *Forest Wealth of Canada*, 24 (1900).

[3] *U.S. Forest Circular*, No. 85 (1907).

[4] *U. elliptica*, Koehne, *Dendrologie*, 136 (1893), Schneider, *Laubholzkunde*, i. 216, fig. 136 (1904), and Ascherson and Graebner, *Syn. Mitteleurop. Flora*, 550 (1911), is *U. fulva*.

[5] This description is drawn up from a specimen in the Kew Herbarium, collected by Markowicz in the Caucasus, which agrees with Koch's description. Markowicz wrote a pamphlet on this elm, which was published at Moscow in 1900. Elwes saw specimens in the St. Petersburg Herbarium which were collected in Abchasia and near Alagir in Ossetia.

[6] Köppen, *Holzgewächse Europ. Russlands und Kaukasus*, ii. 41 (1889), and other Russian botanists regard *U. elliptica* as a dubious species.

[7] Späth, *Catal.* No. 57, p. 4 (1883). Koehne's description of *U. elliptica* was apparently taken from one of these trees, which produced flowers and fruit in Späth's nursery, and not from the typical tree in the Caucasus.

identical in all respects with *U. fulva*. This is remarkable, as I am informed by Mr. Jensen that one of these trees was obtained from Dieck,[1] while another was sent from Tashkend by Koopmann. There must be some error in the account of their origin, as it is improbable that *U. fulva* is either wild or cultivated in Turkestan. (A. H.)

## ULMUS MONTANA, Wych Elm

*Ulmus montana*,[2] Stokes, in Withering, *Bot. Arrange. Veget. Great Brit.* i. 259 (1787); Smith,[3] *Eng. Bot.* t. 1887 (1808); Loudon, *Arb. et Frut. Brit.* iii. 1398 (1838); Mathieu, *Flore Forestière*, 302 (1897).

*Ulmus campestris*, Linnæus, *Sp. Pl.* 225 (1753) (in part); Miller,[4] *Gard. Dict.*, ed. 8, No. 1 (1762); Willkomm, *Forstl. Flora*, 555 (1887).

*Ulmus glabra*,[2] Hudson, *Fl. Angl.* 95 (1762) (not Miller); Rehder, in *Mitt. Deut. Dend. Ges.* 1908, p. 157; Moss, in *Gard. Chron.* li. 217 (1912).

*Ulmus scabra*,[4] Miller, *Gard. Dict.* ed. 8, No. 2 (1762); Schneider, *Laubholzkunde*, i. 216 (1904), 805 (1906); Ley, in *Journ. Bot.* xlviii. 67 (1910); Ascherson and Graebner, *Syn. Mitteleurop. Flora*, iv. 560 (1911).

*Ulmus suberosa*, Michaux, *N. Amer. Sylva*, ii. 244, pl. 129, fig. 2 (1819) (not Moench, Ehrhart, or Smith).

A tree, attaining 120 ft. in height and 20 ft. or more in girth. Bark remaining smooth on the stem and branches for many years, ultimately on the trunk divided by shallow longitudinal fissures into scaly plates. Young branchlets stout, more or less covered with stiff hairs; in the second year smooth or slightly fissured, not showing the fine striation of *U. nitens*. Buds conical, obtuse, with dark brown scales, which are ciliate in margin and densely pubescent on the surface with yellowish brown hair. Leaves (Plate 412, Fig. 13) variable in size and shape, but averaging 3 to 5 in. long, and readily distinguishable from the other species by the short stout densely pubescent petioles, not exceeding $\frac{1}{8}$ in. in length; mostly obovate-elliptic, very unequal at the base, cuspidate-acuminate at the apex, on vigorous branches and coppice shoots often with three cuspidate points; sharply biserrate; lateral nerves fifteen to eighteen pairs, often forked; upper surface scabrous with scattered short hairs; lower surface with a soft white pubescence, dense on the midrib and lateral nerves, forming axil-tufts at their junctions, and scattered on the surface between the nerves.

Flowers twenty to thirty in a cluster, on very short pedicels, regularly pentamerous, hexamerous, and heptamerous; calyx campanulate, contracted towards the base into a narrow wrinkled tubular part, about $\frac{1}{6}$ to $\frac{1}{7}$ in. long; sepals five, six, or

[1] *Neuh. Offer. Nat. Arb. Zöschen*, 1889–1890, p. 22.

[2] The oldest tenable name for this species appears to be *U. glabra*, Hudson; but as *U. glabra*, Miller, has been much used for another species, I retain *U. montana*, Stokes, as a name which has been long in use, and never applied to any other species.

[3] Smith's description applies to *U. montana*; but the leaves are badly drawn in the figure, the stalks being too long. *U. nuda*, Ehrhart, *Beit.* v. 160 (1790), vi. 86 (1791), judging from his specimen in Smith's herbarium at the Linnean Society, is *U. montana*.

[4] Miller, whose account of the elms is very confused, gives two names to *U. montana*, his No. 1 being "the common rough or broad-leaved witch elm, very common in the north-west counties of England, where it is generally believed to grow naturally in woods." His No. 2 is "the witch hazel or rough and very broad-leaved elm," also said by him to "grow naturally in some of the northern counties of England." Woodmen in the Chiltern Hills and in Sussex (cf. p. 1874) at the present day sometimes similarly attempt to make a distinction between the wych elm and the wych hazel.

seven, pinkish; stamens five, six, or seven, with deep pink filaments and red anthers; stigmas red. Fruit, very shortly stalked, oval, about 1 in. long and ¾ in. broad, glabrous except in the interior of the notch, non-ciliate, rounded or pointed at the apex, with a minute notch, usually closed by the incurved stigmas. Seed in the centre of the samara, with its apex distant from the base of the notch.

The seedling[1] of *U. montana* differs from that of *U. nitens* in all the leaves above the first two pairs being alternate in the first year, and considerably larger than in the latter species. The stem is stout and usually bent to one side near the apex, and attains at the end of the first season about 6 to 12 in. in length, developing about nine to thirteen leaves in all.

The flowers of *U. montana*, which is a tree of more northerly distribution than the other European elms, are scarcely ever injured by late frosts; and in consequence ripe seed is produced regularly every year in most parts of Britain, often in great abundance and invariably fertile.

*U. montana* never produces suckers[2] in England, differing in this respect from all the other British elms. There is, however, in the neighbourhood of Cambridge, in hedgerows, a peculiar elm, closely resembling *U. montana*, but producing suckers, though in no great quantity. This elm[3] is a low tree, with wide-spreading branches, forming a globose crown of foliage, and seems to be less vigorous in growth than ordinary *U. montana*. The leaves and buds cannot be distinguished from those of the last, but the branchlets are glabrous, or nearly so. The flowers of one tree which I examined differ from those of *U. montana* in the irregular number of the sepals and stamens; calyx-tube funnel-shaped and scarcely narrowed into a tubular basal part, with five or six lobes; stamens five or six, or often four, one occasionally being aborted and sterile; filaments white, not pink as in *U. montana*. Samaræ similar to those of the last, but rarely ripening, being injured by late frosts. The suckers bear small leaves, 2 to 3 in. in length, with a long acuminate apex, and often one or two additional points on each side of it. This peculiar elm[3] is probably one of the descendants of the Huntingdon elm, in which nearly all the characters of *U. montana* appear; but differing in the irregularity of the flowers, the infertile samaræ, and in the occurrence of suckers.

## Varieties

*U. montana* shows little variation in the wild state; but trees growing in dense woods usually bear smaller and thinner leaves than those which stand in the open.

I. The following geographical form has been described:—

1. Var. *laciniata*, Trautvetter, in Maximowicz, *Fl. Amur.* 246 (1859); Maximowicz, in *Mél. Biol.* ix. 25 (1872); Shirasawa, *Icon. Ess. Forest. Japon*, ii. t. 15, figs. 1-9 (1908).

[1] Cf. *Journ. Linn. Soc.* (*Bot.*) xxxix. 292, pl. 22 (1910).

[2] Mr. A. P. Long saw in 1912 several large trees of *U. montana* producing suckers freely in the Sihlwald, near Zurich. The leaves of these suckers differ in no respect from those produced by the peculiar Cambridge trees here described; but the leaves and branchlets of adult trees are like those of typical *U. montana*.

[3] This hybrid may be named *U. Mossii* after Dr. C. E. Moss, who drew my attention to it.

*Ulmus major*, var. *heterophylla*, Maximowicz and Ruprecht, in *Bull. Acad. Pétersb.* xv. 139 (1857).
*Ulmus laciniata*, Mayr, *Fremdländ. Wald- u. Parkbäume*, 523 (1906).
*Ulmus scabra*, var. *heterophylla*, Schneider, *Laubholzkunde*, i. 218 (1904).

Young branchlets glabrous or with a few scattered hairs. Leaves, at the end of the branchlets, large, 6 to 7 in. long, 3 to 4 in. wide; usually with three, occasionally five, large cuspidate-acuminate lobes; upper surface scabrous, with minute tubercles, each of which bears a short bristle; lower surface covered with a dense soft pubescence. Samara, about $\frac{7}{8}$ in. long, $\frac{1}{2}$ in. wide; narrower than in the type.

This is a common form of *U. montana* in eastern Asia, where it is widely spread, but apparently mixed with the typical form in Amurland, Manchuria, Saghalien, and Japan. Both normal and tricuspidate leaves occur occasionally on the same individual tree. Similar leaves[1] in all respects occur on coppice shoots and on epicormic branches of *U. montana* in Europe, but are rarely seen on ordinary branches. In eastern Asia the peculiar tricuspidate leaves occur normally on adult trees; and young trees raised in the Arnold Arboretum from Japanese seed preserve the remarkable character of the foliage.

Sargent states[2] that in Japan this variety is usually a small tree, about 30 ft. high; but Mayr says that it attains 100 ft. in height in the broad-leaved forests of central Yezo.

The inner bark of this tree is taken off in narrow strips by the Ainos, and after being soaked in water, is woven into a coarse cloth, from which they make their garments. It is also woven into baskets, of which Elwes brought home specimens. The wood is not used at present to any extent; but may possibly become a substitute for Canadian rock elm when the latter becomes scarce.

II. The following varieties have arisen in cultivation :—

2. Var. *fastigiata*,[3] Loudon, *op. cit.* 1399.

*Ulmus Fordii* and *Ulmus exoniensis*, Loudon, *op. cit.* 1399.

Branches directed vertically upwards. Leaves clustered at the ends of short shoots, dark green, small, obovate, more or less uneven and wrinkled on the surface; margin with coarse serrated teeth.

This variety was raised at Exeter by Mr. Ford about 1826, and hence is generally known as Ford's elm or the Exeter elm. The finest specimen which we have seen is one in Canon Ellacombe's garden at Bitton, which was 65 ft. by 12 ft. in 1908. There are two good trees at Bayfordbury, 45 ft. by 5 ft. 8 in., and 35 ft. by 6 ft. 8 in. in 1911. At Drumlanrig Castle, Elwes measured a fine specimen 60 ft. by 8 ft. in 1911. One at Dawyck is reported by Mr. F. R. S. Balfour to be 48 ft. by $7\frac{1}{2}$ ft. in 1911.

[1] Leaves gathered by Elwes in Croatia, and others taken by me from the epicormic branches of a tree at Wyfold Grange, Henley, are identical with specimens collected in Japan by Elwes, who found it growing with *U. japonica* in a virgin forest near Asahigawa in Yezo. Trees with tricuspidate leaves occurring in Galicia have been named *U. montana*, var. *corylifolia*, Zapalowicz, *Consp. Fl. Galic.* ii. 98 (1908). Similar trees are said by Koch, *Dendrologie*, ii. pt. i. 415 (1872), to have been propagated in gardens under the names *U. tricuspis*, *U. tridens*, *U. triserrata*, and *U. intermedia.*

[2] *Garden and Forest*, vi. 323 (1893), and *Forest Flora of Japan*, 57 (1894).

[3] There are several small trees at Kew, with ascending branches, which have been obtained under various names, as var. *etrusca*, var. *gigantea*, and var. *macrophylla fastigiata*; but none of these appear to me to be distinct enough to deserve a special name.

A sub-variety with yellowish foliage, known as var. *fastigiata aurea*, is occasionally seen in botanic gardens. A tree at Kew, 20 ft. high, which was obtained from Lee in 1879, is narrowly pyramidal and not strictly fastigiate in habit.

3. Var. *cinerea*, Lavallée, *Arb. Segrez.* 237 (1877).

*Ulmus cinerea*, Planchon, in De Candolle, *Prod.* xvii. 160 (1873).

Branches stunted and tortuous, the upper ascending, the lower more or less pendulous. Leaves crowded and similar to those of var. *fastigiata*, from which var. *cinerea* appears to differ only in not being fastigiate in habit.

This is represented at Kew by a tree about 25 ft. high, of unknown origin. Elwes found a small tree of it in the late Mrs. Robb's garden at Liphook.

4. Var. *pendula*, Loudon, *op. cit.* 1398.

Var. *horizontalis*,[1] Petzold and Kirchner, *Arb. Musc.* 564 (1864).
*Ulmus pendula*, Loddiges, *Cat.* 1836.
*Ulmus horizontalis*, Loudon, *op. cit.* 1398.

Branches horizontally spreading, branchlets more or less pendulous. This is the common weeping wych elm, which is much planted as an ornamental small tree, and is usually grafted high on a stock of the common species. It occasionally attains 30 ft. in height, and bears flowers and fruit abundantly every year. Booth states[2] that this variety was found in a bed of seedlings in the Perth nursery about 1816. He purchased the plant, from which all the stock, both in England and on the Continent, originated. A fine tree in the Glasnevin Botanic Garden (Plate 393) is grafted at 10 ft. from the ground, and grows to a height of about 30 ft., with a circumference of branches of 54 paces.

5. Var. *pendula Camperdowni*, Hort.

Branches and branchlets pendulous, forming a globose head, in marked contrast to the flat stiff-looking crown of var. *pendula*, Loudon. The original plant grew at Camperdown House, near Dundee; and, according to Mr. Mitchell, of Messrs. R. B. Laird and Sons, Edinburgh, was of considerable age thirty years ago, and quite prostrate in habit, creeping along the ground amongst other elms. There are good specimens of both these weeping elms in the Grange Cemetery, Edinburgh; and the difference in habit is well shown in a photograph by Mr. A. D. Richardson, reproduced in *Gard. Chron.* l. 221, fig. 105 (1911).

6. Var. *crispa*, Loudon, *op. cit.* 1399.

*Ulmus crispa*, Willdenow, *Enum. Pl. Hort. Berol.* 295 (1809).
*Ulmus urticæfolia*, Audibert, *Cat. Hort. Tonn.* 23 (1817).

Leaves very narrow, linear to oblanceolate, wrinkled, laciniate, with numerous incised curved teeth. This variety usually forms a small tree with pendulous branches. There is a fine specimen, about 25 ft. high, at Howden and Company's nursery, Inverness, and smaller ones at Kew, Revesby, and elsewhere.

[1] Var. *horizontalis* is the name generally used in continental nurseries for the common form of the weeping wych elm (var. *pendula*, Loudon); while their var. *pendula* is either identical with or very similar to *U. pendula Camperdowni*.

[2] In Loudon, *Gard. Mag.* 1843, p. 442.

Var. *crispa aurea*, Schelle, *Laubholz-Benennung*, 86 (1903), is a form of the preceding, with yellowish leaves.

7. Var. *libro rubro*,[1] Planchon, in De Candolle, *Prod.* xvii. 160 (1873).

Inner bark of the young branchlets deep red in colour. A tree of this at Kew, about 20 ft. high, was obtained from Van Houtte in 1871. It appears to differ in no respect from the type, except in the colour of the inner bark. This variety is said by Loudon[2] to have been found by M. de Vilmorin in a wood near Verrières some time before 1840. Loudon explains that it was propagated by grafting, and was not a local peculiarity arising from something in the soil, as in the case of the blue flowers of Hydrangea.

8. Var. *nana*, Simon-Louis, *Cat.* 1869, p. 98.

A shrub, attaining about 6 ft. in height, with wide-spreading horizontal branches, stunted branchlets, and small leaves. This variety, the origin of which is unknown, forms at Kew a peculiar slow-growing hemispherical bush, which has not increased appreciably in size for many years.[3]

9. Var. *atropurpurea*, Späth, *Cat.* No. 57, p. 4 (1883-1884).

Leaves dark purple and folded. This was raised from seed in Späth's nursery; but is very similar to or perhaps identical with *U. purpurea*, Koch, *Dend.* ii. pt. i. 416 (1872), of which there was a fine specimen growing in front of Petzold's house at Muskau.

10. Var. *lutescens*, Schelle, *Laubholz-Benennung*, 86 (1903).

Leaves of a beautiful yellow colour. I saw a fine specimen in the Calmpthout Nursery, Belgium, where it was cultivated under the name *U. americana aurea.* Loudon[4] states that in Mr. Jessop's garden at Derby he found a variety, the leaves of which were of a fine yellow colour at the time of expanding in May. This tree was planted by Pontey, and known as the gallows elm, because the original tree grew near a gallows at York.

### Hybrids

*U. montana* is one of the parents in the following hybrids :—

1. *U. vegeta*, Huntingdon elm. See p. 1879.
2. *U. major*, Dutch elm. See p. 1883.
3. *Ulmus Smithii*, Henry. Downton elm.

*Ulmus montana*, var. *Smithii*, Hort. Kew.
*Ulmus pendula*, W. Masters, *Hortus Duroverni*, 66 (1831) (not Willdenow).
*Ulmus glabra*, var. *pendula*, Loudon, *Arb. et Frut. Brit.* iii. 1405 (1838).

A tree with ascending branches, and long pendulous branchlets. Young branchlets stout, more or less pubescent with long hairs. Leaves (Plate 412, Fig. 24) firm in texture, oval, about 3½ in. long and 1¾ in. wide, very unequal at the base, long

[1] *U. campestris rubra*, Simon-Louis, *Cat.* p. 97 (1869), from the description is identical with this variety, which appears in more recent catalogues under the name given above.

[2] *Derby Arboretum*, 52 (1840). Loudon, *op. cit.* 51, states that a similar variety of the white mulberry, *Morus alba*, var. *Morettiana*, was then cultivated in the Jardin des Plantes at Paris; "the soft wood or cambium of the current year's shoot was of a deep red."

[3] Cf. *Woods and Forests*, 1884, p. 482, where an account of this remarkable bush at Kew is given. It may have originated from a witches' broom.

[4] *Gard. Mag.* xv. 449 (1839).

acuminate at the apex; upper surface shining, dark green, glabrous, smooth; lower surface with scattered pubescence and with axil-tufts; margin coarsely biserrate; lateral nerves fourteen to sixteen pairs, very prominent beneath, straight, close, parallel, occasionally forked; petiole $\frac{1}{3}$ in. long, densely pubescent.

Flowers irregular in the number of the sepals and stamens; calyx funnel-shaped, $\frac{1}{10}$ to $\frac{1}{8}$ in. long, with four, five, or six bright pink lobes; stamens, three to five, with whitish filaments and dark red anthers; stigmas bright red. Samara obovate, $\frac{3}{4}$ in. long, $\frac{5}{8}$ in. broad, emarginate at the wide apex, with a shallow rounded notch not closed by the incurved stigmas; seed above the centre of the samara.

This tree has usually slight corky ridges with peculiar brown fissures on the branchlets of the third and fourth years. The leaves are thicker in texture, and with more nerves than those of *U. nitens.*

It was raised in Smith's nursery at Worcester, from seeds obtained from a tree in Nottinghamshire. Some of the seedlings were purchased by Mr. Knight of Downton Castle; and one of them, which turned out to be of weeping habit, was propagated. There are two Downton elms at Kew, about 25 ft. and 35 ft. high. One of these produced ripe fruit in 1909, from which I raised twenty-nine seedlings, very unlike in appearance, twenty-six having alternate leaves, and three having opposite leaves. Mr. Knight informed Loudon [1] that he had raised plants of various kinds from the seed of the Downton elm; and specimens of these seedlings, representing seven remarkably different elms, which were sent from Downton Castle in 1835, are preserved in the Kew Herbarium. A Downton elm, labelled *U. Smithii,* was 25 ft. high at Glasnevin in 1912.

4. *Ulmus belgica,*[2] Burgsdorf, *Anleit. Holzart.* 270 (1805) (*excl. syn.* Du Roi, Miller). Belgian Elm.

*Ulmus latifolia*, Poederlé, *Man. de l'Arbor.* ii. 117 (1792); Petzold and Kirchner, *Arb. Musc.* 561 (1864) (not Moench [3]).
*Ulmus batavina*, Koch, *Dendrologie*, ii. pt. i. 414 (1872).
*Ulmus campestris*, var. *bataviana*, Simon-Louis, *Cat.* 1869, p. 96.
*Ulmus campestris*, var. *belgica*, Lavallée, *Arb. Segrez.* 235 (1877).
*Ulmus campestris*, var. *latifolia*, Gillekens, *Arboric. Forest.* 38 (1891) (not Persoon [4]).
*Ulmus hollandica*, Späth, *Cat.* No. 113, p. 158 (1903-1904) (not Miller [5]).
*Ulmus montana*, var. *hollandica*,[6] Huberty, in *Bull. Soc. Cent. Forest. Belg.* xi. 566 (1904); Aigret, in *Ann. Trav. Publ. Belg.* x. 1230 (1905).

A vigorous tree, with a straight rough-barked stem, and widely extending branches, forming a broad crown of foliage. Young branchlets more or less pubescent with scattered long white hairs, which are usually deciduous in summer,

[1] *Arb. et Frut. Brit.* iii. 1404, paragraph 1 (1838).

[2] The adoption of this name, justified by Burgsdorf's description, "Die hollandische Ulme, le Ypreau de Holland," will avoid the confusion that has been almost universal of this elm with the "Dutch elm" of England, and with *U. hollandica*, Miller. The latter is an uncertain name, and cannot signify the Belgian elm, which has not the corky bark on the twigs, characteristic of Miller's Dutch elm. Cf. p. 1883, note 1.

[3] *U. latifolia*, Moench, *Meth.* 333 (1794), appears to be *U. montana*, to which Moench referred *U. hollandica* as a form. While these sheets were passing through the press, I have been able to consult the *Manuel* by Poederlé, one of whose names for the Belgian elm, *U. latifolia*, is possibly its oldest certain name.

[4] *U. campestris*, var. *latifolia*, Persoon, *Syn. Pl.* i. 291 (1805) is an uncertain name.

[5] Cf. note 2 above.

[6] *U. montana hollandica*, Planchon, in De Candolle, *Prod.* xvii. 160 (1873) is a doubtful name.

but are occasionally retained in the second year. Leaves (Plate 412, Fig. 18) obovate-elliptic, 3½ to 5 in. long, 1½ to 2 in. wide, very oblique at the base, abruptly contracted at the apex into a long serrated point; upper surface slightly scabrous, with a minute scattered pubescence; lower surface densely covered with a soft pubescence, conspicuous on the midrib, and with white axil-tufts; lateral nerves fourteen to eighteen pairs, often forked; margin coarsely biserrate; petiole ⅛ to ⅕ in. long, densely pubescent.

Samaræ differing little from those of *U. montana*, except in their smaller size, about ⅞ in. long and ⅝ in. wide, obovate, rounded and slightly emarginate at the apex, below which is a minute aperture formed by the incurved stigmas. Seed in the centre of the samara, at the base of which the persistent calyx is five-lobed, with remains of five stamens.

This elm is readily distinguishable from *U. montana* by the glabrescent and more slender twigs, and by the narrower leaf, which is usually prolonged at the apex into a long serrated point. The appearance of the stem in mature trees is very characteristic, as it is remarkably cylindrical, being rarely swollen or buttressed at the base, and tapers less than other elms. The tree produces suckers, but not very freely. These have pubescent branchlets, and scabrous leaves, similar in shape to those of the parent tree.

This tree is said by Huberty,[1] who quotes Poederlé's *Manuel de l'Arboriste*, ii. 117 (1792), to have been cultivated in the eighteenth century in the nurseries of the Abbey of Dunes, which was first established at Furnes and later was transferred to Bruges. It was then known by the Flemish name of *Hollander's olm* or elm of Holland, but is now more commonly called *orme gras*.[2] It is undoubtedly a hybrid, as Mr. Springer,[3] who sowed its seed, obtained mixed seedlings, of which I have received specimens. One of these is three weeks later in losing its leaves than the parent. Samaræ which he sent me in 1911 were very variable in size, but uniform as regards the position of the seed in the centre. Those which I sowed failed to germinate. This elm is always propagated in nurseries[4] by layering; and is most extensively cultivated in Belgium and Holland, where it is the principal tree[5] planted along the roads, as well as in parks and avenues. It grows with astonishing vigour, much faster than other elms, according to Mr. Huberty, who states that it yields a soft wood, little used by carriage builders, but valuable for making furniture.

[1] *Bull. Soc. Cent. Forest. Belg.* xi. 564 (1904). This elm is probably referred to in the Dictionary of the French Academy, published in 1694: "*Ypreau*, espèce d'orme à larges feuilles qui est venu premièrement des environs de la ville d'Ypres." Ypres is not far distant from Furnes and Bruges; and it is in this district that the other peculiar elms of Belgium, the Klemmer and Dumont varieties, have originated. Plukenet's *Ulmus hollandica* was possibly the *ypreau*. The word *ypreau*, which is also written *ypereau* and *ipreau*, meaning elm, was first used in 1432, according to Godefroy, *Dict. Anc. Lang. Franc.* x. 873 (1902); and only in the nineteenth century came to mean the grey poplar in the north of France; but it is still used in Flanders for the hybrid elm. Cf. *Bull. Soc. Cent. Forest. Belg.* ii. 817 (1895), and xi. 572 (1906). Gleditsch, *Forstwissenschaft*, i. 285 (1775), thought that the Holland "Yper" originated from the wild elm of North America, but there is no evidence of hybridisation with either *U. americana* or *U. pedunculata*.

[2] It is often called *orme gras de Malines*, because it has been long propagated by the nurserymen of this city; but there are no grounds for believing with Gillekens, *Arboric. Forest.* 39 (1891), that it originated at Malines.

[3] *Mitt. Deut. Dend. Ges.* 1910, p. 272.

[4] Attempts have been made to divide the Belgian elm into numerous varieties; but only two kinds are ordinarily propagated in Belgian and Dutch nurseries, viz., *orme gras* and *orme Dumont*.

[5] Aigret, in *Ann. Trav. Publ. Belg.* x. 477 (1905), gives a table of the numbers of the various species of trees planted along the main roads in Belgium. The elm is the most numerous, 294,725 out of a total of 806,985 trees.

A fine plantation of these elms is reported[1] near Châtelineau, in the Parc de Presles, where on deep fertile soil, resting on limestone, there are sixty-one trees to the acre, averaging 4½ ft. in girth, with a total volume in the round of 3500 cubic ft. These trees are said to be only thirty-four years old. I measured in 1912 a young Belgian elm, which had just been blown down in the Royal Park of Laeken. It was 108 ft. high by 4 ft. 8 in. in girth; but I was unable to count the annual rings. In the poor soil of the Brussels Park, one of these elms averaged[2] six rings to the inch of radius during the first eighty years. Huberty gives[3] a good photograph of two Belgian elms, growing on the road between Jemelle and Rochefort, which, though only forty-eight years old, are 9 ft. in girth. Others near Genappe, thirty-five years old, are 4 to 5 ft. in girth; while some fifty-eight years old are only 7 ft. round.

The elms which were planted in the park at Brussels by Zimmer in 1790, have been described by M. Bommer,[2] who states that they have passed their prime, and that many are suffering from the attack of *Polyporus squamosus.* These elms are very fine trees, growing in sandy soil. The leaves, when seen by Elwes on May 9, 1911, were not fully out, and there were no flowers visible. They have trunks clear of branches to 40 to 60 ft., probably due to pruning, with a total height of 110 ft. to 125 ft.

In the park at Haarlem, and in the "Bosch" at the Hague, there are many fine trees of the same type, on a somewhat similar soil. Some of those at the Hague, which looked quite young, were 120 ft. or more in height; and Elwes was told by Mr. Westbrook, superintendent of the municipal parks and nurseries at the Hague, that the best elms in the country were grown in the large nurseries at Oudenbosch and Calmpthout, north of Antwerp. I was informed at these nurseries in 1912, that the Belgian elm was exported to England on rare occasions; but we have seen no old trees of the kind in this country, except one at Kew, about 35 ft. high, which was obtained under the name *U. Pitteursii* from Lee in 1879. It is possible, however, that it may have escaped the notice of ourselves and of our correspondents, on account of its resemblance to *U. montana.*

Hübner reports[4] that the Belgian elm thrives on the worst soils, and on that account is planted in belts on the margins of pine woods in Prussia. All the indications in Belgium also point to the fact that this elm succeeds better on poor sandy soils than any other variety.

5. *Ulmus belgica*, var. *Dumontii*, Henry. Dumont Elm.

*Ulmus campestris*, var. *Dumontii*, Nicholson and Mottet, *Dict. Prat. Hort.* v. 383 (1898).
*Ulmus montana*, var. *Dumontii*, Aigret, in *Ann. Trav. Publ. Belg.* x. 1231 (1905).

Branches ascending, forming a narrow pyramidal tree. This scarcely differs from the common Belgian elm, except in height, and like it, is propagated by layering.[5] The original tree is said to have been found about 1865 in the park of M. Dumont at Tournai, where it was growing amidst a plantation of ordinary *orme gras*; and was subsequently propagated and sold under the name of *orme Dumont.*

[1] *Bull. Soc. Cent. Forest. Belg.* xvi. 346 (1909).
[2] *Ibid.* iv. 105-109 (1899).
[3] *Ibid.* xi. 571, t. vi. (1904).
[4] *Mitt. Deut. Dend. Ges.* 1908, p. 122.
[5] I have not been able to examine specimens from adult trees. Young nursery plants of *orme gras* and *orme Dumont* are very similar, and resemble *U. montana* in foliage and branchlets, more than the old trees do.

On account of its narrow form, it has become a favourite tree for planting in streets; but is believed not to be so fast in growth as the ordinary Belgian elm. It is much planted in the neighbourhood of Lille, Roubaix, and Tournai; and grows well in the district of Ypres. The finest avenue of this elm is said[1] to be on the road between Waremme and the park of Longchamps; but a specimen brought from there by Elwes, who describes the trees as having the habit of the Wheatley elm, is indistinguishable in branchlets and foliage from *U. montana.* There are peculiar clipped avenues of the Dumont elm in the streets near Moser's nursery at Versailles.

6. *Ulmus Klemeri,*[2] Späth, *Cat.* No. 104, p. 134 (1899); Huberty, in *Bull. Soc. Cent. Forest. Belg.* xi. 495 (1904), and xii. 173, pl. xxi. (1905). Klemmer or Flanders Elm.

*Ulmus campestris,* var. *Klemmer,* Gillekens, *Arboric. Forest.* 40 (1891); Aigret, in *Ann. Trav. Publ. Belg.* x. 1224 (1905).

A tall tree, narrowly pyramidal in habit, with ascending branches, and a straight cylindrical stem, covered with smooth bark. Young branchlets slender, pubescent with short hairs that are deciduous in summer. Leaves ovate, about 3 in. long and 2 in. broad, shortly acuminate at the apex, oblique at the base; scabrous and glabrescent above; lower surface smooth to the touch, covered with a minute pubescence, and with small axil-tufts; lateral nerves about twelve pairs, often forked; margin regularly biserrate. Aigret states that the seed is situated close to the emargination of the samara.

This elm, though similar in some respects to the southern variety of *U. campestris,* is probably a hybrid between *U. nitens* and *U. montana,* but closer to *U. nitens,* in the size of the leaf and the position of the seed in the samara. Leaves on vigorous shoots from the stem are like those of *U. belgica,* in texture and size, and it is possible that the Klemmer elm may be a seedling of the latter, but its origin is unknown.[3]

The Klemmer elm is propagated by layering, and is widely planted in West Flanders, in Belgium, and the adjoining Departement du Nord of France, where it is much esteemed on account of its rapid growth and excellent timber. Its wood is superior to that of *U. belgica* in strength and elasticity, and is preferred by wheelwrights. It grows to a great height, and on account of its narrow form is suitable for planting in streets in towns and along roads in the country. Gillekens states that six trees, which he planted in 1878 amidst a plantation of oaks that were about 20 ft. high, soon outgrew the latter, and in 1891 had attained 60 ft. in height and 2 ft. 9 in. in girth. The finest trees were those recently felled near Alveringhem, which were 10 to 12 ft. in girth. There are good specimens in the avenues adjoining the town of Ypres, which I saw in 1912. The Klemmer elm is sold[4] by French nurseries, and is represented at Kew by two specimens about 15 ft. high, which were obtained from Barbier in 1908.

[1] *Bull. Soc. Cent. Forest. Belg.* ix. 839 (1902).

[2] *Klemmer,* signifying climber in Flemish, is the name given to this elm in Flanders on account of its ascending habit and great height. The name *Klemeri,* implying a supposed person named Klemer, is objectionable, but must be maintained.

[3] Cf. p. 1895, note 1.

[4] I was informed that this elm was also obtainable in nurseries at Ypres and Bruges.

7. *Ulmus Pitteursii*, Petzold and Kirchner, *Arb. Musc.* 566 (1864).

*L'orme Pitteurs*, Morren, in *Journ. Agric. Prat.* 1848, p. 114, and in *Belgique Horticole*, ii. 133 (1852).

This name was given to two varieties which were obtained as seedlings of the *orme gras* in 1845, by M. de Pitteurs at St. Trond near Liège. One variety, described as making annual shoots of 3 ft. in length, and with large leaves, 8 in. long and 7 in. broad, is now little known; but is probably identical with elms propagated by layers, called *orme St. Trond*, which I saw in Looymans' nursery at Oudenbosch in 1912. These were undistinguishable in the young stage from *U. montana*, but bore leaves 5 to 6 in. in breadth; and are perhaps identical with a variety of the latter species, which is occasionally sold as var. *macrophylla*.[1]

The other variety of *U. Pitteursii* apparently differed but little from the ordinary Belgian elm (*U. belgica*) and may be[2] the form of it which is prevalent in the provinces of Liège and Limburg, if this is really distinct. It is said[3] to be represented by a tree so-named in the botanic garden at Liège, which is reputed to have been planted by Morren. This I have had no opportunity of examining.[4]

8. *Ulmus superba*, Henry.

*Ulmus montana superba*,[5] Späth, *Cat.* No. 62, p. 102 (1885-1886).

A narrow pyramidal tree, with smooth bark and very ascending branches. Branchlets glabrous, remaining smooth and without fine striations, corky ridges not being developed. Leaves, similar to those of *U. montana* in shape and numerous nerves, but with long stalks; variable in size,[6] 3 to 4 in. long, 1½ to 2 in. broad, obovate or obovate-elliptic, acuminate at the apex, very oblique at the base; glabrous, dark green, and smooth above; lower surface glabrous, except for minute axil-tufts of pubescence; margin biserrate, non-ciliate; petiole ¼ to ⅓ in. long, with scattered hairs.

Flowers regularly pentamerous; calyx campanulate, like that of *U. montana*, with five pink lobes; stamens five, with pink filaments and red anthers; stigmas white. Samaræ not seen.

This tree is probably identical with the variety[5] of the same name described long ago by Morren; but this is uncertain. Späth states that it is much esteemed at Magdeburg as a street tree. A tree at Kew, obtained from him in 1900, and now about 25 ft. high, is remarkably fast in growth, and peculiar in its narrow pyramidal habit. Another tree at Kew, labelled *U. montana macrophylla fastigiata*,

[1] Possibly identical with *U. campestris*, var. *macrophylla*, Spach, in *Ann. Sc. Nat.* xv. 363 (1841).

[2] Cf. Gillekens, *Arboric. Forest.* 40 (1891).

[3] Aigret, in *Ann. Trav. Publ. Belg.* x. 1230 (1905).

[4] Koch, *Dendrologie*, ii. pt. i. 417 (1872), states that *U. Pitteursii* had scabrous branchlets and brownish leaves. Petzold and Kirchner, *Arb. Musc.* 566 (1864), describe it as having oblong ovate obtuse-toothed dark green leaves, which were reddish brown at the time of unfolding.

[5] *U. montana superba*, Morren, in *Journ. Agric. Prat.* 1848, p. 411, is said to have been introduced into Belgium in 1845 under the name *superba* from Osborne's nursery at Fulham. Morren figures a form of *U. montana* with leaves, up to 10 in. long and 6 in. broad. Morren, who was unacquainted with *U. montana*, which he believed to be identical with *U. pedunculata*, adds to his otherwise correct account of var. *superba* a description of the flowers and fruits of *U. pedunculata*. *U. præstans*, Schoch, ex *Mitt. Deut. Dend. Ges.* 1912, p. 227, appears to be a name, without any description, of *U. superba*, Henry.

[6] Specimens sent by Späth, apparently from young trees, have leaves 4 to 5 in. long, 2½ to 3 in. broad, which are said to fall late in autumn.

the history of which is unknown, is about the same height and is similar in all respects.

9. *Ulmus Dauvessei*, Henry.

*Ulmus montana*, var. *Dauvessei*, Nicholson, *Kew Hand-List Trees*, 139 (1896).

Branches ascending, forming a broad pyramidal tree. Leaves similar to those of *U. montana*, but smaller and thinner in texture, rarely exceeding 4 in. long and 2¼ in. wide, with petioles up to ¼ in. long. Flowers more irregular than in *U. montana*, funnel-shaped or campanulate with a narrowed tubular base; sepals five or six, pink; stamens four to seven, with bright pink filaments and dark red anthers; stigmas pink.

The irregularity of the flowers, the lengthened petioles, the small leaves, and the peculiar habit of this tree, all point to its being of hybrid origin; but its history is unknown to me. It is represented at Kew by a tree, about 40 ft. high and 2½ ft. in girth, which was obtained from Lee in 1879.

## Distribution

*U. montana* is a native of Europe, Asia Minor, and the Caucasus; but seems to be unknown in Siberia,[1] though it occurs in Amurland, Manchuria, and Japan.

In Europe it is a more northerly species than *U. nitens*, and only occurs in the Mediterranean region at high elevations in the mountains, as in the Pyrenees,[2] Apennines, Balkan States, and Greece. It is not known in Portugal, and is limited in Spain[3] to a few mountain woods in Asturias, the Basque Provinces, Aragon, and Navarre. It is the only species in Scandinavia, where it is met with in the wild state as far north as lat. 67° in Norway, and lat. 64° 50′ in Sweden. Its northern limit in Russia[4] extends from southern Finland through Olonetz, Archangel (lat. 62°), and Viatka to Perm (lat. 60° 40′); and it extends southwards to the border of the steppes. In Switzerland, southern Germany, and Austro-Hungary, it is essentially a mountain tree, ascending in the Tyrol and in the Carpathians to 4000 ft. Farther north it descends into the plains, and is common in north Germany on the banks of streams and in alluvial land. It is not known to form pure woods, being always mingled as isolated trees in the mixed broad-leaved forest.

In the British Isles, it is widely spread throughout Ireland and Scotland; and occurs in coppices and woods on hilly ground in most parts of England. In Bucks, Surrey, and Sussex, where it is often seen in coppice, it is commonly known as wych-hazel; while in Scotland and Ireland, it is always known as wych-elm; but the latter name is usually applied to the eastern counties of England, to *U. nitens*, which is a much more common tree there.

Mr. Clement Reid records[5] the remains of elms in interglacial deposits at

[1] Maximowicz, in *Mél. Biol.* ix. 25 (1872).

[2] *U. pyrenaica*, Lapeyrouse, *Hist. Pl. Pyrén. Suppl.* 154 (1818), judging from a specimen in Gay's herbarium at Kew, is *U. montana.*

[3] Willkomm, *Pflanzenverbreit. Iber. Halbinsel*, 126, 202 (1896).

[4] Köppen, *Holzgewächse Europ. Russlands*, ii. 43 (1889).

[5] *Origin of British Flora*, 59, 66, 69, 74, 142 (1899). We are indebted for part of the above information to a recent letter from Mr. Clement Reid.

Grays, Essex, in the form of badly preserved leaves; and in preglacial beds at Happisburg, Norfolk, leaves better preserved, but which cannot be matched with any British elm, being too small for *U. montana*, and having more nerves than the Cornish elm, which it otherwise resembles. He also records elm leaves in a neolithic deposit at Blashenwell, Dorset, and in calcareous tuff of doubtful age at Dursley, Gloucester, and elm wood in Digby Fen, where it was found by Skertchly at a depth of ten feet. The geological evidence [1] throws no light on the present distribution of the elms in Britain. (A. H.)

Sir Herbert Maxwell gives [2] the following account of names of places, derived from *U. montana* in Britain and Ireland:—

"The old Gaelic name for it was *leam* (lam), plural, *leaman*. Ptolemy's *Leamanonius Lacus* is now Loch Lomond, the lake of the elms, out of which flows the Leven, which is the aspirated form *leamhan* (lavan); and it is interesting to find these two forms again side by side in Fife, where are the Lomond Hills overlooking the town of Leven. The two forms come together again in Warwickshire, where, not far from Leamington is Levenhull—*leamhan choill*, elm-wood, and, in the same neighbourhood, a place called Elmdon. The Lennox, a district formerly written Levenax, is the adjectival form *leamhnack* (lavnah), an elm-wood; and in England the river Leam, giving its name to Leamington, the Leven in Cumberland, the Lune in Lancashire (Alauna of Ptolemy), and in Ireland the Laune at Killarney, must all once have been named *amhuinn leamhan*, elm-river. *Leamh chuill* (lav whill), elm-wood, appears as Barluel in Galloway, the hill-top of the elm-wood; the derivative *leamhraidhean* (lavran or lowran), elm-wood, becomes Lowran and Lowring, also in Galloway; and in the same province I have picked up an alternative form to *leamhan*, common in Ireland—namely, *sleamh* (slav) and *sleamhan* (slavvan), whence the names Craigslave and Craigslouan. Yet another derivative, *leamhreach* (lavrah), seems to be the origin of Caerlaverock, *cathair* (caher) *leamhreaich*, fortress in the elm-wood."

### Remarkable Trees

The finest wych elm that I have seen is one at Studley Royal, Yorkshire (Plate 394), which is not mentioned by Loudon, though many trees in the same park were figured in his work. When I measured it in February 1905 it was 105 ft. high by 23 ft. in girth, at 5 feet up, but the roots spread so much below this, that at the ground it measured 37 to 40 ft. This tree is very sound-looking and has lost no branch of any size.

There is an immense wych elm at Cassiobury Park, the seat of the Earl of Essex, which Henry found in 1904 to be about 100 ft. high by 26 ft. 4 in. in girth with two immense branches coming off near the ground and spreading to a diameter of 153 ft. This type of branching is not uncommon in the wych elm and usually leads to premature decay, as the weight of these great limbs tears out a hole in the

[1] Samuel Hassel, in Loudon, *Gard. Mag.* xiii. 477 (1837), states that remains of "the small-leaved elm" are found deep in the bogs of Somersetshire, and also in the foundations of Roman villas.

[2] *Scottish Land-Names*, 110 (1894).

trunk. A wych elm with a remarkably burry trunk, which is growing in the same park, is figured in Plate 400.

In the Park at Stowe, Buckingham, there is a large wych elm with the trunk much buttressed, but a very well-shaped handsome and sound-looking tree. It measures about 110 ft. by 19 ft. 9 in.

At Barham Court, Kent, there is a tree which I have not seen; but Mr. H. Key informs me that it covers a quarter of an acre and girths 21 ft. below the fork, where it divides into several large branches, three of which are rooted in the ground and have thrown up strong upright growths.

The largest wych elm in Gloucestershire, which was blown down some years ago before I began to measure trees, grew close to the church at Rendcombe Park, and was said to have been planted to commemorate the Restoration of Charles II. It was an immense tree of very picturesque shape. A tree at Middle Hill near Broadway, Worcestershire, in 1910 had a girth of 24½ ft., and was 85 ft. high with very wide-spreading branches.

Lees[1] gives a very good account of this species and figures some remarkable old specimens, one of which in Earl Bathurst's park no longer exists. He gives a sketch of it showing great limbs supported by props on each side, and says that it was no less than 36 ft. in girth at 3 ft. from the ground, and 47 ft. round the base. Lees also figures a very remarkable pollard with huge burrs at Cradley, Herefordshire, another at Llanthony, and a curious old stump of great age at Shrawley, which has thrown roots 20 to 30 ft. long to the bottom of the sandstone rock on which it grows. Lees quotes from Dr. Bull who says[2] that in Herefordshire "Weird-like superstitions attach to the wych elm or wych hazel as it is generally called. A spray of wych hazel is at once a potent safeguard against witchcraft, and a wand of awful import in the hand of a witch. It was formerly used as a riding switch, to ensure good luck on the journey. Until quite recently, if not to this very day, not a rural churn was made in the midland districts without a small hole being left in it, for the insertion of a bit of wych elm wood, in order to ensure the quick coming of the butter."

Lord Walsingham sent me a photograph of a very large and handsome wych elm at Rochels near Watton, Norfolk. In Burleigh Park, Notts, I measured in 1903 a tree 110 ft. by 15½ ft., which divided at 10 ft. into three main trunks. At Eaton Hall, Chester, a tree, which I found to be 90 ft. by 20 ft. in 1905, is said to be the finest in Cheshire.

A very fine old tree is growing in Wensleydale, on the village green just before the gate of Bolton Castle, which is said by Spaight[3] to have been planted in 1690. When I saw it in 1906 it was sound and healthy and measured 18 ft. in girth. This is a very characteristic tree of the limestone formation in Wensleydale and Yorkshire generally, but I know of no trees larger than this one. Watson[4] in 1825 mentions as the largest elm which he had seen, a tree growing in the village of Bishop Burton, near Beverley, Yorkshire, which measured 31½ ft. in girth at five feet from the ground, and 44 ft. at the base.

[1] *Gard. Chron.* ii. 102, figs. 20-23 (1874).

[2] *Trans. Woolhope Nat. Field Club*, 1868, p. 83 (1869).

[3] *History of Richmondshire*, p. 293.

[4] *Dendr. Brit.* i. introduction, p. ii. (1825).

As a remarkable instance of the rapid growth of this tree I may mention one on the lawn in front of the late Sir Charles Strickland's house at Hildenley, Yorkshire, which cannot be more than about eighty years old, as he distinctly remembered that when a boy he could step over it. In 1905, when I saw it, it was 100 to 105 ft. high with a bole clean to about 25 ft. and 13 ft. 3 in. in girth. This may become one of the tallest trees of its sort in England. Another instance is a log which I saw in Mr. G. Miles's yard at Stamford in March 1908, which had recently been cut at Clipsham, near Stamford. On a diameter of 6 ft. 3 in. I counted only 110 rings.

In Piercefield Park, near Chepstow, there stands one of the most symmetrical trees of this kind that I know, which measures about 100 ft. by 18 ft. It was covered with half-ripe samaræ on 15th April 1906, and I noticed everywhere in this year that the wych elms were carrying an unusual crop of seeds, which in some places in Wales were so abundant as to give a distinct pink tinge to the trees, seen from a distance in May before they were in leaf. This was particularly noticed at Llandilo on 25th May 1906, when Lord Dynevor showed me what he considers the finest wych elm in that district, which is growing in a meadow by the river in Dynevor Park. This tree measures about 103 ft. by 17 ft., and was covered with nearly ripe seed, though the leaves in this early district were not yet unfolded.

All of these existing trees, however, were far eclipsed by a tree mentioned in Plot's *Natural History of Staffordshire*, p. 210 (1636), which seems so well attested that I quote it as follows. After speaking of some gigantic oaks he says: "But I scarce think either of them held so many, as the prodigious Witch-Elm that grew at Field in this county, and was felled within memory by Sir Harvey Bagot; which, according to an original paper put into my hands by the Right Worshipful Sir Walter Bagot, Bt., the present proprietor, and, as I had it from the mouth of Walter Dixon, yet living, who was a surveyor of the work, was so very great and tall, That 2 able workmen were 5 days in stocking and felling it down; That it fell 120 foot or 40 yards in length; That the stoole was 5 yards 2 foot diameter; That the tree at the butt end was 17 yards in circumference; That it was 8 yards and 18 inches, *i.e.* 25½ foot about by girth measure in the middle; That 14 loads of firewood, each as much as 6 oxen could draw to the house at Field, being not above 300 yards distant, broke off in the fall; That there was 47 loads more of firewood (as large as the former) cut from the top; That they were forced to piece 2 saws together, and put 3 men to each end, to cut the body of it in sunder; That there was cut out of it 80 pair of nathes[1] for wheels, and 8000 feet of sawn timber in boards and planks, after 6 score per cent, which at 5s. per cent came to 12 pounds, All which is attested (as a thing, I suppose, they foresaw in a little while would otherwise become incredible) under the hands of

| | |
|---|---|
| Sir Harvey Bagot. | Lawrence Grews } Cutters. |
| William Cowper, Steward. | Humphrey Chetton } Cutters. |
| Roger Shaw, Baylif. | Francis Marshall } Stockers. |
| Walter Dixon, Surveyor. | Thomas Marsh } Stockers. |

[1] *Nathes* is an old word meaning *naves*.

And so as to the number of Tunns according to the scantlings first above mentioned, they computed it to contain (after their gross country way of measure) 96 tuns of timber, a vast quantity indeed for one tree. But whoever will take the pains to cast it nicely and more artificially, according to the above measured scantlings, will find that it must contain 100 tuns at least of neat timber, a fifth part (which is sufficient in such large butts) being allowed for wast of rind, chips, etc."

Now as one can hardly believe that eight persons would have signed their name to such a statement if they had not believed it to be true, we have dimensions which though they cannot be exact, seem to be unequalled by any hardwood tree on record out of the tropics; and though the quantity of firewood seems incredible, yet if the number of tons be estimated at only 40 ft. to the ton there is something like 4000 cubic ft. of timber, or if the firewood is estimated only at 50 ft. to the load, and the nathes and board at 1000 ft., the volume is 4050 ft., nearly double the contents of the largest tree now standing.

Aubrey,[1] *Natural History of Wilts*, 56 (1685), says:—"At Dunhead, St. Marie's, at the crosse is a wich-hazell not less worthy of remarque . . . for the large circumference of the shadowe that it causeth. When I was a boy, the bowyers did use them to make bowes, and they are the next best to yew."

In Scotland, though the tree is common I have not seen or heard of any of exceptional size. Hunter[2] mentions a good many, but rarely distinguishes between the wych and other elms. Among these the following may be noticed:—A tree at Moncreiffe House, which has layered itself in several places; and a tree at Kinfauns Castle, Perthshire, for which it is said that £50 was offered forty years ago though only computed to contain 460 feet of timber. It was in 1883 about 70 ft. high by 17 ft. in girth at the smallest place, seven feet from the ground. It was figured by Loudon, *Arb. et Frut. Brit.* iii. 1403, fig. 1244 (1838).

In the *Old and Remarkable Trees of Scotland* the same confusion has taken place, the Scotch elm being considered as a variety only and not distinguished in the returns given, but probably most of them relate to the wych elm. The largest of these was a tree then in a decaying condition at Myres in Fifeshire, stated to be 75 ft. by 20 ft. at 9 feet from the ground, dividing into two main stems 16 ft. and 9 ft. in girth. Loudon[3] speaks of an elm near Roxburgh in Teviotdale, called "the Trysting Tree," which, when measured in 1796, was 30 ft. in girth at four feet from the ground. The tallest of which we have any record is a tree measured by Mr. Bean at Dalkeith[4] which was 125 ft. high by 13 ft. 9 in. at four feet.

Strutt, in his *Sylva Scotica*, plate iv, figures a group of four wych elms at Pollok in Renfrewshire, the seat of Sir John Stirling Maxwell, Bart., which were then in extraordinary health and vigour, and of which the largest, measured[5] in 1824, was 85 ft. by 11 ft. 10 in., and contained 669 cubic feet of timber. Mr. Renwick informed me that two of these which remained, were cut down in 1905 and were 90 to 96 ft. high and 12 ft. in girth. The other two were blown down about eleven years

[1] Quoted by Rev. T. A. Preston, in *Trans. Wilts. Arch. and Nat. Hist. Soc.* 1888, p. 269.
[2] *Woods of Perthshire*, 134, 486 (1883). [3] *Arb. et Frut. Brit.* iii. 1402 (1838). [4] *Gard. Chron.* xli. 168 (1907).
[5] Loudon, *Gard. Mag.* xiii. 167 (1837) in an account of the trees at Pollok, states that the largest elm was, in 1836, 90 ft. high and 11 ft. 9 in. in girth close to the ground.

previously. The rings showed the age of the trees to be about 300 years, and there had been practically no growth during the last ten years of their life. Neither the plate in Strutt nor the photograph in *Ann. Anderson. Nat. Soc.* ii. pt. i, frontispiece (1896), show the habit of the wych elm, and I should have supposed these to be English elms except that they produced seed freely every year, even in old age.

Mr. Renwick sends us a list of wych elms, the most remarkable of which are: a hollow tree at Tullichewan Castle, Dumbartonshire, 20 ft. in girth at 2 ft. 5 in. from the ground, the stem being partly covered by burrs; at Woodbank, two trees, 105 ft. by 15 ft. and 100 ft. by 16 ft. 2 in., in 1910; at Strathleven, a tree 90 ft. by 16 ft. 10 in. in 1911; at Ancrum Park, Roxburghshire, a tree, 18 ft. 1 in. in 1893; at Newbattle Abbey, a tree 16 ft. 5 in. in 1896.

In Ireland, *U. montana* is common in the wild state, and is the only species that occurs in rocky and hilly situations in Donegal and Kerry. We have not seen any trees equalling in size those in England or Scotland, the finest specimen probably being one at Charleville, Co. Wicklow, which was 90 ft. high by 17 ft. 4 in. in girth in 1904. In a meadow by the river at Inistioge, Co. Kilkenny, there are two good wych elms, the larger measuring 90 ft. by 14 ft. 11 in. in 1904. At Adare Manor, Limerick, a tree, which had been damaged by a severe gale a short time previously, was 114 ft. by 15½ ft. in the same year.

The largest elm ever known in Ireland was probably one recorded by Hayes,[1] who considered it to be perhaps the finest tree of its species in the world. It grew at St. Wolstans, Co. Kildare, and was supposed to have been planted before the dissolution of the monastery in 1538. It lost two great limbs in 1762, and was blown down in the winter of 1776. Some time before this, the trunk had been carefully measured and was found to be 38 ft. 6 in. in circumference.

Hayes[2] quotes, as showing the extraordinary vigour of the elm in Ireland, a statement given him by Mr. Herbert of Cahirnane, near Killarney:—Six "wyche or native Irish" elms, that were produced by layers from the stool of a tree felled in 1766, measured after twenty-six years' growth, from 3 ft. 11 in. to 5 ft. 1 in. in girth at 5 feet from the ground.

TIMBER, cf. p. 1922. (H. J. E.)

## ULMUS VEGETA, HUNTINGDON ELM, CHICHESTER ELM

*Ulmus vegeta*,[3] Lindley, in Donn, *Hort. Cantab.* 96 (1826); Ley, in *Journ. Bot.* xlviii. 68 (1910).
*Ulmus glabra*, Miller, var. *vegeta*, Loudon, *Arb. et Frut. Brit.* iii. 1404 (1838).
*Ulmus americana*, W. Masters, *Hortus Duroverni*, 130 (1831) (not Linnæus).

A tree, attaining about 100 ft. in height and 15 ft. in girth, with a straight bole and long ascending straight branches. Bark similar to that of *U. montana.* Young branchlets stout, with a few scattered hairs, glabrous and occasionally striated in the

[1] *Practical Treatise on Planting*, 135 (1794). [2] *Ibid.* 162 (1794).

[3] There is no doubt that Lindley meant by this name the Huntingdon elm, although he erroneously gave its habitat as North America, relying on Masters, nurseryman at Canterbury, who called it the American elm, a name by which it is still known in some nurseries.

second year. Buds with minutely pubescent ciliate scales. Leaves (Plate 412, Fig. 16) oval, $3\frac{1}{2}$ to 5 in. long, $2\frac{1}{2}$ to 3 in. broad, very unequal at the base, abruptly contracted at the apex into a long serrated point; upper surface smooth, glabrous; lower surface with slight tufts of pubescence in the axils, glabrous elsewhere, but dotted with numerous minute brown glands; lateral nerves 14 to 18 pairs, often forked; margin coarsely biserrate, non-ciliate; petiole $\frac{1}{4}$ to $\frac{3}{8}$ in. long, more or less pubescent with scattered hairs.

Flowers, twenty to thirty in a cluster, on very short pedicels, very irregular in size and in the number of the sepals and stamens; calyx campanulate with a narrowed wrinkled tubular part at its base, or funnel-shaped, with four or five pink lobes; stamens, three, four, or five, with pale pink filaments and bright red anthers; stigmas bright red. Samaræ on very short pedicels, obovate-oval, about $\frac{7}{8}$ to 1 in. long and $\frac{2}{3}$ to $\frac{3}{4}$ in. broad, glabrous, non-ciliate, rounded at the apex with a short notch closed by the incurved stigmas; seed a little above the centre of the samara, with its apex extending nearly to the base of the notch; seed-cavity long and pointed at both ends.

The Huntingdon elm suckers very freely. Corky wings are never developed on the branchlets. It bears in favourable seasons a great abundance of seed, which is remarkably fertile. Sowings made at Cambridge in 1909, showed that this tree is a hybrid, one of the parents being *U. montana*, and the other uncertain, but probably *U. nitens*. I need not repeat here, the full account of the seedlings of the Huntingdon elm, which appeared in *Journ. Linn. Soc.* (*Bot.*) xxxix. 292-293 (1910). It is interesting to note that the fact that the Huntingdon elm does not come true from seed was established[1] at the Oxford assizes in 1847, when a nurseryman brought an action against Mr. Rivers of Sawbridgeworth, for supplying seedlings of the Huntingdon elm, which were expected to be the same as grafted plants, but which turned out to be very different.

The Huntingdon elm was raised in Wood and Ingram's nursery at Huntingdon about 1750, from seed, which is said[2] to have been gathered from some old trees in Hinchingbrooke Park, near Huntingdon. Mr. John Ingram, who wrote an account[2] of its origin, in 1847, states that these trees were at that time still living, and were the true English or field elm; but as he wrote one hundred years after the original tree was raised, no reliance can be placed on his identification of the parent tree, which was more likely to have been *U. nitens*, which is still common in Hinchingbrooke Park.

The name, Chichester elm, which was given to this tree as early[3] as 1829, cannot be explained. It was also supposed to be of American origin, and is occasionally sold by some nurseries as *U. americana*. (A. H.)

I visited Hinchingbrooke Park, the seat of the Earl of Sandwich, on 30th October 1911, with Mr. M. D. Barkley, agent for the estate, to find if possible the trees from

[1] Cf. *Gard. Chron.* 1847, p. 507. Elm seedlings with opposite and with alternate leaves were noticed and figured by Carrière, in *Rev. Hort.* xlvii. 286, figs. 47, 48 (1875); but he was unable to explain their significance.

[2] *Gard. Chron.* 1847, pp. 507, 526.

[3] Lindley, *Syn. Brit. Flora*, 227 (1829). Cf. Loudon, *Arb. et Frut. Brit.* iii. 1404 (1838).

which the original Huntingdon elm was raised. I found four distinct forms of elm. The oldest, which appear to have been growing in hedgerows before the park was enclosed about 1750, are of no great height, though one of them near the north lodge is 20½ ft. in girth. These are an inferior type of Huntingdon elm, and had shed most of their leaves without colouring well. Some taller and straighter trees which are true English elms grow near them, and had their leaves still on and quite green.

About one hundred yards north of the Japanese garden, there is an old tree 80 ft. by 16½ ft. of the true Huntingdon character, having its trunk split to the ground and the leaves nearly all fallen. On the edge of the grove was a group of bushy ill-shaped trees with small glabrous leaves.

Near these, within the grove which is fenced off, are the tallest and finest elms which I saw in the park, one of them being 120 ft. by 17½ ft., the leaves of which, differing from the Huntingdon type, were still on and quite green. Its tall straight trunk had a large wound where a branch had been torn out by the wind. A bed of suckers surrounded this group of trees, some of which have been transplanted to the Huntingdon nursery, in the hope of preserving a better type of Huntingdon elm than the trees now sold, which, as I am informed by Mr. Perkins, proprietor of this nursery, are usually budded on wych elm stocks. This fine elm is probably a seedling of the Huntingdon elm, and may be called the "Hinchingbrooke" elm.

The Huntingdon elm, though a favourite tree among nurserymen on account of its very rapid growth, and often making an ornamental wide-spreading tree, should not, in my opinion, be planted as an avenue or park tree. Its habit of forking tends to split the trunk in a way that other elms do not show; and out of a large number of trees which I have felled at Colesborne, at about sixty years after planting, hardly one was free from defects caused by this bad habit. I calculated the loss in measurement of the timber on these trees was from 15 to 25 per cent, and the timber, though fit for tinplate boxes, is pale in colour and soft in texture, compared with English or wych elm. It grows, however, with such rapidity, that it might pay to plant in woods, or if care is taken to prune all the branches when young, up to 40 or 50 ft. As a rule, it loses its leaves a month before the English elm and colours badly, but in the remarkable season of 1911, a row of elms of this variety growing at Colesborne on dry soil with a southern aspect, turned a brilliant yellow, just before the leaves fell in October.

The largest elm I have ever seen, and the largest tree of any kind in Great Britain, grew in the grove of Magdalen College, Oxford, but had never been noticed by arboriculturists,[1] until Mr. W. Baker, Curator of the Botanic Garden, showed it me in July 1905. Though generally supposed to be a wych elm, it was undoubtedly identical in habit and leaves with *U. vegeta*. On October 14th of the same year I saw it again, when its leaves had fallen, whilst the English elms in the same park were

[1] H. A. Wilson, in a *History of Magdalen College*, 286 (1899), states: "Most of the trees in the grove are English elms dating from the Restoration period. Two wych elms were planted probably about the same time as the others. The girth of one of these trees in 1831, at five feet, is stated to have been 21 ft., in 1866 23 ft. 9 in. In June 1899 it measured 26 ft. 5 in. and its height was approximately 130 feet. R. T. Günther, *Oxford Gardens*, 218 (1912), gives a reproduction of a photograph of the great Magdalen elm, which was taken in 1899. He states that there is an error in Wilson's account, as the measurement given of 21 ft. girth in 1831 was really that of another tree which fell in 1861. The latter was estimated to contain 1092 cubic feet.

still green. At my suggestion Mr. G. E. Baker, then Bursar of Magdalen College, removed a scrubby tree which stood near it, in order to allow a photograph (Plate 395) to be taken soon after by Mr. Foster of Burford. I took great pains to measure it accurately from four different positions, and found its height to be 140 ft., and its girth at five feet, 27 ft. I estimated its contents when standing at over 2000 cubic feet, but when I showed it later to some of the most experienced judges of timber in the English Arboricultural Society, their estimates were all lower than mine. On 5th April 1911 this splendid tree was blown down, and on hearing of this I went to Oxford at once with Mr. Foster to photograph and measure it on the ground. I made the total height 142 feet. With the help of my forester, J. Irvine, I made a series of thirty-three measurements of the various pieces and limbs as accurately as possible, Mr. Carter, Bursar of Magdalen, booking them as we measured; and found that if no allowance was made for bark, which was three to four inches thick on the trunk, or for the hollows caused by decay, the total contents were 2787 feet. Professor Somerville,[1] and Sir W. Schlich, F.R.S., afterwards measured the tree on the ground, and agree that my calculation of its cubic contents is nearly accurate. The timber of this tree was much redder than that of any tree of *U. montana* which I have ever seen. No suckers were noticed by me, and none have come up since the fall of the tree. I understand that the President and Fellows of Magdalen College have decided to allow the remains to lie undisturbed as a memorial of the fallen giant.

This tree was so rotten in the interior that the annual rings could not be counted. Judging from its great size, it was probably 200 to 300 years, and cannot be of the same origin as the ordinary Huntingdon elm, if the story of the origin of the latter in 1750 is correct. The independent origin of this hybrid at various times and in different places is not improbable.

Another tree, similar in habit and foliage, still survives in the Grove of Magdalen College, and in 1912 measured about 130 ft. in height, and 23 ft. in girth.

Elsewhere I have seen no old trees except in Hinchingbrooke Park; but there is a fine specimen[2] in the Fellows' garden of Trinity College, Cambridge; and another in Victoria Park, Bath. There is at Cambridge, an avenue of Huntingdon elms, known as Brooklands Avenue, which is said to have been planted by Mr. Richard Foster about 1830. (H. J. E.)

[1] Cf. *Quart. Journ. Forestry*, v. 279, fig. (1911).

[2] Figured in *Journ. Linn. Soc.* (*Bot.*) xxxix. pl. 20 (1910). According to C. W. King, this tree was planted about 1814, in the presence of Adam Sedgwick. It is now often called Sedgwick's elm.

## ULMUS MAJOR, Dutch Elm

*Ulmus major*,[1] Smith, *Eng. Bot.* t. 2542 (excl. syn.) (1814); Lindley, *Syn. Brit. Flora*, 226 (1818); Loudon, *Arb. et Frut. Brit.* iii. 1396 (1838); Ley, in *Journ. Bot.* xlviii. 71 (1910).
*Ulmus hollandica*,[1] Miller, *Gard. Dict.* ed. 8, No. 5 (1762) (?).
*Ulmus hollandica*, Moss, in *Gard. Chron.* li. 217 (1912).
*Ulmus fungosa*, Aiton, *Hort. Kew.* i. 319 (1789) (?).
*Ulmus scabra*, Miller, var. *major*, Gürke, in Richter and Gürke, *Plant. Europ.* ii. pt. i. p. 73 (1897); Schneider, *Laubholzkunde*, i. 218 (1904).

A tree, attaining in England over 100 ft. in height and 15 ft. in girth, but usually smaller, with a short bole and irregular wide-spreading branches. Bark of the trunk dark coloured, deeply fissured. Young branchlets glabrous or with a few scattered hairs; finely striated, glabrous and bright reddish-brown in the second year. Buds ovoid, with minutely pubescent ciliate scales. Leaves (Plate 412, Fig. 15) broadly oval, 3 to 5 in. long, 2 to 3 in. wide, very unequal at the base, contracted at the apex into a rather long serrated point; upper surface dark green, shining, nearly smooth, with a scattered minute pubescence; lower surface light green, with conspicuous white axil-tufts, prolonged along the midrib between the insertions of the lower lateral nerves, and with a scattered minute pubescence and numerous minute glands; nerves about twelve to fourteen pairs, mostly forked; margin deeply serrate, non-ciliate; petiole $\frac{1}{4}$ to $\frac{3}{8}$ in. long, pubescent.

Flowers, often very numerous (twenty to fifty) in the cluster, on extremely short pedicels, mostly tetramerous, with four calyx-lobes and four stamens (but often irregular with five calyx-lobes and four stamens, five calyx-lobes and five stamens, etc.): calyx, funnel-shaped, lobes pink and unequal; stamens with filaments tinged with pink and with red anthers; stigmas bright red. Samaræ, on very short pedicels, obovate-oval, when mature $\frac{3}{4}$ to 1 in. long, and $\frac{11}{16}$ to $\frac{3}{4}$ in. broad, full and rounded at the apex, which is emarginate, with a short notch usually closed by the incurved stigmas below the emargination; seed in the upper half of the samara, with its apex close to the base of the notch.

*U. major* may almost always be recognised by the large corky ridges, which are developed only on the epicormic branches of the trunk.[2] It produces suckers freely, the stems of which often have bright reddish-brown corky ridges, whilst their leaves and branchlets are more pubescent than those of the adult tree.

*U. major* rarely bears fertile seed; but in 1909, small lots of seed from trees at Brocklesby, Belton, Bayfordbury, and Cambridge produced in each case two seedlings. These eight seedlings are very variable in appearance, those raised from the

[1] *U. major* was described and figured by Smith from specimens gathered in England by E. Foster, and is the first certain name. We have not adopted Miller's name, *U. hollandica*, as his description is uncertain; moreover, it implies a foreign origin for a tree which is undoubtedly indigenous in England. Miller quotes as a synonym of his elm, *Ulmus major hollandica*, Plukenet, *Alm.* ii. 393 (1696), which is described as "*angustis et magis acuminatis samaris, folio latissimo scabro.*" Plukenet's elm cannot be determined, but his description excludes *U. major*. Cf. page 1869, note 2. The usage of "Dutch elm" as a name for *U. major* apparently began in error about 1730.

[2] In *U. nitens*, var. *suberosa*, which is the true cork-barked elm, the branchlets in the crown of the tree are all corky.

seed of the tree at Belton being exactly like *U. montana*; and I have little doubt that *U. major* is a hybrid, with *U. montana* and *U. nitens* as the parent species.

### VARIETIES

The following varieties are assigned with some doubt to *U. major*:—

1. Var. *serpentina*, Henry.

*Ulmus serpentina*, Koch, *Dendrologie*, ii. pt. i. 417 (1872).
*Ulmus montana*, sub-species *major*, var. *serpentina*, Dippel, *Laubholzkunde*, ii. 29 (1892).

A small tree, with curved and twisted pendulous branches, forming a dense pyramidal or globose crown. Leaves and branchlets similar to those of *U. major*.

The origin of this variety is unknown. It is said by Koch to be known in nurseries as the parasol elm, and is represented at Kew by a good specimen about 15 ft. high, obtained from Späth in 1896. Another smaller tree at Kew is labelled *U. campestris pendula nova*.

2. Var. *Daveyi*,[1] Henry.

A wide-spreading tree, with irregular branches and pendulous branchlets, differing from *U. major* in the epicormic branches never developing corky ridges. Young branchlets more or less pubescent with long hairs. Leaves similar in shape to that species, but smaller, 2½ in. long, 2 in. broad, and with 10 to 12 pairs of lateral nerves; upper surface shining, smooth, with a scattered minute pubescence; lower surface similar, covered with a dense soft pubescence, and with conspicuous white axil-tufts; petiole densely pubescent. Flowers with the sepals and stamens irregular in number, and the stigmas white or tinged with pink. Samaræ not seen.

This occurs as a rare tree in Cornwall, in the valleys mixed with *U. major*, which it resembles in its wide-spreading habit, but has very pendulous branchlets. It is apparently never very large in Cornwall, where I saw it at Coldrenick and near Perranporth; but in Norfolk and Cambridge, where it is also rare, it reaches 80 ft. in height.

3. A tree at Kew, about 25 ft. high, pyramidal in habit, with ascending branches, which was obtained from Späth in 1896 as *U. campestris*, var. *modiolina*,[2] has leaves like those of *U. major*, except that they are very scabrous above.

### DISTRIBUTION

*U. major* is widely distributed throughout England, occurring from Cornwall to Essex and Norfolk, and extending as far north as Yorkshire. It ascends to about 1000 ft. altitude in mountain valleys in South Wales.[3] In many districts it is the commonest tree in hedgerows; and in Cornwall, where it is associated with *U. nitens*, var. *stricta*, these two elms grow in every valley and are certainly indigenous.

*U. major* is said to occur in France, Germany, and Russia; but I have seen no

[1] Named after Mr. F. Hamilton Davey, whose researches into the flora of Cornwall are well known.

[2] This tree is different from *U. modiolina*, Dumont de Courset. See p. 1894.

[3] Ley, in *Journ. Bot.* xlviii. 72 (1910).

specimens, and doubt if it is really known by Continental botanists, who apply the name to forms of *U. montana* with large leaves.[1]

*U. major* is usually called "Dutch elm" by foresters and carpenters, and has been supposed by Miller and subsequent writers to have been introduced into England from Holland in the reign of William III. In all probability the elm which was then introduced was not *U. major*, but a vigorous form of *U. montana*. It is difficult to see how the latter could have been *U. major*, a tree which is apparently unknown in Holland.[2] (A. H.)

### Cultivation

Though it is difficult to describe the habit of this elm, which is very inferior to that of the true English elm, yet it is easy to recognise, even at a distance, as I have repeatedly been able to foretell by the form alone before reaching a tree, that it would have corky twigs on the trunk. It has been largely planted in the neighbourhood of London, where, as elsewhere, it loses its leaves three weeks or a month earlier than the true English elm. In Kensington Gardens, where most of the elms appear to belong to this species, on 30th October 1911, there was hardly a leaf left on the majority of the elms, whereas throughout the Thames valley true English elms were still green in the hedgerows, and on my own place a week later, after 16° of frost, they were only beginning to turn golden, a colour which the Dutch elm seldom or never assumes. An immense deal of trouble and expense and obloquy has fallen on those responsible for the care of the London Parks because a wholesale lopping of the old elms was considered necessary for the safety of passers-by, after several accidents had occurred from falling branches. I am able through the courtesy of the Right Hon. Lewis Harcourt, who, in 1905, was the Minister responsible for the London Parks, to give an illustration (Plate 396) showing the effect of this lopping after a period of some years. I think it must be allowed that, however ugly they seem at first, they throw out young branches with great rapidity and soon improve in appearance.

This tree in some seasons produces a large quantity of seeds which are nearly always infertile. On 13th May 1906, I noticed three trees growing in a row near the North Lodge at Gatton Park, Surrey, their leaves being much more backward than those of *U. campestris*, but none of the samaræ which were sent me from these trees seemed to have any perfect seed.

[1] Cf. Mathieu, *Flore Forestière*, 302 (1897). *U. major*, Reichenbach, *Icon. Fl. Germ.* xii. 13, t. 665 (1850), is *U. montana*. Michaux, *N. Amer. Sylva*, ii. 224, plate 129, fig. 2 (1819), describes and figures ordinary *U. montana* under the name *U. suberosa* as the Dutch elm.

[2] In the large collection of specimens of wild and cultivated elms of Holland, which are preserved in the herbarium of the Dutch Botanical Society at Haarlem, I found none identical with *U. major* of England. Most of the supposed wild elms in Holland, as those on the dunes near Haarlem, are *U. nitens*, var. *suberosa*. Mr. Springer sent me a branch of a solitary large elm growing on these dunes, which he supposed to be *U. major*; but the leaves differ in being very scabrous on the upper surface, and belong to a hybrid intermediate between *U. montana* and *U. major*.

*U. major* may be expected to occur on the continent; but I have not seen any trees of the typical English form, either wild or cultivated, in France, Belgium, and Holland. I have, however, specimens from a wild tree in a wood, near Gray, Haute Saône, intermediate between *U. major* and *U. nitens*.

*U. corylifolia*, Host, *Fl. Austr.* i. 239 (1827), judging from the description and a type specimen in the Kew Herbarium, is allied to, but not identical with *U. major*. This is said to grow on the banks of streams in mountain woods in Austria.

One of the tallest I know which may be referred to *U. major* is a tree in the heronry at Dallam Tower, Westmoreland, which, being drawn up in a thick wood, has attained a height of about 130 ft., and is clear of branches for 60 or 70 ft.; but is only 10 ft. in girth. A very large tree grows by the lake at Dodington Park, Gloucestershire; the main trunk, 22 ft. in girth, is broken, but a branch remains which, in 1910, was 115 ft. high; the branchlets are very corky. A tree at Laverstoke Park, Hants, 113 ft. by 18 ft., is of a superior type to the common Dutch elm, and may be one of the hybrids. There are trees of *U. major* at Hampton Court, some of large size; at Syon; at Aldenham, a large tree on the east front of the house; at Beaulieu Abbey; and at many places in the south of England.

Many of the large elms in Kew Gardens belong to this species, and one, which was felled in December 1911, was 91 ft. high by 13 ft. 5 in. in girth. Mr. Bean counted on the stump about one hundred and fifty rings, so that this tree was probably planted about 1759, when the botanic garden was formed at Kew.

There are many Dutch elms at Boughton; and I was assured by Mr. Neil, the forester here, that their timber is as good as that of the English elm. It is generally believed by carpenters and wheelwrights to have much more brittle and less valuable timber than the redder wood of the English elm.

Many elms in the eastern counties belong to this species; and though there are some fine true English elms in Essex, yet near the coast of this county, and in Suffolk and Norfolk, we rarely see an elm of large size, which does not seem to be *U. major*. On light and sandy soils it is a scrubby tree, and does not carry its leaves so late or turn such a bright colour in autumn as the English elm.

The avenue of elm at Castle Howard in Yorkshire, which is perhaps the best I know so far north, appears to be, and is considered by Mr. Fell, the forester there, to be, of this type, which in some parts of England is often called by carpenters a "bastard" elm.

Boutcher says[1] that in order to get the true variety he imported it from Rotterdam, to make the mother plants of those he raised in his nursery; and adds that though inferior in beauty and elegance of form, to what he calls the French elm,[2] it is still a very valuable tree in the climate of Scotland, as "it will succeed in wet obstinate clay, where no tree I know of equal use, and few but aquatics, will grow freely; but in such places it will soon become a stately tree, and though the wood is not equal to the other mentioned kinds, it is still a useful wood, and is often indiscriminately sold to the carpenter with them from their near resemblance."

In all probability Boutcher here referred to the hybrid elm, *U. belgica*, which was cultivated then in Holland; but we have had no specimens from Scotland that can be referred to this. True *U. major* is rare in Scotland, where most of the elms which are not *U. montana* are *U. nitens* of French origin.[2]

In Ireland *U. major* is the common form in some parts of the country. Many of the elms in the Phœnix Park seem to be Dutch elms, but I have not examined them carefully. It occurs at Loughrea, Co. Galway, and is common at Lismore in the Blackwater valley, Co. Waterford. Here it is known to timber merchants as

[1] *Treatise on Forest Trees*, 19 (1775).

[2] Cf. p. 1897, last paragraph.

"Irish elm," and the wood is used by wheelwrights. The habit of the tree is generally branchy, and inferior to that of *U. nitens*, which is here erroneously called English elm. (H. J. E.)

## ULMUS NITENS, Smooth-leaved Elm

*Ulmus nitens*, Moench, *Meth.* 333 (1794); Rehder, in *Mitt. Deut. Dend. Ges.* 1908, p. 157; Moss, in *Gard. Chron.* li. 217 (1912).

*Ulmus glabra*, Miller, *Gard. Dict.* ed. 8, No. 4 (1768) (not Hudson[1]); Smith, *Eng. Bot.* t. 2248 (1811); Loudon, *Arb. et Frut. Brit.* iii. 1403 (1838); Reichenbach, *Icon. Fl. Germ.* xii. 13, t. 664 (1853); Willkomm, *Forst. Flora*, 553 (1887); Schneider, *Laubholzkunde*, i. 219 (1904) (in part); Ley, in *Journ. Bot.* xlviii. 69 (1910).

*Ulmus foliacea*,[2] Gilibert, *Exercit. Phyt.* ii. 395 (1792); Sargent, *Arnold Arboretum Bull. Pop. Inform.* No. 11 (1911), and in *Gard. Chron.* l. 202 (1911).

*Ulmus campestris*, var. *laevis*, Spach, in *Ann. Sc. Nat.* xv. 362 (1841); Planchon, in *Ann. Sc. Nat.* x. 273 (1848).

*Ulmus campestris*, var. *glabra*, Hartig, *Naturg. Forstl. Kulturpfl.* 458, 460 (1851); Planchon, in De Candolle, *Prod.* xvii. 157 (1873); Ascherson and Graebner, *Syn. Mitteleurop. Flora*, iv. 553 (1911).

A tree, with a straight bole, and wide-spreading branches, with usually pendulous branchlets. Bark grey, deeply fissured in old trunks. Young branchlets slender, glabrous or with a few scattered hairs, usually with the upper margin of the stipule-scars fringed with a tuft of hairs. Buds with minutely pubescent ciliate scales. Leaves (Plate 412, Fig. 23) oval or obovate, 2 to 3½ in. long, 1 to 2 in. broad, very unequal at the base, acuminate at the apex; upper surface dark green, shining, smooth to the touch, in spring pubescent with scattered minute hairs, in autumn glabrescent; lower surface, with conspicuous white axil-tufts, and covered with minute reddish brown glands, in spring pubescent with scattered minute hairs, in autumn glabrescent; margin biserrate, non-ciliate; lateral nerves about twelve pairs, often forked; petiole ¼ to ½ in. long, pubescent.

Flowers, twenty to thirty in a fascicle, on very short pedicels (less than $\frac{1}{25}$ in. in length); tetramerous or pentamerous, but often irregular in the number of sepals and stamens; calyx funnel-shaped, about $\frac{1}{12}$ to ⅛ in. long, with four or five pink short lobes; stamens, four or five, occasionally three, with pink filaments and red anthers; stigmas white, or rarely pale pink. Samaræ, on very short pedicels, glabrous, non-ciliate, obovate with a cuneate base, about ¾ in. long, and ½ in. broad; broad and rounded at the slightly emarginate apex, with the notch closed by the incurved stigmas; seed in the upper part of the samara, with its apex nearly touching the base of the notch.

Seedling: The two cotyledons are raised above the ground on a short

[1] *U. glabra*, Hudson, *Fl. Angl.* 95 (1762), is the tree almost universally known as *U. montana*; and being earlier than Miller's name, renders the latter inapplicable to the smooth-leaved elm.

[2] *Ulmus foliacea*, Gilibert, *Exercit. Phyt.* ii. 395 (1792), was founded on a specimen of an elm, which was said to be frequent about Grodno in Lithuania. The description is very imperfect, but probably applies best to *U. nitens*, which occurs in Lithuania. There are no grounds for resuscitating a name like *U. foliacea*, which cannot be identified with certainty.

pubescent caulicle, which ends in a long tap-root; and are sub-orbicular to obovate, $\frac{1}{3}$ to $\frac{2}{5}$ in. long, broad and rounded at the apex, deeply cordate at the base, green above with scattered short bristles, whitish and glabrous beneath; indistinctly veined; margin entire and ciliate; stalklets very short, pubescent. Leaves, arising from the stiff erect pubescent stem in opposite pairs, ovate, sharply serrate or biserrate, scabrous above with papillæ and numerous short bristles, the latter being scattered on the under surface; ciliate in margin; petioles short, pubescent. Seedlings, sown in June, attain in October 6 to 12 in. in height, and bear six to eight pairs of leaves as a rule. The normal alternate foliage of the adult plant appears in the second year; but develops at once on any branch that may be formed in the first year. The pubescent twigs and rough hairy leaves are preserved for a considerable period, the normal foliage and branchlets not appearing till the trees are about ten years old.[1]

### VARIETIES

This species, being distributed over a wide area (cf. p. 1896), is variable in the wild state, in the amount of pubescence on the branchlets and leaves, and in the presence or absence of corky ridges on the twigs and branches. In the ordinary form in England, the twigs are not suberose.

1. Var. *suberosa*, Henry. Cork-barked Elm.

*Ulmus hollandica*, Pallas, *Fl. Ross.* 76 (1784) (not Miller).
*Ulmus suberosa*, Moench, *Verz. Weissenst.* 136 (1785); Reichenbach, *Icon. Fl. Germ.* xii. 13, t. 663 (1850); Hartig, *Naturges. Forst. Cult.* 459, t. 56 (1851) (not Ehrhart,[2] Smith, Lindley, or Loudon).
*Ulmus tetrandra*, Schkuhr, *Bot. Handb.* i. 178, t. 58 b (1791).
*Ulmus campestris*, var. *suberosa*, Wahlenberg, *Fl. Carpat.* 71 (1814).
*Ulmus glabra*, var. *suberosa*, Gürke, in Ritter and Gürke, *Pl. Europ.* ii. 72 (1897).

Branchlets of the second to the tenth year furnished with corky wings. Leaves and samaræ as in the type (Plate 412, Fig. 19).

This variety is occasionally seen in England, as on trees at Hatfield; and is represented at Kew by four trees of no great size, but of considerable age, which were long labelled erroneously *U. alata*.[3] It appears to be a common variety in the forests of central Europe, as in the oak forests on the banks of the Save in Slavonia, where I saw it in 1909, and Elwes in 1910. Elwes also gathered a specimen at Gisselfelde in Denmark in 1910.

A form of var. *suberosa* with small leaves, about an inch in length, is cultivated at Kew, as var. *microphylla pendula*.[4]

2. Var. *stricta*, Aiton, *Hort. Kew.* i. 319 (1789); Schneider, *Laubholzkunde*, i. 220 (1904); Ley, in *Journ. Bot.* xlviii. 70 (1910). Cornish Elm.

[1] Cf. our account of French seedlings of this species, p. 1897. A young seedling is figured in *Journ. Linn. Soc.* (*Bot.*) xxxix. 292, pl. 22 (1910).

[2] *U. suberosa*, Ehrhart, *Beit.* vi. 87 (1790), was a general name, applied to all the elms in Europe that were not *U. montana* or *U. pedunculata*. Ehrhart's specimen of *U. suberosa* in Smith's herbarium at the Linnean Society is the "English elm," *U. campestris*, L.

[3] Cf. *Gard. Chron.* xix. 453, fig. 66 (1896).

[4] Distinguished as *U. glabra*, var. *propendens*, Schneider, *Laubholzkunde*, i. 220 (1904).

*Ulmus campestris*, var. *cornubiensis*, Loudon, *Arb. et Frut. Brit.* 1376 (1838).
*Ulmus stricta*, Lindley, *Syn. Brit. Flora*, 227 (1829); Moss, in *Gard. Chron.* li. 234 (1912).

A tree, attaining about 80 ft. in height and 15 ft. in girth, rather variable in habit, but usually with a narrow crown, with the upper branches short and ascending, the lower branches spreading and curving upwards at the ends. Bark light grey, fissuring into small plates. Young branchlets often densely pubescent at the insertions of the leaves, elsewhere with scattered hairs. Buds minute, with glabrous scales, slightly ciliate in margin. Leaves (Plate 412, Fig. 20) firm in texture, obovate to oval, 2 to 2½ in. long, 1 to 1½ in. broad, unequal at the base, acuminate at the apex; upper surface dark green, shining, glabrous, smooth to the touch, lower surface lighter green, with conspicuous tufts of pubescence at the junctions of the midrib and lateral nerves, and at the forks of the latter, and with slight tufts usually near the margin at the base of the teeth, elsewhere glabrous or with minute scattered pubescence; lateral nerves about twelve pairs, often forked; margin crenately biserrate, non-ciliate; petiole about ⅓ in. long, pubescent.

Flowers, fifteen to twenty in small clusters, on very short pedicels, irregular in the number of sepals and stamens, but mostly tetramerous; calyx funnel-shaped, minute, about $\frac{1}{12}$ in. long, with pink lobes; stamens four or five, with pink filaments and dull red anthers; stigmas pink. Samaræ rarely ripening, but when mature, similar to those of typical *U. nitens*, but narrower, obovate, cuneate at the base, about ⅔ in. long and ⅜ in. wide; imperfect samaræ are usually broadly obovate, less than ½ in. long.

From ten lots of seed, sent me from different localities in Cornwall, I raised in 1909, thirty-eight seedlings, all of which bore opposite leaves. These plants are now all uniform in appearance; and in all probability the Cornish elm is a pure species.

In Abbeyleix Park, Ireland, there are a few elms, growing with oak trees on the alluvial flat of the river Nore, which are probably indigenous. These are similar in foliage to the Cornish elm; and have similar small tetramerous flowers, few in a cluster, but differing in having white stigmas. These trees produced a few ripe samaræ in 1909, similar to those of *U. nitens*, but scarcely ½ in. long. From these I raised two seedlings, with opposite leaves, which in 1912 resembled the seedlings of the Cornish elm.

The suckers of the Cornish elm, which are freely produced, have pubescent stems, and small leaves, scabrous above with scattered minute tubercles and short bristles. The epicormic branches are usually smooth, but occasionally develop corky ridges, which are, however, never seen on the normal branches in the crown of the tree.

In Cornwall, the tree is slow in growth, but it produces a remarkably tough wood, which is used by wheelwrights for naves, felloes, and framework of waggons; and was formerly utilised as staves for the casks in which cement and china clay were exported, and also for making boxes in which gunpowder was compressed by an hydraulic press, as no other wood was found to bear great pressure so well.

The Cornish elm is undoubtedly indigenous in Cornwall and south Devon, where it grows abundantly in the hedge rows, reproducing itself regularly by suckers, and possibly by seed in rare years as in 1909. Mr. F. Hamilton Davey tells me that it is plentiful also in parts of Somerset, and says regarding its distribution in Cornwall, that it is fairly common from the Tamar to a little to the west of Penzance. From the extreme north-east of the county to Land's End, it does not approach the coast nearer than a mile or two. It is also rare along the Lizard peninsula, and near the range of granite hills, which run from east to west through the centre of the county. It is a tree of low altitudes, the finest specimens being always found in sheltered valleys, especially in those which run to the south coast.

The Cornish elm was reported[1] by the late Rev. A. Ley to be common in Brittany, and this is confirmed by Dr. C. E. Moss, who observed it in most parts of Brittany, though good specimens were rare on account of the local practice of lopping. (A. H.)

I measured two large trees (Plate 397) of characteristic habit in the lower part of the entrance drive at Coldrenick in Cornwall, the suckers of which were of precisely similar habit. One was 100 ft. by 11 ft. 4 in., the other 92 ft. by 13 ft. 5 in. On the approach to Menabilly House from Fowey, I also saw fine Cornish elms, some over 100 ft. high. I measured one, 95 ft. by 6 ft. 2 in. At Scorrier, near Truro, there are many of these elms, though I saw none of great size; and Mr. John Williams told me that they resist the sea wind, which is here very strong, better than any other tree; and that he plants suckers, which are abundant. The Cornish elm seldom attains a great girth; but a tree at Enys, which Henry saw in 1911, measured 24 ft. in girth at the ground, 17 ft. at five feet up, and 16 ft. at seven feet up; but it was only about 65 ft. high. In the vicarage garden at Perranarworthal, Henry saw a fine specimen 80 ft. by 11 ft.

In Devonshire I have seen no Cornish elm remarkable for size.

At Shawford, near Winchester, there are four fine trees, not quite of the typical Cornish form, the largest of which measured 125 ft. by 10 ft. 10 in. in 1907. At Bisterne Park, near Ringwood, I saw in 1896 three trees, differing somewhat in habit. One had a very erect trunk and pointed top; and measured 86 ft. by 13 ft. The second was more rounded at the top, and was 90 ft. by 14 ft. The third tree could scarcely be called a true Cornish elm, as it had the pendulous branches of an American elm, and measured 98 ft. by 13½ ft.

At Cowdray, near the ruins, there are three fine trees, of the same habit as the second tree at Bisterne, and not so regular as the Wheatley elm. These measured in 1906, 113 ft. by 13 ft., 113 ft. by 13½ ft., and 110 ft. by 11 ft. 9 in. respectively.

At Bagshot Park, on May 20, 1907, I measured a very fine tree, which had not yet come into leaf, though some seeds were nearly ripe. It measured 98 ft. by 11½ ft. At Arley Castle, there is a tree, No. 351 in the catalogue, which measured 75 ft. by 8 ft. 8 in. in 1907.

In Wales the Cornish elm grows well on the south coast; and at Singleton, the property of Lord Swansea, there is an avenue of them. Mr. Harris, formerly

[1] In *Journ. Bot.* xlviii. 70 (1910).

gardener here, considered this to be the best variety for seaside planting, as it bears the wind well.

In the Edinburgh Botanic Garden there are two trees, nearly equal in size, about 65 ft. by 5 ft. 4 in. in 1908, with small leaves, which seem to be a variety of the Cornish elm.

In Ireland, especially in the south, this tree comes to great perfection. In the Blackwater valley, I measured in 1910, in a meadow two miles east of Lismore, a splendid tree, about 100 ft. by 15 ft. In the avenue to the castle, there are some very tall and slender elms, one of which was 110 ft. by 10 ft. At Mallow Park, the seat of Mrs. Norris, I saw in 1909, some fine elms of similar type, one of which measured 95 ft. by 12 ft. 10 in.; another 95 ft. by 10 ft.; and a third, a very straight tree, about 90 ft. by 6 ft. The Cornish elm is probably native in the south of Ireland, where there are very many old trees of this variety. (H. J. E.)

3. Var. *Wheatleyi*, Simon-Louis, *Cat.* 1869, p. 98. Wheatley or Jersey Elm.

(?) *Ulmus sarniensis*, Loddiges, *ex* Loudon, *Arb. et Frut. Brit.* iii. 1376 (1838).

A pyramidal tree with stiff ascending long branches, and a narrow pointed crown. Leaves (Plate 412, Fig. 21) similar to those of the Cornish elm, but broader in proportion to their length, and glandular on the petiole and on the midrib, veins, and surface beneath, with less conspicuous axil-tufts. Flowers as in var. *stricta*, but with white stigmas. Fruit rarely ripening, but when mature similar to that of *U. nitens*, with the apex of the seed touching the small circular closed notch. From seed collected by Mr. J. F. Rayner from a tree in Southampton Cemetery, I raised forty-three plants in 1909, half with opposite, and half with alternate leaves. These now look a mixed lot, differing in the size of the leaves and in the presence or absence of corky ridges on the stem; and are not uniform like the seedlings of the Cornish elm.

This tree is generally regarded as a form of the Cornish elm, of which it is probably a seedling; but it differs in the characters noted. It is now generally sold in nurseries as the Wheatley elm; and is occasionally known in Germany and Holland under the erroneous name, *U. campestris*, var. *monumentalis*.[1] The Wheatley elm is so commonly known as the Jersey or Guernsey elm, that in all probability it is identical with the tree propagated by Loddiges as *U. sarniensis*, but Loudon's description of this is inadequate, and points rather to some form of the English elm.

A form of the Wheatley elm with leaves of a fine yellow colour, said to last till autumn, originated in 1900 in the Chester nurseries, and is now sold by Messrs. Dicksons as the "golden Cornish elm." (A. H.)

The finest Wheatley elm[2] is probably one growing in the public garden on

[1] The Wheatley elm is well figured by Springer in *Mitt. Deut. Dend. Ges.* 1910, p. 271, fig. 273; where Beissner points out that the true var. *monumentalis*, Rinz, *ex* Petzold and Kirchner, *Arb. Musc.* 554 (1864), is a columnar tree, with a few upright main branches and numerous short twigs bearing dense crowded dark green leaves, which was propagated by Rinz at Frankfort from a sucker of *U. nitens*, var. *suberosa*. A beautiful narrow pyramidal tree in Späth's nursery, called *U. campestris cornubiensis*, bears leaves similar in size and appearance to a common form of *U. nitens*, and differs from the Wheatley or the true Cornish elm.

[2] A good specimen at Kew is figured in *Gard. Chron.* xli. 150, fig. 67 (1907).

the terrace at Richmond, overlooking the Thames, which is over 90 ft. in height (Plate 398).

In Messrs. Rogers' nursery at Southampton, large numbers of this elm are propagated; and I noticed that the grafts have rough leaves at first, and do not produce the typical smooth leaves until they are older. Trees of this variety retain their leaves till late in the season; and at Colesborne, colour well in the autumn. In the Isle of Wight, there are many trees of this type, and an avenue of small ones at Barton Farm, Osborne.

At Stowe, near Buckingham, there are some fine trees of this variety, here called Jersey elm, which produce many suckers. The best that I measured was 85 ft. by 8 ft. 2 in. in 1905. At Merton Hall, Norfolk, Henry measured a fine tree, 86 ft. by 8 ft. 10 in.; and there is a good specimen in Sir Hugh Beevor's grounds at Hargham.

At Monreith, Sir Herbert Maxwell has a number, which are called Southampton elms. These are from 52 to 57 ft. high, fifty-six years after planting, and are nearly as tall as some English elms planted about 100 years ago. These trees, however, are not so well-shaped as in the south of England, and looked as if the climate was too damp for them. In Queen's Park, Glasgow, there are twenty Wheatley elms, which were planted about 1859; they are 46 to 61 ft. in height, and average 6 ft. in girth. (H. J. E.)

4. Var. *italica*, Henry (*var. nova*). Mediterranean Elm.

Leaves (Plate 411, Fig. 9) coriaceous, similar to those of *U. nitens* in shape and size, mainly differing in the numerous lateral nerves, which are never less than fourteen pairs, and often as many as eighteen pairs. Branchlets and samaræ as in the type. This variety usually has leaves smooth and glabrous above, and glabrescent beneath in autumn, with conspicuous axil-tufts; petiole ¼ in. long, pubescent.

This elm appears to be common in Italy, Spain, and Portugal, and also in Algeria. I observed it wild on Montserrat and in the Guadarrama mountains in Spain, where at 3000 to 4000 ft. elevation it attains a height of 70 to 80 ft. Dr. Henriquez sent to Kew on loan a specimen of this elm from Coimbra, in Portugal.

The small elms on which the vines are trained in Italy, north of the Apennines from Parma to Ravenna, are probably a form of *U. nitens*; but as these trees are lopped annually and kept low in stature, their leaves are irregular in size, have usually few nerves, and show a varying amount of roughness and pubescence on different individuals. Near Parma these elms produced good seed in 1911, from which I raised seedlings at Cambridge.

In the courtyard of the Villa Paveri-Fontana at Collecchio, near Parma, there is a remarkable old elm of this variety, about 60 ft. high, with a short, very burry, and rugged bole, 20 ft. in girth, and dividing above into three or four stems. Between Turin and the ancient royal palace at Stupinigi there is an avenue of elms, about six miles long, which was planted in 1781. Prof. Mattivoli, who kindly sent me specimens, tells me that the trees have been badly lopped during the last fifty years, and are not so large or so old as those in the gardens of the palace. The largest which he measured in this avenue varied from 50 to 70 ft. in height, with a girth ranging from 7 to 11 ft. The largest elms in Turin are in the La Marmora garden, and measure about 100 ft. in height and 12½ ft. in girth. These are said to have been planted in 1706.

5. Var. *umbraculifera*, Trautvetter, in *Act. Hort. Petrop.* ii. 590 (1873).

A tree with a dense globose crown of foliage, which is commonly cultivated in Persia, and only differs from ordinary *U. nitens* in its peculiar habit. This elm attains occasionally an enormous size, and is much prized on account of its dense shade and beautiful form. It is said to have been known in Persia for centuries, and has been introduced into the Caucasus and Armenia.[1] Regel, in *Gartenflora*, xxx. 3. t. 1034 (1881), figures a beautiful specimen, apparently of great age and size, which was then growing near Erivan in Russian Armenia. This peculiar elm is known to the Persians as the *narwan*,[2] and is always propagated by grafting. It was introduced in 1878 by Späth, who received it from a German gardener in the employment of the Shah, and is said to be planted in some of the streets in Berlin.[3] There is a fine specimen about twenty years old, grafted at 6 ft. from the ground, in Desfossé's nursery at Orleans.

Var. *umbraculifera gracilis*, Späth, *Cat.* No. 100, p. 121 (1897-1898).

This originated in Späth's nursery from a shoot of the preceding variety, which had been grafted on a standard, and is said to differ only in forming a long oval and not a globose crown. At Kew, a shrub, labelled var. *umbraculifera*, obtained from Simon-Louis in 1904, has numerous ascending branches from the base, no main stem being developed, and forms a globose bush about 6 ft. high.

6. Var. *Rueppellii*,[4] Späth, *Cat.* No. 73, p. 124 (1888-1889).

A pyramidal tree, with a distinct stem and numerous ascending branches, forming a globose or ovoid crown, and closely resembling var. *umbraculifera* in habit. Branches slightly corky, branchlets pubescent. Leaves similar to those of the Cornish elm in size and shape, but scabrous above. This is represented at Kew by two trees, about 10 ft. high, that were obtained from Barbier in 1902.

7. Var. *pendula*,[5] Rehder, in Bailey, *Cycl. Amer. Hort.* 1882 (1902).

Branches and branchlets very pendulous. There is a good specimen[6] of this

[1] Radde, *Pflanzenverb. Kaukasus*, 305 (1899). A fine tree in the city of Bokhara is figured in *Mitt. Deut. Dend. Ges.* 1910, p. 73. *Ulmus densa*, Litwinow, in *Schedæ Herb. Fl. Ross.* vi. 163 (1908), is the name given to the wild elm with a dense crown of foliage, which grows in the mountains of Turkestan and Ferghana. Litwinow considers var. *umbraculifera* to be probably a graft of *U. densa*.

[2] This Persian word is also transcribed *narwand*, *narband*, and *narbun*, and primarily means a pomegranate tree. Probably it has been transferred to this peculiar elm on account of the fancied resemblance of the globose crown to a pomegranate.

[3] *Garden and Forest*, ii. 516 (1889). Cf. also *Gartenflora*, xxxvi. 643 (1887). E. Morren, in *Belg. Hort.* 1879, p. 269, states that Späth received this variety from M. Scharrer at Tiflis.

[4] The origin of this variety is unknown; but it was probably named after T. Rueppell, owner of P. Smith and Co.'s nursery at Hamburg from 1862 to 1899.

[5] A tree cultivated at Kew as *U. glabra pendula nova* is a common form of the species with drooping branches, which do not form a dense crown of foliage as in true var. *pendula*.

[6] The outer lower branches of this tree bear abnormal leaves, some of which have one or two small supernumerary leaflets at the base. Other leaves are large and broad, as if composed of two ordinary leaves, and are often cleft from the apex to the base. Some leaves form ascidia. Similar leaves occur on a single lower branch of a tree of *U. nitens* in the Cambridge Botanic Garden. Cf. Worsdell, in *Gard. Chron.* l. 285 (1911), and in *Journ. Roy. Hort. Soc.* xxxvii., *Proc.* ccxiii. (1912). Similar abnormal leaves are present on a specimen at Kew of *U. nitens*, gathered from a large tree in Persia by Dr. Stapf. Delavaud, in *Bull. Soc. Bot. France*, viii. 144 (1861), gives an explanation of the same kind of malformation, which he had observed on an elm at Rochefort.

variety on the lawn in front of the palace at Kew. It is grafted high, and resembles in appearance the weeping form of *U. montana.*

A weeping form of *U. nitens* in Victoria Park, Bath, is known as the Scampston elm, var. *scampstoniensis*, Petzold and Kirchner, *Arb. Musc.* 560 (1864). Schneider states that what he has seen under this name is ordinary *U. nitens.* Loudon says that the Scampston elm came from a place of that name in Yorkshire, and that a tree, 18 ft. high in 1834, which bore this name in the Chiswick garden, was clearly some variety of the glabrous elm, and differed little from the species. Elwes saw the decayed stump of the original tree at Scampston Hall in Yorkshire; and so far as we can ascertain, this weeping variety, if it ever was distinct, is no longer known in nurseries in England.

8. Var. *Dampieri*, Henry.

*Ulmus Dampieri* and *Ulmus montana Dampieri*, Petzold and Kirchner, *Arb. Musc.* 563 (1864).
*Ulmus campestris Dampieri*, Späth, *Cat.*
*Ulmus campestris*, var. *plumosa*, Lavallée, *Arb. Segrez.* 236 (1877).

A tree, fastigiate in habit, forming a narrow pyramidal crown. Branches curved. Leaves, crowded on short glabrous branchlets, broadly ovate, 2 to 2½ in. long, 1½ to 1¾ in. broad, smooth and glabrous above, glabrous beneath except for conspicuous axil-tufts; margin incised with serrated teeth.

This beautiful tree, which has been much confused with Ford's elm (the similar variety of *U. montana*), is probably of continental origin.[1] The finest specimens which I have seen are two trees in Antwerp Park, about 45 ft. high, and one in the Leyden Botanic Garden, which was 40 ft. by 4½ ft. in 1912.

A sub-variety of this with yellowish leaves is occasionally seen in botanic gardens, where it is known as *U. Dampieri aurea* or *U. Dampieri Wredei.*[2]

9. Var. *modiolina*, Henry.

*Ulmus modiolina*, Dumont de Courset,[3] *Bot. Cult.* vi. 384 (1811).
*Ulmus tortuosa*, Loddiges, *Cat.* 1836, *ex* Loudon, *Arb. et Frut. Brit.* iii. 1376, 1379 (1838) (not Host).

This is *l'orme tortillard* of the French,[4] described as a tree with a twisted

[1] Nothing is known of Dampier, after whom this elm is named.

[2] Named after Wrede, curator of the Arboretum at Alt-Geltow, near Potsdam, where the yellow form originated in 1875 as a branch on a tree of *U. Dampieri.* It received a certificate from the Royal Horticultural Society in 1893, as *U. Wredei aurea.*

[3] Dumont de Courset described *U. modiolina* as a pyramidal tree with crowded branches and small leaves. Du Roi, *Harbk. Baumz.* ii. 501 (1772), identified *l'orme tortillard*, which was then much valued in France, with *U. glabra*, Miller (*U. nitens*, Moench). Michaux, *Arb. Amer.* iii. 274 (1813), who advocated its introduction into the United States, said that it could be propagated by layering. Loudon states that it was reported to come true from seed frequently. M. de Vilmorin tells me that his grandfather planted several lines of this variety at Les Barres, which still exist but are growing in poor soil. His specimens include two different trees, one being ordinary *U. pedunculata*, while the other is a peculiar elm with moderately sized leaves, scabrous on both surfaces. The tree in the Jardin des Plantes, Paris, which is now labelled *U. campestris modiolina*, is probably not the original tree, as it is identical with the hybrid elm (*U. belgica*) of Holland and Belgium. The elm cultivated as var. *modiolina* by Späth is closely allied to *U. major* (see p. 1884). Huberty, in *Bull. Soc. Cent. Forest. Belg.* xi. 492 (1904) and xv. 788 (1908), considers this elm to be a special variety with small leaves, which grows slowly even in good soil, but produces the most suitable wood for making naves of wheels; but I have seen no specimens in Belgium. Aigret, in *Ann. Trav. Public. Belg.* x. 1225 (1905), assigns to var. *tortuosa* some hybrid elms with large leaves, which are growing with twisted trunks at Louveigné in the province of Liège.

[4] The earliest account is by Duhamel, *Exploit. des Bois*, i. 294 (1764), who says that *le tortillard* has not only the most useful wood of all the elms, but has also fine foliage; and adds that it can be raised by seeds, grafts, or layers. Poederlé, *Man. de l'Arbor.* i. 116 (1792) identifies Duhamel's *l'orme tortillard* with the elm called *orme maigre* in Belgium, which is *U. nitens.*

arrangement of the bark and of the fibres of the wood, the stem when old being covered with burrs. This is stated by old writers to be a distinct variety, which produced wood valued by wheelwrights. Koch[1] states that it is a fine tree, with burry excrescences on the trunk, much planted in avenues in the north of France.

It is somewhat doubtful, however, whether this elm was really a distinct variety, as the description points to abnormal growth, which might occur in any kind of elm. M. Hickel informs me that the term *l'orme tortillard* is now applied to small elms of *U. nitens* with twisted and knotty stems, growing as a rule in poor soil.[2]

10. Var. *variegata*, Dumont de Courset, *Bot. Cult.* vi. 384 (1811); Loudon, *Arb. et Frut. Brit.* iii. 1405 (1838).

Leaves variegated with white, elliptic, 1 to 3 in. long, long-acuminate at the apex, smooth to the touch above, on long petioles. Branchlets glabrous. This occasionally bears branches with green leaves, similar to the ordinary form of *U. nitens.*

This[3] was cultivated in 1838 in the Chiswick garden, and differs from the variegated form of the English elm. Elwes saw a very fine variegated smooth-leaved elm at The Mote, Maidstone, 90 ft. by 9 ft. in 1911; another at Leaton Knolls, Shrewsbury, 62 ft. by 5 ft. 8 in. in 1910; one at Woburn, 60 ft. by 5 ft. in 1908; and a fine tree by the roadside at Little Parndon rectory, Essex, which was 85 ft. by 10 ft. in 1907. This was said by the Rev. P. Deedes to be at its best in July, when the leaves are of a golden colour; but they are shed early in autumn. There are also specimens at Beauport, at Stanway, and in Victoria Park, Bath.

Another variegated variety, which is probably derived from *U. nitens*, is in cultivation at Kew, where it forms a wide-spreading tree about 20 ft. high, obtained from Smith of Darley Dale in 1879 under the name *U. campestris variegata nova.* It bears leaves, which are often much reduced in size and entirely whitish. Occasionally branches are produced bearing leaves of normal size, with the variegation confined to the margin, while one or two other branches bear pure green leaves larger than is usual in typical *U. nitens.*

11. Var. *Webbiana*, Lee, *ex* Simon-Louis, *Cat.* 1869, p. 97.

A tree, pyramidal in habit, with ascending branches and sparse foliage. Leaves folded longitudinally, so that most of the upper surface is concealed; in other respects similar to those of var. *stricta.* There are two good specimens at Kew, about 35 ft. high, which were planted in 1871.

This variety is said[4] to have been raised in Lee's nursery about 1868; but it seems to be identical with *U. campestris*, var. *concavæfolia*, Loudon, *Arb. et Frut. Brit.* iii. 1378 (1838), which is insufficiently described.

[1] *Dendrologie*, ii. pt. i. 410, where it is referred to *U. montana.* Koch's description possibly refers to *U. Klemeri.* See p. 1872.

[2] This was the view taken by Planchon, in De Candolle, *Prod.* xvii. 158 (1873). M. Jouin showed me in the nursery of Simon-Louis at Metz, an avenue of what he called "*orme tortillard*," which he said were much liked by wheelwrights. They seemed to be a bad stunted elm, with swellings on the crooked trunks, and are propagated by grafting on stocks of *U. nitens.*—H. J. E.

[3] This is probably identical with var. *argenteo-variegata*, Rehder, in Bailey, *Cycl. Amer. Hort.* 1882 (1902), which is the same as var. *folio argenteo-marginata*, Petzold and Kirchner, *Arb. Musc.* 557 (1864). It appears to have been first mentioned by Miller, *Cat. Plant.* 86 (1730), and *Gard. Dict.* ed. 1, No. 7 (1731), as *Ulmus folio glabro eleganter variegata*, the striped witch elm.

[4] *Gard. Chron.* 1868, p. 918.

*Ulmus viscosa*, Koch, *Dendrologie*, ii. pt. i. 416 (1872), appears to differ very little from var. *Webbiana*. It is represented at Kew by a stunted specimen, obtained from Booth in 1871.

12. Var. *virens*, Henry. Kidbrook Elm.

*Ulmus campestris*, var. *virens*, Loudon, *Arb. et Frut. Brit.* iii. 1376 (1838).

A tree, narrowly pyramidal in habit, with long ascending branches, retaining its foliage till December. Branchlets slender, glabrous. Leaves oval, about 3 to 4 in. long, and 1½ to 2 in. wide, long-acuminate at the apex, coarsely biserrate, glabrous and smooth above, slightly scabrous beneath, with numerous glands; lateral nerves, about twelve pairs, remote, often forked. Flowers similar to those of the Huntingdon elm. Samaræ similar, but smaller.

The only specimen of this tree which I have seen is one about 40 ft. high at Ashwell Bury, near Baldock. In habit it somewhat resembles the Huntingdon elm, but is more narrowly pyramidal. It retained its leaves in 1910 till the first week in December. It is probably identical with Loudon's tree, of which he says: "Notwithstanding its name of Kidbrook elm, a place in Sussex, it is a Cornish variety," probably on account of its pyramidal habit. Var. *virens* appears to be unknown in nurseries at the present day.

13. Var. *betulæfolia*, Loudon, *Arb. et Frut. Brit.* iii. 1376 (1838).

*Ulmus betulæfolia*, Loddiges, *Cat.* 1836.

A tree, pyramidal in habit, with ascending branches. Leaves ovate-oblong, up to 3½ to 4 in. long and 1½ in. broad, long-acuminate at the apex, tapering to a cuneate but unequal base, coarsely biserrate; lateral nerves 12 to 14 pairs.

This tree, which is readily distinguished by its foliage and habit, is represented by a good specimen in the Cambridge Botanic Garden. It is occasionally seen in hedgerows in Herts and Huntingdonshire.

## Distribution

*U. nitens* is a native of central, western, and southern Europe, Algeria, Asia Minor, Caucasus, Armenia, Persia, and Turkestan. In eastern Asia it is replaced in Manchuria, Korea, and Japan by the closely allied species, *U. japonica*.

In Europe, this species has a more southerly distribution than *U. montana*, and is unknown in Denmark, Sweden, and Norway as a wild tree, though it is said to occur, possibly planted, on the island of Gothland, in the Baltic. In Russia, it is limited to the southern provinces, its northern limit passing through Grodno, Volhynia, Chernikof, Tambof, Saratof, Samara, and Orenburg. It is especially common in the south-west, occasionally forming pure woods, one of which is said by Köppen[1] to extend for fifteen versts in the valley of a tributary of the river Ingul in Kherson. It occurs in the Crimea, and in the Caucasus, where it grows from sea-level to an altitude of 6000 feet.

[1] *Holzgewächse Europ. Russlands*, ii. 34 (1889).

In Germany the northern limit of this species is not accurately known; but it is a rare tree in the north, except in the river valleys.[1] It is widely spread through Austria-Hungary, the Balkan States, Greece, Italy, Spain, and Portugal.

*U. nitens* is rare, either wild or cultivated, in Belgium and Holland; but it grows on the dunes near Haarlem. I saw no trees of it in the Ardennes,[2] where the wild elms are *U. pedunculata* and *U. montana*.[3] I found it in a nursery at Malines, where it was called *orme maigre*.

In France it is widely distributed, except in Provence, where it is replaced by the southern variety of *U. campestris*. It is common, however, only in some parts, and rarely if ever forms a prominent feature in the landscape, such as it does in East Anglia. It is apparently wild in Normandy, where it grows often in hedges as a small tree with corky branches and numerous suckers. It is indigenous in certain forests, mixed with oak, notably in the Bois de Vincennes near Paris, where it is associated with *Quercus lanuginosa*. I noticed a few trees growing in the forest of Marly. It constitutes a small percentage of the forest of Thétieu, through which the Adour flows near Dax. The soil here is liable to inundation once or twice a year, and the main species is the pedunculate oak. I estimated many elms in this remarkable forest to be about 70 ft. in height and 5 to 6 ft. in girth. Near Angoulême I noticed this elm in hedges on the road to the forest of La Braconne, one fine tree being 90 ft. in height and 7 ft. in girth. It is also common on the outskirts of the forest of Orleans, where it grows in hedgerows around cultivated fields in better soil than that of the interior of the existing woodland. M. Mathey gives an interesting account[4] of the remarkable woods, known as *vaivres*, which occur on the banks of the Saône from Port-sur-Saône near Vesoul to twenty-eight kilometres north of Macon. These moist woods, with an alluvial soil which is frequently flooded, are mainly composed of oak, ash, and glabrous elm, with a sprinkling of aspen, alder, hornbeam, maple, and blackthorn. The herbaceous vegetation is characterised by moisture-loving plants like *Carex brizoides*, *Panicum Crus-Galli*, and *Iris Pseudacorus*. Most of these woods are treated as coppice with standards, the number of the latter that are reserved being often in the proportion of 15 oaks, 2 ashes, and 1 elm; but as the elm is the least valuable, it is not maintained in the overwood, although it increases by suckering in the underwood. The late M. Broilliard stated[5] that the woods richest in elms are near Gray (Haute Saône), Pontailler, Heuilley (Côte d'Or), Ecuelles, Boyer (Saône-et-Loire), and Truchère (Ain).

The seedling elms[6] that are imported into England, under the name

[1] Willkomm, *Forst. Flora*, 554 (1887), states that, on the alluvial land along the Elbe in north Germany, small pure woods of this species are not uncommon.

[2] Reputed wild trees of small size on the side of the river near Rochefort were, I found on enquiry in 1912, elms that had been imported as seedlings twenty-five years previously from a French nursery in Calvados.

[3] Specimens of wild elm in the Bois de Colfontaine near Mons, sent me as "*U. campestris*" by M. Quievy, are also *U. montana*.

[4] *Bull. Soc. Forest. Franche-Comté*, iv. 494 (1898), and *Bull. Soc. Cent. Forest. Belg.* vi. 87 (1899). I am indebted for specimens of this elm to M. Mourlot, Inspector of Forests at Gray, and to Mdme. Broilliard.

[5] *Bull. Soc. Cent. Forest. Belg.* xii. 53 (1905).

[6] On account of their rough pubescent leaves and branchlets, these seedlings have been much confused with *U. campestris*, the true English elm; but they are undoubtedly *U. nitens*, and agree perfectly with seedlings that I have raised from seed of the latter species gathered in England. Sargent sent me in 1911 seedlings of his "*U. foliacea*," raised from seed gathered in Hungary in 1905; these have corky pubescent twigs and rough hairy leaves and are also *U. nitens*.

*U. campestris*, from French nurseries at Orleans, Ussy, and elsewhere, are raised from the seed[1] of *U. nitens*. This importation has continued probably without interruption from the time of Evelyn, who refers[2] to the French elm as having "glabrous and smooth" leaves, and mentions in 1664 "a cloyster of the right French elm in the little garden near to her Majesties the Queen Mothers Chapel at Somerset-house, which were (I suppose) planted there by the industry of the F.F. Capuchines." In all probability many of the glabrous elms that are now growing in parks and pleasure grounds in England and Scotland are of French origin.

In England, *U. nitens* is limited to the southern, midland, and eastern counties. Throughout the east of England, where it is much commoner than *U. montana*, it is usually known, when of wide-spreading habit, as the wych elm. *U. nitens* is unknown, except as a planted tree, in Scotland; but it is probably wild in the south of Ireland, where it is frequent in hedge-rows; and at Abbeyleix, var. *stricta* is scattered amidst oaks on the alluvial flat of the river Nore. Cf. p. 1889.

Many elms may be found in the south and west of England, which differ somewhat in foliage and habit, both from typical *U. nitens*, and its well-marked varieties, described above. It seems unnecessary to describe these in detail or to give them special names. In several cases, in which I sowed seeds of trees, slightly abnormal in foliage and habit, the seedlings produced are not uniform, and show the characters of opposite and alternate leaves in the first year in Mendelian proportions; and are presumably of hybrid origin. Some of these elms have leaves thicker in texture, larger in size, often longer with more numerous nerves; whilst others show peculiar serrations, less obliquity at the base, etc. A tree at Colesborne, from which numerous seedlings were raised in 1909, is a typical instance of this class of elm.[3] (A. H.).

## Remarkable Trees

Among the most remarkable trees of this species is an elm (Plate 399) at Sharpham, near Totnes, the seat of O. Durant Parker, Esq., which was figured as a wych elm in *Gard. Chron.* xxxix. 152 (1906). I found this tree to be a glabrous-leaved elm of unusual habit, with immense branches spreading to a distance on one side of 104 ft., and covering a total area of a quarter acre. Its height when I saw it in 1906 was 80 to 90 ft., and its girth 17 ft. The lateral branches were covered with small spray and ferns, which are favoured by the damp climate.

In Dyrham Park, Gloucestershire, there is an old and decayed glabrous-leaved elm, which in 1906 measured 19½ ft. in girth, but was throwing out much young spray on the live branches.

From West Dean, in Sussex, Mr. F. Arthur sent me specimens of this elm, which is there known as wych elm; its timber sells locally for 1s. per foot.

At Godinton, Kent, a fine tree which has the habit of the weeping Hertfordshire elm, measured in 1907 about 100 ft. by 12 ft.

[1] M. de Vilmorin informs me that the seeds of this elm, which are sold by him as "*U. campestris*," are usually gathered in the neighbourhood of Soissons (Aisne) and Le Mans (Sarthe).

[2] *Sylva*, 19 (1664).

[3] Referred to in *Journ. Linn. Soc.* (*Bot.*) xxxix. 294 (1910).

Some trees of the same habit, which possibly originated in the nursery at Canterbury, where Masters raised elms of this type, are at Betteshanger, near the church, and in 1907 were 70 to 80 ft. by 7 to 8 ft. in girth.

At Boughton Park, in the broad avenue, which is a hundred yards wide and stretches for a mile in front of the house, are a number of elms, which like those at Cassiobury, and some of those at Hinchingbrooke, are remarkable for their burry growths and stunted habit. Whether this is due to their having been propagated from suckers of trees having this peculiarity I cannot say, but there are much better-shaped trees in the same avenue which are free from burrs, and have attained a much greater height. Two which I measured were 107 ft. by 18 ft. 4 in., and 104 ft. by 18½ ft. in 1908. Boughton is celebrated for its extensive avenues of elms, planted by "John the Planter," Duke of Montagu, who died in 1749. Most of them are true English elms, and one which I saw in the timber yard had 170 rings, of which only twenty were sapwood, in a diameter of 3 ft. Elm timber at Boughton averages about 1s. per foot, though in an exceptional case it has realised 2s. 6d. per foot. It is much liked for chair seats, on account of its rich colour.

In Hertfordshire and on the western borders of Essex, the most graceful form of this tree may be seen in perfection. Among the largest in this district are the following:—At Bayford Lodge, Bayfordbury, a fine weeping tree, very distinct from an English elm growing beside it, in 1909 was 92 ft. by 11 ft. 9 in. At Amwell Bury, near Ware, Herts, there is a magnificent tree of *U. nitens*, of which Mr. H. Clinton-Baker sends us the following particulars. It is situated on Mr. E. S. Hanbury's property, and in November 1911 measured 133 ft. in height, with a trunk 23½ ft. in girth at five feet, above which it divides into two stems, 15 ft. 2 in. and 14 ft. 6 in. in girth respectively.

At Eastwick rectory, in the Stort valley, the best of a group of four trees near the church was 119 ft. by 10 ft. in 1909; while another was 116 ft. by 6½ ft. A little higher up in a narrow grass field there is a fine tree of the same habit, 113 ft. by 13½ ft. Suckers from these trees are numerous; and Mr. Rivers, of the Sawbridgeworth nurseries, has propagated a number of this type. At Briggins Park, the seat of the Hon. H. C. Gibbs, two splendid trees (Plate 401) stand together by a small lake; the larger was 128 ft. by 12 ft. 10 in. in 1909. On the road from Dunmow to Easton Park, there are a number of weeping elms, which vary a good deal in habit, and appear to belong to this species, but I have not seen them in leaf. At Aldenham, in Sawyer's Lane, there are good trees of this type which, as usual, have rough-leaved suckers; the best measured in 1908 about 108 ft. by 7 ft. 11 in.

At Great Saling, Essex, on the village green, there is a handsome weeping tree (Plate 402), with an immense witches' broom near the top, which in 1907 measured 110 ft. by 20 ft. 9 in. Loudon says[1] that Mr. Jukes, who made drawings of the large elms at Studley Royal, pronounced this elm to be the most handsome that he had seen. It measured, in 1841, 114 ft. in height and 17½ in girth at five feet from the ground.

[1] *Gard. Mag.* xvii. 356 (1841).

At Ryston, near Downham, Norfolk, in 1908, I measured one in a hedgerow which was 95 ft. by 16 ft. 9 in.

At Belton, Grantham, there are large glabrous elms in the grounds; and at Barholm, not far from Stamford, there are two remarkable trees, very wide spreading, with a rounded crown of foliage, of no great height, but 16 ft. 3 in., and 17½ ft. in girth in 1910.

In Gloucestershire there are several old elms on my own property which have similar leaves to *U. nitens*, but which keep their leaves green later in autumn and turn to a brighter colour, which may be due to the locality, for I have noticed that beech, oak, elm, and other trees colour better in the Cotswold hills than in the eastern counties. The timber, though hard and tough, is not so red as that of the English elm, and their habit is more spreading owing probably to their hybrid origin. From one of these just outside my park, which does not lose its leaves till late in November, I have raised numerous seedlings, which vary extremely in habit and vigour, but all have rough leaves at four years old like the suckers of the parent tree. At Toddington Manor there is a fine spreading tree near the house, which in 1910 measured 100 ft. by 15 ft. 2 in.; but most of the elms here and generally in the Vale of Gloucester are true English elms.

In Scotland I have not noticed any large trees of this species; but Mr. Renwick measured a tree at Loudon Castle, 107 ft. by 15 ft. 4 in. in 1910.

In Ireland, Mr. R. A. Phillips informs us that it is common in hedgerows between New Ross and Waterford, and along the river Barrow. At Killarney, Henry measured one in Lord Kenmare's grounds, which was 91 ft. by 12 ft. in 1903.

On the Continent there are many large elms, referred by Continental botanists to *U. campestris*, which are often *U. nitens*. The most famous elm in Germany is a tree standing in the market-place of Schimsheim, near Worstadt in Hesse, which is figured by Seidel in *Woods and Forests*, 1884, p. 577. Willkomm[1] states that this tree is 500 or 600 years old, about 100 ft. high, and 44 ft. in girth at 3 ft. 4 in. above the ground. The trunk is hollow, but otherwise the tree is quite healthy, with abundance of foliage. Neither Seidel nor Willkomm identify the species of this elm. The elm at Worms under which Luther preached is said by Willkomm[1] to be probably *U. nitens*, and is reported by Seidel to be much taller than the Schimsheim elm, and 25 ft. in girth at 8 ft. above the ground. Mr. Springer has sent me a photograph of a wide-spreading smooth-leaved elm at Windesheim Castle, near Zwolle in Holland. This tree measured, in February 1913, 75 ft. in height and 23 ft. 9 in. in girth, with a crown of foliage 106 feet in diameter. (H. J. E.)

[1] *Forstliche Flora*, 550, note (1887).

## ULMUS MINOR, Goodyer's Elm

*Ulmus minor*, Miller,[1] *Gard. Dict.* ed. 8, No. 6 (1768); Reichenbach,[2] *Icon. Fl. Germ.* xii. 12, t. 660 (1850).

*Ulmus glabra*, Miller, var. *minor*, Ley, in *Journ. Bot.* xlviii. 70 (1910).

*Ulmus campestris*, Smith, *Eng. Bot.* t. 1886 (1808) (not Linnæus); Lindley, in Rees, *Cyclopædia*, xxxvii. No. 1 (1819).

*Ulmus sativa*, Moss, in *Gard. Chron.* li. 216 (1912) (not Miller).

*Ulmus sativa*, Miller, var. *Lockii*, Druce, *List of British Plants*, 63 (1908).

*Ulmus Plotii*,[3] Druce, in *Northampton Nat. Hist. Journ.* 1911, p. 88, and in *Gard. Chron.* l. 408 (1911).

A small tree, 40 to 90 ft. in height, with the stem usually curving at the summit, and a few short stout ascending branches, and pendulous branchlets, forming a narrow crown of peculiar appearance. Young branchlets slender, with a scattered minute pubescence, glabrous and finely striate in the second year. Leaves (Plate 411, Fig. 3) firm in texture, obovate or elliptical, about 1½ to 2½ in. long and ¾ to 1½ in. broad, unequal and often cordate at the base; acute, acuminate, or occasionally rounded at the apex; upper surface dull, scabrous with scattered minute tubercles and minute hairs; lower surface scabrous and densely pubescent with short hairs in spring, later glabrescent, but with conspicuous axil-tufts; biserrate; lateral nerves, eight to ten pairs, often forked; petiole ⅕ in. long, pubescent.

Flowers, twenty to twenty-five, in small clusters, on very short pedicels, irregular in the number of sepals and stamens; calyx funnel-shaped, about ⅛ in. long, with four, five, or six pink lobes; stamens, three, four, or five, with deep pink filaments and red anthers; stigmas pink. Samaræ rarely ripening, but in 1909 a few were produced, narrowly obovate, ½ in. long, emarginate at the apex, with a triangular open notch; seed in the upper half of the samara.

This tree produces suckers freely. It is possibly a hybrid, as a small packet of seed produced in 1909 fifty-seven alternate-leaved and twenty-eight opposite-leaved seedlings.

[1] Miller, *Gard. Dict.* ed. 8, No. 6 (1768), describes *U. minor* as "the smooth narrow-leaved elm, by some called the upright elm," and adds that "the leaves are narrower and more pointed than the English elm and are smoother; they are later in coming out in the spring than these, but continue later in autumn." He identifies it with *Ulmus minor folio angusto glabro*, which is the elm described by him in *Gard. Dict.* ed. 3, No. 6 (1737) and in ed. 6, No. 10 (1752), as "very common in some parts of Hertfordshire and in Cambridgeshire, where there is scarce any other sort of elm to be seen. This makes a very handsome upright tree, and retains its leaves as late in the autumn as the common small-leaved elm, which is called the English elm by the nurserymen near London; but it doth not come out so early in the spring." As Miller's *U. sativa* is undoubtedly the English elm (our *U. campestris*, L.), there is little doubt that Miller's *U. minor* is Goodyer's elm. The latter was called *U. minor* by Parkinson, *Theat. Bot.* 1405 (1640). Plot's *Ulmus folio angusto glabro* was not cited by Miller, and is *U. campestris*, var. *viminalis*. If *U. minor*, Miller, is objected to, on account of the uncertainty of the description, the tree may be styled *U. minor*, Reichenbach.

[2] Reichenbach's figure agrees, as regards leaves, flowers, and samaræ with English specimens of *U. minor*, and he cites correctly Smith, *Eng. Bot.* t. 1886. It is probable that *U. tortuosa*, Host, *Fl. Austr.* i. 330 (1827), is identical with *U. minor*, although Host's specimen in the Kew herbarium is *U. nitens*. *U. tortuosa* was said to be a low tree with a twisted trunk and small leaves, which grew in hilly districts in Hungary.

[3] This is not, as Druce supposes, the elm described by Plot, *Nat. Hist. Oxfordshire*, 158 (1677), which is identical with *U. campestris*, var. *viminalis*. See p. 1906. I have examined specimens of the tree at Banbury referred to *U. Plotii* by Druce, and they agree exactly with *U. minor*, as here described. Cf. Moss, in *Gard. Chron.* li. 234, figs. 104, 105, and 106 (1912).

This elm was first described by Goodyer, in Johnson's edition of *Gerarde's Herball*, 1478 (1633), as *Ulmus minor folio angusto scabro.* He says that he saw it once, growing in hedgerows between Lymington and Christchurch, where it has been lately found by Dr. Moss. There are specimens in the British Museum, collected by Buddle and other early botanists; but it has apparently escaped the notice of later writers. So far as I know, it has never been cultivated in nurseries; and nearly all the trees which we have seen, grow in hedgerows and similar situations, where they may be looked upon as indigenous.

*U. minor* occurs on the Continent, as it is recorded by Reichenbach for Hungary; and there are specimens of it in the Haarlem herbarium from trees growing in Holland; but its distribution has not been worked out—French, German, and Russian botanists having paid little attention to the different kinds of wild elms.

This elm, which is of a remarkably distinct appearance, is common on the Madingley road, Cambridge, where the trees grow, from which I have drawn up my description. It is widely spread in the eastern counties, from Lincoln [1] through Norfolk and Suffolk to Essex, and occurs also in Northamptonshire, Huntingdonshire, Bedfordshire, Oxfordshire, and Hampshire. The usual habit of the tree is well shown in a photograph,[2] taken by Mr. Druce, of a tree growing near Fineshade Abbey, Northamptonshire. He has also photographed [2] a tree near Banbury, which is over 80 ft. in height. (A. H.)

There are many elms of this variety in the eastern and midland counties, which are readily known by their smaller leaves, inferior size, and irregular habit. They are known sometimes in these districts as "lock" elms.[3] A great many elms on the sandy land of west Norfolk are of this character, and, generally speaking, the hedgerow elms of this county are very inferior in size and shape to those of Essex, Herts, and the south generally. Similar trees grow in the hedges between Grantham and Belvoir Castle.

The finest specimen which we know of *U. minor* is a tree growing in the park at Weston Birt (Plate 403), which was 97 ft. high and 7 ft. in girth in June 1912. At Studley College, Warwickshire, there are about a dozen trees in a field, which are 70 to 80 ft. high, the three largest being 9 ft. 4 in., 9 ft. 6 in., and 9 ft. 9 in. in girth. Mr. C. C. Rogers, who has kindly sent specimens, tells us of a tree at Hagnaby Priory, Spilsby, Lincolnshire, nearly 90 ft. in height and 7 ft. 7 in. in girth. (H. J. E.)

[1] Ley mentions a tree at Skellingthorpe in south Lincoln.

[2] Reproduced in *Gard. Chron.* l. 408, figs. 165, 166 (1911).

[3] This name indicates the toughness of the wood of this elm, which is difficult to work with tools like the saw or plane, which become "locked," as carpenters say. Sir J. E. Smith, *Eng. Flora*, ii. 20, 23 (1824), states that the wood of this species "is greatly preferred in Norfolk to any other, and sells for nearly double the price, serving more especially for the naves of wheels."

## ULMUS CAMPESTRIS, English Elm

*Ulmus campestris*, Linnæus, *Sp. Pl.* 225 (1753) (in part), and *Flora Anglica*,[1] 11 (1754) (not Smith[2]); Loudon,[3] *Arb. et Frut. Brit.* iii. 1374 (1838) (in part); Planchon, in De Candolle, *Prod.* xvii. 156 (1873) (in part); Moss, in *Gard. Chron.* li. 199 (1912).

*Ulmus sativa*,[4] Miller, *Gard. Dict.* ed. 8, No. 3 (1768).

*Ulmus suberosa*, Smith, *Eng. Bot.* t. 2161 (1810) (not Moench or Ehrhart); Loudon, *Arb. et Frut. Brit.* iii. 1395 (1838) (in part); Lindley, *Syn. Brit. Flora*, 226 (1829).

*Ulmus atinia*, Walker, *Essays Nat. Hist.* 70 (1812).

*Ulmus surculosa*, Stokes, var. *latifolia*, Stokes, *Bot. Mat. Med.* ii. 35 (1812); Ley, in *Journ. Bot.* xlviii. 72 (1910).

A tree attaining 130 ft. in height and 20 ft. in girth, with a tall straight stem, and spreading or ascending branches, rather variable in habit, but usually forming in the open a broad or narrow oval crown. Bark dark-coloured, deeply fissured. Young branchlets densely pubescent with short erect white hairs, more or less retained in the second year, when the twigs become finely striate. Buds ovoid, with minutely pubescent ciliate scales. Leaves (Plate 412, Fig. 14) broadly oval or ovate, about 2 to 3 in. long, and 1½ to 2 in. wide, very oblique at the base, shortly acuminate at the apex; upper surface dark green, scabrous, with a scattered minute pubescence, conspicuous on the midrib; lower surface pale green, with broad[5] conspicuous white axil-tufts prolonged along the sides of the midrib, and covered between the nerves

[1] Linnæus, *Flora Anglica*, 11 (1754), identifies his *U. campestris* with one of the four elms mentioned by Ray, *Syn. Meth.* 468 (1724), namely, No. 1, *U. vulgatissima folio lato scabro*, Gerarde, *Emac.* 1480, which is the English elm. According to the rules of the Vienna Congress, the correct name, as it is the earliest, of the English elm must then be *U. campestris*, Linnæus, which cannot be applied to any other species. Continental botanists usually mean by *U. campestris*, all the European elms, except *U. montana* and *U. pedunculata*. *U. campestris*, Miller, *Gard. Dict.* ed. 8, No. 1 (1768), described as "the common rough or broad-leaved wild elm . . . very common in the north-west counties of England, where it is generally believed to grow naturally in the woods," is *U. montana*, and not the English elm, as asserted in *Gard. Chron.* li. 199 (1912).

[2] Smith, *Eng. Bot.* t. 1886 (1808), is *U. minor*. See p. 1901.

[3] Loudon's account of the English elm is very confused, and his reference to Smith, *Eng. Bot.* t. 1886, is erroneous. He also fails to recognise that *U. suberosa*, Smith, *Eng. Bot.* t. 2161, is the English elm, and is quite different from *U. suberosa*, Moench, which is the corky-twigged variety of *U. nitens*.

[4] Miller's *U. sativa*, which he calls "the small-leaved or English elm," is in my opinion the tree which is still named English elm and which is described above as *U. campestris*, L. The name "*sativa*" implies that it was commonly cultivated; and Miller states that it was planted near London. No other elm can have been meant, as it is certain that *U. campestris* was known near London under the name of English elm in Miller's time. The Society for the Encouragement of Arts offered a gold medal in 1765 (and also in subsequent years) "for properly planting the greatest number of the small-leaved English elms for raising timber, commonly used for keels of ships and water-works." [Cf. *Museum Rusticum*, iv. 380 (1765).] The elm that was used for making water-pipes was *U. campestris*, L., as is shown by T. H. W.'s clear account of the species around London, in *Gentleman's Magazine*, lv. 453 (1785). John Harrison, nurseryman at Cambridge, plainly means *U. campestris* by the English elm, in his *New Method*, 33 (1766): "That which we call the English elm, is peculiar only to the southern part of this island, where it grows wild in hedgerows; there is not a tree of that kind to be seen in the northern counties, but what has been planted within seventy or eighty years; and these are either in avenues or some plantations near gentlemen's seats." Cf. also Hunter, *Evelyn's Sylva*, 124 (1776). Miller quotes as a synonym Gerarde's *Ulmus minor folio angusto scabro*, of which he took a different view from Ray, as is clearly shown in his *Gard. Dict.* ed. 1, No. 3 (1731), where this elm of Gerarde's is identified with "the common English elm, the timber of which is generally preferred to the rest, and is the largest tree when planted on a kindly soil." It is impossible to suppose that *U. sativa*, Miller, can be the tree which is called by us *U. minor*, as the latter has never been called English elm, has never, so far as we know, been planted anywhere, and never attains a great size. Sir J. E. Smith, *Eng. Flora*, ii. 21 (1824), took the same view as I do regarding Miller's *U. sativa*.

[5] The axil-tufts are peculiar, broad, and diffused in this species.

with a dense soft pubescence and numerous minute glands; lateral nerves ten to twelve pairs, often forked; petiole $\frac{1}{5}$ in. long, densely pubescent.

Flowers, about twenty in a cluster, on very short pedicels, irregular in the number of stamens and sepals; calyx funnel-shaped, with four or five or six red sepals; stamens three, four, or five, with red filaments and red anthers; stigmas white. Samaræ on very short pedicels, nearly orbicular, $\frac{1}{2}$ in. in diameter, glabrous, non-ciliate, emarginate at the apex, with a short notch closed by the incurved stigmas; seed in the upper part of the samara, with its apex touching the base of the notch.

The English elm occasionally, but rarely, produces epicormic branches with corky wings; but, as a rule, the branches of all parts of the tree, as well as the suckers, are not suberose. The suckers are produced very freely, with leaves and twigs more pubescent than those of adult trees. It rarely bears[1] good seed in England; but in 1909, out of about twenty lots of seed gathered in different places in the south of England, I raised one seedling from a tree at Cambridge, two seedlings from a tree at Bayfordbury, and one seedling from a tree in the Isle of Wight.[2]

### Varieties

1. Var. *australis*, Henry.

A tree, often pyramidal in habit, with short branches. Branchlets similar in pubescence to the type. Leaves (Plate 412, Fig. 17) thick and firm in texture, oval, 2 to 3 in. long, $1\frac{1}{4}$ to $1\frac{3}{4}$ in. broad, with a longer and more cuspidate acuminate apex than in the type; scabrous and pubescent above; lower surface densely pubescent, the pubescence conspicuous on the midrib and nerves, with axil-tufts not so well developed as in the typical form of the species; lateral nerves about twelve pairs, very prominent beneath; coarsely biserrate in margin; petiole up to $\frac{1}{4}$ in. long, pubescent. Flowers similar to those of the English elm, but with pale pink sepals, anthers, filaments, and stigmas. Samaræ more obovate than orbicular, but rounded at the base and otherwise similar to the typical form.

This elm differs mainly from the English elm in the thicker texture of the leaves, which have more prominent nerves beneath. It appears to be distributed in the wild state in south-eastern France, Switzerland, and the Riviera. The small elms along the stream on the golf course at Cannes, some on the hills above La Mortola, and others of which I gathered specimens at Pertuis, in Provence, more closely resemble the English elm than any others which I have seen, in branchlets, leaves, and samaræ, but still are not identical. I have also received from Lord Walsingham specimens of this variety from trees growing at Venice;

[1] The English elm has long been known to produce unfertile seed. W. Watkins, *Forest Trees*, 24 (1753), says "I could never find any seed worth gathering on the English elm"; and proceeds to give directions for layering, which he preferred to propagation by root-suckers.

[2] Loudon, *Arb. et Frut. Brit.* iii. 1384 (1838), states that Masters raised in 1817, from the seed of a common English elm at Lea Park, near Canterbury, a considerable number of seedlings, which comprised about twenty distinct varieties. In all probability this tree was a hybrid. All the elms, however, now standing in Lea Park, judging from copious specimens which I received in 1910, appear to be English elms.

and it probably exists farther to the south and east, as it seems similar to an elm[1] found by Baldacci in mountain woods near Spizza in the extreme south of Dalmatia.

Var. *australis* is occasionally planted in botanic gardens in France, as at Le Mans and Bordeaux; but it seems rare in cultivation. I refer to this variety the elms in the avenue of the Cours-la-Reine, Rouen, which is said to have been planted in 1649 by the Duke de Longueville. The best of these trees still remaining were about 90 ft. by 9 ft. in 1912.

2. Var. *variegata*, Dippel, *Laubholzkunde*, ii. 25 (1892).

*Ulmus campestris*, var. *foliis variegatis*, Loddiges, *Cat.* 1836, *ex* Loudon, *Arb. et Frut. Brit.* iii. 1376 (1838).

Leaves striped and spotted with white. This is a very ornamental tree, no doubt of English origin,[2] as it agrees with the English elm in all its essential characters. So far as I know it never produces fruit. There are two good trees about 50 ft. high at Kew. A fine specimen at Moor Park, Rickmansworth, which was 65 ft. by 7½ ft. in 1910, produces suckers with variegated leaves. Elwes saw, in 1909, a tree 83 ft. by 7½ ft. at Campsea Ashe in Suffolk, also with variegated suckers. These are said to lose their colour if transplanted. Another at Kenwood, without suckers, was 75 ft. by 7 ft. 2 in. in 1909.

*U. picturata*, Cripps, *ex* Simon-Louis, *Cat.* 1880, p. 66, of which there is a small tree at Kew, does not appear to differ much from the preceding variety; but has larger leaves.

3. Var. *Van Houttei*, Schneider, *Laubholzkunde*, i. 220 (1904).

Leaves scabrous above and beneath, tinged with yellow. Branchlets pubescent with long hairs.[3] The variety,[4] which is known in catalogues as *U. campestris*, "*Louis van Houtte*," is represented at Kew by several trees about 20 ft. high, obtained from Waterer in 1894.

4. Var. *purpurea*, Petzold and Kirchner, *Arb. Musc.* 558 (1864).

Leaves 2 to 2½ in. long, tinged purple, often folded, irregularly serrate. This is probably of hybrid origin, as it resembles *U. montana* in texture and roughness of surface, but has slender twigs and small leaves. It is grown at Kew under the name *U. montana*, var. *purpurea*.

5. Var. *purpurascens*, Schneider, *Laubholzkunde*, i. 220 (1894).

*U. campestris myrtifolia purpurea*, Louis de Smet, *Cat.* No. 10, p. 59 (1877).

Leaves small, about an inch long, scabrous above, pubescent beneath, tinged with a purple colour. Branchlets pubescent with long hairs. This is represented at Kew by a grafted tree, about 20 ft. high, obtained from Waterer in 1885.

[1] Var. *dalmatica*, Baldacci, in *Malpighia*, v. 79 (1891).

[2] Plot, *Nat. Hist. Oxfordshire*, 172 (1677), mentions a striped elm in Dorsetshire.

[3] In *Gard. Chron.* xi. 368, fig. 77 (1887), an instance is given of the influence of scion upon stock, where this variety was grafted on *U. campestris* with ordinary green leaves, and the stock subsequently produced a variegated shoot below the graft.

[4] The origin of this elm is unknown; but it may be the "yellow-leaved elm" referred to by Miller, *Cat. Plant.* 86 (1730), and *Gard. Dict.* ed. 1, No. 8, as "*Ulmus minor foliis flavescentibus.*"

6. Var. *Berardii*, Simon-Louis. *Cat.* 1869, p. 96.

A small tree or shrub, with minute leaves, ½ to ¾ in. long, firm in texture, deeply incised with a few teeth, almost glabrous; petioles, as well as the young branchlets, pubescent. This peculiar variety, with leaves somewhat resembling those of *Zelkova Verschaffeltii*, but smaller, was raised by Simon-Louis from seeds gathered in 1863 from the large elms growing on the ramparts at Metz. There is a specimen at Kew, about 10 ft. high, which was obtained from Späth in 1902

At Kew there is a small tree raised from seed collected by me off a tree at Nancy in 1903, which has similar minute leaves, but scabrous above and pubescent beneath, and biserrate in margin. This is one of a number of seedlings raised from the same lot of seed, which differed extraordinarily in the size and other characters of the leaves, showing that the parent tree was a hybrid.

In the Kew herbarium there are remarkable specimens of an elm with minute foliage, which were gathered in Jersey by Oliver in 1874 and 1880, but of which I can learn no particulars. Mr. Miller Christy sent me in 1911 a branch with minute leaves of an elm, like a bush and only 4 ft. high, which grows close to the shore at the foot of the cliff at West Mersea, Essex. Elwes collected a similar specimen in a hedgerow between Bisterne and Ringwood.

7. Var. *viminalis*, Loudon, *Arb. et Frut. Brit.* iii. 1376 (1838).

*Ulmus viminalis*, Loddiges, *Cat.* 1836.
*Ulmus antarctica*, Petzold and Kirchner, *Arb. Musc.* 552 (1864).

A tree, with ascending branches, pendulous branchlets, and sparse foliage. Leaves (Plate 412, Fig. 22) incised on the margin with deep serrated teeth: about 1½ to 2½ in. long, obovate-elliptic or narrowly elliptic, often nearly equal at the base, long-acuminate at the apex, scabrous above, slightly pubescent beneath, with conspicuous axil-tufts. Young branchlets slightly pubescent.

This variety is said by Loudon to have been raised by Masters in 1817 from seeds of the common English elm; but it appears to me to be identical with the elm described by Plot[1] in 1677 as occurring in avenues at Hanwell, "where there is a whole walk of them planted in order, besides others that grow wild in the coppices of the park." Plot describes this elm as having a narrow leaf with a peculiar kind of pointed ending, and his figure is unmistakably that of var. *viminalis.* Mr. Druce states[2] that the avenue at Hanwell is now composed of English elm.

The best tree of this variety that we know of is at Milton Abbey, Dorsetshire, the seat of Sir A. E. Hambro. It measured in 1906 71 ft. by 6 ft. 4 in., and resembled in habit the Cornish elm. A tree in the Cambridge Botanic Garden is about 70 ft. high. There are three trees[3] at Kew, two about 35 ft. high, which are labelled var. *viminalis*, and are of unknown origin, but evidently of considerable age; and one precisely alike in foliage, but labelled *U. campestris antarctica*, which was obtained

[1] *Nat. Hist. Oxfordshire*, 158, plate x. fig. 1 (1677).

[2] In *Gard. Chron.* l. 408 (1911). Druce's identification of Plot's elms with the tree (which he calls *U. Plotii*) now growing at Banbury and elsewhere is, in my opinion, erroneous. Cf. Moss, in *Gard. Chron.* li. 234, figs. 104-106 (1912). Cf. also p. 1901, note 3.

[3] Another tree at Kew, 18 ft. high, which was obtained from Osborne in 1879 and is labelled var. *betulaefolia*, scarcely differs from var. *viminalis.*

from Volxem in 1879, and is now about 35 ft. high. There are also fair specimens at Bradwell Grove (Oxon), Highnam, Tortworth, and Weston Birt. At Gisselfelde, Denmark, a fine tree is about 60 ft. high.

8. Var. *viminalis aurea*, Henry.

*Ulmus Rosseelsii*, Koch, *Dendrologie*, ii. pt. i. 412 (1872).
*Ulmus campestris*, var. *aurea*, Morren, *Belg. Hort.* 1866, p. 356, coloured plate; Lemaire, *Illust. Hort.* 1867, t. 513.
*Ulmus campestris*, var. *antarctica aurea*, Nicholson, in *Kew Handlist Trees*, ii. 135 (1896).

This is a sub-variety[1] of var. *viminalis*, in which the leaves are variegated with yellow. It appears to have originated about 1865 in Rosseel's nursery at Louvain; and is represented at Kew by a grafted tree about 20 ft. high.

9. Var. *viminalis marginata*,[2] Petzold and Kirchner, *Arb. Musc.* 556 (1864).

Var. *viminalis variegata*, Nicholson, in *Kew Handlist Trees*, ii. 137 (1896).

This is similar to var. *viminalis*, but the leaves are variegated with white. There are good specimens at Bayfordbury and Beauport, and a shrub at Kew about 10 ft. high. At Hamwood, Co. Meath, Elwes saw a tree about 30 ft. high in 1910.

## Distribution

The "English elm," by which name this species is usually known, is a native of southern England, growing in hedgerows, where it reproduces itself only by suckers. It is common in the Thames valley, extending southwards to the Isle of Wight, where it is abundant, and westwards to Devonshire, whence it ascends the Severn valley through Somerset and Gloucester to Worcester and Warwickshire, and the Wye valley to Herefordshire. It is unknown, except as a planted tree in the east of England, where it is replaced mainly by *U. nitens*; and is totally absent from Cornwall, where it is replaced by the Cornish elm.

This tree on account of the rarity with which it produces fertile seed, has been supposed to be not indigenous; and some writers, without any evidence, have asserted that it was introduced by the Romans. It is unknown in Italy, where the elm on which the vines are trained is quite a distinct variety.[3] So far as I can ascertain, it has not been seen in the wild state anywhere but in England; though an allied form occurs in the south-east of France.

The English elm has been largely planted in royal parks and public gardens in Spain. Evelyn,[4] speaking of this tree, states: "Those incomparable walks and vistas of them at Aranjuez, Casa del Campo, Madrid, the Escurial, and other places of delight belonging to the King and grandees of Spain, are planted with such, as they report Philip the Second to be brought out of England; before which (as that most honourable person the Earl of Sandwich, lately his Majesty's Ambassador

[1] It was exhibited at the London Horticultural Society in 1868 by Lee (*Gard. Chron.* 1868, pp. 914, 1038).
[2] A specimen of this in the Sherard Herbarium, Oxford, was gathered in the Chelsea Physic Garden in 1713.
[3] Cf. *U. nitens*, var. *italica*, p. 1892.
[4] *Sylva*, 33 (1679). The plantation of elms at Aranjuez is further described by Evelyn, *Sylva*, 303 (1706).

Extraordinary at that Court writ to me) it does not appear that there were any of those trees in all Spain. In that Princely Seat it is, that double rows of them are planted in many places for a league together in length, and some of them forty yards high." Philip II. created Aranjuez a royal residence in 1575, and probably planted the elms about this date.

I visited the Royal Park at Aranjuez in 1911, and found many fine avenues of elms, the largest trees measuring 100 to 120 ft. in height, and 12 to 16 ft. in girth. These differed in no respects from the English elms in Windsor Park or other places in England; and their deeply-furrowed dark-coloured bark has caused this species to be known in Spain as *olmo negro.* None of the trees which I saw looked old enough to date back to the reign of Philip II.; but I was informed that a very old tree which died and was removed in 1910, had a trunk 30 ft. in girth. This was probably one of the three ancient elms[1] noticed here by Lady Holland in 1803.

At Aranjuez, situated at a low altitude in a warm climate, on the banks of the Tagus, the English elm produces every year fertile seed in great abundance; and numerous seedlings were observed by me in a nursery in the park. At Madrid, on a plateau, 500 feet higher, with a cold temperature in spring, there are extensive groves of these elms in the Retiro Park; but the samaræ were empty and contained no seed, as is almost invariably the case in England; and M. Hickel noticed the poor seed on the elms of the same species at Toledo. Farther south in Spain, the English elm probably produces as good seed as at Aranjuez; and the prevalence of this tree in Spain must be due to the ease with which seedlings can be raised.

A large quantity of seed from the English elms in the Royal Park at Aranjuez was sent to me in May 1911, and sown immediately at Cambridge. In spite of the great drought, I raised numerous seedlings, which are very uniform in character, all having opposite leaves, indicating a pure species.

It is possible that the tradition of the introduction of the English elm into Spain from England is incorrect; and that this tree may be a true native of Spain, indigenous in the alluvial plains of the great rivers, now almost completely deforested. As explained under *U. nitens*, var. *italica*, the elm, which occurs wild in the mountains of Spain has different foliage, and it is readily distinguished by its light grey-coloured bark. (A. H.)

## Cultivation

The English elm is, so far as we are able to judge after careful study of the genus at home and abroad, a variety peculiar to the southern parts of England,[2] where it

[1] Cf. Earl of Ilchester, *Spanish Journal of Elizabeth, Lady Holland*, 73 (1910), where mention is made of three venerable trees ("either elms or oaks") in front of a small hunting villa built by Charles V. in the garden at Aranjuez. These, according to oral traditions, were said to have been planted by the Emperor Francis I. during his captivity, and by Philip II. Two were flourishing, but one was in a piteous state in 1803.

[2] We have never seen any specimens from the Continent which are identical with this species, except those planted in Spain which are mentioned above.

has grown in hedgerows from the earliest historical times in great abundance, and has propagated itself by suckers only. We can distinguish it from the other elms by its erect habit, by the rarity of suberose branches, by its dark red heartwood, and especially by the late period to which it holds its leaves, and by the bright golden colour which they assume, in seasons when the elms of all other varities are comparatively dull in colour. This latter character is constant on all soils and, so far back as I can remember, in all seasons. Two reasons may be suggested to account for its rarely producing seed. One is that reproduction by suckers during a very long period has diminished its floral fertility.[1] The other that being here at the northern limit of its range (if it is a true native of Spain), the seasons are rarely warm enough. Whatever may be its origin, it is one of the finest and most characteristic trees of those parts of England where it thrives; and both from its economic and ornamental value deserves to be propagated in the only way it can be kept true, either by suckers or by layers from stocks of the best type.

So far as we know there is no nursery in this country which has so produced it for many years past, and the results are only too evident in the elms which are now growing in almost all modern places. Soil no doubt has a material influence on the growth of this as of other trees, and a rather heavy and deep soil is necessary to bring it to perfection, but even on the thin dry oolite of the Cotswold hills, the true English elm retains its characteristics, and though slower in growth attains a greater height and bulk than any other tree, except perhaps the beech and the wych elm.

Its true value as a landscape tree may be best estimated by looking down from an eminence in almost any part of the valley of the Thames, or of the Severn below Worcester, during the latter half of November, when the bright golden colour of the lines of elms in the hedgerows, is one of the most striking scenes that England can produce.

Its economic value is also much greater than has been generally realised, for a tree that will, on grassland, of only moderate quality, without detriment to the adjoining pasture, and without any outlay but a moderate attention in trimming the lower branches, produce a log worth from £4 to £5 in a hundred years or less, cannot be ignored as an important element in the value of all grazing districts.

No author that I know has written on the propagation of the elm with so much personal knowledge as William Boutcher, nurseryman at Comely Garden, Edinburgh (now Comely Bank Nursery), whose *Treatise on Forest Trees*, first published in 1775, contains more exact observation on nursery work than most recent books. The elm was a great favourite of his, and I believe that if his advice had been more generally followed, the elms produced during the last century by nurserymen would be much finer than they generally are. Though observant woodmen know that the best and

[1] This is very unlikely, as the English elm in Spain produces ripe and fertile seed in abundance, yet suckers there very freely. Moreover, *U. nitens*, which suckers quite as much as the English elm, produces good seed in abundance in favourable years.

straightest timber trees are those naturally produced in hedges from suckers, yet the practice of budding or grafting elms on the stock of the wych elm has become so general, that I do not know a single commercial nursery in England to-day where true layered English elms can be procured; and though nurserymen assert that the budded trees will produce equally fine specimens, yet one has only to compare the younger trees planted in the last hundred years with the older ones, to see how inferior they generally are.

This practice, however, was recommended as long ago as Miller's time on the grounds mentioned by Hunter, in Evelyn's *Silva* 124 (1776), as follows:—"The practice of grafting will be found a valuable improvement of the English elm, if we consider the nature of the wych elm on which it is grafted. First the wych elm will not only grow to the largest size of all the sorts, but will grow the fastest. This is not to be wondered at, if we examine the root, which we shall find more fibrous than in any of the other elms. Now as all roots are of a spongy nature to receive the juices of the earth for the nourishment and growth of the tree, that tree must necessarily grow the fastest, whose root is most spongy and porous; and therefore the English elm being set upon the root of the wych will draw from the earth a greater quantity of nutriment. The English elm on this basis, will arrive at timber many years sooner than those raised by layers, and be also forced to a greater size."

Boutcher says:[1] "The English elm grafted on the Scots makes both a beautiful and valuable tree, yet it is still inferior in regularity of form, and loftiness of stature, to those raised from their own mother, and as every tree must in some measure partake of the stock on which it is grafted, so this has a near resemblance of the Scots elm in its bark even when young, and when old, like them, grows more loose spreading and less erect than the true English, though when young they are extremely beautiful. Here it may be necessary to observe a practice extremely common among ignorant nurserymen, which is cutting their English elm grafts from those on Scotch stocks, and which, indeed, have the fairest and plumpest buds (a plain indication from whence they immediately proceed, the buds of the Scotch being larger and more turgid than those of the English), but these gentlemen do not regard the quality of the plants they sell, so they are paid for them, or are ignorant that by repeating this practice the English elm may be brought so far to degenerate, as in many graftings this way, to differ very little from the Scots; therefore, whatever kind the stocks are on which you graft the English, let the grafts be taken from trees of the true kind, raised by layers of their own mother. This, however little attended to, nature plainly dictates." Further, he goes on to say: "It may also be proper to notice here that all elms planted in gardens and by the sides of walks, lawns, or avenues, ought to be on Scots stocks, as these produce no suckers, which the English, French, or Dutch do in such quantities as to make it very troublesome and expensive keeping such places clear of them and in good order."

The whole of Boutcher's article on the elm is so valuable that it should be studied by anyone wishing to plant elms.

[1] *Treatise on Forest Trees*, 12-13 (1784).

Many persons at the present time are anxious when planting to obtain quick results; yet the planting of an avenue is an operation which cannot be hurried, or let by contract to persons who have no future interest in it. No doubt the nature of the soil has much influence on the growth and habit of the tree, but even in poor, dry calcareous soils the difference between the old trees grown from suckers in hedges and the trees bought from nurseries is so great that no one can mistake them.

It also frequently happens that elms are budded on stocks raised from seedlings of foreign origin, and these vary so much in their habit and origin, and are usually so much inferior to the true English elm that the results are what we usually see in modern plantings. There is no better proof of this than the young trees planted to fill the gaps in the Long Walk at Windsor, which are a lamentable illustration of carelessness in the propagation of elms.

No tree possesses the power of suckering[1] to a greater extent than the English elm, the roots often extending 50 yards or more; and in order to procure a quantity of young plants, it is only necessary to shut up a small field surrounded by elms of good type, and transplant the most vigorous suckers into a nursery, where they can be pruned and cultivated until 6 to 10 ft. high; or if desired to have them larger they may be transplanted every two or three years and safely moved when as much as 15 or 20 ft. high.

Mr. Knight published[2] in 1840 a method of propagating elms by using, as cuttings, slender shoots which were pulled out from the trunk near the ground, and then reduced to about an inch in length, with a single leaf at the apex.

If, however, it is desired to produce new varieties, the raising of elms from seed is a very simple matter, provided that the seeds are sown as soon as they are ripe, when they germinate in a few days, and make strong plants in the first year. The variation of most elms that I have sown, except the wych elm, is very great, and natural cross-fertilisation no doubt accounts for this.

The leaves do not show their true character at first, and it would be difficult to judge of their fitness for planting until they have attained a considerable size. If good timber trees are wanted, no variety surpasses the true English elm on its own roots in its own district. For the maritime climate of south-western England, Scotland, and Ireland I would recommend the true Cornish elm on its own roots. For Scotland, and the North of England, Boutcher was probably right in preferring trees of a good local variety budded on the stock of seedling wych elms, but never on that of imported seedlings.

The bark of dead or sickly elms will usually be found to contain the elm-bark beetle (*Scolytus destructor*) in one stage or another. Whether this insect attacks healthy trees has never yet been satisfactorily determined; and the remarks on this point of Dr. T. A. Chapman, an entomologist whose actual experience of Scolytidæ

[1] In *Gard. Chron.* 1872, p. 603, fig. 504, an article by the late W. Ingram, gardener at Belvoir Castle, has a figure showing the extraordinary root-development of an elm growing on the edge of a quarry.

[2] Loudon, *Gard. Mag.* xvi. 474 (1840).

is probably greater than that of all other English writers on the subject, seem worth quoting. He says:[1] "I do not remember having seen a *felled* elm trunk that *S. destructor* had not attacked, frequently whilst still trying to throw out shoots; yet I have never seen a trace of it in healthy growing trees; these are supposed to resent and repel the attacks of the *Hylesinidæ* by pouring out sap into their burrows; and, in the case of *S. pruni*, I have observed burrows less than an inch long, some of which, containing a few eggs already laid, had been abandoned uncompleted by the beetles, apparently on account of the presence of a fluid which must have been sap, as no rain had fallen to account for it; these burrows had been formed in bark that was still nearly healthy, though near some dying bark which had doubtless attracted the beetles." The following is an account of Mr. J. Edwards' experience. Some years ago a large and apparently sound limb of a big elm at Colesborne was broken off, and it was allowed to remain where it fell. This fallen limb in due course showed traces of attack by *Scolytus*, of which, judging from the number of holes in the bark, there must have been some thousands. Up to the present no signs of *Scolytus* can be seen on the tree, which could not have been the case if any of the beetles bred in the bark of the fallen limb had established themselves either in the trunk or the main branches. As a preventive measure it has been recommended[2] that the bark be smeared with a fermenting mixture of cow-dung, slaked lime, bullock's blood, and tobacco; but I am inclined to think that steps calculated to maintain or increase the vigour of the tree would be more likely to succeed than any bark-dressing whatever. With the view of lessening the numbers of this pest, sickly trees and felled timber, as well as fallen limbs, should be got rid of without delay.

The branches of old elm trees are liable to fall[3] without warning in calm weather, especially after heavy rains. Though in some cases the branches have been weakened by fungoid attacks, there is no doubt that in other cases the tree was quite sound when the branches fell. Many fatal accidents have been reported from this cause, as at Powis Castle[4] in 1899, and in Kensington Gardens a few years ago. The English elm seems to be more liable to drop its branches than any of the other kinds.

## Elm as a Woodland Tree

The elm has seldom been considered by writers on forestry as a woodland tree in this country, and is rarely found in plantations[5] except scattered among other trees. I have seen, however, at several places in the Midlands small areas of

[1] *Entomologist's Monthly Magazine*, vi. 127 (1869-1870). An interesting account of the ravages of this beetle was given by W. S. Macleay, in an article entitled "Abstract of a Report on the State of the Elm Trees in St. James's and Hyde Parks," which appeared in *Edin. Phil. Journ.* xi. 123 (1824).

[2] W. R. Fisher, Schlich's *Manual of Forestry*, iv. 278 (1907).

[3] Cf. correspondence in *Gard. Chron.* xxxviii. 119, 134, 252, 268, 331 (1905), and xxxix. 11 (1906).

[4] *Gard. Chron.* xxv. 340 (1899).

[5] In the Forest of Dean, a mixed larch and oak plantation, aged 39 years, in which there were a few English elms, showed the following average measurements taken by Mr. A. P. Long in 1911: larch, 59 ft. high, 41 in. girth; English elm, 54 ft. high, 37 in. girth; oak, 40 ft. high, 26 in. girth.

almost pure elm, probably grown from suckers, which convince me that on suitable soil and near a good market it might prove to be one of the most profitable forms of woodland.

At Brampton, near Huntingdon, I saw on the roadside a belt of elms of moderate quality which stood thick on the ground, and am indebted to M. D. Barkley, Esq., of Huntingdon, for the following particulars:—"The elm belt at Brampton contained 1 acre and 30 poles; there were 330 trees containing 8750 feet of timber. These were left much too thick on the ground to develop properly." I should expect on land worth 20s. to 25s. per acre, in the vale of Gloucester, or on the Oxford clay, about 6000 to 8000 cubic feet per acre from elms grown from suckers that had been properly trimmed, and this should be produced in from 70 to 100 years, according to the quality of the land. Assuming only 9d. per foot to be the price, such a crop would realise £250 to £300 per acre at 100 years, and no other tree except black Italian poplar would be likely to approach this result.

A good system on those properties which are managed systematically in elm-growing districts is to leave in every chain of hedgerow two or three of the best suckers to grow into timber; and these are carefully trimmed every few years to prevent their branches from spreading, and are felled when they contain 50 to 100 feet of timber. Such hedgerow trees are not detrimental to grass land, and produce a sum equal in some cases to 15 or 20 per cent of the rent of the land adjoining.

With regard to the elm as a forest tree on the Continent, Huberty is of opinion that it can only be cultivated profitably on fresh fertile soils, and in situations where it can extend both its branches and its roots without hindrance. *U. montana* is less exacting in its demands on the soil than the other species, and is extremely hardy, having been unaffected by the severe spring frost ($-4°$ Cent.) which severely injured, in the Ardennes on May 26-27, oak, ash, chestnut, Robinia, and larch. *U. montana* keeps pace in youth with the beech, though eventually it is suppressed by the latter in the Forêt de Soignes. The elm is liable to be split by severe frost, when the trunk is exposed by a fall of timber. It requires light and room when it passes the youthful stage. Mouillefert says[1] that the elm, which he calls *U. campestris* (but which is *U. nitens*), forms nearly pure groves in the woods attached to the Agricultural School at Grignon; while Fliche states[2] that it forms a considerable proportion of the forests in the valleys of the Saône and Adour.

## Remarkable Trees

Loudon says that the oldest trees on record are perhaps those at Mongewell Park in Oxfordshire, once the property of the Bishop of Durham, but the figure of these in Strutt's work (*Sylva Britannica*, plate XVI.) shows that they were neither

[1] *Essences Forestières*, 160 (1903).
[2] In Mathieu, *Flore Forestière*, 300 (1897). Cf. p. 1897.

very large nor very old. Of other very old trees mentioned by Loudon the Crawley elm, also figured by Strutt (plate XXII.) is perhaps the finest remaining. It stands in the Brighton road at one end of the town, and was in 1838, 70 ft. high and 61 ft. in girth at the ground. It was then hollow, and measured 31 ft. round the inside, which was closed by a door, and it is said that a poor woman once gave birth to a child in the hollow tree. Now it is a venerable wreck with some living branches; but the great swelling at the base which formed its immense girth is much decayed, and owing to the suckers which have sprung from the ground round it, it is difficult to measure. Miss Smith, who lives opposite the tree, told me that in her youth she had seen twelve people seated at tea in the hollow trunk. The other English elm figured by Strutt, which grew at Chipstead in Kent, was not a very large tree, and died soon after he drew it. None of the other elms mentioned by Loudon seem to have been as large as several which I have measured, as will be seen in the following list drawn up in the alphabetical order of the counties:—

In Bedfordshire there are fine trees at Woburn Abbey and Wrest Park, but most of the big elms in this county are *U. nitens.*

In Berks the best that I have seen are in Windsor Park, where the "Long Walk," leading from the Castle gates to the statue, is still one of the finest and most imposing avenues in the world. I am indebted to Mr. W. C. Squires, who has charge of the trees in Windsor Park, for reading and correcting when necessary, the following account:—The planting of the Long Walk is ascribed by Menzies[1] to the example of Evelyn, who states that in 1664 he planted some land at Says Court with elms, "being the same year that the elmes were planted by His Majesty in Greenwich Park." In 1670 Evelyn visited Windsor. "King Charles II.," he says, "passed most of his time in hunting the stag and in walking in the Park, which he was now planting with rows of trees." In 1678, 1679, and 1680, Evelyn was at Windsor again. In 1680 a survey of the land between the Castle and the Great Park was made, and the intervening fields were purchased at a cost of £1242:4:9. We may safely presume that the planting of the Long Walk was at once commenced. The distance from the Castle to the statue is 2¾ miles. There were originally 1652 trees. The distance between the two inner rows is 50 yards. The trees are 30 feet apart from each other in two lines also 30 ft. apart, perhaps rather too close for the health of the trees. At the end nearest the Castle the soil is loam from 10 to 15 feet deep overlying chalk, and here the trees have done well and are mostly still sound. Where they have room enough, as in the adjoining part of the Park, there are many really fine trees up to 120 ft. in height and 15 ft. or 16 ft. in girth; but farther on where the soil is clay overlying gravel, and beyond the double gates where it is heavy clay, the trees are much smaller, and many have died at various periods and been replaced with elms of varied character, many of which now are, and never will be anything but an eyesore. Previous to 1861 the condition of this part of the avenue had attracted attention, and a good deal of correspondence on the subject had taken

[1] *History of Windsor Great Park and Forest*, 17 (1864).

place, which will be found in the Report of the Commissioners of Woods to Parliament in March 1861.

Along the Thames valley throughout Berks there are abundance of splendid elms of the best type, but as far as I know none so large as exist farther west.

In Bucks, the same remarks apply to all the alluvial soil; and those in the playing fields at Eton must recall happy days to numbers of our readers, though like some of the older elms in the country they are decaying and being replaced by inferior grafted trees. The two best trees in 1907 were, 113 ft. by 17 ft. 8 in. and 115 ft. by 17 ft. 3 in., as measured by Henry.

In Cambridgeshire we are out of the region of native English elms, but there are some of good size in the College Backs at Cambridge. Two of the largest, which grew in the grounds of St. John's College, and were known as the Sisters, were blown down in the great gale of 14th October 1881. These were recorded at the time by Mr. J. W. Clark,[1] as 10 ft. in girth, and about 130 ft. in height to the topmost branches. Prof. Hughes[2] counted 218 rings on the base of one of the fallen trunks, showing that these elms were planted a few years after the College grounds were laid out, as is generally supposed, in 1630. Perhaps the finest English elm now at Cambridge is in the grounds of King's College, which measured 130 ft. by 13 ft. in 1906.

In Cheshire the only fair elms that I have seen are in the park at Eaton Hall; but I must confess that I have unduly neglected this county, where, perhaps on account of the soil, trees do not generally attain a large size.

In Cornwall true English elms are rare, if not entirely absent, all those which I have seen being of other species.[3]

In Cumberland, and in Derbyshire we are too far north to see this tree at home, but there are some fine trees in Kedleston Park.

In Devonshire there are many splendid elms, among which I have seen none finer than those at Powderham. Plate 404 shows a tree which, in 1902, when it was photographed for this work, measured 125 ft. by 22 ft., but when I revisited Powderham in 1906, it had lost a large limb.

In Dorsetshire the finest elms I know of are at Melbury, where a row of seven, known as the Seven Sisters, stand in front of the house, on what was described by the gardener as a thin, but evidently a very fertile soil. The centre tree in this row measured 130 ft. by 17 ft. 10 in. in 1906.

In Essex[4] there are great numbers of this elm, which may be called the prevalent tree of the county, and it is difficult to select the finest. Mr. J. C. Shenstone

[1] In *Cambridge Review*, 26th October 1881.

[2] In *Comm. Antiq. Soc. Cambridge*, xxiv. page xxxix (1884).

[3] Cf. pp. 1884, 1889.

[4] Holinshed, *Chronicles*, ii. cap. 22 (1586), says:—"Of elme we have great store in everie high waie and elsewhere, yet have I not seene there of anie together in woods or forrests, but where they have beene first planted and then suffered to spread at their own willes. Of all the elms that ever I saw, those in the south side of Dover Court, in Essex, near Harwich, are the most notable, for they grow in such a crooked maner, that they are almost apt for nothing else but navie timber, great ordinance and beetels: and such thereto is their naturall qualitie, that being used in the said behalfe, they continue longer, and more long than anie the like trees in whatsoever parcell else of this land, without cuphar, shaking, or cleaving as I find."

tells me that in the avenue at Earls Colne he measured a hollow tree 124 ft. by 24 ft. 4 in., but I cannot be sure that this is a true English elm. At Boreham House there is a fine avenue leading to the main road from London to Chelmsford, in which Mr. Shenstone in 1890 measured a tree 132 ft. by 20 ft. which he considered to be the tallest in Essex; but when I was there in 1907 I could find none taller than a double-stemmed tree of which the one half was 125 ft. by 17 ft. and the other broken off at about 25 feet was 23 feet in girth. Nearer the coast elms grow to a great size on the fertile brick earth, but usually are either *U. major* or *U. nitens.*

Gloucestershire holds the record for the tallest elms that I know of, which are at Forthampton Court, near Tewkesbury, the seat of the late J. Reginald Yorke, Esq. A tree, of which I saw the stump, was blown down in March 1895, and was reported by Mr. French, the gardener, in the *Gardener's Magazine* for 6th April of that year, to have been 150 ft. by 20 ft., as measured on the ground with a tape. Mr. Yorke confirmed this statement, assuring me that it was considerably taller than one which I measured in 1906 at the same place, and found to be about 140 ft. by 20 ft. A butt covered with ivy, about 23 ft. in girth over bark, is broken off about 30 ft. high; and its top is said to have been 40 yards long as it lay.

At Badminton there are two immense elms close to the kennels, the largest of which measures 30 ft. in girth or 26 ft. if taken below a large burr. An immense limb has broken off low down; and the top of this tree as well as that of its neighbour, which is 24 ft. in girth, were blown off many years ago. There are many other fine elms and oaks in the park here, but I saw none of equal girth to the one above mentioned.

A splendid old elm, of which Col. Thynne has sent me a photograph, grows near Thornbury Castle, and measures 93 ft. by 27½ ft. Elms containing from 600 to 1000 cubic feet of timber are mentioned by timber merchants as having been felled on several occasions in the vale of Gloucester. The tree mentioned by Loudon under the name of Piffe's Elm, which he says was in 1783 the largest tree in Gloucestershire, was quite small in comparison with many now living. I saw two blown down at Sandywell Park, the seat of C. W. Lawrence, Esq., which each contained over 600 feet.

In the valley of the Coln, close to an old Manor House called Compton Casey, belonging to the Earl of Eldon, there are eleven (formerly twelve) fine elms growing in a row only 90 yards long, which average 13½ ft. in girth and about 115 ft. high. The largest is 16 ft. 9 in. in girth.

At Cirencester, at Ampney, at Williamstrip, at Barnsley, and other places, there are many large trees, which have, as they grow old, a tendency to become very burry at the base, but on the more brashy soil of the Cotswolds they rarely exceed about 110 ft. by 15 ft.

At Huntly Manor, Gloucestershire, the seat of B. St. John Ackers, Esq., there is a remarkable tree of which the text-plate gives a better idea than any description. The watercourse when I saw it in September 1910 was nearly dry, but runs freely most of the year. Mr. Ackers tells me that if you put a ferret into the root on one side, it can pass through and come out on the other, which shows that the butt is partly hollow, though the tree looks sound and is about 100 ft. high. It is difficult to explain the origin of this curious freak.

ACTUAL SPAN OF THE TWO FEET IS 5½ FEET.
DISTANCE FROM A TO B. 12½ FEET.

In Hampshire the fertile valleys are full of big trees; but on the chalk they are of a more stunted character, and are often *U. major*. The largest true English elm in this county was recorded in 1887 at Broadlands, near Romsey, and, according to Mr. J. Smith of that town, was 110 ft. by 24½ ft.

Herefordshire, though not so remarkable for the size of its elms as for its oaks, is full of fine trees, some of which were recorded by Dr. Bull in the volume of the *Woolhope Transactions* for 1868; and in later volumes other elms are mentioned at Croft Castle and Longworth. The largest I have measured is in the deer park at Hampton Court, a tree with two trunks from the same root, girthing 20 ft. and 16 ft. respectively, and 125 ft. in height in 1908.

The elms of Hertfordshire are mostly of the glabrous-leaved species, and are described under *U. nitens*; but the finest English elm I have seen is in Hatfield Park, and measured in 1905 120 ft. by 20½ ft. The girth in 1911 was found by Mr. Barton to be 21½ ft. At Bayford Church there is a fine tree, apparently quite sound, which was measured by Mr. H. Clinton-Baker in 1911 as 110 ft. by 18 ft. 9 in. Near Bayford Lodge another is 93 ft. by 21 ft. 9 in.; and a record exists that this tree had a girth of 17 ft. 1 in. in 1813.

Kent has a great variety of elms, which appear to be mostly of a different type to those of the Thames and Severn valleys, and as a rule are planted in parks and pleasure grounds, and not natural in the hedgerows. The best that I have measured is one at Eastwell Park, which in 1907 was 115 ft. by about 20 ft. in girth.

In Lancashire, Leicestershire, and Lincolnshire I have not seen any English elms remarkable for size or beauty. The majority of the trees in these counties look more like hybrid seedlings of foreign origin than true English elms.

On November 23, 1910, I saw at Well Vale, Lincolnshire, the seat of Mr. W. H. Rawnsley, some English elms (Plate 405) which had preserved their leaves quite green after several nights of frost, when the thermometer sank to 17° Fahr. These elms, one of which measured 108 ft. by 8 ft. 2 in., seem to have been grown from suckers; and though they retain their leaves abnormally late in the season—in 1911 till the first week in December—they are English elms, and differ entirely from the Kidbrook elm referred to, p. 1896.

In Middlesex, and all round London, the majority of the elms appear, like those of the London parks, to be foreign seedlings or hybrids; and Loudon did not record a single one of any great size which seems to exist at present, though he gave many details of trees at Hampstead, Fulham, and elsewhere. Though there are many large English elms at Hampton Court, Richmond, Syon, and in Kew Gardens, they seem to be suffering more or less like those in Hyde Park from smoke and old age, and none of the younger trees which have been supplied from nurseries to fill their place appear likely to develop into first-class elms. The tallest that I have measured is at Chiswick House, and this in 1904 was a healthy sound tree 137 ft. by 12½ ft. Another in the grounds of Fulham Palace is 120 ft. by 18 ft. Henry saw one at Hampton Court, with a broken top, girthing 27 ft. 9 in. in 1910; and another at Osterley Park, 108 ft. by 16 ft. 4 in. in 1907.

In Monmouthshire and in South Wales generally, though elms are common enough they are no longer a characteristic hedgerow tree.

In Norfolk I cannot remember to have seen a single really first-class English elm, most of the hedgerow trees being either *U. nitens* or the small-leaved elm (*U. minor*), which, so far as I have seen, is usually a comparatively small and ill-shaped though picturesque tree. Marshall[1] was, I think, not far wrong when he said that there was not, generally speaking, a good elm in the county of Norfolk; and the only elm mentioned by Grigor of great size was what he calls an English elm over 20 ft. in girth.

In Northamptonshire, where oak and ash grow so well, I have seen few very striking elms; and the same may be said of Nottinghamshire, where in the beautiful parks of Thoresby, Rufford, Welbeck, and Clumber they are not so conspicuous a feature as the oaks. At Althorp there are many splendid trees, three of which were measured in 1893 by Mr. F. Mitchell[2] as follows:—(No. 1) In the pleasure grounds, 117 ft. by 20 ft., with two stems, containing 924 ft. (No. 5) By the carriage drive, 105 ft. by 19 ft. 8 in., also divided, and containing 841 ft. (No. 17) West of Harleston House, 110 ft. by 18½ ft., containing 715 ft. I verified these measurements in 1904, and found that in eleven years they had made but little increase.

[1] *Planting*, ii. 431 (1796). Cf. Loudon, p. 1383.

[2] *Trans. Roy. Scott. Arb. Soc.* xiii. 90 (1893).

Oxfordshire may be called *par excellence* the county of the elm, and it would be hard to find a place in it, except in the barest parts of the Cotswolds, where they do not form a conspicuous feature of the landscape. Among the parks where they grow best the following may be mentioned:—Barrington, Blenheim, Cornbury, Fawley Court, Heythrop, Thame. At Brightwell Park, among a number of large old elms, I measured in 1905 two trees, 119 ft. by 17½ ft. and 110 ft. by 19 ft. respectively; and probably larger ones can be found. One of the oldest in this county is the "Tubney Tree" at a roadside meet of the Old Berkshire hounds, eight miles from Oxford. This has larger leaves than usual, but no suberose branches. Its trunk is 27 ft. in girth, and still fairly sound, and at about 20 ft. from the ground divides into five ascending limbs, which were topped about twenty-five years ago, and have thrown out many healthy branches. The Rev. H. J. Bidder of St. John's College has a water-colour drawing of this tree, made before it was topped. In Magdalen College Park at Oxford there are some very tall English elms, one being, in 1905, 134 ft. by 13 ft. 5 in.

In Shropshire, notwithstanding the splendid soil, the elms are not so large as the oaks; and in Somersetshire[1] I have neither seen nor heard of any elms of larger size than one in the park at Dunster Castle, which in 1904 measured 120 ft. by 22 ft. The same remarks apply to Staffordshire and Suffolk.

Surrey and Sussex are more remarkable for conifers than for elms, and my local knowledge is not enough to enable me to indicate where the finest elms are to be found. Among the innumerable residences with which these counties are so thickly studded, no part of England would better repay the researches of any one interested in trees; and we have probably missed some of great rarity, though we have visited a great number of places in these counties. Henry measured in 1905 in Betchworth Park, near Dorking, an English elm 122 ft. in height and 23 ft. in girth.

Warwickshire and Worcestershire are, like Oxford and Gloucester, renowned for large and fine elms. Loudon recorded at Coombe Abbey a tree, 150 ft. high and 28½ ft. in girth, which I fear no longer exists, as I could find none nearly as large now standing at that place.

In Mr. Berkeley's park at Spetchley, near Worcester, there are great numbers of large trees, about 200 years of age. I measured one 108 ft. by 21½ ft.; another 127 ft. by 14 ft.; and in front of the house, one with very large spreading branches, 125 ft. by 20 ft. The largest in girth was 22 ft. 8 in., broken off at about 20 ft. up. At Ombersley Court, 6 miles north of Worcester, Lord Sandys showed me the remains of a huge tree, only the ivy-covered stump of which is left, and measures about 23 ft. in girth. From it about 1700 ft. of timber was sold; and I was assured by Mr. Groom of Hereford that this tree contained 2000 ft. of timber. A little way off I measured a tree, which is perhaps the largest sound English elm now standing, and which I found to be 130 ft. high by 23½ ft. in girth; another in the park was 125 ft. by 21½ ft.

[1] Babbage, in Loudon, *Gard. Mag.* xvii. 356 (1841), and xviii. 488 (1842) gave particulars of English elms at Nettlecombe Court—one, only eighty years old, having a clean trunk 32 ft. long, with a middle girth of 12 ft., containing 200 cubic ft. of timber; and another, sixty-nine years old, containing 360 ft. in the trunk, and 120 ft. in the top and branches.

Wiltshire, like the adjoining counties, is full of elms, which attain a large size where the soil is deep. The best that I have measured is at Corsham Court, close to the house, and was 123 ft. by 20 ft. in 1905.

In Yorkshire I have seen no remarkable English elms, even at Castle Howard or Studley, where the wych elm grows to perfection; but in the finer soils of this great county there are no doubt many good trees which we have not heard of.

In Wales the best place for English elms that I know is at Maeslwych, near the borders of Herefordshire, in the rich valley of the Wye, where there is an avenue containing many fine trees, which attain 110 to 120 ft. in height. In south Wales, where the demand for elm timber is probably greater than anywhere in England, I have seen none worthy of special notice; and in central and north Wales the wych elm is the prevalent species.

In Scotland we know few English elms[1] of importance; but Mr. Renwick measured two trees at Loudon Castle, Ayrshire, 107 ft. by 15 ft. 4 in., and 105 ft. by 16 ft. 4 in., in 1908; and at Milton Lockhart, Lanarkshire, a tree 90 ft. by 15 ft. 1 in. in 1911. Most of the elms in Scotland are *U. montana*.

In Ireland I find no records in my journals of true English elms, except at Adare, the seat of the Earl of Dunraven, where, on the river bank, the finest English elm I have seen in Ireland, was about 120 ft. by 13½ ft. in 1909. There is a beautiful tree at Riverview, Ferrybank, Waterford, of which Mr. T. A. Penrose sent us a photograph. It is growing in deep loamy soil, and measured 91 ft. in height and 14 ft. in girth in 1912. Mr. R. A. Phillips records an English elm at Loughrea, 100 ft. by 12 ft. Most of the hedgerow elms in the east are scrubby and corky twigged. In the south of Ireland the glabrous and Cornish varieties are more generally grown, but only in the demesnes of large landowners, and the timber is here so little valued that when required for large works it is sometimes imported from England.

## Timber

After the oak and ash, elm is the most important of the non-coniferous timbers grown in England, but its value per foot is inferior to either of them. It varies immensely in different districts, and according to the distance from a sawmill; but it may be put at about 1s. per foot for the best quality of trees, containing from 50 to 200 cubic feet, and 8d. to 10d. for old, faulty, and ill-shaped trees.

Large old elms are often defective in some part of their trunk on account of their liability to lose large limbs from wind; and where limbs have been blown off, or cut off, the stump generally decays, and its decay extends into the body of the tree, unless the wound has been dressed with tar, creosote, or some preservative, and covered with lead or cement.

The removal of elm trees, often weighing many tons, is expensive and difficult; and though engines are now used for the purpose, it is often found economical, where a number are to be felled, to erect a sawmill close by, and cut them up

[1] Walker, *Essays Nat. Hist.* 70 (1812), says: "We have no English elms in Scotland of an old date or of a large size."

into such sizes as they are fit for on the spot; and sometimes large trees are sawn on the ground lengthways into halves so as to make them more portable.

Laslett[1] says that if used where it is constantly under water, or in situations where it is always kept dry, elm is one of the most durable of timbers, but it decays rapidly under any other circumstances. I have seen, however, many old barns and sheds in Hants, Wilts, and Essex covered with elm weather-boards, which, when tarred, have remained sound for a very long period, perhaps over a century.

Elm timber should not be left lying long after it is cut; and when converted into boards must be carefully stacked to dry under cover, as, though it does not crack so much as oak or ash, it twists and warps much more.

Until iron water-pipes were introduced, elm was the favourite wood for this purpose, and old water-pipes are often dug up in London and elsewhere which have remained sound for 200 years or more. I saw one of these at Chirk Castle, which was found at Clerkenwell in 1898, by the new River Water Company, of which Sir Hugh Myddleton, ancestor of the present owner, R. Myddleton, Esq., was the founder. This is said to have been laid down in 1613, and measures 8 ft. long by 14 in. in diameter at the big end. The bore, which is much encrusted by a limy deposit, is 9 in. in diameter at one end and 6 in. at the other. Similar water-pipes are preserved in the museum at Kew and in the Surveyors' Institution in London.

Elm was also formerly much used for the blocks and keels of ships; and most of the large elms cut in Essex are used, I believe, for piles, groynes, wharves, and for the timbering of the sluices, which carry off the flood-water of the Essex marshes, as well as for weather and coffin boards.

In all the counties nearest to south Wales the greater part of the elms which are blown down or felled, are used for making boxes for tin-plates, as it has been found that this wood does not discolour the plates, and there is less waste in converting for this purpose than for any other, the boards being small and thin. The price in these counties therefore depends very much on the prosperity of the tin-plate trade.

A very large quantity of elm is also used for coffin-boards though, according to "Acorn," whose account of this timber is well worth consulting, this branch of trade has lately suffered like all our home industries from foreign competition. The sizes required for this purpose, he says, are 6½ to 7 ft. long by 11 to 14 in. wide, and ¾ to 1 in. thick, and cost about half as much as oak, though the figure is in some cases very handsome when polished.

For the seats of Windsor chairs, elm is also largely used. The greater part of the tip-waggons used by railway and harbour contractors were formerly made of elm, though now in many cases iron and imported timber are preferred. Barge-builders, box-makers, and wheelwrights also consume a large quantity of elm, the latter preferring it to oak for the stocks or hubs of wheels, as well as for felloes of large size; and here again Canadian rock elm (*Ulmus racemosa*) is now a formidable competitor, which tends to keep down the price of home-grown timber.

With regard to the best quality of elm for wheelwrights I have made many inquiries among country tradesmen, whose opinions differ according to the locality.

[1] *Timber and Timber Trees*, 218 (1875).

Wych elm is generally considered, even in the south, the toughest and best for wheel stocks, trees of small size being cut into lengths and bored through the heart to keep them from splitting as they dry. Most country wheelwrights who have a reputation for the durability and soundness of their waggons prefer to select the trees when growing in their own district; but I have not been able to find any agreement as to what they consider the best timber.[1]

As one goes north, however, the comparative value of wych elm as compared with English and smooth-leaved elm, increases; and it may be said generally that whilst in the southern and midland counties, English elm is worth a third more than wych elm; in the north, where the former is falling off in size and quality, they are equally valued, and in Scotland wych elm, the native tree of the country, is preferred; and when speaking of elm in Scotland, wych is generally implied. Many wheelwrights speak of "bastard elm," a term which they apply to any trees which do not show in their timber the colour and character which is associated with English elm.

In France the same uncertainty and difference of opinion seems to exist; and I was told by an experienced waggon-builder in Savoy that he preferred the elm stocks imported from Angoulême, to any that were grown in the east of France.

For ornamental purposes or furniture, elm is little used in England, probably because it is liable to the attacks of wood-boring larvæ. Large tables were sometimes made of a solid elm plank, but few examples remain of ancient furniture made from this wood. One of the secondary staircases at Hatfield House is made of elm, which has not warped and shows handsome figure, but it does not seem to be of great antiquity. The best ornamental use, however, to which elm wood can be put is as veneer, cut from the mottled, waved, and curiously veined burrs, which are found not uncommonly on the trunks of old trees, especially of the wych elm. These when cut into board, crack, shrink, and warp so badly that they can hardly be used in cabinetmaking, but in the form of veneer they are sometimes of very great beauty. In rare cases, the outside slab of an elm tree is found full of markings like that of bird's-eye maple, which are due to undeveloped buds. Mr. E. Gimson, of Daneway, near Cirencester, whose reputation as an artistic worker in fine woods is great, showed an elm cabinet with a very rare mottled figure, made from a tree grown in the neighbourhood, which I have never seen equalled.

In Scotland I procured very fine examples of wych elm burr which, when polished, show a deep rich reddish-brown colour, and from which all the lower part of a most handsome bookcase was made for me.

When visiting the palace of Fontainebleau, I was shown the cradle presented by the city of Paris to Napoleon I. for his son and heir. I was assured by the guardian that this was made from a rare and costly foreign wood, but though overloaded with gilt metal-work, I recognised an elm burr, no better than many which are allowed to rot in our English parks, from one of which it probably came. Very handsome solid table-tops are also made from similar burrs, but more often they are too full of holes and rindgalls to be of any value. (H. J. E.)

[1] Cf. p. 1886 for the wood of the Dutch elm, p. 1889 for that of the Cornish elm, and p. 1902, note 3, for that of the small-leaved elm.

## ULMUS JAPONICA

*Ulmus japonica*, Sargent, *Trees and Shrubs*, ii. 1, t. 101 (1907).
*Ulmus campestris*, var. *japonica*, Sargent *ex* Rehder, in Bailey, *Cycl. Amer. Hort.* iv. 1882 (1902).
*Ulmus campestris*, var. *lævis*, Fr. Schmidt, in *Mém. Acad. Sci. St. Pétersb.* xii. 174 (1868) (in part).
*Ulmus campestris*, var. *vulgaris*, Shirasawa, *Icon. Ess. Forest. Japon*, ii. t. 15, figs. 10-21 (1908).
*Ulmus campestris*, Komarov, *Flora Manshuriæ*, ii. 82 (1903) (not Linnæus).

A tree, attaining in Japan about 100 feet in height, and 12 feet in girth. Young branchlets pale brown, often roughened with minute tubercles or ridges, and covered with dense soft pubescence, more or less retained on the branchlets of the second year, which are fissured and roughened with slight corky ridges; in some specimens,[1] prominent corky ridges are developed in the second and third years. Leaves (Plate 411, Fig. 4) obovate or elliptic, about 3 to 4 in. long, and 1½ to 2½ in. wide, oblique at the base, acuminate at the apex; upper surface scabrous with numerous tubercles and scattered bristle-like hairs; lower surface pale green, pubescent throughout with white short hairs, conspicuous on the midrib and nerves, and forming slight axil-tufts at their junctions; lateral nerves twelve to sixteen pairs, prominent beneath, occasionally forking before reaching the margin, which is coarsely biserrate; petiole $\frac{1}{6}$ to ¼ in. long, densely pubescent.

Flowers nearly sessile, regularly tetramerous, with four sepals and four stamens. Fruit narrowly obovate-oblong gradually tapering to the base, glabrous, non-ciliate, about ¾ in. long, and ⅜ in. wide near the apex; notch open, triangular, with the stigmas slightly incurved; seed touching at its apex the base of the notch.

This species is readily distinguished by the peculiar fawn colour of the branchlets in their first season.

This species, which is closely related to the European *U. nitens*, is a native of Japan, Manchuria, and Amurland. Komarov states that it is common throughout Manchuria, growing along rivers and on hill-sides, usually solitary, but occasionally forming small woods. (A. H.)

In Japan, this elm is said by Sargent,[2] to occur in the mountain ranges of Hondo at 3000 to 5000 feet, where it is a small tree with the branchlets often conspicuously winged. In Hokkaido this tree is much more abundant, growing on the plains almost at sea-level, and on the lower slopes of the mountains. In the streets and environs of Sapporo it is the most conspicuous tree, and attains 80 to 90 ft. in height by 10 to 12 ft. in girth. I measured a tree close to the station at Iwamigawa 110 ft. by 11 ft., with a clean stem 40 to 50 ft. high; but as a rule it has a more branching and pendulous habit, which reminded me, as it did Sargent, of the American elm, but which is as variable as the habit of *U. nitens*. I found a tree of this species growing close to one of *U. montana* in the virgin forest near

[1] Collected by Elwes, in virgin forests at Asahigawa in central Yezo. Cf. also Shirasawa's figure, *Icon. Ess. Forest. Japon*, ii. t. 15 (1908).

[2] In *Garden and Forest*, vi. 323, fig. 50 (1893), and *Forest Flora of Japan*, 57, t. 18 (1894), where a tree growing near Sapporo is figured.

Asahigawa, and noted that though the bark of the two was indistinguishable, the leaves were very distinct. I did not, however, notice that it produced suckers.

*U. japonica* was introduced into the Arnold Arboretum by seeds sent in 1895 from Sapporo by Professor Miyabe, and has grown there rapidly. Professor Sargent informs us that in October 1911, the trees which were raised vary from 17 to 30 ft. in height, and 1 ft. 2 in. to 2 ft. 4 in. in girth. Some of them produced flowers in the spring of 1907. This elm is perfectly hardy in eastern Massachusetts, where it promises to become an ornamental tree of great value. It was subsequently introduced into Europe by Späth[1] in 1900.

There are two trees at Kew, one obtained from the Arnold Arboretum in 1897, which is now about 12 feet high; and another obtained from Späth in 1900. There is also a good specimen in the Edinburgh Botanic Garden, about 15 feet high. (H. J. E.)

## ULMUS ALATA, Wahoo, Winged Elm

*Ulmus alata*, Michaux, *Fl. Bor. Am.* i. 173 (1803); Loudon, *Arb. et Frut. Brit.* iii. 1408 (1838); Sargent, *Silva N. Amer.* vii. 51, t. 313 (1895), and *Trees N. Amer.* 291 (1905).
*Ulmus pumila*, Walter, *Fl. Carol*, 111 (1788) (not Linnæus).

A tree, attaining in America 50 ft. in height and 5 ft. in girth, with thin scaly bark. Young branchlets minutely pubescent, furnished in the second or third season with two peculiar thin and wide corky wings which persist for many years. Buds ovoid, glabrous, sharp-pointed. Leaves (Plate 411, Fig. 12) deciduous in autumn, thin in texture, 1½ to 2¼ in. long, averaging ¾ in. broad, oblong-lanceolate or narrowly elliptic, slightly unequal or subcordate at the base, acute or acuminate at the apex; smooth above; lower surface without axil-tufts and glabrescent, except for hairs on the midrib; lateral nerves eight to ten pairs, rarely forking before reaching the margin, which is biserrate and fringed with minute cilia, only visible with a good lens.

Flowers, appearing in spring before the leaves, few in a fascicle, on long pedicels; calyx five-lobed; ovary tomentose, stalked. Samara narrowly elliptic, ⅓ in. long, contracted at the base into a long slender stalk, tipped at the apex with long incurved stigmas, covered on the surface with long white hairs, which are most numerous on the thickened margin.

This species is readily distinguishable by the corky wings on the branchlets, which differ from those on the other corky elms (except *U. crassifolia*) in being regularly two in number, very thin, and of considerable width (Plate 411, Fig. 12).

*U. alata*, which is sometimes known by the native Indian name *wahoo*, is indigenous in the warmer parts of the eastern United States, occurring from southern Virginia southward to Florida, and westward to southern Indiana, southern Illinois, Missouri, Arkansas, Indian Territory, and the valley of the Trinity river, Texas. It is usually too small in size to be of any value for timber.

[1] Späth, *Cat.* No. 106, p. 124 (1900-1901).

It is said by Loudon to have been introduced into England in 1820; but it has apparently not thriven in our climate, as we have never found any old trees. It appears to be very rare, if it exists at all, in collections on the Continent. It has lately been reintroduced by Späth of Berlin; and young trees which were obtained from him in 1909, are now to be seen at Kew, where they are already beginning to show the corky branchlets that are characteristic of the species. (A. H.)

## ULMUS CRASSIFOLIA

*Ulmus crassifolia*, Nuttall, in *Trans. Amer. Phil. Soc.* v. 169 (1837); Sargent, *Silva N. Amer.* vii. 57, t. 315 (1895), and *Trees N. Amer.* 294 (1905).
*Ulmus opaca*, Nuttall, *Sylva*, i. 35, t. 11 (1842).

A tree, attaining in America 80 ft. in height and 9 ft. in girth. Bark deeply divided by interrupted fissures into broad flat scaly ridges. Buds ovoid, acute, ⅛ in. long, with minutely pubescent scales. Young branchlets slender, covered with a short erect pubescence; glabrous and striated in the second year, when they often begin to develop two shining brown corky wings. Leaves (Plate 411, Fig. 2) oval, about 1 to 2 in. long, and ½ to 1 in. broad; acute or rounded and never acuminate at the apex; slightly oblique with both sides rounded and usually subcordate at the base; upper surface light green, scabrous, with scattered minute hairs; lower surface paler green, with a scattered minute pubescence, conspicuous on the mid-rib, and not forming axil-tufts; margin serrate or occasionally biserrate, with triangular spreading and not incurved teeth, minutely ciliate; lateral nerves, eight to eleven pairs, usually forked, not regularly parallel; petiole ⅛ in. long, covered with a minute pubescence. Stipules triangular-lanceolate, ¼ in. long, pubescent, clasping the stem by their broad bases, persistent till the leaves are fully developed in May.

Flowers opening in autumn, on slender pedicels (⅓ to ½ in. long), three to five in a fascicle; calyx deeply divided below the middle into five to eight narrow pointed lobes; ovary pubescent. Samara, ⅓ in. long, ovate, deeply notched at the pointed apex, with the stigmas incurved and nearly touching at their tips, pubescent on both surfaces, densely ciliate in margin; seed occupying more than two-thirds of the samara.

*U. crassifolia*, which is known in North America as the cedar elm, is distributed from the valley of the Sunflower river, Mississippi, through southern Arkansas and Texas to Nuevo Leon in Mexico. In Arkansas, it grows usually on river cliffs and low hill-sides; but is most common in Texas, where it ranges in the west from the valley of the Pecos river to the coast, growing both in deep alluvial soil and on dry limestone hills, and attaining its largest size in the alluvial flats of the Guadalupe and Trinity rivers.

This species is very rare in Europe, the only specimen which I have seen being a tree at Kew, about 15 ft. high, which was obtained from Sargent in 1870. The twigs die off annually; and evidently this species is unsuitable to our climate, and can only be grown as a botanical curiosity in favoured districts in the south of England. (A. H.)

## ULMUS PUMILA

*Ulmus pumila*, Linnæus,[1] *Sp. Pl.* 226 (1753); Trautvetter, in Maximowicz, *Prim. Fl. Amur.* 248 (1859); Planchon, in De Candolle, *Prod.* xvii. 159 (1873); Franchet, *Pl. David.* i. 268 (1884).

*Ulmus pumila*, var. *transbaicalensis*, Pallas, *Fl. Ross.* i. 77, t. 48, A, B, C, and *e* (1784).

*Ulmus humilis*, Gmelin, *Fl. Sibir.* iii. 105 (1768).

*Ulmus microphylla*, Persoon, *Syn.* i. 291 (1805).

*Ulmus campestris*, var. *parvifolia*, Loudon, *Arb. et Frut. Brit.* iii. 1377 (1838) (not *U. parvifolia*, Jacquin).

*Ulmus campestris*, var. *pumila*, Ledebour, *Fl. Ross.* iii. 647 (1851); Maximowicz, in *Mél. Biol.* ix. 23 (1872).

A tree attaining 50 ft. in height in Eastern Asia, but often shrubby. Bark furrowed and scaly, as in *U. nitens*. Young branchlets slender, clothed with a dense short white pubescence, persistent more or less on the second year's branchlets, which are fissured and finely striate. Buds minute, ovoid, pubescent. Leaves (Plate 411, Fig. 1) membranous and thin in texture, ovate to ovate-lanceolate, about 1 to 1½ in. long, and ½ to ¾ in. broad; acute or shortly acuminate at the apex; nearly equal, occasionally subcordate, at the base; upper surface dark green, glabrous or scabrous and minutely pubescent; lower surface lighter green, with a scattered minute pubescence, often conspicuous on the midrib, and with minute often obsolete axil-tufts at the junctions of the midrib and nerves; nerves about ten pairs, often forked; margin simply and regularly serrate; petiole $\frac{1}{16}$ to ⅛ in. long, pubescent.

Flowers five to six in a fascicle, appearing in spring, on very short pedicels; tetramerous or pentamerous. Samara orbicular, about $\frac{7}{16}$ in. in diameter, glabrous, non-ciliate, with a deep notch at the apex, usually closed by the overlapping incurved stigmas; seed in the centre of the samara, with its apex close to the base of the notch.

### Varieties

1. Var. *pinnato-ramosa*, Henry.

*Ulmus pinnato-ramosa*, Dieck, *ex* Späth, *Cat.* No. 95, p. 113 (1895), and Dieck, *Verk. Verzeich. Zöschen*, 1897, p. 20; Koehne, in Fedde, *Rep.* viii. 74 (1910) and in *Mitt. Deut. Dend. Ges.* 1910, p. 92.

*Ulmus turkestanica*, Regel, in Dieck, *Hauptkat. Baumeschul. Zöschen*, 1883, p. 36, and in *Gartenflora*, xxxiii. 28 (1884).

A tree of straggling habit, giving off remarkably long shoots, 2 to 3 ft. in length; leaves (Plate 411, Fig. 6) ovate-lanceolate, acuminate.

This elm, which is said to have been introduced by Dieck from western Siberia,

[1] Linnæus founded his species on *U. humilis*, Gmelin, in Ammanus, *Stirp. Ruth.* 180 (1739), and on *U. pumila*, Plukenet, *Alm.* 393 (1696), the latter being described as an elm in Siberia with small leaves. Litwinow, in *Schedæ Herb. Fl. Ross.* vi. 166 (1908), states that the specimen of *U. humilis*, Gmelin, in the St. Petersburg Herbarium, is *U. pumila*. He has also verified at St. Petersburg the type specimens of *U. pumila*, var. *transbaicalensis*, Pallas, and of *U. microphylla*, Persoon.

appears to differ little from *U. pumila*, and is identical with specimens[1] collected lately in Turkestan by Mr. M. P. Price. It was raised both by Späth and by Von Sivers from seed sent from Turkestan; and produces flowers at Berlin in the middle of April, ripening its fruit in the end of May. Three trees at Kew, labelled *U. pinnato-ramosa*, which were obtained from Späth in 1900, are now about 25 ft. high, and are very vigorous in growth. A small tree named *U. turkestanica*, obtained from the Arnold Arboretum in 1907, is identical with those labelled *U. pinnato-ramosa.*

2. *Ulmus arbuscula*, Wolf, in *Mitt. Deut. Dend. Ges.* 1910, p. 286, was raised in 1902 from seeds gathered from a large tree of *U. montana* in the St. Petersburg Botanic Garden. It bears leaves 1 to 3 in. long, and is supposed to be a hybrid between *U. montana* and *U. pumila.* I have seen no specimens of this tree.

3. *Ulmus Koopmannii*, Lauche, *ex* Späth, *Cat.* No. 62, pp. 6 and 101 (1885-1886). This is a form of *U. pumila* with small ovate leaves, 1 to 1¼ in. in length. It is represented at Kew by a poor tree, about 15 ft. high, which was obtained from Transon in 1896. Lauche, *Deutsche Dendrologie*, 349 (1883), states that Koopmann sent seeds of the small-leaved elm from Margilan in Turkestan to the Berlin Botanic Garden, where this variety is represented by a tree with a dense oval crown of foliage like *U. nitens*, var. *umbraculifera* in habit; and the same form is now sold by Späth. Koopmann informed Ascherson and Graebner[2] that this elm is frequently planted in cemeteries in Turkestan, where it is often of great size.

## Distribution

*U. pumila*, which was described by Pallas from shrubby specimens gathered in Dahuria, appears to be widely distributed in eastern Siberia, Manchuria, and northern Korea, where, according to Komarov,[3] it attains 50 ft. in height, and usually grows solitary in river valleys on stony or sandy ground. Bretschneider[4] states that it is the common elm in the Peking plain, where it grows very rapidly, forming a stately tall tree. It is much valued by the Chinese for its timber, which is used in making carts. A nourishing white meal, containing mucilage, is obtained from the thick inner bark, and is used as food by the people in the mountains. The use of this meal is ancient in China. *U. pumila* also occurs in Turkestan and in western Tibet, where Thomson collected specimens[5] at Nubra at 10,000 feet elevation.

*U. pumila*, though mentioned by Loudon, does not seem to have been introduced in his time; and the only specimens which I have seen are at Kew. One of these, labelled *U. pekinensis*, was obtained from the Arnold Arboretum in 1908. Purdom also sent this species from north China to Messrs. Veitch in 1910.

(A. H.)

I received from Peking in 1908 a section of a branch of this tree about 2 in. in diameter, as that of a poplar, but on potting it leaves were produced; and I have

[1] It also agrees with specimens in the Kew Herbarium, labelled *U. campestris*, var. *pumila*, Ledebour, which were collected by Regel in Turkestan in 1878.

[2] *Syn. Mitteleurop. Flora*, iv. 557 (1191).

[3] *Flora Manshuriæ*, ii. 88 (1903).

[4] *Bot. Sinic.* ii. 128 (1892).

[5] Identified with *U. parvifolia*, Jacquin, by Hooker, *Flora British India*, v. 481 (1888).

now two healthy trees about 5 ft. high, which are quite hardy, and seem to grow well at Colesborne. I have rarely heard of an elm being struck from cuttings.[1]

(H. J. E.)

## ULMUS PARVIFOLIA

*Ulmus parvifolia*, Jacquin, *Hort. Schoenbr.* iii. 6, t. 262 (1798); Maximowicz, in *Mél. Biol.* ix. 25 (1872); Franchet et Savatier, *Enum. Pl. Jap.* i. 431 (1875); Forbes and Hemsley, in *Journ. Linn. Soc.* (*Bot.*) xxvi. 448 (1894); Shirasawa, *Icon. Ess. Forest. Japon*, i. text 68, t. 37, figs. 1-9 (1900).

*Ulmus chinensis*, Persoon, *Syn.* i. 291 (1805).

*Ulmus virgata*, Roxburgh, *Fl. Ind.* ii. 67 (1832).

*Ulmus campestris*, var. *chinensis*, Loudon, *Arb. et Frut. Brit.* iii. 1377 (1838).

*Planera parvifolia*, Sweet, *Hort. Brit.* 464 (1830).

*Microptelea parvifolia*, Spach, in *Ann. Sci. Nat.* xv. 359 (1841).

A tree, attaining in China and Japan about 40 ft. in height and 4 ft. in girth Bark scaling off in small plates, showing the reddish brown cortex beneath. Young branchlets slender, sparingly covered with a short wavy white pubescence, retained more or less on the branchlets of the second year, which are fissured, but not finely striate; on older branchlets corky ridges are not produced. Buds minute, ovoid, pubescent. Leaves (Plate 411, Fig. 5) sub-evergreen, persisting till December or January, ovate- or obovate-lanceolate, thick in texture, 1 to 1¾ in. long, ½ to ¾ in. broad, acute at the apex, nearly equal at the base; upper surface dark green, shining, glabrous, smooth to the touch; lower surface lighter green, glabrous except for axil-tufts at the junctions of the midrib with the basal nerves; nerves ten to twelve pairs, usually forked; margin crenately and simply serrate, non-ciliate; petiole, ⅛ to ⅕ in. pubescent.

Flowers appearing in autumn, in clusters of two to five in the axils of the leaves, on very short pedicels, tetramerous; calyx deeply cleft into four segments; ovary minutely pubescent; stigmas white. Fruit ovate to almost orbicular, about ⅓ in. long, shortly cleft at the apex with convergent densely pubescent stigmas; surface minutely pubescent; seed in the centre of the samara.

*U. parvifolia* is a native[2] of China, Tongking,[3] Formosa,[4] and Japan. In Japan, it is confined to the southern parts of Hondo, Skikoku, and Kiusiu, where it is usually a small tree, often only 15 to 20 ft. in height.

In China, it is known as *lang-yü*, and is widely spread throughout the provinces of the Yangtze valley, and southwards, extending to Tongking. The wood is hard, heavy, and tough, but difficult to cleave, and seldom large enough for planking.

*U. parvifolia* is said to have been introduced into France in the reign of Louis

[1] In *Gard. Chron.* xxxix. 35, fig. 20 (1906), a case is illustrated of elm posts, put in the ground for a pergola, which took root and produced abundant foliage at Redworth, Totnes.

[2] The western Tibetan specimens referred to this species by Hooker, *Flora British India*, v. 481 (1888), are all barren branches, and appear to be *U. pumila*. Mayr, *Fremdländ. Wald- u. Parkbäume*, 524 (1906), confuses *U. parvifolia* and *U. pumila*, stating erroneously that the former is a native of the cold regions of Manchuria and North China.

[3] A specimen in the Kew Herbarium was collected by Balansa in the mountains of Tongking at 2700 feet altitude.

[4] Collected by me at Bankinsing.

XV. by Abbé Gallois, who supposed it to be the tea-plant, and for a long time it was known as "thê de l'Abbé Gallois."

It appears to have been introduced into England by Mr. James Main[1] in 1794, who brought home some plants from China, which were cultivated in a garden at Hackney. It is, however, very rare in cultivation, though it seems to be perfectly hardy, and is very ornamental, retaining its foliage in England usually till late in December or early in January. A tree[2] at Kew, about 35 ft. high and 3 ft. in girth, died in 1912; it used to produce flowers occasionally in November, but never set fruit. Another at Beauport, Sussex, grafted on the common elm, was 40 ft. by 3 ft. 8 in. in 1911. There is also a tree at Enys, in Cornwall, about 20 ft. high.

At Verrières,[3] near Paris, its growth is rapid, a specimen only twelve years old being 25 ft. in height in 1906. This produces flowers and fruit, but the latter is usually destroyed by frost. There are also good specimens at Segrez and Grignon. A fine tree in the Jardin des Plantes at Paris, about 45 ft. high and 3½ ft. in girth, was just about to open its flowers on 5th September 1902. This tree is remarkable for its large leaves, up to 3 in. long and 1¼ in. wide; and possibly constitutes a distinct variety, which I have not been able to match exactly with any native specimens in herbaria. The finest specimen in Italy is probably one on Isola Bella in Lake Maggiore, which produces good fruit, and measured in 1909 about 40 ft. by 2½ ft.

The largest and finest specimen in the United States is growing in Central Park, New York, near the 72nd Street entrance from Fifth Avenue. It was introduced in 1865 by Thomas Hogg.[4] (A. H.)

[1] Main was sent as a collector to China in 1791-1794 by Mr. Gilbert Slater of Low Layton, Essex, noted for its extensive gardens and rare plants. Bretschneider, *Hist. Europ. Bot. Disc. China*, 214 (1898), refers to Slater's introductions; but omits all mention of Main. The latter gives an interesting account of his voyage in *The Horticultural Register*, v. 62 (1836).

[2] The tree at Kew was figured under the erroneous name of *U. pumila* in *The Garden*, February 20, 1904, p. 133. This, tree suffered from the disease known as "slime-flux," the trunk exuding a sweet sap, which attracted a large number of wasps during the summer and autumn of 1911. Mr. Bean in *Nature*, vol. lxxxvii. p. 516 (1911), states that this affection is probably due to a yeast, which finds its way to the cambium layer, by means of a wound, and there sets up decomposition of the cells, forming sugary products, which exude from the trunk in solution, and partly ferment into alcohol. Cf. also *Gard. Chron.* l. 323 (1911). Fraser, in *Gard. Chron.* xlix. 59 (1911), also alludes to exudation of sap from elm trees, which may continue for many years, causing the bark to become perfectly white, owing to the death of the green alga, which usually lives upon it.

[3] *Hort. Vilmorin.* 52 (1906).

[4] *Garden and Forest*, i. 231, 312 (1888).

# KOELREUTERIA

*Koelreuteria*, Laxmann, in *Nov. Comm. Acad. Petrop.* xvi. 561, t. 18 (1772); Bentham et Hooker, *Gen. Pl.* i. 396 (1862).

DECIDUOUS trees and shrubs belonging to the order Sapindaceæ. Branchlets, with numerous lenticels, showing in winter elevated leaf-scars, marked in the centre with numerous dots, and girt with a projecting rim-like margin. Buds all axillary, no true terminal bud being developed, each covered externally by two opposite scales, which are united at first by their apices, but ultimately gape apart, exposing the villous interior of the bud. Leaves alternate, without stipules, either unequally and simply pinnate or equally bipinnate; leaflets opposite, sub-opposite, or alternate, with lobed, toothed, or serrate margins.

Flowers polygamous, irregular, in large terminal panicles. Calyx deeply divided into five unequal lobes. Petals five, yellow, unequal, each with a bifid gland on the base of the lamina, above the woolly claw. Disc oblique, with three to five lobes alternating with the petals. Stamens five to eight, inserted within the disc, exserted, declinate; filaments woolly. Ovary three-angled, pubescent, with an elongated style and a trifid stigma, three-celled, each cell containing two ovules. Fruit, an inflated capsule, membranous, three-winged, splitting loculicidally into three valves. Seeds, two or three in each capsule, usually only one developing on the centre of the septum in the middle of each valve, subglobose, blackish, without an aril; embryo spirally convolute.

Four species are known, natives of China and Formosa, distinguishable as follows:—

I. *Leaves simply pinnate with an odd number of leaflets.*

1. *Koelreuteria minor*, Hemsley, in Hooker, *Icon. Plant.* t. 2642 (1900).

Leaflets, fifteen to twenty-five, small, not exceeding 1½ in. long and ½ in. broad, crenate. Capsule, with orbicular valves, ¾ in. in breadth; seed ⅛ in. in diameter.

This is a rare shrub, which was discovered on the Lienchow river in Kwangtung, China, by Ford in 1887. It has not been introduced, and probably would not be hardy in England.

2. *Koelreuteria paniculata*, Laxmann. See p. 1932.

Leaflets, nine to thirteen, 1½ to 3 in. long, toothed or lobed in margin; occasionally some of the leaflets are so deeply cut that the lobes are distinct, making the leaf incompletely bipinnate. Capsule, with ovate acuminate valves, 1½ in. long, and ¾ in. wide; seed ¼ in. in diameter.

II. *Leaves bipinnate, with an even number of pinnæ, which bear an even or odd number of leaflets.*

3. *Koelreuteria Henryi*, Dümmer, in *Gard. Chron.* lii. 148 (1912).

Leaflets alternate, five to eleven, 3 in. long, glabrous except for axil-tufts beneath, serrate. Branchlets glabrous. Capsule with broadly oval or orbicular valves, about 1¼ in. across; seed ⅕ in. in diameter.

This is a tree, attaining about 50 ft. in height, which was discovered by me in 1894, near Bankinsing in the mountains of central Formosa. It has not yet been introduced, but might possibly be hardy in the milder parts of England and Ireland.

4. *Koelreuteria bipinnata*, Franchet, in *Bull. Soc. Bot. France*, xxxiii. 463 (1886), in *Rev. Hort.* lx. 393, fig. 93 (1888), and *Pl. Delav.* 143, pl. 29, 30 (1889).

Leaflets opposite, sub-opposite, or alternate, nine to thirteen, 3 in. long, pubescent on the midrib and veins, serrate. Branchlets pubescent. Capsule with oval or elliptic valves, rounded at the apex, about 2½ in. long and 1½ to 2 in. broad; seed ¼ in. in diameter.

This is a tree, attaining 90 ft. in height in mountain woods in central and south-western China. It was discovered in 1887 by Delavay near Tapintze, north-east of Tali in Yunnan, at 6000 ft. elevation. It was subsequently found by me near Mengtse in the same province, and in the hills north of Ichang in Hupeh. Faber collected it in the mountains near Ningpo in Chekiang.

In 1887 Delavay sent home seed, which germinated freely in the Jardin des Plantes at Paris. One of the seedlings, which has been in the Temperate House at Kew since 1889, is now about 40 ft. high, but has not as yet borne flowers. This species has not apparently been tried in the open air in this country, but is likely to prove hardy in mild districts like Cornwall. It is rare[1] on the Continent, but is reported to be growing at Les Barres[2] and at the Villa Thuret, Antibes.[3] (A. H.)

[1] Simon-Louis, *Cat.* 1908, p. 44, mentions it as very tender to frost at Metz.

[2] *Frut. Vilmorin.* 42 (1904).

[3] Pardé, in *Bull. Soc. Dend. France*, 1911, p. 259.

## KOELREUTERIA PANICULATA

*Koelreuteria paniculata*, Laxmann, in *Nov. Comm. Acad. Petrop.* xvi. 561, t. 18 (1772); Loudon, *Arb. et Frut. Brit.* i. 475 (1838); Hemsley, in *Journ. Linn. Soc.* (*Bot.*) xxiii. 138 (1886); Masters, in *Gard. Chron.* ii. 563, fig. 111 (1887); Mottet, in *Rev. Hort.* lxxviii. 466, fig. 181 (1906).
*Koelreuteria paullinioides*, L'Héritier, *Sert. Angl.* 18, t. 19 (1788).
*Koelreuteria chinensis*, Hoffmannsegg, *Verz. Pfl.* 70 (1824).
*Sapindus chinensis*, Linnæus, *Syst. Veg.* 315 (1774).
*Sapindus sinensis*, Gmelin, *Syst. Veg.* 642 (1796).

A tree attaining about 60 ft. in height and 6 ft. in girth. Bark[1] smooth at first, becoming scaly on old trunks. Branchlets glabrous. Buds ovoid, about $\frac{3}{8}$ in. long; outer scales two, glabrous without, villous within, gaping apart in winter; leaf-scars triangular to semi-orbicular, elevated on projecting pulvini. Leaves unequally pinnate, 6 to 12 in. long; rachis pubescent on the upper side; leaflets nine to thirteen, opposite or sub-opposite, ovate, $1\frac{1}{2}$ to 3 in. long, sub-sessile; variously toothed or lobed, the basal lobes occasionally separated by sinuses extending to the midrib; both surfaces slightly pubescent on the midrib and veins.

Panicles terminal, large, 6 to 9 in. long, with pubescence slight on the principal and secondary axes, and dense on the pedicels and calyx; flowers numerous, yellow, clustered in threes, about $\frac{1}{4}$ in. wide. Fruiting capsules, ripe in autumn, with ovate valves, which are acuminate at the apex, and about 2 to $2\frac{1}{4}$ in. long and $1\frac{1}{4}$ in. wide. Seeds globose or slightly pyriform, about $\frac{1}{4}$ in. in diameter, blackish, shining, only two or three maturing in each capsule. Seedling[2] with two strap-shaped cotyledons raised above ground, followed on the stem by alternate primary leaves, which have three irregularly toothed or lobed leaflets.

### Varieties

The large deeply cut leaflets,[3] borne by some trees, are probably associated with vigorous growth due to soil and similar conditions, and scarcely indicate a distinct variety.

A specimen branch with variegated leaves, which does not seem to have been propagated, was sent to Kew in 1885 by Major Alcock Beck from Easthwaite Lodge, Hawkside, Ambleside.

### Distribution

*K. paniculata* is a native of northern China, where it is common in the hills around Peking; and has also been found in the mountains of Shensi, Kansu, and

[1] Schneider, *Dend. Winterstud.* 40, fig. 47 (1903), depicts the scaly bark on an old stem.

[2] Cf. Kerner, *Nat. Hist. Plants*, Eng. trans. i. 9, fig. 1 (1898).

[3] Wyman, in Bailey, *Cycl. Amer. Hort.* ii. 861 (1900), mentions under the name *K. japonica*, Siebold, a form with deeply cut leaflets. Beissner, in *Mitt. Deut. Dend. Ges.* 1898, p. 424, refers to *K. japonica*, a shrub with a "bipinnate leaf, pubescent beneath," growing in the nursery at Plantières, near Metz, but this is not recognised as a distinct variety in the catalogue of Simon-Louis. Cf. also *Hortus Vilmorin.* 338 (1906).

western Szechwan. It was known by the Chinese as the *luan* tree in classical times, when it was planted around the graves of Ministers of State; but is now called *mu-lan-tze* at Peking, where the leaves are used as a black dye and the seeds as beads.[1] It is not recognised by Japanese botanists as a native of Japan,[2] where it was introduced at an early period by the Buddhist monks, but it is frequently cultivated and occasionally naturalised[3] in Hondo.

It was cultivated under glass at St. Petersburg in 1752, and was probably raised from seed sent to Paris about 1747–1751 from Peking by D'Incarville, who introduced about this time other trees from North China, like *Sophora japonica* and *Ailanthus glandulosa.*[4] *K. paniculata* is said[5] to have been introduced into England by the Earl of Coventry in 1763. It is perfectly hardy, forming a small ornamental tree, which produces abundant panicles of yellow flowers about midsummer. The leaves, which are elegant in form, often turn a beautiful crimson colour in autumn. It is easily propagated by seeds, by layers in autumn, by cuttings of young branches in spring, or by root-cuttings.

The finest specimen in England is probably one in Waterer's nursery at Knaphill, Woking, which was 40 ft. high by 6 ft. in girth in 1911. Elwes saw a fine tree[6] at the east corner of the upper north terrace of Windsor Castle in July 1912, when it was covered with a rich crop of flowers. This measured about 40 ft. by 5½ ft. It ripens seeds at Bitton, near Bath, which come up naturally.

In France[7] it reproduces itself naturally by seed in the neighbourhood of Montpellier; and there is a fine specimen at Verrières, which is considered by Mottet[8] to be one of the original seedlings, dating perhaps from 1751. In 1906, it measured 62 ft. high and 6 ft. 11 in. in girth, with a spread of branches about 60 ft. in diameter. It produces fruit freely every year. M. Hickel tells me that there is a large tree at Heidelberg; but it does not seem to be hardy[9] in other parts of Germany where the winter is very severe, as in Silesia. In the United States[10] it is perfectly hardy as far north as Massachusetts, but is liable after a hard winter to have single limbs die back to the trunk. The seeds readily germinate where they fall to the ground, so that in some places in North America it is becoming naturalised. (A. H.)

[1] Bretschneider, *Bot. Sinic.* ii. 381 (1892), and iii. 491 (1895), and *Hist. Europ. Bot. Disc. China*, i. 159, ii. 850 (1898).

[2] *K. japonica*, Hasskarl, *Cat. Pl. Hort. Bog. Alt.* 226 (1844), was a name given to a shrub of *K. paniculata* that was introduced from Japan into Java.

[3] Franchet and Savatier, *En. Pl. Jap.* i. 85 (1875), record a specimen as wild in a wood in Hondo, which was undoubtedly naturalised.

[4] Cf. vol. i. p. 32. A tree of *Ailanthus glandulosa* in the garden of the Master of Trinity College, Cambridge, is said to have come as a seedling from London in 1758, and is undoubtedly one of the original trees raised from seed sent by D'Incarville in 1751. The old *Sophora japonica* tree (vol. i. p. 42) at Cambridge in all probability dates from the same year.

[5] Aiton, *Hort. Kew.* ii. 7 (1789).

[6] A branch with fruit of a tree at Windsor is figured in *Gard. Chron.* ii. 563, fig. 111 (1887).

[7] Pardé, in *Bull. Soc. Dend. France*, 1909, pp. 103, 114.

[8] *Rev. Hort.* lxxviii. 466, fig. 181 (1906).

[9] *Mitt. Deut. Dend. Ges.* 1903, p. 8. A fine specimen, growing in the Royal Garden at Friedrichshafen on the Lake of Constance, is figured in *Mitt. Deut. Dend. Ges.*, 1912, p. 310.

[10] *Garden and Forest*, vii. 305 (1894), and x. 49 (1897).

*Printed by* R. & R. CLARK, LIMITED, *Edinburgh.*

Plate 372.

SMALL-LEAVED LIME AT SPROWSTON, NORWICH

PLATE 373.

WHITE LIME AT ALBURY

PLATE 374.

WEEPING WHITE LIME AT HATHEROP CASTLE

PLATE 375.

CHUSAN PALM AT LAMORRAN

PLATE 376.

ACACIA DEALBATA AT DERREEN

Plate 377.

HOLLY AT GORDON CASTLE

Plate 378.

WHITETHORN AT HETHEL

PLATE 379.

WEEPING WILLOW AT CHELTENHAM

Plate 380.

WHITE WILLOW AT HAVERHOLME

PLATE 381.

CRICKET-BAT WILLOW AT HERTFORD

PLATE 382.

GREY POPLAR AT COLESBORNE

Plate 383.

FEMALE LOMBARDY POPLAR IN BRUNSWICK

Plate 384.

CAROLINA POPLAR AT DANNY PARK

PLATE 385.

FASTIGIATE BLACK POPLARS IN BELGIUM.

PLATE 386.

BLACK ITALIAN POPLAR AT BELTON

Plate 387.

BALSAM POPLAR AT BUTE HOUSE, PETERSHAM

PLATE 388.

WESTERN BALSAM POPLAR IN VANCOUVER ISLAND

PLATE 389.

EUROPEAN WHITE ELM AT SYON

PLATE 390.

EUROPEAN WHITE ELM AT UGBROOKE

Plate 391.

AMERICAN WHITE ELM IN MASSACHUSETTS

PLATE 392.

AMERICAN WHITE ELM AT HARGHAM

Plate 393.

WEEPING WYCH ELM AT GLASNEVIN

PLATE 394.

WYCH ELM AT STUDLEY ROYAL

Plate 395.

HUNTINGDON ELM AT MAGDALEN COLLEGE, OXFORD

PLATE 396.

ELMS IN KENSINGTON GARDENS

PLATE 397.

CORNISH ELMS AT COLDRENICK

Plate 398.

WHEATLEY ELM AT RICHMOND

PLATE 399.

SMOOTH-LEAVED ELM AT SHARPHAM

PLATE 400.

WYCH ELM AT CASSIOBURY

PLATE 401.

SMOOTH-LEAVED ELM AT BRIGGINS

PLATE 402.

SMOOTH-LEAVED ELM AT SALING

Plate 403.

GOODYER'S ELM AT WESTON BIRT

PLATE 404.

ENGLISH ELM AT POWDERHAM

PLATE 405.

ENGLISH ELMS AT WELL VALE

## FASTIGIATE BEECH

WE are indebted to Mr. F. R. S. Balfour for the picture of a very remarkable beech which he showed me in 1908, growing close to Dawyck House, Peeblesshire. So far as we know, the habit of this tree is unique; though, according to Koch,[1] a fastigiate beech formerly existed in the nursery of Simon-Louis at Metz. The tree at Dawyck was 48 ft. by 4 ft. 2 in. in 1912; and is supposed by Mr. Balfour to be about forty years old, but to my eye it is much older. It has lately been described and illustrated by H. A. Hesse[2] of Weener, Hanover, under the name *Fagus silvatica Dawycki*. Young plants have been propagated by layering and by grafting, and two of these are growing well at Colesborne.

[1] *Dendrologie*, ii. pt. ii. 17 (1873). Simon-Louis, who called this form var. *fastigiata*, do not advertise it now in their catalogues. Var. *pyramidalis*, Petzold and Kirchner, *Arb. Musc.* 662 (1864), described as being pyramidal in habit, is unknown to us.

[2] In *Mitt. Deut. Dend. Ges.* 1912, p. 366.

Plate 406.

FASTIGIATE BEECH AT DAWYCK

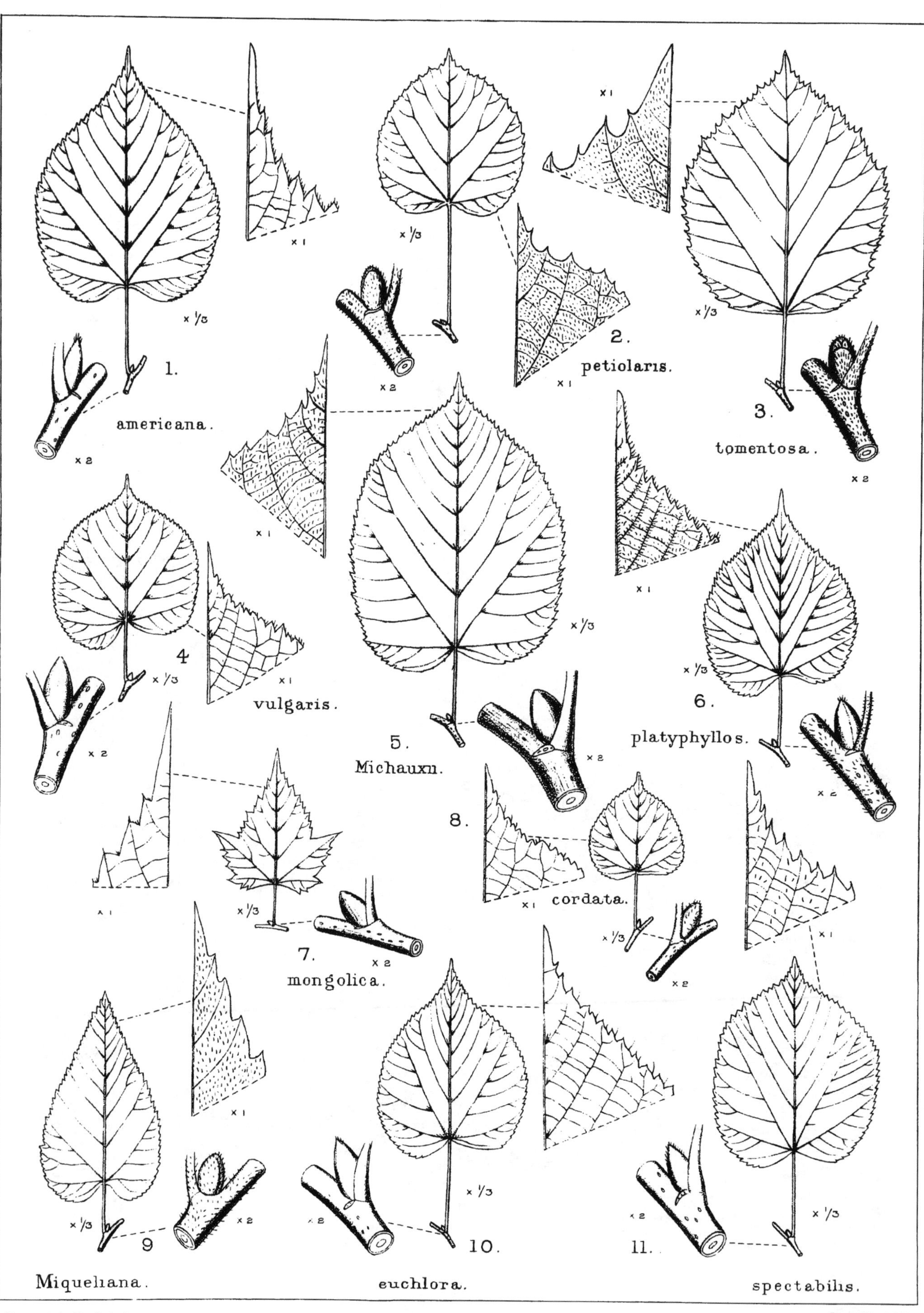


Huitt del., Huth lith.

PLATE 407.

TILIA.

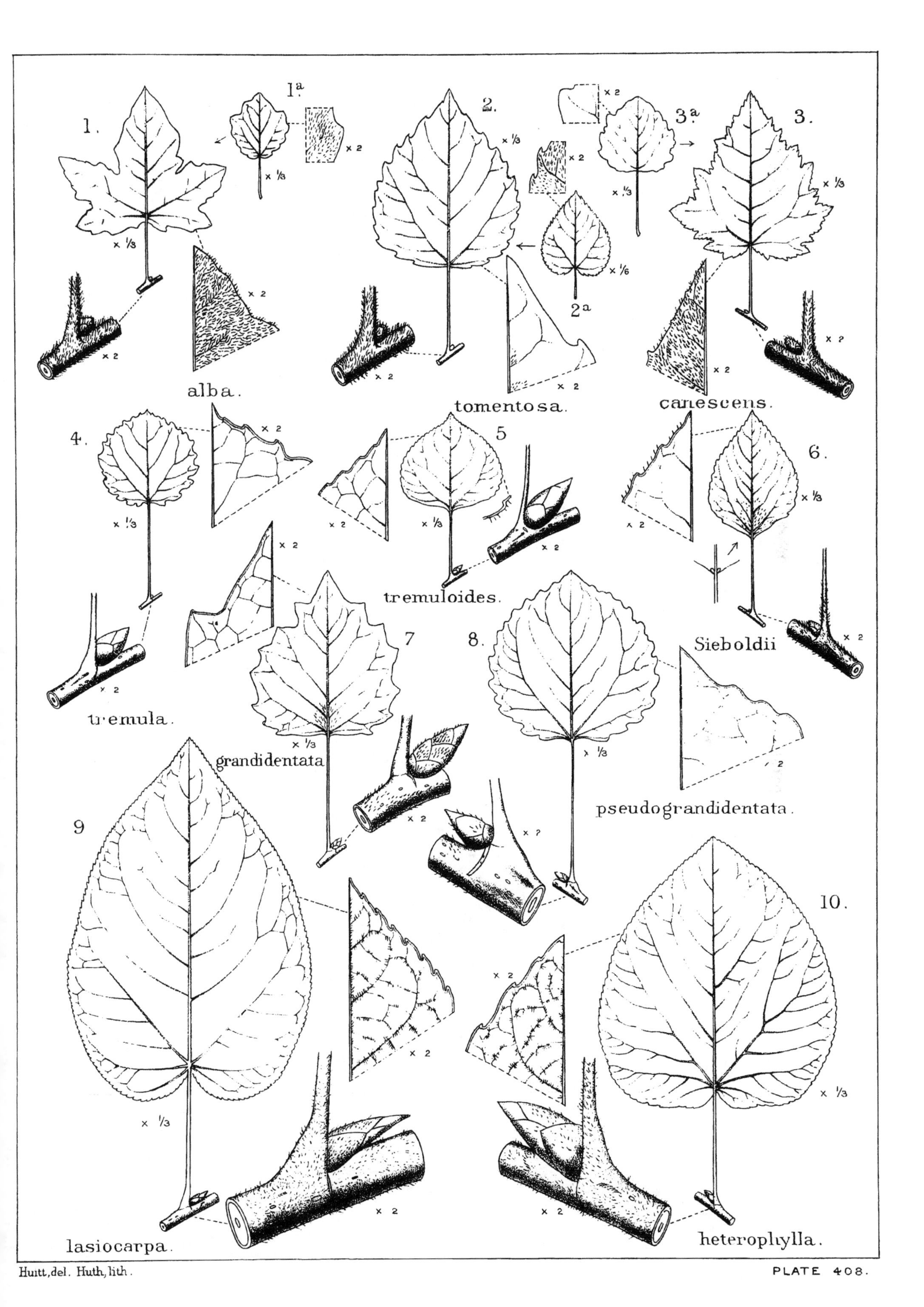


Huitt, del. Huth, lith.

PLATE 408.

POPULUS.

Huitt, del. Huth, lith.

PLATE 409.

POPULUS

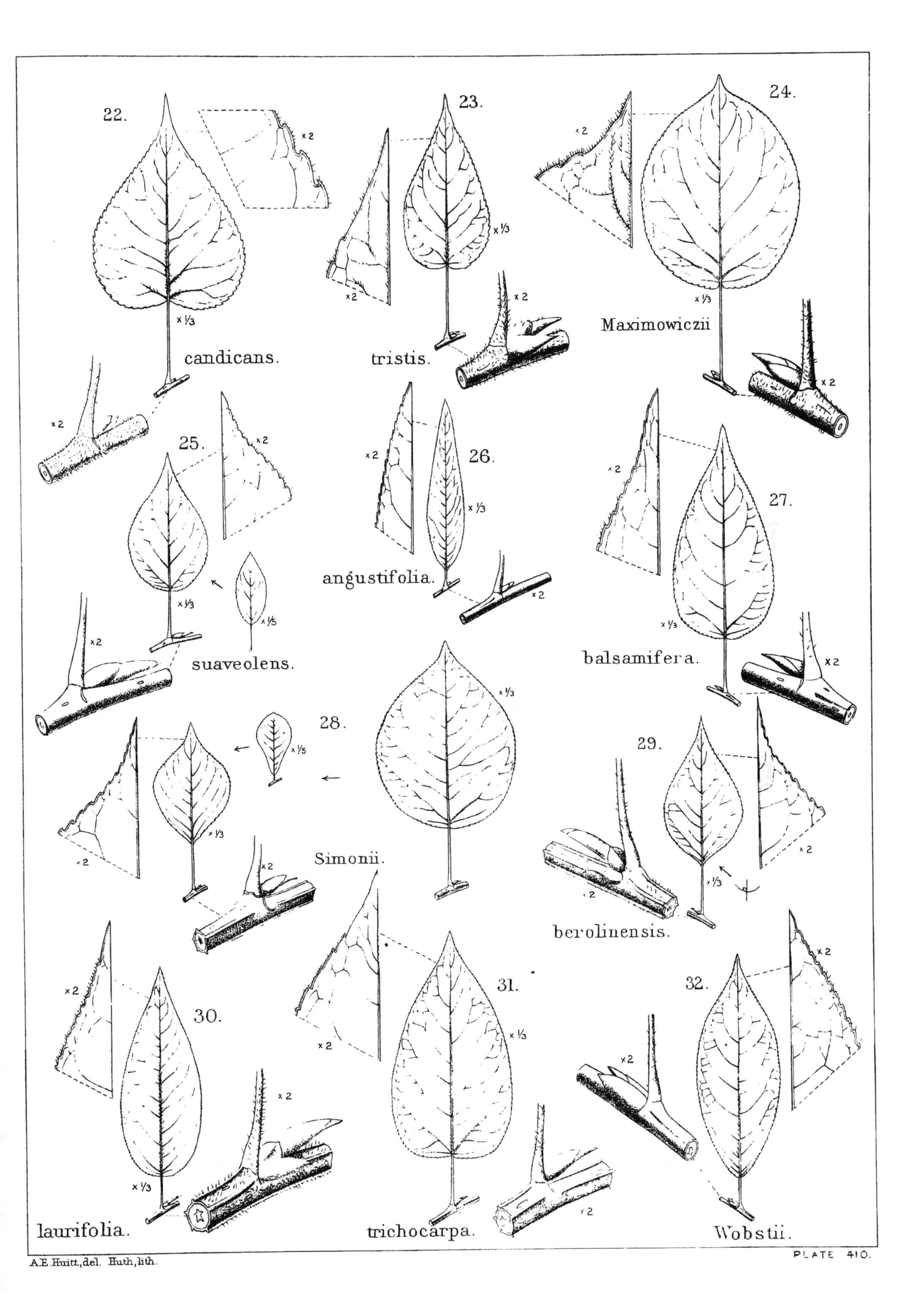
22.
candicans.
23.
tristis.
24.
Maximowiczii
25.
suaveolens.
26.
angustifolia.
27.
balsamifera.
28.
Simonii.
29.
berolinensis.
30.
laurifolia.
31.
trichocarpa.
32.
Wobstii.
x 2
x 1/3
x 1/5
A.E. Huitt, del. Huth, lith.
PLATE 410.

POPULUS.

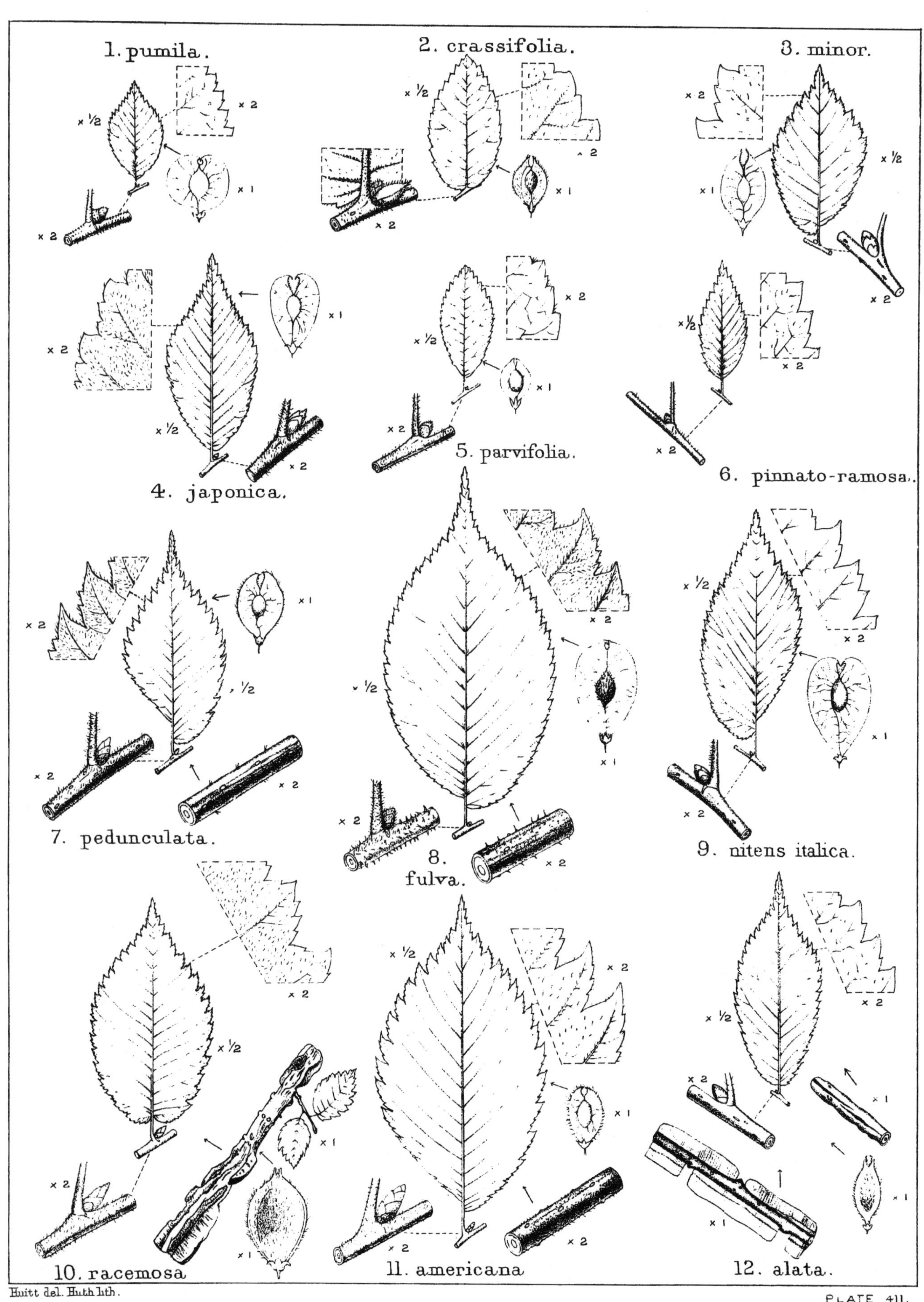


Huitt del. Huth lith.

PLATE 411.

ULMUS.

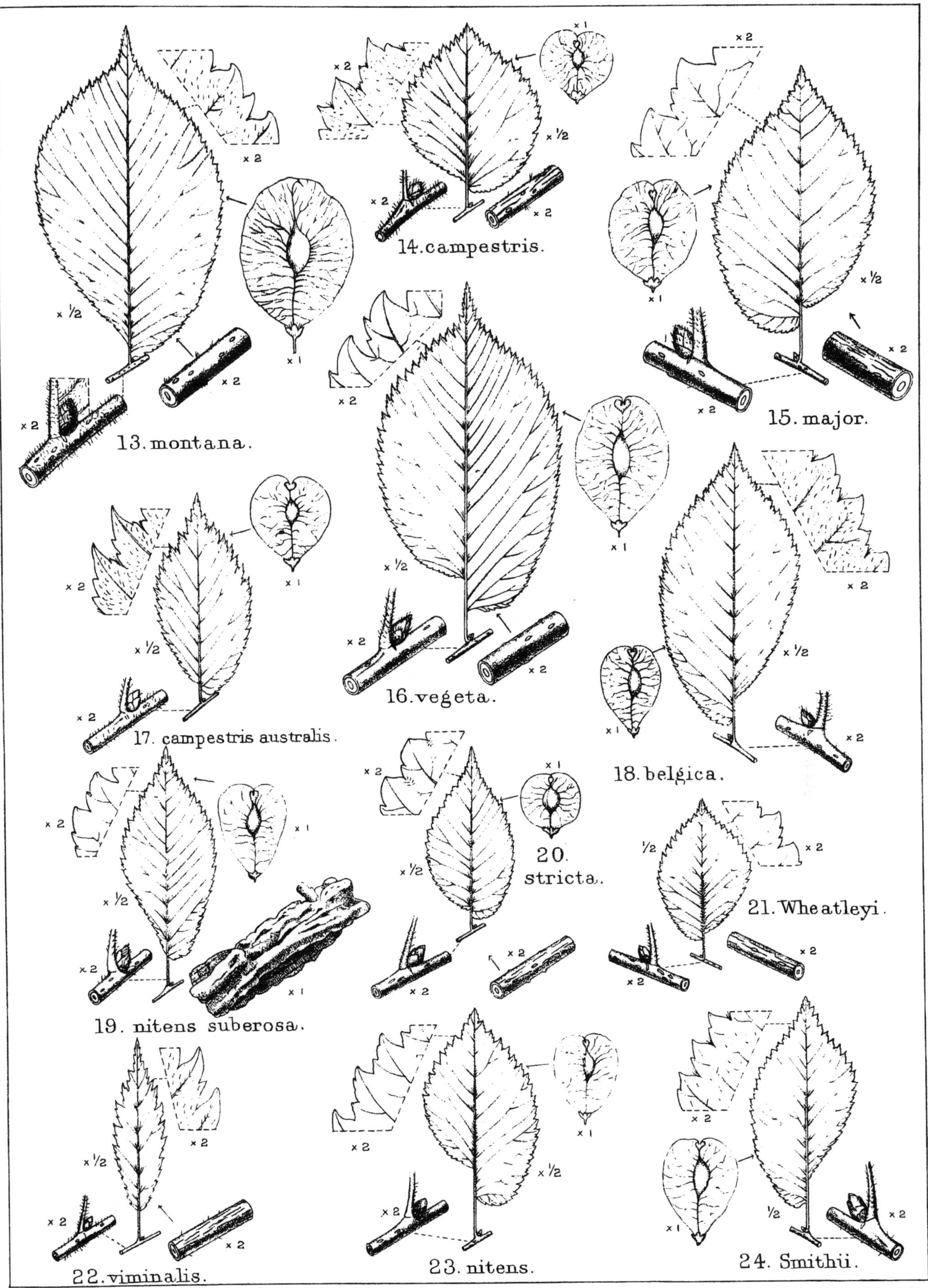


Hutt del. Huth lith

PLATE 412.

ULMUS.

THE TREES OF GREAT BRITAIN AND IRELAND

PRINTERS
PRINTERS
EDINBURG

# THE TREES
OF
# GREAT BRITAIN AND IRELAND

BY

HENRY JOHN ELWES, F.R.S.

AND

AUGUSTINE HENRY, M.A.

## INDEX, Etc.

EDINBURGH: PRIVATELY PRINTED

MCMXIII

# CONTENTS

# LIST OF SUBSCRIBERS

Ackers, B. St. John, Esq., Huntley Manor.
Acland, Sir C. T. D., Bart., Killerton, Exeter.
Addington, Lord, 24 Princes Gate, London.
Ailsa, The Marquess of, Culzean Castle, Maybole, Scotland.
Andrews, Hugh, Esq., Toddington Manor, Winchcombe.
Ashton Court Estate, The, Long Ashton, Bristol.
Avondale Forestry Station, Rathdrum, Co. Wicklow.

Backhouse, R. O., Esq., Sutton Court, Herefordshire.
Bacon, R. C., Esq., Willingham by Stow, near Gainsborough.
Bagot, Lord, Blithfield, Rugeley.
Baird, H. R., Esq., Durris House, Drumoak, Aberdeenshire.
Baker, H. Clinton, Esq., Bayfordbury, Herts.
Balfour, F. R. S., Esq., 39 Phillimore Gardens, London.
Barclay, F. H., Esq., The Warren, Cromer.
Barclay, R. L., Esq., Gaston House, Bishop Stortford.
Barrymore, Lord, Fota, Queenstown, Ireland.
Bath, The Marquess of, Longleat, Warminster.
Bathurst, Earl, Cirencester.
Battersea, the late Lord, The Pleasaunce, Overstrand, Cromer.
Battye, A. Trevor-, Esq., Ashford Chase, Petersfield, Hants.
Bazley, the late Gardner S., Esq., Hatherop Castle, Fairford, Gloucestershire.
Bazley, Sir T. S., Bart., Kilmorie, Torquay.
Beauchamp, the late Sir R. P., Bart., Langley Park, Norfolk.
Beaufort, The Duchess of, Badminton.
Bedford, The Duchess of, The Abbey, Woburn.
Bedford, The Duke of, K.G., The Abbey, Woburn (two copies).
Benson, Lieut.-Col. L., Whinfold, Hascombe, Godalming.
Bentham Trustees, The, Royal Botanic Gardens, Kew (two copies).
Berkeley, The Earl of, Foxcombe, near Oxford.
Biddulph, Lord, Ledbury, Herefordshire.
Birkbeck, Robert, Esq., 20 Berkeley Square, London.
Bosanquet, Percival, Esq., Ponfield, Herts.
Bowles, E. A., Esq., Myddelton House, Waltham Cross, Herts.
Bradford, The Earl of, Weston Park, Shifnal.
Brassey, Albert, Esq., Heythrop, Chipping Norton.
Brodie of Brodie, Brodie Castle, Forres, Scotland (two copies).
Brownlow, Earl, Belton House, Grantham.
Bubb, Henry, Esq., Ullenwood, Cheltenham.
Buchanan, J. Hamilton, Esq., Leny, Callander, Scotland.
Burroughes, T. H., Esq., 16 Lower Berkeley Street, London.

Carlile, W. W., Esq., Gayhurst, Newport-Pagnell.
Carlton Club Library, Pall Mall, London.
Castletown of Upper Ossory, Lord, Doneraile Court, Ireland.
Cator, Mrs., Trewsbury, Coates, Cirencester.
Cawdor, the late Earl, Stackpole Court.
Chadwick, the late Major, Findhorn House, Forres, Scotland.

Chambers, the late B. E. C., Esq., Grayswood Hill, Haslemere.
Champion, W. N. L., Esq., Riddlesworth Hall, Thetford, Norfolk.
Charterhouse School, The, Godalming.
Cheylesmore, Lord, K.C.V.O., 16 Princes Gate, London.
Chichester, Major C. H., Hall, Barnstaple.
Childe, Mrs. Baldwyn, Kyre Park, Tenbury.
Cirencester, The Royal Agricultural College.
Clark, Kenneth M., Esq., Sudbourne Hall, Orford, Suffolk.
Clarke, Harvey R. G., Esq., Brook House, Hayward's Heath, Sussex.
Clarke, Stephenson R., Esq., C.B., Borde Hill, Cuckfield, Sussex.
Clifford, Henry F., Esq., The Grange, Frampton-on-Severn, Gloucestershire.
Clinton, Lord, Heanton Satchville, Dolton, N. Devon.
Cockburn, N. C., Esq., Harmston Hall, Lincoln.
Colley, T. H. Davies-, Esq., 71 Princess Street, Manchester.
Conant, E. W. P., Esq., Lyndon Hall, Oakham.
Cookson, Hugh, Esq., Trelissick, Truro.
Cornewall, the late Rev. Sir George H., Bart., Moccas Court, Hereford.
Cowdray, Lady, Dunecht House, Aberdeenshire.
Cranbrook, the late Earl of.
Crewdson, Theodore, Esq., Spurs, Styal, Handforth, Cheshire.
Crichton, The Hon. Arthur, 8 Southwick Crescent, London.

Dalrymple, The Hon. Hew, Lochinch, Castle Kennedy, Wigtownshire.
Darnley, The Earl of, Cobham Hall, Kent.
Davies, Capt. H. Kevill-, Croft Castle, Kingsland, Herefordshire.
Dawnay, The Hon. Mrs., Ruston, Wykeham, Yorkshire.
De Mauley, Lord, Langford House, Lechlade, Oxfordshire.
Dent, Major John W., Ribston Hall, Wetherby, Yorkshire.
De Vesci, Viscount, Abbey Leix, Ireland.
Dimsdale, R., Esq., Eastleach, Lechlade, Oxfordshire.
Dorington, the late Rt. Hon. Sir John E., Bart.
Drummond, Mrs. Hay, Cromlix, Dunblane.
Duchesne, M. C., Esq., Farnham Common, Slough, Bucks.
Ducie, The Earl of, Tortworth Court, Gloucestershire (two copies).
Dunsmuir, The Hon. James, Victoria, B.C.
Dyer, Sir William T. Thiselton-, F.R.S., Great Witcombe, Gloucester.

Eldon, The Earl of, Stowell Park, Gloucestershire.
Elwes, the late Arthur H. S., Esq., Congham House, King's Lynn.
Elwes, the late V. D. H. Cary-, Esq., Billing Hall, Northamptonshire.
Enys, the late John D., Esq., Enys, Penryn, Cornwall.
Evans, John D. D., Esq., Ffrwdgrech, Brecon.

Fenwick, Mark, Esq., Abbotswood, Stow-on-the-Wold.
Fenwick, Walter L., Esq., Witham Hall, Bourne.
Firebrace, C. W., Esq., Danehurst, Uckfield, Sussex.
Fletcher, W. H. B., Esq., Aldwick Manor, Bognor.
Flux, the late William, Esq.
Ford, the late Mrs., Pencarrow, Cornwall.
Foster, P. S., Esq., Canwell Hall, Sutton Coldfield.
Foster, W. R., Esq., The Granville, Ilfracombe.
Fothringham, W. Steuart, Esq., Murthly Castle, Perthshire.
Francis, T. Musgrave, Esq., Quy Hall, near Cambridge.

Garnett, W., Esq., Quernmore Park, Lancaster.
Gascoigne, Col., Lotherton, Aberford, Leeds.
Gibbs, the late Antony, Esq., Tyntesfield, Bristol.
Gibbs, The Hon. Herbert C., 9 Portman Square, London.

Gibbs, The Hon. Vicary, Aldenham House, Elstree.

Gilbey, Lieut.-Col. Alfred, Twyford Lodge, Winchester.

Glasgow, The Mitchell Library.

Glasnevin, The Royal Botanic Gardens.

Godman, F. Ducane, Esq., F.R.S., South Lodge, Horsham.

Goold, John C., Esq., Killeen Glebe, Dunshaughlin, Co. Meath.

Greaves, R. M., Esq., Wern, Portmadoc.

Grosvenor, Lady Hugh, 9 Southwick Crescent, London.

Gull, Sir W. Cameron, Bart., Frilsham House, Yattendon, Newbury.

Hamilton, The Duke of, Hamilton Palace.

Hanbury, Cecil, Esq., Queen Anne's Mansions, London.

Hardy, Laurence, Esq., Sandling Park, Hythe, Kent.

Harford, H. W. L., Esq., Horton, Chipping Sodbury, Gloucestershire.

Harford, John C., Esq., Falcondale, Lampeter.

Harlech, Lord, Brogyntyn, Oswestry.

Havelock, W. B., Esq., The Nurseries, Brocklesby Park, Lincolnshire.

Hawkshaw, John C., Esq., Hollycomb, Liphook, Hampshire.

Hazlerigg, Sir Arthur G., Bart., Noseley Hall, Leicester.

Hindmarsh, the late W. T., Esq., Alnwick.

Holford, Sir George L., K.C.V.O., Weston Birt House, Tetbury.

Horlick, James, Esq., Cowley Manor, near Cheltenham (three copies).

Howard, Sir E. Stafford, K.C.B., Thornbury Castle, Gloucestershire.

Hudson, Edward, Esq., 15 Queen Anne's Gate, London.

Izquierdo, Señor Don Salvador, Santiago de Chile.

James, A. J., Esq., Edgeworth Manor, Cirencester.

James, Henry A., Esq., Hurstmonceaux Place, Hailsham, Sussex.

James, the late William D., Esq., West Dean Park, Chichester.

Joicey, James, Esq., Poulton Priory, Cirencester.

Kekewich, T. H., Esq., Peamore, Exeter.

Kesteven, Lord, Casewick, Stamford.

Keyser, Charles E., Esq., Aldermaston Court, near Reading.

Kingscote, Thomas, Esq., Watermoor House, Cirencester.

Lawrence, C. W., Esq., Sandywell Park, Gloucestershire.

Leatham, A. W., Esq., Miserden Park, Gloucestershire.

Leicester, The Earl of, Holkham, Norfolk.

Lennard, Sir Henry A. H. F., Bart., Wick ham Court, West Wickham, Kent.

Lewis, John B., Esq., 65 Goulbourn Avenue, Ottawa, Canada.

Llewellyn, Sir John T. D., Bart., Penllergaer, Swansea.

Loder, Sir Edmund G., Bart., Leonardslee, Horsham.

Loder, G. W. E., Esq., Wakehurst Place, Ardingly, Hayward's Heath.

Longchamps, Baron de Selys-Longchamps, Waremme, Belgium.

Lucas, C. J., Esq., Warnham Court, Horsham.

Luttrell, the late G. F., Esq., Dunster Castle, Somerset.

M'Clellan, Frank C., Esq., 38, The Avenue, Kew Gardens, Surrey.

Mackenzie, Sir Kenneth, Bart., of Gairloch, Conan House, Ross-shire.

Mackenzie, W. D., Esq., Fawley Court, Henley-on-Thames.

M'Laren, H. D., Esq., 43 Belgrave Square, London.

Malcolm, Col., C.B., of Poltalloch, Lochgilphead.

Manvers, Earl, Thoresby Park, Nottinghamshire.

Mason, W. H., Esq., Morton Hall, East Retford.

Maxwell, The Rt. Hon. Sir Herbert W., Bart., Monreith, Whauphill, Wigtownshire.

Maxwell, Sir John M. Stirling, Bart., Pollok House, Pollokshaws, Scotland.
Menzies, the late Sir Neil, Bart., Castle Menzies, Aberfeldy.
Messel, L., Esq., Nymans, Handcross, Sussex.
Methuen, Field-Marshal Lord, G.C.B., Corsham Court, Wiltshire.
Middleton, Lord, Birdsall House, York.
Mitchell, Mrs. A. C., Highgrove, Tetbury.
Moiser, Cyril, Esq., Heworth Grange, York.
Moncrieffe, Sir Robert D., Bart., Moncrieffe House, Perthshire.
Montana, U.S.A., The University Library, Missoula.
Moore, Sir F. W., Royal Botanic Gardens, Glasnevin, Dublin.
Moore, G. F., Esq., Chardwar, Bourton-on-the-Water.
Moore, Dr. Norman, Hancox, Battle, Sussex.
More, T. J. M., Esq., 31 St. George's Road, London.
Munich, Die K. B. Forstliche Versuchanstalt.

Nelson, T. A., Esq., Ach-na-Cloich, Connel, Argyllshire.
Newcastle, The Duke of, Clumber Park, Worksop.
Newman, Sir Robert L., Bart., Mamhead Park, Exeter.
Newman, Robert L., Esq., 11 Cadogan Square, London.
Nicholl, Mrs., Merthyr Mawr, Bridgend.
Northbourne, Lady, Betteshanger, Kent.
Norwich, The Free Library.

Ogilvie, F. M., Esq., 72 Woodstock Road, Oxford.
Oxford, Magdalen College.

Palmer, Charles, Esq., Stewkley Grange, Leighton Buzzard.
Palmer, the late General William J., Colorado Springs, Colorado, U.S.A. (seven copies).
Parsons, Alfred, Esq., A.R.A., 54 Bedford Gardens, Kensington, London.
Peckover of Wisbech, Lord, Bank House, Wisbech.
Peirse, E. Beresford-, Esq., Gredington, Whitchurch, Salop.
Pember, the late E. H., Esq., Vicar's Hill, Lymington, Hants.
Pembroke, the late Earl of, Wilton House, Wiltshire.
Phillimore, Lady, Cam House, Campden Hill, London.
Phillips, Lady Faudel, Balls Park, Hertford.
Pinchot, Gifford, Esq., Washington, D.C., U.S.A.
Platt, Col. Henry, C.B., Gorddinog, Llanfairfechan, N. Wales.
Portland, The Duke of, K.G., Welbeck Abbey, Worksop.
Potts, W. Trumperant, Junr., Esq., Correen Castle, Ballinasloe, Co. Roscommon.
Powis, The Earl of, Powis Castle, Welshpool.
Prain, Sir David, F.R.S., Royal Botanic Gardens, Kew.
Pratt, E. R., Esq., Ryston Hall, Norfolk.
Price, M. P., Esq., Tibberton Court, Gloucester.
Price, T. P., Esq., Mark's Hall, Essex.
Price, W. R., Esq., Pen Moel, Chepstow.
Probyn, General The Rt. Hon. Sir Dighton M., G.C.B., Buckingham Palace.
Pryor, Marlborough R., Esq., Weston Park, Stevenage, Herts.

Quaritch, Mr. Bernard, 11 Grafton Street, London.

Radnor, The Earl of, Longford Castle, Wiltshire.
Ralli, Pandeli, Esq., Alderbrook, Cranleigh, Surrey.
Ralli, Pantia, Esq., Ashtead Park, Surrey.
Rayleigh, Lady, Terling Place, Witham, Essex.
Redesdale, Lord, 1 Kensington Court, London.
Richmond and Gordon, The Duke of, K.G., Goodwood, Chichester.
Rogers, C. Coltman, Esq., Stanage Park, Brampton Bryan, Herefordshire.
Rogers, Lieut.-Col. J. M., Riverhill, Sevenoaks.
Rolle, the late Hon. Mark.
Rolleston, Sir John F. L., Glen Parva Grange, Leicester.
Rothschild, The Hon. Walter, F.R.S., Tring Park, Herts.

St. Oswald, Lady, Appleby Hall, Doncaster.
Sanderson, Finlay, Esq., University College, Oxford.
Sawyer, Charles, Esq., 74 New Oxford Street, London.
Schlich, Sir William, F.R.S., 29 Banbury Road, Oxford.
Sherborne, Lord, Sherborne House, Gloucestershire.
Soames, A. G., Esq., Sheffield Park, Sussex.
Stanhope, The Hon. Richard, Revesby Abbey, Boston, Lincolnshire.
Stern, Capt. H. J. J., Bective House, Navan, Co. Meath.
Stewart, Lady Alice Shaw-, Ardgowan, Greenock, Scotland.
Stewart, Lady Octavia Shaw-, Fonthill Abbey, Tisbury, Wiltshire.
Stirling, J. A., Esq., 41, The Pryors, East Heath Road, Hampstead, London.
Stooke, J. E. Hellyar, Esq., 2 Palace Yard, Hereford.
Strickland, Algernon, Esq., Apperley Court, Tewkesbury.
Strickland, the late Sir Charles W., Bart.

Talbot, G. J., Esq., 36 Wilton Crescent, London.
Toronto, Canada, The University Library.
Tottenham, Col. C. G., Ballycurry, Ashford, Co. Wicklow.
Trench, R. C., Esq., Penrhyn Estate Office, Bangor, North Wales.
Trotter, John, Esq., Brickendon Grange, Hertford.
Turner, T. Warner, Esq., Langwith Lodge, Mansfield, Nottinghamshire.

Underdown, H. C. B., Esq., Buckenham Hall, Mundford, Norfolk.

Victoria, B.C., The Provincial Library.
Vienna, Die Dendrologische Gesellschaft für Oesterreich-Ungarn.
Vilmorin, M. Maurice L. de, 13 Quai d'Orsay, Paris.
Vilmorin, M. Philippe L. de, 23 Quai d'Orsay, Paris.
Vines, Prof., F.R.S., Botanic Garden, Oxford.

Waldo, E. G. B. Meade-, Esq., Stonewall Park, Edenbridge, Kent.
Watney, Daniel, Esq., 33 Poultry, London.
Watney, Herbert, Esq., M.D., Buckhold, Pangbourne, Berks.
Watney, Vernon J., Esq., Cornbury Park, Charlbury, Oxfordshire (two copies).
Wilding, E. H., Esq., Wexham Place, Stoke Poges, Buckinghamshire.
Wilkinson, C. J., Esq., Sebergham Castle, Cumberland.
Willmott, Miss, F.L.S., Warley Place, Great Warley, Essex (three copies).
Wilson, Sir Maurice Bromley-, Bart., Dallam Tower, Milnthorpe, Westmorland.
Woodward, Robert, Esq., Arley Castle, Bewdley.
Wythes, E. J., Esq., Copped Hall, Epping.

Yarborough, The Earl of, Brocklesby Park, Lincolnshire.

## PRESENTATION COPIES

His late Majesty King Edward VII.
King Edward VII. Convalescent Home for Officers of the Navy and Army, Osborne, Isle of Wight.
The Royal Society, London.
The Linnean Society, London.
The Athenæum, Pall Mall, London.
The Bodleian Library, Oxford.
The University Library, Cambridge.
Harvard University, U.S.A.
The Imperial Japanese Bureau of Forestry.
Beevor, Sir Hugh R., Hargham, Norfolk.
Elrod, Prof. M. J., Missoula, Montana, U.S.A.
Elwes, A. L., Esq., Staple Grove, Beckenham.

# POSTSCRIPT

BY


## HENRY JOHN ELWES, F.R.S.

AND

## AUGUSTINE HENRY, M.A.


In the Introduction to the first volume of this work, published in 1906, we stated the objects which we had in view, and which, during the seven years which have now elapsed, we have endeavoured to carry out to the best of our power. If the patience of our Subscribers has been unduly taxed, we can only say that the magnitude of the task which we set before us was even then hardly realised; and that the difficulty of discovering, identifying, describing, and figuring the rare and remarkable trees in Great Britain is one which grew with our knowledge, and with each succeeding volume. The want of order in this work, on which some of our reviewers have remarked, has been really of the greatest service; for by leaving the more difficult and little-known genera to the last, we have been able to make the work more complete and accurate than it would have been if every genus had been taken in its accepted botanical sequence; and though a few additions may be made to the earlier volumes, we know of no really important omissions in them.

Though some local botanists and arboriculturists have studied the native trees of their own counties with more or less care, British botanists, until recently, have taken little notice of the trees which form so conspicuous a feature in the vegetation and scenery of England; and in many counties, whose flora, birds, and insects have been most carefully and accurately studied by local naturalists, we have found no one who, apparently, knew or cared for the trees, and have had to depend largely on our own observations.

An immense quantity of foreign as well as British literature has been referred to, as evidenced by almost every page of the work; but a general bibliography seems unnecessary, as the references are fully given; as well as the authority for nearly every fact, opinion, or observation not made by ourselves.

As it is possible that in the future, questions may arise as to the correct nomenclature of some of the numerous trees that we have described and figured, it is well to say that the herbarium accumulated by us in the course of our work is preserved at Cambridge and at Kew, so that the actual specimens from which the descriptions and identifications were made can be referred to in case of doubt.

It is impossible to acknowledge in full detail the help that we have received from landowners and their agents, foresters, gardeners, and friends in all parts of Great Britain which we have visited during the course of the work; but we have felt on many occasions that without this help the work would have been impossible.

Amongst those to whom we are most indebted for help, I must especially mention the following:—

Sir William T. Thiselton-Dyer, F.R.S., who has carefully read the proofs of all the volumes except the first, and whose suggestions and advice we have followed in many difficult questions; Mr. Charles Palmer of Stewkley Grange and Manchester, who has also read the proofs with extraordinary care; the Director and staff of the Royal Gardens at Kew, who from first to last have shown a personal interest in our work which has been of the greatest support and assistance. We must here call attention to the fact that the collection of living trees at Kew is, and we hope always will be, so far as its soil and climate will allow, the most complete, correctly named, and well-cared-for in Europe; while its unrivalled library and herbarium, where much of our work has been prepared, and the references checked, have been indispensable in connection with the living specimens.

The staff of the Botanical Department of the British Museum of Natural History, as well as the Directors of the Botanical Gardens at Edinburgh, Glasnevin, Oxford, and Cambridge have afforded us every facility for studying both the trees and herbaria in their charge, and we gratefully acknowledge their assistance.

In the United States we have received much help from the officials of the National Bureau of Forestry, and from many private individuals, amongst whom I must mention Prof. Elrod of the Montana University, Mr. Gifford Pinchot, the late Chief Forester of the United States, and especially Mr. Charles S. Sargent of the Arnold Arboretum, who has given us much valuable information and help on many occasions.

In Canada Messrs. J. M. and W. T. Macoun of Ottawa have been on many occasions most helpful and obliging.

In France we have to thank numerous friends for shewing us many of the finest trees and forests in all parts of the country; amongst them we are especially grateful to MM. Maurice and Philippe de Vilmorin of Paris, Prof. Flahault of Montpellier, M. Leon Pardé, Inspecteur des Eaux et Forêts of Beauvais, M. Hickel of Versailles, M. Guinier of Nancy, M. Jouin of Plantières near Metz, and Mr. Cecil Hanbury of La Mortola in Italy.

In Holland Mr. L. A. Springer of Haarlem, in Belgium Prof. Bommer of Brussels, M. Huberty, Inspector of Forests, Verviers, and the late Baron de Selys-Longchamps, in Germany the late Herr Späth of Berlin and the late Prof. Blasius of Brunswick, and the Directors of the Botanic Gardens we have visited in Germany, Russia, Italy, Sweden, Norway, Spain and Portugal, Bulgaria, and Servia, have all given us most valuable notes and assistance.

In Denmark we have also received much help and information from Forest Inspectors Mundt and Bramsen, who personally conducted us through many interesting forests and private arboreta.

In Japan I have on two occasions received unusual attention and assistance in visiting the most interesting forests, and in procuring good photographs of the trees, for which I shall be ever grateful both to the past and present Ministers of Agriculture, and especially to Dr. H. Shirasawa of Tokyo, and numerous other officials and friends.

Amongst the English landowners, who have afforded us the greatest assistance and kindness on all occasions, I must specially mention His late Majesty King Edward, to whom our work is dedicated, and whose fine trees at Windsor, Sandringham, Balmoral, and Osborne have been frequently mentioned in our pages. The Earl of Ducie, who during a long life has had the pleasure of seeing many rare trees planted by himself come to maturity, and who, perhaps better than any English landowner, has realised the importance of attending to his trees after they were planted. The late Sir Charles Strickland, who often told me that the pleasure of watching the growth of trees which he had himself raised and planted, was the one interest in life which remains undiminished in extreme old age. Sir Hugh Beevor, who has inherited a taste for arboriculture and forestry, and on many occasions has contributed most valuable notes and measurements of remarkable trees in many places. The Dukes of Argyll, Bedford, Northumberland, Portland, Richmond, and Wellington. The Marquesses of Bath, Lansdowne, Ripon, and Waterford. The Earls of Annesley, Bathurst, Bradford, Brownlow, Cawdor, Coventry, Darnley, Fortescue, Ilchester, Leicester, Manvers, Pembroke, Portsmouth, Powis, Radnor, Selborne, Spencer, and Yarborough. The Viscounts Falmouth and Powerscourt. The Lords Bagot, Barrymore, Clinton, Dynevor, Kesteven, Llangattock, Lovat, Methuen, Northbourne, Peckover of Wisbech, Penrhyn, Rayleigh, Redesdale, Sackville, Scarsdale, Sherborne, and Walsingham. Sir C. T. D. Acland, Sir E. Stafford Howard, Sir George Holford, Sir E. Loder, Sir John Stirling Maxwell, Sir Herbert Maxwell, Sir Frederick Moore, Sir John Ross-of-Bladensburg. The Hon. Vicary Gibbs, Mr. H. Clinton Baker, Mr. R. Birkbeck, Mr. F. R. S. Balfour, Brodie of Brodie, Mrs. Baldwyn Childe, Capt. D. Cameron of Lochiel, Major Dent, Mr. W. Steuart Fothringham, Major Lloyd, Col. Malcolm of Poltalloch, Mr. E. R. M. Pratt, Mr. C. Coltman Rogers, Dr. Herbert Watney, and Mr. R. Woodward, Jr.

For special information respecting the trees of Scotland we are indebted to Mr. John Renwick of Glasgow, and for notes on Irish trees to Mr. R. A. Phillips of Cork and Sir F. W. Moore.

The conditions which determine the successful cultivation of exotic trees in different parts of Great Britain are so complicated by local variations of soil, climate, and elevation that after many attempts I have failed to construct a map which would divide the country into arboricultural regions. The best guide to the possibility of growing any particular species in any given locality is to know whether it has succeeded reasonably well in any place of similar soil and climate; and as we have recorded the accumulated experience of our predecessors, it may usually be presumed that if no degree of success has been attained it is only a waste of time and money to plant species which have died out or remained in a stunted condition. Generally speaking it may be said that there are four principal types of climate in Great Britain.

The first is characterised by a high summer and a low winter temperature, combined with a low average rainfall; and this includes all those districts in which wheat is the most important agricultural crop, namely the eastern counties as far north as the Humber, and the southern counties as far west as Dorset and the Severn Valley. In many parts of this region where, owing to the influence of sea air or elevation, late spring and early autumn frosts rarely occur, the best climatic conditions for the growth of most exotic trees are found.

Secondly, the south-western counties and the maritime districts of Wales, the greater part of Ireland and western Scotland, where the summer temperature is lower and the winters shorter and milder, and where the rainfall and the humidity of the air is much greater. In this region alone a great many of the rarer trees and shrubs thrive wherever suitable soil and shelter from wind are found; but the trees which require a high summer temperature and abundant sunshine to ripen their wood, such as walnuts, hickories, Catalpas, and many North American species, are not often successful, and do not attain large dimensions or ripen their seeds in normal seasons.

Thirdly, the north-eastern and midland counties and those districts of southern England where the summers are shorter, where late and early frosts are prevalent, or where the soil is too heavy or too wet to suit a great many exotic trees; but many exceptions will be found in this region, especially near the sea, and in this district may be included parts of Scotland bordering on the Moray Firth, where the climate is distinctly more suitable to arboriculture than in the northern and midland counties of England generally.

Fourthly, the mountains of central and northern England, Wales, Scotland, and Ireland, at elevations over 1000 feet, where only a few of the hardier conifers can be grown with much hope of success.

Next to climate, the geological formation and depth of the soil is the most important factor in the successful growth of trees; and this again is variable even in the districts where climate is favourable; so that we find the best instances of arboriculture are scattered in all parts of the country. Generally speaking, the soils which produce the finest hardwoods are on the old red sandstone and lower greensand formations; and though some species attain a large size on other soils, yet wherever heavy clay or thin rocky soils prevail, especially on limestone, many exotic trees will not grow at all, or become stunted and unhealthy as soon as their roots get into the subsoil. As a result of careful soil preparation trees may appear to thrive for a number of years, yet they eventually become sickly or die if nature has not provided a suitable subsoil. Want of knowledge or of attention to these two factors, has led in the past to immense waste of time and money in planting trees which are quite unsuited to their surroundings, and we have seen numerous instances in all parts of Great Britain which prove how little man can do where nature does not favour his efforts.

Another point which has been insufficiently realised by most planters, but which my own experience has repeatedly proved, is the remarkable variation in the individual constitution of trees of the same species, even when raised from seeds of the same tree under precisely similar conditions. In a state of nature there is a continual struggle for existence among individuals under which only a few of the strongest survive, and these are the seed-bearers from which future generations arise; but when we sow a number of seeds under the most favourable conditions which we can devise, and protect them artificially in their youth against their natural enemies, or when we propagate trees by other means, which are often adopted not because they are the fittest but because they are the quickest and cheapest means of reproducing them, we obtain a variable percentage of weaklings which thrive only under the best conditions, and which when transplanted to less favourable environment are sure sooner or later to succumb. For this reason I have always advocated the raising of forest trees when possible from seed of known healthy parents growing in the same or a similar locality to that where they are to be planted, and rejecting in the nursery all but the most vigorous. This may prove a slower and more costly method than that of buying the trees from a nursery, but I believe it to be the most economical in the end wherever conditions are not very favourable for their growth. With regard to exotic trees which only ripen their seed in this climate in very favourable seasons, I have often found that the seedlings raised from imported seed are more vigorous than those raised from home-grown seed, but my observations, though they have been carried on since 1900, are insufficient to enable me to express a decided opinion on this subject.

When we consider how remarkably variable our climate is, it is clear that a great deal of the success of planting any trees which are more exacting as regards heat, sunshine, or moisture[1] than indigenous species, must depend on a good series of seasons in their youth; for if trees are severely checked by drought, or by late spring or early autumn frosts when young, they suffer much more than when older and better established. Therefore in planting exotic trees it is wise not to depend on one or two individuals, but to plant several of the same species in a group, with the object of selecting the most vigorous and well-shaped when they begin to crowd each other.

Another point which is often forgotten by planters of ornamental trees is the fact that in nature these grow gregariously or mixed with other trees, in more or less thick or shady woods, and not isolated in grass. To use a gardener's expression, "a well-furnished tree" such as a cedar, a Sequoia, or a beech with spreading branches resting on the ground, may be a very beautiful object on a lawn or in a park; but it is not the usual natural shape of these trees; and if the trunk is always

[1] As an instance I may say that at Colesborne the rainfall of June, July, and August, during which months most of the growth of the majority of trees is made, was, in 1911, 3.39 inches, with only 21 days on which rain fell, and in 1912, 22.30 inches, with only 21 days without rain.

sacrificed to the branches, we rarely get such magnificent boles as these trees are capable of producing when planted under more natural conditions.

Another point too often neglected by English arboriculturists is early and regular pruning, which, unless they are planted in close order, is necessary in the case of most broad-leaved and some coniferous trees, until they have attained a considerable size and age. Though the lower branches should not be cut off close to the trunk before it is thick enough to support a well-shaped crown, yet the earlier it is done after the tree has become established, the quicker and better is the wound healed; and the careful planter must continually watch that no branches are allowed to attain undue proportions at the expense of the trunk. With some trees pruning and shortening of large branches is best done in early spring before the sap rises; with others, in July or August, when the tree is in vigorous growth; but though art may do a great deal to form a well-shaped tree, yet unless the soil is deep enough and fertile enough to keep it in health, stunted and stag-headed trees will be the result.

The influence of grass on the roots of trees is a subject upon which a great deal of light has been thrown by the experiments carried on by the Duke of Bedford and Mr. Spencer Pickering.[1] Though opinions differ on the extent to which their conclusions apply to trees universally, and how far they are due to particular soils, yet there can be no doubt that as a general rule the soil should be kept free from grass for a distance of about three feet all round the trunk for some years at least. On the peculiar soil of Colesborne I have found that some trees, among which Scots and Corsican Pine are conspicuous, do not suffer when planted in a thick sward, and actually seem to grow faster and to be more healthy. I am inclined to believe that wherever the soil is deficient in fertility, grass over the roots of trees, though it may check their growth in dry seasons, is better in most cases than a bare soil which has been impoverished by cultivation. Good soil will overcome almost all other obstacles to the growth of trees, and this fact leads me to speak of another question which planters and foresters often insufficiently consider.

Is it more economical and profitable to plant land, which, like many of the natural woods and plantations in Great Britain, has been allowed to remain as woodland only because it was not thought good enough to cultivate; or to plant land which is producing more or less profitable crops? We rarely see the dimensions that trees are capable of attaining in Great Britain, except on the estates of families which for centuries have been rich enough to plant good land, and to leave trees standing after they have attained maturity; and though the largest trees are not always the most valuable, yet the highest quality of timber, which usually fetches a higher price per foot, is rarely produced on inferior land, whilst the quantity per acre that can be grown in a short time is the most important factor in the profit or loss of planting.

In forestry as in agriculture the best land is the most profitable and therefore the cheapest. But when I speak of good land I do not mean that the best land for farming is the best for trees, because trees root so much more deeply than agricultural plants that they can penetrate and feed where the roots of plants cannot reach; and on steep hill-sides especially, trees seem to thrive wherever the rock is suffi-

[1] *Thirteenth Report of the Woburn Experimental Fruit Farm*, London, 1911. See also *Ninth Report*, 1908.

ciently broken up, provided that it contains the chemical elements of fertility ; so that the agricultural or grazing value of any particular spot often affords no indication of its value for tree-planting.

Though the world has now been nearly everywhere explored, and the number of trees capable of growing in Great Britain has been more than doubled by fresh introductions since Loudon wrote in 1838, it is very surprising how few, if any, of the newer introductions seem likely to supersede or even to equal our long-tried native and introduced trees. Though no one can be a stronger advocate than I am, both in principle and in practice, of the planting of a great variety of exotic trees, and though in localities where somewhat unusual conditions are found, some of these species—such as Japanese larch, Sitka spruce, Lawson and Nootka cypress, and Douglas fir—do seem likely to be more profitable than older introductions like the European larch, Silver and Spruce firs, yet I can hardly think of any broad-leaved tree, except some poplars which are not yet fully proved, and possibly the grey alder, which seems likely to have much economic value or to supersede our native oak, beech, ash, elm, and sycamore, which to timber merchants are at present the only trees really worth looking at, and which will grow to a large size in almost all parts of our islands. Whether time will prove this to be a fact or not, we have done our best to describe every tree as yet introduced, with the exception of some of the latest introductions from China.

When trees have passed the age of maturity and are beginning to decline in health they usually show it by the death of the upper branches, which is particularly noticeable on very heavy or very dry soils after a long hot summer. If there is no serious decay in the trunk, this decline may often be checked, for a considerable time at any rate, by a top-dressing of leaf-mould mixed with old rotten manure and fertile soil, spread about three inches deep over the area covered by the branches ; but such dressing must not be too thick or too strong, especially in the case of conifers, for which pure leaf-mould is perhaps the best manure. When wounds caused by broken branches, fungi, or other injuries appear on the trunk, it is important to fill them as soon as possible. The same principles adopted in filling decayed teeth are perfectly applicable to trees, namely, to clear out all decayed wood as far as possible, to apply an antiseptic to the exposed surface, to fill up the cavity with cement, and to cover with lead any cracks or holes by which water can get into the trunk. When branches become dangerous or inclined to split off, they are often supported by iron rods, bands, and chains ; and I believe that when this is done in such a way that the band can be loosened as the branch thickens, it is a safer and more permanent method than passing a rod through a hole in the branch and putting a nut on it to keep it in place.

There are many other special points in the cultivation of trees which only experience can teach, though the general principles are given in works on forestry and arboriculture ; and in many cases these special points are dealt with in the pages of our work. Opinions vary and always will vary as to the best systems of planting, and the best mixture to adopt, which depend on the peculiar conditions we have to deal with ; but in tree-planting, whether for economic or other objects, it cannot be too strongly emphasised that a study of local conditions is always essential. The more a man knows of the risks and difficulties which planters have to contend with in most parts of the country, the less willing will he be to offer advice, or to form estimates of cost and expected profits based on experience which is not local. For this reason I look with suspicion on hastily considered working plans and estimates of costs and results which are not supported by local knowledge. Even in old-established forests abroad, where the expenses and profits are based on the experience of centuries, and where nothing is done without the approval of Government foresters, changes are constantly being made in practice to meet the changing conditions of the times; and when we remember that our long-established systems of woodcraft in England have been completely revolutionised in the last fifty years by changes in economic conditions, the duration of which cannot be foreseen, it seems evident that what we think right to-day may turn out to be wrong long before the trees we plant are mature.

Before concluding I should again like to offer a warning word on the question of planting and management for profit, as contrasted with planting for ornament, sport, and shelter. English landowners are sometimes reproached with ignorance and neglect of the principles of pure forestry; but it must always be remembered that the economic, social, and in many districts the climatic conditions of Great Britain, make planting for profit a very uncertain and often a very risky investment. In our work we have endeavoured to show the possibility of cultivating a great number of exotic trees which have not, and probably never will have, any economic value from the forester's point of view, because we believe that when planted in small quantities with sufficient knowledge of their cultural requirements, they will add greatly to the interest, beauty, and residential value of British country homes; and we have illustrated the finest examples of most of the trees which exist in Britain. We ought to aim as far as possible at planting those species which local experience has tested, in situations which are most suitable to them. This warning must apply not only to the rarer species, but to some which have been very extensively planted of late years by enthusiasts in arboriculture, who look only at the successes of others and refuse to look at the failures. I have myself learnt more from my failures than from my successes; and have never been able to understand why in so many cases people are unwilling to show or to write of their failures, when these are due—as they often are—to natural obstacles rather than to want of care or knowledge.

The future of arboriculture in Great Britain is a brilliant one, if landowners are not deterred from planting by ill-considered or hostile legislation; but the future of pure forestry—in England at least—is very problematical. For though there are districts where the land may—under State foresters working on a larger scale than private owners—produce a more profitable return under timber than when used for other purposes, yet I believe that these districts are so few and far between that the establishment of a State industry, financed by taxation, to compete with the long-established private industry of timber-growing would not be justified by any advantage that would result to the country. (H. J. E.)

# Postscript

In concluding our work there are a few points about which I think it advisable to add some explanatory remarks. First, with regard to its scope, it was intended at the outset to include only those trees which attain timber size in the British Isles. It was found, however, impossible in practice to draw a rigid line between the timber trees and the smaller trees, which like shrubs are cultivated for ornament or curiosity. In the case of genera, which comprise both large and small trees as well as shrubs, our treatment has not been logically uniform. All the species of oak in cultivation have been described in detail, because in this genus shrubs are of exceptional occurrence. On the other hand, only the larger maples have been the subject of separate articles, as it was evident that a brief notice would suffice for the shrubby species. In the case of genera of exceptional interest, as Pinus, Juniperus, Cephalotaxus, all the introduced species, even those of small size, have been treated in full.

The keys for the identification of species are based upon the characters of the twigs, leaves, and buds, and not upon those of the flowers and fruits, as has been usual in botanical works. The latter characters are often not available in the case of trees, the determination of which may be required when they are in the young state, or at some period of the year when flowers and fruit are not present.

When the preparation of the seventh volume was drawing near a close, we saw that it would be impracticable, without unduly deferring the completion of the work, to include many of the new species that of late years have been introduced from China and Japan. On this account, Eucommia, Tetracentron, Cercidiphyllum, Pistacia, Phellodendron, Idesia, Poliothyrsis, Davidia, etc., have been necessarily omitted. For obvious reasons, I was unable to take up the complete study of such genera as Cratægus, which is almost exclusively composed either of shrubs or small trees that are merely ornamental in character; and I have limited my account in this case to a full description of the two indigenous species. Generally speaking, the Conifers have been described exhaustively in our book; but certain rare kinds of which there are only a few specimens of small size in Cornwall and in the mild districts of Ireland, have been left untouched, as Podocarpus, Callitris, Tetraclinis, Widdringtonia, Dacrydium, and Phyllocladus.

Though, as just explained, all the species of trees in cultivation in England, Scotland, Wales, and Ireland, are not included, yet a vast number have been described, as is evidenced by the Index, which extends over 80 pages. The Index has been compiled with great care, and should prove of service to our readers, as it embraces, in addition to the common names and the usually accepted scientific names, nearly all the appellations which have been applied to the various species and their varieties and sports in countless lists, catalogues, and books. The compilation of this synonymy has been a heavy labour and a thankless task. The choice of the correct name of each species has not been always easy.

As some of the reviews of the published volumes criticise certain names which I have adopted, it will be well for me to explain my views on the vexed question of nomenclature. That in a work of this magnitude, I have refrained from the invention of new names, except in one case for a species[1] and in two or three cases for hybrids,[2] gives me a claim to be heard on the subject. In the earlier volumes, I followed the Kew practice of the time, that of selecting as the correct name the oldest one, which had been used under the genus to which the species is now referred. In the latter volumes, I have followed, except in a few instances, the Vienna Rule, that of adopting the oldest specific name, no matter to what genus it had been attached. Most German and American botanists follow this rule "blindly," as one of them remarked to me.

As a result names which have been current for a century, not only in scientific books but in popular literature, have now to give way to supposed earlier names, which have been resuscitated from the works of writers whose descriptions are often so ambiguous as to render it impossible to say what species was actually meant. The strict application of the Vienna Rule is sometimes so difficult that the best authorities disagree thereon. As an example, we are bidden to change the name *Alnus glutinosa*, Gaertner, which has been used for the common alder for over a century in every botanical text-book and Flora. Certain botanists substitute for it the name *Alnus rotundifolia*, Miller, whilst one authority puts forth a strong claim for *Alnus vulgaris*, Hill.[3] This displacement of a well-known scientific name by an obscure one is a pedantic and harmful practice. Its absurdity is shown by the fact that some recent writers, who adopt under all circumstances the Vienna Rule, are obliged, in order to specify clearly the species which they mean, to use two names, the second of which is the old-established name in brackets. The common oak is called by these writers[4] "*Quercus Robur*, Linn. (*Q. pedunculata*)"; while the wych elm is cumbered[5] with the appellation "*Ulmus glabra*, Huds. = *U. montana*." The confusion of the new practice is doubled, when the Vienna name for a species happens to be (as in the last case) the same as the old-established name for another species of the same genus. Thus, no one now knows without some explanation which species of elm is meant by the term *Ulmus glabra*.

In order to avoid such confusion, I have preserved in certain cases the old-established name, if its use involves no ambiguity. I have thus kept up *Larix europæa*, *Larix americana*, *Larix leptolepis*, *Abies pectinata*, *Quercus pedunculata*, *Betula pubescens*, *Betula verrucosa*, a series of names which have been consecrated by long usage in books on botany, arboriculture, and forestry. If I am wrong in using these names, I err in company with nearly all the writers who have mentioned these trees during the last century.

The postponement to the last volume of the more difficult genera has enabled me to devote time, labour, and travel to their study; and has resulted in a clearer knowledge of the numerous cultivated kinds of elms and poplars, the systematic position of which has been the despair of botanists. The results of the experimental sowings of the seeds of various elms, which I made in 1909, together with a study

[1] Cf. *Populus Maximowiczii*, vol. vii. 1838.
[2] Cf. *Populus Lloydii*, vol. vii. 1830, *Ulmus Mossii*, vol. vii. 1865, note 2.
[3] Cf. Schneider, *Laubholzkunde*, ii. 890 (1912).
[4] Cf. Schneider, *Laubholzkunde*, i. 197 (1904).
[5] Cf. Tansley, *British Vegetation*, 148 (1911).

of the history of the Lucombe Oak and its descendants, have thrown much light on a class of trees, which though common in cultivation on account of their vigour of growth, are unknown in the wild state.

These trees undoubtedly originated as chance seedlings, due to accidental cross-fertilisation of two distinct species, and were immediately selected by observant nurserymen as desirable varieties to propagate. Trees like the Black Italian Poplar and the London Plane, which in botanical characters appear to be first crosses in each case, between an American and a European species, can be traced back to 1700, about which date the introduced species was long enough in the country to produce flowers and to fertilise the native species growing beside it.

Other trees, which I consider to be also first crosses, on account not only of their botanical characters but because they are unknown in the wild state, are:— the Common Lime, Huntingdon Elm, and Cricket-Bat Willow, the parents in their case being closely allied European species. All the preceding hybrids were produced accidentally; and similar cases occur periodically in certain nurseries, as at Plantières, near Metz, where two splendid hybrid poplars (*Populus Eugenei* and *P. robusta*) have been picked out of the seed-bed. The first to produce artificial crosses between forest trees was Klotzsch in 1845; and his experiments, though on a small scale, were successful in indicating that extra vigour of growth was obtainable with certainty.

Practically nothing, however, has been done, since his time, to improve the breeds of forest trees; and foresters have never even thought of the possibilities in this direction, though gardeners and farmers have shown the way for centuries. I suggested in a paper read on 7th April 1910 before the Linnean Society that artificial crossing should be tried in the Ash and Walnut, as the quality of the wood of these two valuable trees would be improved by more rapid growth.

During the past three years I have carried on cross-pollination of trees of certain kinds at Cambridge, in Kew Gardens, and near Exeter and Gloucester. These experiments have shown that there are obstacles in the way of obtaining successful results, owing to the difficulty of manipulation on trees swaying in the wind, and to the spring frosts which often injure the pollinated flowers. Moreover, suitable exotic trees, especially of the broad-leaved sorts, are hard to find, as those that exist rarely produce good flowers. I am convinced that such experiments could be carried on much more successfully in stations like Montpellier in France, or Washington in the United States, where suitable trees and a warm climate can be readily found. Nevertheless, I have raised a considerable number of hybrid seedlings, which are now under observation. First-crosses once obtained can readily be reproduced by cuttings or by layers; and the cost of propagation would be very moderate. That this is feasible, is shown by the splendid hybrid elms in Belgium and Holland, all of which are raised by layering in nurseries.

Many interesting problems, which had to be laid aside for the time, arose in the course of the researches which were undertaken in the work of preparation of the seven volumes. To some of these problems, especially those connected with the origin of sports and varieties, I hope to return. The belief is rapidly gaining ground that all species of trees are comparable, in a greater or less degree, as regards their variation, to the remarkable instance of the Douglas Fir, the Rocky Mountain variety of which differs so much from that growing on the Pacific Coast, while between these two extremes lie a series of intermediate forms. Similar variation is well known in the case of *Pinus Laricio*, the geographical forms of which behave very differently in cultivation. Such differences probably exist in all species with a wide distribution; and in the future, care in the selection of seed may be the most important point in sylviculture.

All the bibliographical references in the work have been checked independently, and no effort has been spared to indicate with accuracy the source of our information, where it is not the result of our own observations. Though in the course of our labours, we have had the benefit of mutual criticism, it should be clearly understood that each of us is only responsible for the parts which are signed by his initials. It would be inaccurate to quote both our names, when only one of us has verified the fact or studied the question at issue. As an example,—"Elwes, in Elwes and Henry, *Trees of Great Britain*, v. 1179," is the correct citation for the account of *Cupressus lusitanica* in Portugal; whereas, "Henry, in Elwes and Henry, *Trees of Great Britain*, v. 1183," is responsible for the article on *Cupressus arizonica*. It will readily be seen, that while the purely botanical part (including the identification of specimens) has been done by me, that the other part, dealing with distribution, history, and cultivation, has been divided in varying proportions between the two authors.

(A. H.)

# ERRATA AND ADDENDA

Vol. i. p. vii, line 15. The English Elm was not introduced from Italy. Cf. vol. vii. p. 1907.

i. p. 2, line 32. **For** Dryander, in Ait. **read** Aiton.

The leaves of the North American beech are figured in vol. iii. Pl. 202, Fig. 6.

Rehder, in *Rhodora*, ix. 113 (1907), and in *Mitt. Deut. Dend. Ges.* 1907, p. 70, states that the correct name of the American beech is *Fagus grandifolia*, Ehrhart, *Beit.* iii. 22 (1788), and describes three varieties: (1) var. *typica*, Rehder; (2) var. *pubescens*, Fernald and Rehder; and (3) var. *caroliniana*, Fernald and Rehder, the latter having a sub-variety *mollis*.

i. p. 2, line 39. **For** north **read** the north.

i. p. 4, line 1. **Add** *Fagus Hohenackeriana*, Palibin, in *Bull. Herb. Boiss.* viii. 378 (1908), is probably a form with large leaves of *F. orientalis*, Lipski.

i. p. 6, note 2, line 2. **Omit** the hazel, the cotyledons of which are not aerial. Cf. vol. iii. p. 521.

i. p. 8, line 40. This is var. *Rohanii*, Masek, in *Mitt. Deut. Dend. Ges.* 1905, p. 196 and 1908, p. 140, described as a purple beech with deeply cut leaves like those of var. *quercoides*, from a tree of which it is said to have arisen through pollination by a purple beech that stood near.

i. p. 10, line 21. The Weeping Beech at Endsleigh is figured in vol. ii. Pl. 58A.

i. p. 13, line 31. **For** Lyons-le-Forêt **read** Lyons-la-Forêt.

i. p. 17, line 42. **For** Buckholt **read** Buckhold.

i. p. 29, line 13. **For** *Vilmoriana* **read** *Vilmoriniana*.

i. p. 30, note 1. **For** 14 **read** 15.

i. p. 31, note 1. **Add** vol. ii. Pl. 126, Fig. 3.

i. p. 32, line 2. **For** *Japan* **read** *Japon*.

i. p. 38. Concerning *Sophora japonica*, var. *pendula*—

Bretschneider, in *Journ. N. China Br. R. Asiat. Soc.* xv. 15 (1880), states: "The Chinese produce this tree artificially by causing two young trees of *Sophora japonica*, growing close together, to join by grafting, and then turning upwards the roots of one of them." De Vries, *Mutation Theory*, 101 (1911), says, however, that the Weeping Sophora originated in Joly's nursery at Paris in 1800.

i. p. 38, note 1. This note is to be deleted, as the plate mentioned has not been published.

Vol. i. p. 44. *Araucaria araucana*, Koch, is the correct name of the Chilean Araucaria, according to the Vienna Rules.

i. p. 44, line 2. **For** iv. 2432 (1844) **read** iv. 2432 (1838).

i. p. 44, line 4. **For** Rich. **read** Richard.

i. p. 44, line 5. **For** Mirb. **read** Mirbel.

i. p. 57, line 31. **For** develops **read** develop.

i. p. 64, line 4. **For** var. ? *chinensis* **read** var. ? *chinense*.

i. p. 65, line 2. **For** *tulipifera* **read** *Tulipifera*.

i. p. 65, line 8. **After** leaves **insert** (Vol. iii. Pl. 204, Fig. 7).

i. p. 66, line 30. **After** buds **insert** (Vol. ii. Pl. 126, Fig. 1).

i. p. 76, line 33. **Omit** opening red.

i. p. 76, line 37. **Omit** the buds open green, and

i. p. 77, line 1. *Picea morindoides* is fully described under the correct name, *Picea spinulosa*, Henry, in vol. vi. p. 1392. Its native country is Sikkim and Bhutan.

i. p. 80, line 37. **For** Serajevo **read** Sarajevo.

i. pp. 85, 89. If, as is probable, both the Ajan spruce (*Picea ajanensis*) and the Hondo spruce (*Picea hondoensis*) constitute only one species, its correct name is *Picea jezoensis*, Carrière.

i. p. 85, line 16, and i. p. 89, line 13. It is doubtful if the distinctions here noted, as regards the colour of the opening buds, are valid. In the specimens of *P. hondoensis* which I have examined, the buds open with a slightly reddish tinge.

i. p. 88, line 10. **For** *tremula* **read** *Sieboldii*.

i. p. 92, line 26. The specimens in the Kew Herbarium with pubescent branchlets, from the Columbia river, which are mentioned as a variety of *Picea sitchensis*, must be referred to *Picea Engelmanni*.

i. p. 93, line 3. **After** 1892 **insert** xiv. 184.

i. p. 99, line 6. **For** *Pruminopitys* **read** *Prumnopitys*.

i. p. 100, line 32. *Taxus canadensis*, Marshall, is undoubtedly a distinct species, and is peculiar not only in its shrubby habit but in having monœcious flowers.

i. p. 107, line 21. **For** 53 **read** 31.

i. p. 107, line 22. **For** the Hokkaido **read** Hokkaido.

i. p. 110, line 32. **For** 58 **read** 34.

i. p. 111, lines 13, 14. **For** *Chesthuntensis* **read** *Cheshuntensis*.

i. p. 114, line 18. **After** mentioned above **add** (p. 107).

i. p. 117, line 24. **For** Low **read** Lowe.

i. p. 117, line 32. R. J. Moss, in *Scient. Proc. R. Dublin Soc.* xii. 92 (1909), found 0.6 per cent of taxine in the leaves of the Irish yew (var. *fastigiata*) as compared with 0.12 per cent in the leaves of the female common yew and 0.18 per cent in those of the male common yew, recorded by Thorpe and Stubbs in *Journ. Chem. Soc. Trans.* 1902, p. 874. Moss refers to a case, in which pheasants, which are believed to eat common yew with impunity, were poisoned by the leaves of the Irish yew.

i. p. 120, line 35. **For** 54, 55 **read** 32, 33.

Vol. i. p. 122, line 22. **For** Low **read** Lowe.

i. p. 123, line 9. **For** Whittinghame **read** Whittingehame.

i. p. 132, line 34. The article on Distribution in China should be signed (A. H.).

i. p. 139, line 1. **For** Hempsted **read** Hemsted.

i. p. 139, line 28. **For** Worcestershire **read** Herefordshire.

i. p. 142, line 17. The species of the sections *Aucuparia* are treated in detail in vol. vi. p. 1574.

i. p. 142, line 25. **For** *Pyrus thianschanica*, Regel, **read** *Pyrus tianschanica*, Franchet.

i. p. 142, line 30. **For** Torrey and Gray **read** De Candolle. See vol. vi. p. 1574.

i. p. 142, line 32. *Pyrus sambucifolia*. **Add** the following note :—Cf. vol. vi. p. 1574, note 3.

i. p. 144, line 32. **For** Linnæus **read** Ehrhart.

i. p. 148, line 19. **For** Mountmorris **read** Mountnorris ; **for** rector **read** late vicar.

i. p. 148, line 23. **For** Lee's **read** Lees.

i. p. 149, line 33. **For** Connaught **read** Albany.

i. p. 149, line 34. **For** Snell **read** Smelt.

i. p. 159, line 3. **For** 915 **read** 912.

i. p. 175, line 26. **For** Seeman **read** Seemann.

i. p. 177, last line. **For** *uniflora* **read** *aquatica* (cf. vol. iii. 513).

i. p. 182, and note 1. *Thuja*, the spelling adopted by Linnæus, *Sp. Pl.* 1002 (1753), is to be preferred to that of *Thuya*.

i. p. 184, line 8. **For** *non* A. Murray **read** not Balfour in A. Murray.

i. p. 192, lines 6 and 15, and i. 198, line 16. **For** *Retinospora* **read** *Retinispora*. Regarding the correct spelling of the word, see vol. v. p. 1146, note 2.

i. p. 195. The correct name, according to the Vienna Rules, of *Thuya japonica*, Maximowicz, is *Thuya Standishii*, Carrière.

i. p. 196, line 1. **For** Komaror **read** Komarov.

ii. p. vi, last line. **Insert** a comma between Ailanthus and Cladrastis.

ii. p. 202, line 36. **For** *Cryptomera* **read** *Cryptomeria*.

ii. p. 202, line 4. **For** *Thoujopsis* **read** *Thujopsis*.

ii. p. 203, last line. **Read** *Thujopsis Hondai*, A. Henry.

ii. p. 208, line 41. **For** *A*. **read** *Æ*.

ii. p. 218, line 10. **For** *A*. **read** *Æ*.

ii. p. 225, line 6. **Add** the note :—*Pavia rubra*, Poiret, in Lamarck, *Ency*. v. 94 (1804), is *Æsculus Pavia* (cf. p. 207).

ii. p. 225, line 17. **For** *A*. **read** *Æ*.

ii. p. 228, line 34. **For** Maximowicz **read** Masters.

ii. p. 234. It will probably be better to adopt for this species the name *Tsuga heterophylla*, Sargent, as the correct one under the Vienna Rules.

ii. p. 240, line 9. **For** *loc. cit.* **read** *Timber Trees and Forests oj North Carolina*, 1898.

ii. p. 244. The correct name for this species is *Tsuga dumosa*, Sargent.

ii. p. 282. Moss, in *Journ. Bot.* 1910, p. 6, and Schneider, *Laubholzkunde*, i. 197 (1904), adopt the name, *Quercus Robur*, Linnæus, for the pedunculate or common oak.

ii. p. 282, line 25. **After** Branchlets **insert** (Pl. 78, Fig. 1).

Vol. ii. p. 283, line 3. **After** Leaves **insert** (Plate 79, Fig. 1).

ii. p. 285, line 12. **For** (1844) **read** (1838).

ii. p. 285, line 15. **For** *comptonæfolia* **read** *comptoniæfolia.*

ii. p. 288, line 30. **For** Boenn. **read** Boenninghausen. A fuller account of the hybrid oak, *Quercus intermedia,* Boenninghausen, is given by Dr. C. E. Moss, in *Journ. Bot.* 1910, pp. 1, 34, plate 502.

ii. p. 291, line 14. *Quercus sessilis,* Ehrhart, cannot be adopted as the correct name of the sessile oak, as it was unaccompanied by any description. Cf. Moss, in *Journ. Bot.* 1910, p. 2, and Schneider, *Laubholzkunde,* i. 196 (1904), and ii. 901 (1912).

ii. p. 291, line 22. **After** Branchlets **insert** (Pl. 78, Fig. 2).

ii. p. 291, line 25. **After** Leaves **insert** (Pl. 79, Fig. 8).

ii. 294, line 2. **Read** *Quercus lanuginosa,* Lamarck, *Fl. Franc.* ii. 209 (1778); Thuillier. **Insert** the following reference as a new line:—*Quercus Robur,* var. *lanuginosa,* Lamarck, *Encyc.* i. 717 (1783).

ii. p. 376, line 18. **For** unlikely **read** likely.

ii. p. 379, title. **Add** Dahurian Larch.

ii. p. 382, line 5. **Insert** A remarkable tree at Henham Hall, Suffolk, which according to the Earl of Stradbroke was planted between 1790 and 1803, is grafted on a common larch stock, forming a trunk about 6 ft. high and 7 ft. 4 in. in girth. It divides into numerous branches, which extend horizontally for a distance of about 100 ft., supported on larch poles. None of the branches rise more than two or three feet above this level; and when Elwes saw the tree in August 1909, they were covered with new and old cones.

There are two good specimens of the Dahurian larch at Stanage Park, Herefordshire, which measured about 80 ft. by 5 ft. in 1910. They were planted as *Larix pendula* in 1836.

ii. p. 383, title. **Add** Kurile Larch.

ii. p. 384, title. **Add** Japanese Larch.

ii. p. 393, note. **Insert** reference [1].

ii. p. 398, line 27. **For** *albertiana* **read** *Albertiana.*

ii. p. 409, line 18, and ii. 410, line 18. The form of *Pinus Laricio,* which occurs in the mountains of the central, north-eastern and south-eastern provinces of Spain is identical with that of the Cevennes and Pyrenees, and is to be referred to var. *tenuifolia.* The best account of the *Pinus Laricio* of Spain is given by S. E. Cook, *Sketches in Spain,* ii. 228, 234, 237, 244 (1834); and the name *P. hispanica,* Cook, *op. cit.* 234, may be added to the synonymy of this tree. Laguna, *Flora Forestal Española,* i. 80 (1883), gives full details of its distribution.

ii. p. 424, line 6. **Omit** rarely. **For** 6 **read** 8.

ii. p. 428, line 23. **For** *Guilandina diocus* **read** *Guilandina dioica.*

ii. p. 433, line 6. **After** regular **insert** full stop.

ii. p. 435, line 11. **Insert** A tree standing close to Weston Birt House, measured by Elwes in October, 1912, was 47 ft. by 3½ ft.

Vol. ii. p. 437, line 18. **For** DIPTERA (*sectio nova*) **read** DIPTERA, A. Henry (*sectio nova*).

ii. p. 438, line 26. **For** *Juglans fraxinifolium* **read** *Juglans fraxinifolia.*

iii. p. 452, line 23. **Add** Interesting details concerning the flowers, cones, and seedlings, as well as the rate of growth of the cedars, are given by Hutchison, in *Trans. Roy. Scott. Arbor. Soc.* xiii. 200 (1893).

iii. p. 452, line 24. The seedling here described was that of *Pinus halepensis*; and this paragraph should be transferred to vol. v. p. 1100.

iii. p. 471, line 12. **For** last January **read** in January 1907.

iii. p. 475, line 25. **After** now **add** (1907).

iii. p. 480, line 23. **For** *albertiana* **read** *Albertiana.*

iii. p. 485, line 25. **For** Endlichler **read** Endlicher.

iii. p. 489, line 10. **For** 1888 **read** 1870.

iii. p. 489, line 27. **For** *Thuya Craigana*, Murray, **read** *Thuja Craigana*, Balfour, in Murray. The correct name of the Incense Cedar of North America is *Libocedrus Craigana*, Low, *ex* R. Brown, in *Trans. Bot. Soc. Edin.* ix. 373 (1868). Cf. A. Henry, in *Gard. Chron.* liii. 325 (1913).

iii. p. 490, line 30. **For** Murray **read** Balfour, and **for** Thuya **read** Thuja.

iii. p. 492, line 35. **For** Park **read** Place.

iii. p. 494, note 2. The new species from Formosa has been named *Cunninghamia Konishii*, Hayata, in *Gard. Chron.* xliii. 194 (1908). Cf. Clinton-Baker, *Illust. Conif.* iii. 84 (1913), and Elwes, in *Quart. Journ. Forestry*, vi. 274 (1912). It has not yet been introduced.

iii. p. 513, note 1. **For** 1904 **read** for 1903. **Add** Cf. vol. i. p. 177 and note 3.

iii. p. 540, line 29. **For** *Ostrya Ostrya*, Sargent, **read** *Ostrya Ostrya*, Karsten, *Deutsch. Fl.* 20 (1895); Sargent.

iii. p. 545, lines 1 and 23; 546, lines, 8, 11, 21, 24, and 37; 547, lines 9, 11, and 16; 548, line 13; 549, line 1; 550, line 18; 553, line 29. **For** Oerstedt **read** Oersted.

iii. p. 553, line 21. **After** now **add** (October, 1907).

iii. p. 553, line 33. **For** Foster **read** Forster.

iii. p. 558, note 1. *Arbutus canariensis* is said, by W. Fitzherbert in *Gard. Chron.* lii. 44 (1912), to be 30 ft. high at Abbotsbury. It is perfectly hardy at Rostrevor, where I saw a shrub 10 ft. high in May 1913.

iii. p. 563, line 33. **For** *Brit.* 1119 **read** *Brit.* ii. 1119.

iii. p. 564, line 28. **For** Klotzch **read** Klotzsch.

iii. p. 565, line 31. **For** *procera*, Lindley, **read** *procera*, Douglas, *ex* Lindley.

iii. p. 573, line 27. **After** var. *scotica* **add** A. Henry.

iii. p. 576, note 2. **Add** Cf. vol. v. p. 1130, note 2.

iii. p. 586, line 5. **For** 162 **read** 131.

iii. p. 595, line 23. **For** *Primer* **read** *Pruner.*

iii. p. 599, line 5. **For** Schneider *ex* **read** Schneider, *Laubholzkunde*, i. 803 (1906);

iii. p. 600, line 17. **For** *cordifolia*, Schneider, *ex* **read** *cordiformis*, Schneider, *Laubholzkunde*, i. 803 (1906), and ii. 1008 (1912);

iii. p. 601, line 32. **For** Schneider *ex* **read** Schneider, *Laubholzkunde* i. 803 (1906);

Vol. iii. p. 604, note. **For** Schneider by **read** Schneider, *Laubholzkunde*, i. 803 (1906) and ii. 872 (1912);

iii. p. 605, line 29. **For** Schneider *ex* **read** Schneider, *Laubholzkunde*, i. 804 (1906);

iii. p. 620, line 18. The evidence is now almost conclusive that *Platanus acerifolia*, Willdenow, the London Plane, is of hybrid origin, the parents being *P. occidentalis*, Linnæus, and *P. orientalis*, Linnæus.

iii. p. 627, line 5. **For** *hybridus* **read** *hybrida*.

iii. p. 640, note. **For** *mandschuricum* **read** *mandshuricum*.

iii. p. 642, line 31. **For** *villosa* **read** *villosum*.

iii. p. 652, line 22. **For** *austriaca* **read** *austriacum*.

iii. p. 654, line 30. **After** paces **add** (Plate 185).

iii. p. 657, line 38. **For** sugar maple **read** common sycamore.

iii. p. 664, note 3. **For** 1889 **read** 1899.

iii. p. 669, line 20. **For** M. E. Louis **read** M. Jouin.

iii. p. 677, line 24. **For** *nigra* **read** *nigrum*.

iii. p. 690, line 31. **For** trees, **read** trees (Plate 195).

iii. p. 703, note. **Add** *Silva of California*, 143 (1910).

iii. p. 706, line 7. **For** he wrote **read** this was written.

iii. p. 706, line 35. **After** now **add** (1907).

iii. p. 707, line 5. **For** 106 **read** 196.

iii. p. 709, line 27. **After** now **add** (1907).

iii. p. 709, line 28. **After** now **add** (1907).

iv. p. 714, line 23. **For** all other conifers by **read** all other conifers except Keteleeria by. Cf. vol. vi. 1473.

iv. p. 728, line 25. **Insert** The Marquess of Bath informed Elwes in 1910 that Col. Thynne had measured a silver fir near the house at Longleat which was 144 ft. by 17 ft. 4 in. in 1909.

iv. p. 731, note 2. **For** 1896 **read** 1894.

iv. p. 739, line 21. **For** *Panachaica* **read** *panachaica*.

iv. p. 747, line 33. **For** a **read** at.

iv. p. 756, line 15. **For** inch **read** inches.

iv. p. 782, line 2. **For** 1840 **read** 1839.

iv. p. 812, line 37. **For** *Pseudotsuga japonica*, Sargent, **read** *Pseudotsuga japonica*, Beissner, in *Mitt. Deut. Dend. Ges.* 1896, p. 62; Sargent.

iv. p. 817, line 22. **For** in the Yellowstone Park in **read** from the Yellowstone Park in Wyoming to.

iv. p. 818, line 11. **For** 3 **read** 1.

iv. p. 824, line 10. **For** pine **read** fir.

iv. p. 854, lines 9, 10. **For** *Icon. Forest. Japon.* **read** *Icon. Ess. Forest. Japon.*

iv. p. 854, line 12. **For** Seeman **read** Seemann.

iv. p. 860, line 20. **For** 898 **read** 900.

iv. p. 862, line 1. **For** 907 **read** 905.

iv. p. 862, line 5. **For** 904 **read** 912.

iv. p. 862, line 19. **For** *longicuspis*, Blume, **read** *longicuspis*, Siebold and Zuccarini.

Vol. iv. p. 862, line 36. **For** 906 **read** 907.

iv. p. 862, note 1. **For** 21 **read** 25.

iv. p. 863, line 15. **For** 905 **read** 906.

iv. p. 876, line 2. **For** 124 feet **read** 97 feet.

iv. p. 884, line 5. Schneider, *Laubholzkunde*, ii. 832 (1912) adopts the earlier name, *Fraxinus obliqua*, Tausch, in *Flora*, xvii. 521 (1834) for *F. Willdenowiana*, Koehne; and states that Lingelsheim identifies it with a wild specimen found by Bornmüller in Anatolia.

iv. p. 904, line 13. **For** Hildenley at **read** Hildenley in.

iv. p. 912, line 33. **For** 265 **read** 264.

iv. p. 915, line 21. **For** *Pittcursii* **read** *Pitteursii.*

iv. p. 918, line 8. **For** Hertfordshire **read** Herefordshire.

iv. p. 918, line 38. **For** University Park **read** Christ Church Meadows.

iv. p. 919, line 10. **For** 61 feet in girth **read** 61 feet in height.

iv. p. 922, line 25. **For** Stukeley **read** Stewkley.

iv. p. 940, line 17, and note 2. The variety in Japan has been described as a distinct species, *Alnus serrulatoides*, Callier, in Fedde, *Repert.* x. 229 (1911).

iv. p. 946, line 24. **For** *montrosa* **read** *monstrosa.*

iv. p. 952, line 16. **For** *multinervis*, Schneider, **read** *multinervis*, Callier, *ex* Schneider.

iv. p. 953, line 24. **Delete** comma **after** *Flora.*

iv. p. 963, note 2. **For** *Eriophes* **read** *Eriophyes.*

iv. p. 971, line 21. **For** 225 **read** 255.

iv. p. 989, line 6. **For** *dahurica* **read** *davurica.*

iv. Plate 264, Fig. 16. **For** lancelotata **read** lanceolata.

v. p. 1013, line 15. **For** Bewdley **read** Bromsgrove.

v. p. 1029, note 1. **For** *Rusticanum* **read** *Rusticum.*

v. p. 1031, line 24. **For** Burwood House, near Cobham, Surrey, **read** Bearwood, near Wokingham.

v. p. 1080, note 7. **For** saved **read** sowed.

v. p. 1100, line 24. **For** *Pithyusa* **read** *pithyusa.*

v. p. 1100, line 31. **For** *Eldarica* **read** *eldarica.*

v. p. 1102, line 38. **After** *P. halepensis* **insert** (Plate 287).

v. p. 1103, note 3. **For** Strangways **read** Steven. Cf. p. 1100 and note.

v. p. 1107, line 32. **For** Solander in Aiton **read** Aiton.

v. p. 1113, line 12. **For** Solander in Aiton **read** Aiton.

v. p. 1124, line 36. **For** Burwood House, Surrey, **read** Bearwood, Berks.

v. p. 1130, note 2. **For** Klotsch **read** Klotzsch.

v. p. 1134, line 26. **For** *Mackintoshiana* **read** *Macintoshiana.*

v. p. 1139, line 23. **For** *Aboretum* **read** *Arboretum.*

v. p. 1140, line 1. **For** Solander in Aiton **read** Aiton.

v. p. 1142, line 5. **For** 5 ft. 3 in. **read** 5.23 in.

v. p. 1142, line 6. **For** 6 ft. 4 in. **read** 6.4 in.; **for** 8 ft. 3 in. **read** 8.3 in.

Vol. v. p. 1150, line 9. *Cupressus Hodginsii*, Dunn, has been made the type of a new genus, and is now known as *Fokienia Hodginsii*, A. Henry and H. H. Thomas, in *Gard. Chron.* xlix. 66, 84, figs. (1911). Cf. Clinton-Baker, *Illust. Conif.* iii. 85 (1913), who gives particulars regarding its introduction into cultivation.

v. p. 1179, note 1, line 4. **For** Parlatore **read** Antoine.

v. p. 1188, line 18. **For** Maichi **read** Imaichi.

v. p. 1200, line 28. **Insert** (H. J. E.).

v. p. 1203, note 2. **For** *pygmea* **read** *pygmæa.*

v. p. 1229, note 5. **For** *schochiana* **read** *Schochiana.*

v. p. 1235, note 2. **After** Small, **read** in *Bull. Torrey Bot. Club*, 1901, p. 157.

v. p. 1257, line 21. **For** 317 **read** 318.

v. p. 1257, line 24. **For** 318 **read** 317.

v. p. 1289, between lines 19 and 20. **Add** as a synonym, *Quercus austriaca sempervirens*, Hort., a name by which this oak is still occasionally known in some nurseries.

v. p. 1315, line 17. **For** *cerquhino* **read** *cerquinho.*

vi. p. 1372. Two varieties of the Hondo spruce, vars. *reflexa* and *acicularis*, have recently been described by Shirasawa and Koyama, in *Tokyo Bot. Mag.* xxvii. 129, pl. ii. figs. 1-17 (1913).

An allied species, *Picea Koyamai*, Shirasawa, was published in *Tokyo Bot. Mag.* xxvii. 128, pl. ii. figs. 28-35 (1913).

vi. p. 1374, line 3. **Add** Shirasawa and Koyama describe and figure *Picea Maximowiczii* in *Tokyo Bot. Mag.* xxvii. 130, pl. ii. figs. 18 to 27 (1913).

vi. p. 1380, line 25. **For** Solander in Aiton **read** Aiton.

vi. p. 1411, line 3. **For** Timbal **read** Timbal-Lagrave.

vi. p. 1481, line 3; 1483, line 6; 1485, line 11; 1487, line 22; 1489, lines 12 and 23; 1490, line 29. **For** *Bull. Soc. Dend. France*, i. **read** *Bull. Soc. Dend. France*, ii.

vi. p. 1487, line 18 and note 2. It will be advisable to adopt for this species the oldest name, *Catalpa ovata*, Don, and to abandon that of *C. Kaempferi.* Rehder, in Sargent, *Plant. Wilson*, ii. 304 (1912) agrees with me that this species is truly wild in Central China, and that *C. Henryi*, Dode, cannot be maintained even as a variety.

vi. p. 1489, line 24. *Catalpa Duclouxii*, Dode, and *C. sutchuenensis*, Dode, in *Bull. Soc. Dend. France*, ii. 204 (1907), are considered by Rehder, in Sargent, *Plant. Wilson.* ii. 304 (1912), to be identical, and constitute a glabrous variety of *C. Fargesi*, Bureau, which varies much in pubescence.

vi. p. 1569, line 11. **For** Solander in Aiton **read** Aiton.

vi. p. 1580, note. **Add** The rowan (*Pyrus Aucuparia*) is called quicken tree in Wicklow and other parts of Ireland.

vi. p. 1585, line 1. Schneider, *Laubholzkunde*, i. 806 (1906), adopts for this species the name *Magnolia virginiana*, Linnæus, *Sp. Pl.* 535 (1753), instead of *Magnolia glauca.*

vi. p. 1616, note 2. **For** *E. Gunnii* **read** *E. coccifera.*

vii. p. 1736, note 3. **For** Chevallier **read** Blin. Cf. *Bull. Dend. Soc. France*, 1913, p. 122.

(A. H.)

# INDEX

THE END

*Printed by* R. & R. CLARK, LIMITED, *Edinburgh*

Printed in the United States
By Bookmasters